# GPS · Theory, Algorithms and Applications

Guochang Xu · Yan Xu

# GPS

Theory, Algorithms and Applications

Third Edition

With 62 Figures

 Springer

Guochang Xu
Shandong University at Weihai
Institute of Space Sciences
Weihai
China

Yan Xu
GFZ German Research Centre for Geosciences
Geodesy
Potsdam
Germany

ISBN 978-3-662-50365-2     ISBN 978-3-662-50367-6   (eBook)
DOI 10.1007/978-3-662-50367-6

Library of Congress Control Number: 2016937942

© Springer-Verlag Berlin Heidelberg 2003, 2007, 2016
This work is subject to copyright. All rights are reserved by the Publisher, whether the whole or part of the material is concerned, specifically the rights of translation, reprinting, reuse of illustrations, recitation, broadcasting, reproduction on microfilms or in any other physical way, and transmission or information storage and retrieval, electronic adaptation, computer software, or by similar or dissimilar methodology now known or hereafter developed.
The use of general descriptive names, registered names, trademarks, service marks, etc. in this publication does not imply, even in the absence of a specific statement, that such names are exempt from the relevant protective laws and regulations and therefore free for general use.
The publisher, the authors and the editors are safe to assume that the advice and information in this book are believed to be true and accurate at the date of publication. Neither the publisher nor the authors or the editors give a warranty, express or implied, with respect to the material contained herein or for any errors or omissions that may have been made.

Printed on acid-free paper

This Springer imprint is published by Springer Nature
The registered company is Springer-Verlag GmbH Berlin Heidelberg

*To*
*Liping, Jia, Yuxi, and Pan*

# Preface to the Third Edition

After the first edition of this book was published at the end of 2003, Springer requested a revision for the second edition around 2006, and that was published at the end of 2007. The original edition was written based on experience with the software design of KSGsoft (Xu et al. 1998) and KGsoft (Xu 1999), and related research and practice in Germany and Denmark. The second edition benefited from the design of the multi-functional GPS/Galileo software (MFGsoft, Xu 2004). The new model of solar radiation for satellite orbit determination discovered through GPS research led to an attempt to solve the equations of satellite motion affected by second-order disturbances, which resulted in the book *Orbits*, published at the end of 2008, with a second edition in 2013, combined with intensive research (cf. Xu et al. 2010a,b, 2011; Xu and Xu 2012, 2013a,b). During that time, the author was further involved in GPS activities through his role in the supervision of Ph.D. studies and software development (cf. Wang et al. 2010; He 2015). In 2011, the second edition of *GPS* was translated into Chinese by the Peking Institute of Satellite Controlling and Telecommunication and was published by Tsinghua University Press Peking, and was sold out around 2015. In 2014, *GPS* was translated into Persian, and that edition was published by Dhahran University. The preparation of the third edition of *GPS* was contracted with Springer several years ago, but was interrupted by work on the books *Orbits* (2008, 2013) and *Sciences of Geodesy* (2010, 2012), as well as other scientific endeavours. In 2014 the author moved from Germany to a position at Shandong University in Weihai to create a new area of study comprising navigation, remote sensing, and celestial mechanics. The rapid development of the GPS, GLONASS, Galileo, and BeiDou systems, and the realisation of net-based multi-system real-time civil GNSS applications at GFZ in Germany and at Shandong University, provided the impetus for this most recent revision of the book. Dr.-Ing. Yan Xu, who has played an important role, was asked to be the second author.

A fully new Chap. 12 has been added, which highlights the newly developed singularity-free theory of analytical satellite orbits. A brief historical review of the singularity problem is presented, and the solution is derived mathematically. The study was conducted by the first author of this book with his friends from 2007 to 2015, and the most elegant formulas—the Lagrange-Xu and Gauss-Xu equations of motion—were derived rigorously and mathematically and are described here for the first time. Three singularity criteria with clear geometric meaning are defined. This theory opened up a new area of study in orbit determination using analytical theory and is believed to be extremely important with respect to GNSS applications for onboard orbit determination.

Chapters that have been intensively revised and supplemented are the first, "Introduction"; the fifth, "Physical Influences of GPS Surveying"; the eighth, "Cycle Slip Detection and Ambiguity Resolution"; and the ninth, "Parameterisation and Algorithms of GPS Data Processing". The others are minorly revised. These are reviews of the works of scientists around the world. The supplemented content concerns the following 25 points.

1. Review of GPS modernisation
2. Review of the key developments of GLONASS
3. Review of the development of the Galileo system
4. Review of the development of the Compass (BeiDou) system
5. Review of progress in ionospheric studies
6. Review of tropospheric research
7. Research concerning the GPS clock
8. Review of the use of an external clock
9. Introduction of water vapour measurements
10. Review of activities involving GPS altimetry
11. Review of the study of GPS equivalence algorithms
12. The evaluation of ambiguity search criteria
13. Review of research progress in adaptive filtering
14. Research on GPS data processing by combined and uncombined methods
15. Research on reference satellite changes in the GPS difference algorithm
16. Research on tropospheric models in airborne kinematic positioning
17. Research on reference station changes in the GPS difference algorithm
18. Review of float ambiguity fixing
19. Outline of the progress in precise point positioning
20. GPS software introduction
21. Review of satellite orbital theory
22. Review of numerical orbit determination
23. Summary of GEO satellite orbit determination
24. Research on independent parameterisation
25. Summary of problems remaining in GPS research

With such intensive content supplementation, the authors hope that this new edition of *GPS* can better serve as a reference for GNSS research and study. The theoretical contributions and new findings in the original edition of this book can be summarised as follows:

1. The soft equivalence of differencing and undifferenced GPS algorithms
2. The unified GPS data processing algorithm

3. The general ambiguity search criterion
4. The equivalent ambiguity search criterion
5. The diagonalisation algorithm
6. Yang's filter—adaptive robust Kalman filtering
7. Theory of numerical orbit determination using GPS
8. The algebraic solution of the variation equation
9. The problem of the ambiguity function criterion

The new findings supplemented in the second edition are as follows:

1. The equivalence of combined and uncombined GPS algorithms
2. The independent parameterisation method
3. The equivalence theorem of GPS algorithms
4. The optimal differential GPS baseline network
5. The new model of solar radiation
6. The new model of atmospheric drag

New theoretical content in this third edition can be described as follows:

1. The equivalent equations of triple differences
2. The idea of an intelligent Kalman filter
3. Float ambiguity fixing in the case of ionosphere-free combination
4. The singularity-free Lagrange-Xu equations of motion
5. The singularity-free Gauss-Xu equations of motion
6. The criteria of singularity and their geometric meaning

The extended content is derived in part from published international papers and has been subjected to individual review. We thank Academician Yuanxi Yang of the Institute of Surveying and Mapping in Xian, Prof. Ta-Kang Yeh of Taibei University of Taiwan, Prof. Wu Chen of Hong Kong Polytech University, Prof. Yunzhong Shen of TongJi University, and Prof. Zhiping Lv of the Institute of Surveying and Mapping in Zhengzhou for their valuable review of portions of the supplemented content of this book.

The first author extends sincere thanks to Prof. Dr. D. Lelgemann of TU Berlin for supervision of the author's Ph.D. study and work many years ago. Thanks go to the directors Prof. Dr. Ch. Reigber, Prof. Dr. Markus Rothacher, and Prof. Dr. Harald Schuh of GFZ for their support and faith in the author during his approximately 20 years of research activities at GFZ. Prof. Ming Li of the China Academy of Space Technology (CAST) is thanked for the unwavering support since the author's involvement at CAST as Thousand Talents expert. Of course, Shandong University at Weihai is thanked for the opportunity to develop a new study area and form an international team, as well as obtaining internal and external funding support. The second author is thanked for taking over most of the text processing work for this new edition.

The navigation/remote sensing team members of Shandong University at Weihai are thanked for their warm support. Special thanks go to Thousand Talents expert Prof. Hermann Kaufmann of Germany, Guest Prof. Pierre Rochus of Belgium,

Guest Prof. Luisa Bastos of Portugal, Guest Prof. Anna Jenssen of Sweden, scientist Nina Boesche, and guest engineer Nan Jiang of Germany, senior engineers Wenlin Yan and Zhangzheng Sun, engineers Chunhua Jiang and Fangzhao Zhang, post-doctoral researchers Yujun Du and Fang Gao, and Ph.D. candidate Wenfeng Nie.

January 2016

Guochang Xu
Yan Xu

## References

He KF (2015) GNSS Kinematic Position and Velocity Determination for Airborne Gravimetry. PhD Thesis, (Scientific Technical Report; 15/04). Potsdam: Deutsches GeoForschungsZentrum GFZ. http://doi.org/10.2312/GFZ.b103-15044

Wang Q, Xu G, Chen Z (2010) Interpolation Method of Tropospheric Delay of High Altitude Rover Based on Regional GPS Network. Geomatics and Information Science of Wuhan University, 35(12), 1405–1408

Xu G (1999): KGsoft – Kinematic GPS Software – Software User Manual, Version of 1999, Kort & Matrikelstyrelsen (National Survey and Cadastre – Denmark), ISBN 87-7866-158-7, ISSN 0109-1344, 35 pages, in English

Xu G (2004) MFGsoft – Multi-Functional GPS/(Galileo) Software – Software User Manual, (Version of 2004), Scientific Technical Report STR04/17 of GeoForschungsZentrum (GFZ) Potsdam, ISSN 1610-0956, 70 pages, www.gfz-potsdam.de/bib/pub/str0417/0417.pdf

Xu G, Schwintzer P, Reigber Ch (1998) KSGSoft – Kinematic/Static GPS Software – Software user manual (version of 1998). Scientific Technical Report STR98/19 of GeoForschungsZentrum (GFZ) Potsdam

Xu G, Xu J (2012) On the Singularity Problem of Orbital Mechanics, MNRAS, 2013, Vol.429, pp1139-1148

Xu G, Xu J (2013a) On Orbital Disturbance of Solar Radiation, MNRAS, 432 (1): 584-588 doi:10.1093/mnras/stt483

Xu G, Xu J (2013b) Orbits – 2nd Order Singularity-free Solutions, second edition, Springer Heidelberg, ISBN 978-3-642-32792-6, 426 pages, in English

Xu G, Xu TH, Chen W, Yeh TK (2010b) Analytic Solution of Satellite Orbit Perturbed by Atmospheric Drag, MNRAS, Vol. 410, Issue 1, pp 654-662 87.

Xu G, Xu TH, Yeh TK, Chen W (2010a) Analytic Solution of Satellite Orbit Perturbed by Lunar and Solar Gravitation, MNRAS, Vol. 410, Issue 1, pp 645-653

Xu Y, Yang Y, Zhang Q, Xu G (2011) Solar Oblateness and Mercury's Perihelion Precession, MNRAS, Vol. 415, 3335-3343

# Preface to the Second Edition

After the first edition of this book was published at the end of 2003, I was very happy to put the hard work of book writing behind me and concentrate with my small team on the development of multi-functional GPS/Galileo software (MFGsoft). The experiences from the practice and implementation of the theory and algorithms into high-standard software caused me to strongly feel that I would like to revise and to supplement the original book, to modify some of the content, and to report on the new developments and knowledge. Furthermore, with the EU Galileo system now being realised and the Russian GLONASS system under development, GPS theory and algorithms needed to be re-described so that they would be valid for the Galileo and GLONASS systems as well. Therefore, I am grateful to all of the readers of this book, whose interest led Springer to ask me to complete this second edition.

I remember that I was in a hurry during the last check of the layout of the first edition. The description of a numerical solution to the variation equation in Sect. 11.5.1 was added to the book at the last minute, and comprised exactly one page. Traditionally, variation equations in orbit determination (OD), geopotential mapping, and OD Kalman filtering, have been solved by integration, which is complicated and compute-intensive. This marks the first time in the history of OD that the variation equation is not integrated, but is solved by a linear algebraic equation system. However, this was mentioned neither in the preface nor at the beginning of the chapter. The high precision of this algebraic method is verified by a numerical test.

The problems discussed in Chap. 12 of the first edition are largely solved and are now described by the so-called independent parameterisation theory, which points out that in both undifferenced and differencing algorithms, the independent ambiguity vector is double-differencing. With the use of this parameterisation method, the GPS observation equations are regular equations, and can be solved without any a priori information. Many conclusions may be derived from this new knowledge. For example, synchronisation of the GPS clocks may not be realised by the carrier-phase observables because of the linear correlations between the clock error parameters and the ambiguities. The equivalence principle is extended to show that the equivalences are valid not only between the undifferenced and differencing algorithms, but also between uncombined and combining algorithms and their mixtures. In other words, the GPS data processing algorithms are equivalent under the same parameterisation of the observation model. Different algorithms are beneficial for different data processing purposes. One consequence of the equivalence theory is that a so-called secondary data processing algorithm is created. Thus the complete GPS positioning problem may be separated into two steps (first to transform the data to the secondary observables and then to process the secondary data). Another consequence of the equivalence is that any GPS observation equation can be separated into two sub-equations, which is very advantageous in practice. Furthermore, it shows that the combinations under traditional parameterisation are inexact algorithms compared with those under independent parameterisation.

The additional content features a more detailed introduction, which includes not only GPS developments, but also those of the EU Galileo and Russian GLONASS systems, as well as the combination of the GPS, GLONASS, and Galileo systems. The book thus covers the theory, algorithms, and applications of the GPS, GLONASS, and Galileo systems. The equivalence of GPS data processing algorithms and the independent parameterisation of GPS observation models is discussed in detail. Other new content includes the concept of optimal network formation, application of the diagonalisation algorithm, and adjustment models of radiation pressure and atmospheric drag, as well as discussions and comments of what are currently, in the author's opinion, key research problems. Application of the theory and algorithms in the development of GPS/Galileo software is also outlined. The content concerning the ambiguity search is reduced, whereas the content regarding ionosphere-free ambiguity fixing is cancelled out, although it was reported by Lemmens (2004) as new. Some of the content of various sections has also been reordered. In this way, I hope this edition may better serve as a reference and handbook of GPS/Galileo research and applications.

The extended content is partly the result of the development of MFGsoft, and has been subjected to individual review. I thank Prof. Lelgemann of the TU Berlin, Prof. Yuanxi Yang of the Institute of Surveying and Mapping in Xian, Prof. Ta-Kang Yeh of ChingYun University of Taiwan, and Prof. Yunzhong Shen of TongJi University for their valuable reviews. I am grateful to Prof. Jiancheng Li and Dr. Zhengtao Wang of Wuhan University, as well as Mr. Tinghao Xiao of Potsdam University, for their cooperation during the development of the software from 2003 to 2004 at the GFZ.

I sincerely thank Prof. Dr. Markus Rothacher for his support and faith in me during my research activities at the GFZ. I also thank Dr. Jinghui Liu of the educational department of the Chinese Embassy in Berlin, Prof. Heping Sun and Jikun Ou of IGG in Wuhan, and Prof. Qin Zhang of ChangAn University for their warm support during my scientific activities in China. The Chinese Academy of Sciences is thanked for the Outstanding Overseas Chinese Scholars Fund. During this work, several interesting topics have been carefully studied by some of my students. I am grateful to Ms. Daniela Morujao of Lisbon University, Ms. Jamila Bouaicha of TU Berlin, Dr. Jiangfeng Guo and Ms. Ying Hong of IGG in Wuhan, and Mr. Guanwen Huang of ChangAn University. I am also thankful for the valuable feedback from readers and from students through my professorships at ChangAn University and the IGG CAS.

June 2007                                                         Guochang Xu

# Preface to the First Edition

The contents of this book cover static, kinematic, and dynamic GPS theory, algorithms, and applications. Most of the content comes from the source code descriptions of the Kinematic/Static GPS Software (KSGsoft), which was developed at GFZ before and during the EU AGMASCO project. The principles described here have been largely applied in practice, and are carefully revised from a theoretical perspective. A portion of the content is dealt with on a theoretical basis and applied to the development of quasi-real-time GPS orbit determination software at GFZ.

The original purpose in writing this book was, indeed, to have it for myself as a GPS handbook and as a reference for a few of my friends and students who worked with me in Denmark. The desire to describe the theory in an exact manner comes from my mathematical education. My extensive geodetic research experience has led to a detailed treatment of most topics. The comprehensiveness of the content reflects my nature as a software designer.

Some of the results of research carried out in GFZ are published here for the first time. One example is the unified GPS data processing method using selectively eliminated equivalent observation equations. Methods including zero-, single-, double-, triple-, and user-defined differential GPS data processing are unified in a unique algorithm. This method has advantages of both un-differential and differential methods, in that the un-correlation property of the original observations is retained, and the unknown number may be greatly reduced. Another example is the general criterion and its equivalent criterion for integer ambiguity search. A search using this criterion can be carried out in ambiguity or coordinate domains, or both. The optimality and uniqueness properties of the criterion are proved. Further examples include the diagonalisation algorithm of the ambiguity search problem, the ambiguity-ionospheric equations for ambiguity and ionosphere determination, and the use of the differential Doppler equation as system equation in the Kalman filter.

The book includes 12 chapters. After a brief introduction, the coordinate and time systems are described in the second chapter. Because orbit determination is also an important topic of the book, the third chapter is dedicated to Keplerian satellite orbits. The fourth chapter deals with the GPS observables, including code range, carrier phase, and Doppler measurements.

The fifth chapter covers all physical influences on GPS observations, including ionospheric, tropospheric, and relativistic effects, earth tide and ocean loading tide effects, clock errors, antenna mass centre and phase centre corrections, multipath effects, anti-spoofing, and historical selective availability, as well as instrumental biases. Theories, models, and algorithms are discussed in detail.

The sixth chapter first covers GPS observation equations, including their formation, linearisation, related partial derivatives, and linear transformation and error propagation. Useful data combinations are then discussed, particularly with respect to the introduction of the concept of ambiguity-ionospheric equations and related weight matrix. The equations include only ambiguity, ionospheric, and instrumental error parameters and can also be solved independently in kinematic applications. Traditional differential GPS observation equations, including the differential Doppler equations, are also discussed in detail. The method of selectively eliminated equivalent observation equations is proposed to unify the un-differential and differential GPS data processing methods.

The seventh chapter covers all adjustment and filtering methods that are suitable and necessary for GPS data processing. The main adjustment methods described are classical, sequential, block-wise, and conditional least squares. The key filtering methods include classical and robust filtering, as well as adaptively robust Kalman filters. In addition, a priori constraints, a priori datum, and quasi-stable datum methods are discussed for dealing with rank-deficient problems. The theoretical basis of equivalently eliminated equations is derived in detail.

The eighth chapter is dedicated to cycle slip detection and ambiguity resolution. Several cycle slip detection methods are outlined, with emphasis on deriving a general criterion for integer ambiguity search in ambiguity or coordinate domains, or both. Although the criterion is derived from conditional adjustment, in the end, it has nothing to do with any condition. An equivalent criterion is also derived, and shows that the well-known least squares ambiguity search criterion is just one of the terms of the equivalent criterion. A diagonalisation algorithm is proposed for use with ambiguity search, which can be done within one second after the normal equation is diagonalised. The ambiguity function and float ambiguity fixing methods are also outlined.

The ninth chapter describes GPS data processing in static and kinematic applications, and data pre-processing is outlined. Emphasis is given to solving the ambiguity-ionospheric equations and single point positioning, relative positioning, and velocity determination using code, phase, and combined data. The equivalent un-differential and differential data processing methods are discussed, and a method of Kalman filtering using velocity information is described. The accuracy of the observational geometry is outlined at the end of the chapter.

The tenth chapter covers the concepts of kinematic positioning and flight-state monitoring. The use of the IGS station, multiple static references, airport height information, kinematic tropospheric modelling, and the known distances of the multiple antennas on aircraft are discussed in detail. Numerical examples are also given.

The 11th chapter deals with the topic of perturbed orbit determination. Perturbed equations of satellite motion are derived, and perturbation forces of satellite motion are discussed in detail, including the earth's gravitational field, earth tide and ocean tide, the sun, moon, and planets, solar radiation pressure, atmospheric drag, and coordinate perturbation. Orbit correction is outlined based on the analytical solution of $C_{20}$ perturbation. Precise orbit determination is also discussed, including its principle and related derivatives, as well as numerical integration and interpolation algorithms.

The final chapter is a brief discussion regarding the future of GPS and comments on some remaining problems.

The book was subjected to an individual review of chapters and sections or according to content. I am grateful to reviewers Prof. Lelgemann of the Technical University (TU) Berlin, Prof. Leick of the University of Maine, Prof. Rizos of the University of New South Wales (UNSW), Prof. Grejner-Brzezinska of Ohio State University, Prof. Yuanxi Yang of the Institute of Surveying and Mapping in Xian, Prof. Jikun Ou of the Institute of Geodesy and Geophysics (IGG) in Wuhan, Prof. Wu Chen of Hong Kong Polytechnic University, Prof. Jiancheng Li of Wuhan University, Dr. Chunfang Cui of TU Berlin, Dr. Zhigui Kang of the University of Texas at Austin, Dr. Jinling Wang of UNSW, Dr. Yanxiong Liu of GFZ, Mr. Shfaqat Khan of KMS of Denmark, Mr. Zhengtao Wang of Wuhan University, and Dr. Wenyi Chen of the Max-Planck Institute of Mathematics in Sciences (Leipzig, Germany). The book was subjected to a general review by Prof. Lelgemann of TU Berlin. A grammatical check of technical English writing was performed by Springer-Verlag Heidelberg.

I offer my sincere thanks to Prof. Dr. Ch. Reigber for his support and faith in me throughout my scientific research activities at GFZ. Dr. Niels Andersen, Dr. Per Knudsen, and Dr. Rene Forsberg at KMS of Denmark are thanked for their support for starting work on this book, and Prof. Lelgemann of TU Berlin for his encouragement and help. During this work, many valuable discussions were held with many specialists. My thanks go to Prof. Grafarend of the University Stuttgart, Prof. Tscherning of Copenhagen University, Dr. Peter Schwintzer of GFZ, Dr. Luisa Bastos of the Astronomical Observatory of University Porto, Dr. Oscar Colombo of Maryland University, Dr. Detlef Angermann of German Geodetic Research Institute Munich, Dr. Shengyuan Zhu of GFZ, Dr. Peiliang Xu of the University Kyoto, Prof. Guanyun Wang of IGG in Wuhan, Dr. Ludger Timmen of the University of Hannover, and Ms. Daniela Morujao of Coimbra University. I thank Dr. Jürgen Neumeyer of GFZ and Dr. Heping Sun of IGG in Wuhan for their support. Dipl.-Ing. Horst Scholz of TU Berlin is thanked for redrawing a portion of the graphics. I am also grateful to Dr. Engel of Springer-Verlag Heidelberg for his advice.

I thank my wife Liping, son Jia, and daughters Yuxi and Pan for their loving support and understanding, as well as for their help on part of the text processing and graphing.

March 2003

Guochang Xu

# Contents

| 1 | **Introduction** | 1 |
|---|---|---|
| | 1.1 A Key Note on GPS | 2 |
| |     1.1.1 GPS Modernization | 4 |
| | 1.2 A Brief Message About GLONASS | 7 |
| |     1.2.1 The Development of GLONASS | 7 |
| | 1.3 Basic Information on Galileo | 9 |
| |     1.3.1 The Development of Galileo | 10 |
| | 1.4 Introduction of BeiDou | 11 |
| |     1.4.1 The Development of BeiDou | 12 |
| | 1.5 A Combined Global Navigation Satellite System | 13 |
| | References | 14 |
| 2 | **Coordinate and Time Systems** | 17 |
| | 2.1 Geocentric Earth-Fixed Coordinate Systems | 17 |
| | 2.2 Coordinate System Transformations | 21 |
| | 2.3 Local Coordinate System | 22 |
| | 2.4 Earth-Centred Inertial Coordinate System | 24 |
| | 2.5 IAU 2000 Framework | 28 |
| | 2.6 Geocentric Ecliptic Inertial Coordinate System | 32 |
| | 2.7 Time Systems | 33 |
| | References | 36 |
| 3 | **Satellite Orbits** | 37 |
| | 3.1 Keplerian Motion | 37 |
| |     3.1.1 Satellite Motion in the Orbital Plane | 40 |
| |     3.1.2 Keplerian Equation | 44 |
| |     3.1.3 State Vector of the Satellite | 46 |
| | 3.2 Disturbed Satellite Motion | 49 |
| | 3.3 GPS Broadcast Ephemerides | 49 |
| | 3.4 IGS Precise Ephemerides | 51 |
| | 3.5 GLONASS Ephemerides | 52 |

|  |  |  |  |
|---|---|---|---|
| | 3.6 | Galileo Ephemerides. | 53 |
| | 3.7 | BDS Ephemerides | 53 |
| | References | | 53 |
| **4** | **GPS Observables** | | **55** |
| | 4.1 | Code Pseudoranges | 55 |
| | 4.2 | Carrier Phases | 57 |
| | 4.3 | Doppler Measurements | 59 |
| | References | | 61 |
| **5** | **Physical Influences of GPS Surveying** | | **63** |
| | 5.1 | Ionospheric Effects | 63 |
| | | 5.1.1 Code Delay and Phase Advance | 63 |
| | | 5.1.2 Elimination of Ionospheric Effects | 66 |
| | | 5.1.3 Ionospheric Models | 69 |
| | | 5.1.4 Mapping Functions | 73 |
| | | 5.1.5 Introduction of Commonly Used Ionospheric Models | 76 |
| | 5.2 | Tropospheric Effects | 80 |
| | | 5.2.1 Tropospheric Models | 81 |
| | | 5.2.2 Mapping Functions and Parameterisation | 85 |
| | | 5.2.3 Introduction of Commonly Used Tropospheric Models | 88 |
| | | 5.2.4 Tropospheric Model for Airborne Kinematic Positioning | 91 |
| | | 5.2.5 Water Vapour Research with Ground-Based GPS Measurement | 93 |
| | 5.3 | Relativistic Effects | 94 |
| | | 5.3.1 Special Relativity and General Relativity | 94 |
| | | 5.3.2 Relativistic Effects on GPS | 97 |
| | 5.4 | Earth Tide and Ocean Loading Tide Corrections | 99 |
| | | 5.4.1 Earth Tide Displacements of GPS Stations | 99 |
| | | 5.4.2 Simplified Model of Earth Tide Displacements | 101 |
| | | 5.4.3 Numerical Examples of Earth Tide Effects | 103 |
| | | 5.4.4 Ocean Loading Tide Displacement | 105 |
| | | 5.4.5 Computation of the Ocean Loading Tide Displacement | 108 |
| | | 5.4.6 Numerical Examples of Loading Tide Effects | 109 |
| | 5.5 | Clock Errors | 110 |
| | | 5.5.1 Introduction of Commonly Used Clock Error Models | 112 |
| | | 5.5.2 Impact of Frequency Reference of a GPS Receiver on the Positioning Accuracy | 114 |

|  |  |  |  |
|---|---|---|---|
| 5.6 | Multipath Effects | | 115 |
| | 5.6.1 | GPS Altimetry, Signals Reflected from the Earth's Surface | 117 |
| | 5.6.2 | Reflecting Point Positioning | 117 |
| | 5.6.3 | Image Point and Reflecting Surface Determination | 119 |
| | 5.6.4 | Research Activities in GPS Altimetry | 120 |
| 5.7 | Anti-spoofing and Selective Availability Effects | | 121 |
| 5.8 | Antenna Phase Centre Offset and Variation | | 122 |
| 5.9 | Instrumental Biases | | 126 |
| References | | | 127 |

## 6 GPS Observation Equations and Equivalence Properties ........ 133

| | | | |
|---|---|---|---|
| 6.1 | General Mathematical Models of GPS Observations | | 133 |
| 6.2 | Linearisation of the Observation Model | | 135 |
| 6.3 | Partial Derivatives of Observation Function | | 137 |
| 6.4 | Linear Transformation and Covariance Propagation | | 141 |
| 6.5 | Data Combinations | | 142 |
| | 6.5.1 | Ionosphere-Free Combinations | 144 |
| | 6.5.2 | Geometry-Free Combinations | 145 |
| | 6.5.3 | Standard Phase–Code Combination | 148 |
| | 6.5.4 | Ionospheric Residuals | 149 |
| | 6.5.5 | Differential Doppler and Doppler Integration | 150 |
| 6.6 | Data Differentiations | | 152 |
| | 6.6.1 | Single Differences | 153 |
| | 6.6.2 | Double Differences | 156 |
| | 6.6.3 | Triple Differences | 158 |
| 6.7 | Equivalence of the Uncombined and Combining Algorithms | | 160 |
| | 6.7.1 | Uncombined GPS Data Processing Algorithms | 161 |
| | 6.7.2 | Combining Algorithms of GPS Data Processing | 163 |
| | 6.7.3 | Secondary GPS Data Processing Algorithms | 168 |
| | 6.7.4 | Summary | 171 |
| 6.8 | Equivalence of Undifferenced and Differencing Algorithms | | 172 |
| | 6.8.1 | Introduction | 172 |
| | 6.8.2 | Formation of Equivalent Observation Equations | 173 |
| | 6.8.3 | Equivalent Equations of Single Differences | 175 |
| | 6.8.4 | Equivalent Equations of Double Differences | 179 |
| | 6.8.5 | Equivalent Equations of Triple Differences | 181 |
| | 6.8.6 | Method of Dealing with the Reference Parameters | 182 |
| | 6.8.7 | Summary of the Unified Equivalent Algorithm | 183 |
| References | | | 184 |

# 7 Adjustment and Filtering Methods ........ 187
- 7.1 Introduction ........ 187
- 7.2 Least Squares Adjustment ........ 187
  - 7.2.1 Least Squares Adjustment with Sequential Observation Groups ........ 189
- 7.3 Sequential Least Squares Adjustment ........ 191
- 7.4 Conditional Least Squares Adjustment ........ 193
  - 7.4.1 Sequential Application of Conditional Least Squares Adjustment ........ 195
- 7.5 Block-Wise Least Squares Adjustment ........ 196
  - 7.5.1 Sequential Solution of Block-Wise Least Squares Adjustment ........ 198
  - 7.5.2 Block-Wise Least Squares for Code–Phase Combination ........ 200
- 7.6 Zhou's Theory: Equivalently Eliminated Observation Equation System ........ 201
  - 7.6.1 Zhou–Xu's Theory: Diagonalised Normal Equation and the Equivalent Observation Equation ........ 204
- 7.7 Kalman Filter ........ 206
  - 7.7.1 Classic Kalman Filter ........ 206
  - 7.7.2 Kalman Filter: A General Form of Sequential Least Squares Adjustment ........ 208
  - 7.7.3 Robust Kalman Filter ........ 209
  - 7.7.4 Yang's Filter: Adaptively Robust Kalman Filtering ........ 212
  - 7.7.5 Progress in Adaptively Robust Filter Theory and Application ........ 216
  - 7.7.6 A Brief Introduction to the Intelligent Kalman Filter ........ 218
- 7.8 A Priori Constrained Least Squares Adjustment ........ 218
  - 7.8.1 A Priori Parameter Constraints ........ 219
  - 7.8.2 A Priori Datum ........ 220
  - 7.8.3 Zhou's Theory: Quasi-Stable Datum ........ 222
- 7.9 Summary ........ 224
- References ........ 226

# 8 Cycle Slip Detection and Ambiguity Resolution ........ 229
- 8.1 Cycle Slip Detection ........ 229
- 8.2 Method of Dealing with Cycle Slips ........ 231
- 8.3 A General Criterion of Integer Ambiguity Search ........ 231
  - 8.3.1 Introduction ........ 231
  - 8.3.2 Summary of Conditional Least Squares Adjustment ........ 232
  - 8.3.3 Float Solution ........ 234
  - 8.3.4 Integer Ambiguity Search in Ambiguity Domain ........ 235
  - 8.3.5 Integer Ambiguity Search in Coordinate and Ambiguity Domains ........ 236

|  |  | 8.3.6 | Properties of Xu's General Criterion.............. | 238 |
|---|---|---|---|---|
|  |  | 8.3.7 | An Equivalent Ambiguity Search Criterion and Its Properties.......................... | 239 |
|  |  | 8.3.8 | Numerical Examples of the Equivalent Criterion..... | 242 |
|  |  | 8.3.9 | Conclusions and Comments.................... | 244 |
|  | 8.4 | Ambiguity Resolution Approach Based on the General Criterion............................................ | | 245 |
|  | 8.5 | Ambiguity Function.................................. | | 247 |
|  |  | 8.5.1 | Xu's Conjecture: Maximum Property of Ambiguity Function................................. | 248 |
|  | 8.6 | Ionosphere-Free Ambiguity Fixing...................... | | 251 |
|  |  | 8.6.1 | Introduction.............................. | 251 |
|  |  | 8.6.2 | Concept of Ionospheric Ambiguity Correction....... | 253 |
|  |  | 8.6.3 | Determination of the Ionospheric Ambiguity Correction............................... | 256 |
|  |  | 8.6.4 | Integer Ambiguity Fixing Through Ambiguity-Ionospheric Equations................ | 257 |
|  |  | 8.6.5 | Float Ambiguity Fixing....................... | 257 |
|  | 8.7 | PPP Ambiguity Fixing............................... | | 257 |
|  | References................................................ | | | 259 |
| 9 | **Parameterisation and Algorithms of GPS Data Processing**....... | | | 263 |
|  | 9.1 | Parameterisation of the GPS Observation Model........... | | 263 |
|  |  | 9.1.1 | Evidence of the Parameterisation Problem of the Undifferenced Observation Model.......... | 264 |
|  |  | 9.1.2 | A Method of Uncorrelated Bias Parameterisation..... | 265 |
|  |  | 9.1.3 | Geometry-Free Illustration.................... | 271 |
|  |  | 9.1.4 | Correlation Analysis in the Case of Phase–Code Combinations ............................. | 272 |
|  |  | 9.1.5 | Conclusions and Comments.................... | 273 |
|  | 9.2 | Equivalence of the GPS Data Processing Algorithms........ | | 274 |
|  |  | 9.2.1 | Equivalence Theorem of GPS Data Processing Algorithms ............................... | 275 |
|  |  | 9.2.2 | Optimal Baseline Network Forming and Data Condition ....................... | 277 |
|  |  | 9.2.3 | Algorithms Using Secondary GPS Observables...... | 279 |
|  |  | 9.2.4 | Simplified Equivalent Representation of GPS Observation Equations ....................... | 280 |
|  | 9.3 | Non-equivalent Algorithms........................... | | 287 |
|  | 9.4 | Reference Changing in GPS Difference Algorithm.......... | | 287 |
|  |  | 9.4.1 | Changing Reference Satellite................... | 287 |
|  |  | 9.4.2 | Changing Reference Station.................... | 288 |
|  | 9.5 | Standard Algorithms of GPS Data Processing ............. | | 291 |
|  |  | 9.5.1 | Preparation of GPS Data Processing.............. | 291 |

|  |  | 9.5.2 | Single Point Positioning | 292 |
|---|---|---|---|---|
|  |  | 9.5.3 | Standard Un-differential GPS Data Processing | 297 |
|  |  | 9.5.4 | Equivalent Method of GPS Data Processing | 300 |
|  |  | 9.5.5 | Relative Positioning | 301 |
|  |  | 9.5.6 | Velocity Determination | 302 |
|  |  | 9.5.7 | Kalman Filtering Using Velocity Information | 305 |
|  | 9.6 | Accuracy of the Observational Geometry | | 306 |
|  | 9.7 | Introduction to the Real-Time Positioning System | | 308 |
|  |  | 9.7.1 | Network RTK | 308 |
|  |  | 9.7.2 | PPP-RTK | 311 |
|  | References | | | 311 |
| 10 | **Applications of GPS Theory and Algorithms** | | | 313 |
|  | 10.1 | Software Development | | 313 |
|  |  | 10.1.1 | Functional Library | 313 |
|  |  | 10.1.2 | Data Platform | 318 |
|  |  | 10.1.3 | A Data Processing Core | 320 |
|  | 10.2 | Introduction of GPS Software | | 321 |
|  | 10.3 | Concept of Precise Kinematic Positioning and Flight-State Monitoring | | 323 |
|  |  | 10.3.1 | Introduction | 324 |
|  |  | 10.3.2 | Concept of Precise Kinematic Positioning | 326 |
|  |  | 10.3.3 | Concept of Flight-State Monitoring | 330 |
|  |  | 10.3.4 | Results, Precision Estimation, and Comparisons | 333 |
|  |  | 10.3.5 | Conclusions | 338 |
|  | References | | | 339 |
| 11 | **Perturbed Orbit and Its Determination** | | | 341 |
|  | 11.1 | Perturbed Equation of Satellite Motion | | 341 |
|  |  | 11.1.1 | Lagrangian Perturbed Equation of Satellite Motion | 342 |
|  |  | 11.1.2 | Gaussian Perturbed Equation of Satellite Motion | 345 |
|  | 11.2 | Perturbation Forces of Satellite Motion | | 348 |
|  |  | 11.2.1 | Perturbation of the Earth's Gravitational Field | 348 |
|  |  | 11.2.2 | Perturbations of the Sun, the Moon, and the Planets | 353 |
|  |  | 11.2.3 | Earth Tide and Ocean Tide Perturbations | 354 |
|  |  | 11.2.4 | Solar Radiation Pressure | 358 |
|  |  | 11.2.5 | Atmospheric Drag | 362 |
|  |  | 11.2.6 | Additional Perturbations | 365 |
|  |  | 11.2.7 | Order Estimations of Perturbations | 367 |
|  |  | 11.2.8 | Ephemerides of the Moon, the Sun, and Planets | 368 |
|  | 11.3 | Analysis Solution of the $\overline{C}_{20}$ Perturbed Orbit | | 372 |
|  | 11.4 | Orbit Correction | | 379 |
|  | 11.5 | Principle of GPS Precise Orbit Determination | | 383 |
|  |  | 11.5.1 | Xu's Algebraic Solution to the Variation Equation | 385 |

|  |  |  |
|---|---|---|
| | 11.6 Numerical Integration and Interpolation Algorithms | 387 |
| | 11.6.1 Runge–Kutta Algorithm | 387 |
| | 11.6.2 Adams Algorithms | 391 |
| | 11.6.3 Cowell Algorithms | 394 |
| | 11.6.4 Mixed Algorithms and Discussions | 396 |
| | 11.6.5 Interpolation Algorithms | 397 |
| | 11.7 Orbit-Related Partial Derivatives | 398 |
| | References | 407 |
| **12** | **Singularity-Free Orbit Theory** | **409** |
| | 12.1 A Brief Historical Review of the Singularity Problem | 409 |
| | 12.2 On the Singularity Problem in Orbital Mechanics | 412 |
| | 12.2.1 Basic Lagrangian and Gaussian Equations of Motion | 412 |
| | 12.2.2 Solving Algorithm for the Singularity Problem | 417 |
| | 12.2.3 Xu's Criteria for Singularity | 418 |
| | 12.2.4 Derivation of Lagrange-Xu Equations of Motion | 419 |
| | 12.2.5 Derivation of Gauss Equations from Lagrange Equations | 429 |
| | 12.2.6 Derivation of Gauss-Xu Equations of Motion | 431 |
| | 12.3 Bridge Between Analytical Theory and Numerical Integration | 434 |
| | References | 435 |
| **13** | **Discussions** | **439** |
| | 13.1 Independent Parameterisation and A Priori Information | 439 |
| | 13.2 Equivalence of the GPS Data Processing Algorithms | 441 |
| | 13.3 Other Comments | 442 |

**Appendix A: IAU 1980 Theory of Nutation** ............ **445**

**Appendix B: Numerical Examples of the Diagonalisation of the Equations** ............ **449**

**References** ............ **455**

**Index** ............ **483**

# Abbreviations and Constants

## Abbreviations

| | |
|---|---|
| AF | Ambiguity Function |
| AS | Anti-Spoofing |
| AU | Astronomical Units |
| C/A | Coarse Acquisition |
| CAS | Chinese Academy of Sciences |
| CIO | Conventional International Origin |
| CHAMP | Challenging Mini-Satellite Payload |
| CRF | Conventional Reference Frame |
| CTS | Conventional Terrestrial System |
| DD | Double Difference |
| DGK | Deutsche Geodätische Kommission |
| DGPS | Differential GPS |
| DOP | Dilution of Precision |
| ECEF | Earth-Centred Earth-Fixed (system) |
| ECI | Earth-Centred Inertial (system) |
| ECSF | Earth-Centred Space-Fixed (system) |
| ESA | European Space Agency |
| EU | European Union |
| Galileo | Global Navigation Satellite System of the EU |
| GAST | Greenwich Apparent Sidereal Time |
| GDOP | Geometric Dilution of Precision |
| GFZ | GeoForschungsZentrum Potsdam |
| GIS | Geographic Information System |
| GLONASS | Global Navigation Satellite System of Russia |
| GLOT | GLONASS Time |
| GMST | Greenwich Mean Sidereal Time |
| GNSS | Global Navigation Satellite System |
| GPS | Global Positioning System |

| | |
|---|---|
| GPST | GPS Time |
| GRACE | Gravity Recovery and Climate Experiment |
| GRS | Geodetic Reference System |
| GST | Galileo System Time |
| GXu | Guochang Xu, first author of this and other books |
| HDOP | Horizontal Dilution of Precision |
| IAG | International Association of Geodesy |
| IAT | International Atomic Time |
| IAU | International Astronomical Union |
| IERS | International Earth Rotation Service |
| IGS | International GPS Geodynamics Service |
| INS | Inertial Navigation System |
| ION | Institute of Navigation |
| ITRF | IERS Terrestrial Reference Frame |
| IUGG | International Union for Geodesy and Geophysics |
| JD | Julian Date |
| JPL | Jet Propulsion Laboratory |
| KMS | National Survey and Cadastre (Denmark) |
| KSGsoft | Kinematic/Static GPS Software |
| LEO | Low-Earth Orbit (satellite) |
| LS | Least Squares (adjustment) |
| LSAS | Least Squares Ambiguity Search (criterion) |
| MEO | Medium-Earth Orbit (satellite) |
| MFGsoft | Multi-Functional GPS/Galileo Software |
| MIT | Massachusetts Institute of Technology |
| MJD | Modified Julian Date |
| NASA | National Aeronautics and Space Administration |
| NAVSTAR | Navigation System with Time and Ranging |
| NGS | National Geodetic Survey |
| OD | Orbit Determination |
| OTF | On-the-Fly |
| PC | Personal Computer |
| PDOP | Position Dilution of Precision |
| PRN | Pseudorandom Noise |
| PZ-90 | Parameters of the Earth Year 1990 |
| RINEX | Receiver Independent Exchange (format) |
| RMS | Root Mean Square |
| RTK | Real-Time Kinematic |
| SA | Selective Availability |
| SC | Semicircles |
| SD | Single Difference |
| SINEX | Software Independent Exchange (format) |
| SLR | Satellite Laser Ranging |
| SNR | Signal-to-Noise Ratio |
| SST | Satellite–Satellite Tracking |

| | |
|---|---|
| SV | Space Vehicle |
| TAI | International Atomic Time |
| TD | Triple Difference |
| TDB | Barycentric Dynamic Time |
| TDOP | Time Dilution of Precision |
| TDT | Terrestrial Dynamic Time |
| TEC | Total Electron Content |
| TJD | Time of Julian Date |
| TOPEX | (Ocean) Topography Experiment |
| TOW | Time of Week |
| TRANSIT | Time Ranging and Sequential |
| TT | Terrestrial Time |
| UT | Universal Time |
| UTC | Coordinated Universal Time |
| $UTC_{SU}$ | Moscow Time UTC |
| VDOP | Vertical Dilution of Precision |
| WGS | World Geodetic System |
| ZfV | Zeitschrift für Vermessungswesen |

# Constants

| Symbol | Value | Unit | Explanation | cf. |
|---|---|---|---|---|
| $a_e$ | 6,378,137 | m | Semi-major axis of WGS-84 | §2.1 |
| $f_e$ | 1/298.2572236 | | Flattening factor of WGS-84 | §2.1 |
| $a_p$ | 6,378,136 | m | Semi-major axis of PZ-90 | §2.1 |
| $f_p$ | 1/298.2578393 | | Flattening factor of PZ-90 | §2.1 |
| $a_{eI}$ | 6,378,136.54 | m | Semi-major axis of ITRF-96 | §2.1 |
| $f_{eI}$ | 1/298.25645 | | Flattening factor of ITRF-96 | §2.1 |
| $\varepsilon$ | 84,381."412 | | Obliquity of the ecliptic at J2000.0 | §2.4 |
| JDGPS | 2,444,244.5 | JD | Julian Date of GPS standard epoch (1980 Jan. 6, 0 h) | §2.6 |
| JD2000.0 | 2,451,545.0 | JD | Julian Date of 2000 January 1, 12 h | §2.6 |
| $G$ | 6.67259e–11 | $m^3 s^{-2} kg^{-1}$ | Constant of gravitation | §3.1 |
| $\mu_e$ | 3.986004418e14 | $m^3 s^{-2}$ | Geocentric gravitational constant | §3.1 |
| $\omega_e$ | 7.292115e–5 | $rad s^{-1}$ | Nominal mean angular velocity of the earth | §3.3 |
| $c$ | 299,792,458 | $m s^{-1}$ | Speed of light | §4.1 |
| $\mu_s$ | 1.327124e20 | $m^3 s^{-2}$ | Heliocentric gravitational constant | §5.4 |
| $\mu_m$ | $\mu_e(M/M_e)$ | $m^3 s^{-2}$ | Gravitational constant of the moon | §5.4 |
| $M/M_e$ | 0.0123000345 | | Moon–Earth mass ratio | §5.4 |

(continued)

(continued)

| Symbol | Value | Unit | Explanation | cf. |
|---|---|---|---|---|
| $h_2, h_3$ | 0.6078, 0.292 | | Love number | §5.4 |
| $l_2$ | 0.0847 | | Shida number | §5.4 |
| $f_0$ | 10.23 | MHz | Fundamental frequency of GPS | §8.5 |
| $f_1$ | $154 f_0$ | MHz | First carrier frequency of GPS | §8.5 |
| $\lambda_1$ | 19.029 | cm | Wavelength of $f_1$ | §8.5 |
| $f_2$ | $120 f_0$ | MHz | Second carrier frequency of GPS | §8.5 |
| $\lambda_2$ | 24.421 | cm | Wavelength of $f_2$ | §8.5 |
| $f_5$ | $115 f_0$ | MHz | Third carrier frequency of GPS | §8.5 |
| $\lambda_5$ | 25.482 | cm | Wavelength of $f_5$ | §8.5 |
| $P_s$ | 4.5605e–6 | $Nm^{-1}$ | Luminosity of the sun | §11.2 |
| $a_s$ | 1.0000002AU | m | Semi-major axis of the orbit of the sun | §11.2 |
| AU | 149597870691 | m | Astronomical units | §11.2 |
| $a_m$ | 384,401,000 | m | Semi-major axis of the orbit of the moon | §11.2 |
| $f_{g1}$ | 1602 | MHz | First carrier frequency of GLONASS | §1.2 |
| $\Delta f_{g1}$ | 0.5625 | MHz | First carrier frequency interval of GLONASS | §1.2 |
| $f_{g2}$ | 1246 | MHz | Second carrier frequency of GLONASS | §1.2 |
| $\Delta f_{g2}$ | 0.4375 | MHz | Second carrier frequency interval of GLONASS | §1.2 |

# Chapter 1
# Introduction

The Global Positioning System (GPS) is a navigation system based on satellite technology. Its fundamental technique involves measuring the ranges between the receiver and a few simultaneously observed satellites, and the positions of the satellites are forecasted and broadcasted along with the GPS signal to the user. Through several known positions (of the satellites) and the measured distances between the receiver and the satellites, the position of the receiver can be determined. The position change, which can also be determined, is then the velocity of the receiver. The most important applications of GPS are positioning and navigation.

Through its evolution over the past few decades, GPS has now come to be known even by school children. It has been extensively applied in several areas, including air, sea, and land navigation, low-earth orbit (LEO) satellite orbit determination, static and kinematic positioning, and flight-state monitoring, as well as surveying. Its wide utility has made GPS a necessity for industry, research, education, and daily life.

For example, joggers wishing to determine their location using a GPS watch can do so very simply, merely by pressing a key. The underlying principles of such an application, however, are complex, and include knowledge of electronics, orbital mechanics, atmospheric science, geodesy, relativity theory, mathematics, adjustment and filtering, and software engineering. Many scientists and engineers have devoted efforts toward making GPS theory easier to understand and its applications more precise.

Galileo is the European global positioning system, and GLONASS is the Russian system. In China, the BeiDou Navigation Satellite System has undergone rapid development in recent years. The positioning and navigation principles of these systems are nearly the same as those of the US GPS system. With very few exceptions, GPS theory and algorithms can be directly used for the Galileo, GLONASS, and BeiDou systems. The global navigation satellite system (GNSS) of the future will likely feature a combination of the GPS, GLONASS, Galileo, and BeiDou systems.

In order to describe the distance measurement using a mathematical model, coordinates, and time systems, the orbital motion of the satellite and GPS observations must be discussed (Chaps. 2–4). The physical influences on GPS measurement such as ionospheric and tropospheric effects also must be dealt with (Chap. 5). Linearised observation equations can then be formed using various methods, such as data combination and differentiation as well as the equivalent technique (Chap. 6). The equation system may be full-rank or rank-deficient and may need to be solved in a post-processing or a quasi-real-time way, so the various adjustment and filtering methods shall be discussed (Chap. 7). For precise GPS applications, phase observations must be used, and therefore the ambiguity problem must be dealt with (Chap. 8). The algorithms of parameterisation and the equivalence theorem as well as standard algorithms of GPS data processing can then be discussed (Chap. 9). Applications of GPS theory and algorithms to GPS/Galileo software development will be sequentially outlined, and the concept of precise kinematic positioning and flight-state monitoring from practical experience will be present (Chap. 10). The theory of dynamic GPS applications for perturbed orbit determination based on the above-referenced theory will then be described (Chap. 11). Singularity-free orbits theory will be outlined and discussed for the purpose of a combined analytical and numerical orbit determination (Chap. 12). Discussions and comments will be presented in the final chapter. Thus the contents and structure of this book are organised in a logical sequence.

The book covers kinematic, static, and dynamic GPS theory and algorithms. Most of the content is refined theory that has been applied to the independently developed scientific GPS software, KSGSoft (**K**inematic and **S**tatic **GPS Soft**ware) and MFGSoft (**M**ulti-**F**unctional **GPS**/Galileo **Soft**ware), and which was obtained from extensive research on individual problems. Because of our strong research and application background, we are able to describe complex theories comfortably and with confidence. A brief summary of the contents is given in the preface.

Numerous GPS and GPS-related books are frequently quoted and carefully studied. Some of these are strongly suggested for further reading, e.g., Bauer (1994), Hofmann-Wellenhof et al. (2001), King et al. (1987), Kleusberg and Teunissen (Eds.) (1996), Leick (1995), Liu et al. (1996), Parkinson and Spilker (Eds.) (1996), Remondi (1984), Seeber (1993), Strang and Borre (1997), Wang et al. (1988), Xu (1994), Xu (2003b, 2007, 2008, 2010, 2012), Xu and Xu (2013).

## 1.1
## A Key Note on GPS

The US Global Positioning System was designed and built and is operated and maintained by the US Department of Defense (cf., e.g., Parkinson and Spilker 1996). The first GPS satellite was launched in 1978, and the system was fully operational by the mid-1990s. The GPS constellation consists of 24 satellites in six orbital planes, with four satellites in each plane. The ascending nodes of the orbital planes are equally spaced 60° apart, and the orbital planes are inclined at 55°. Each GPS satellite is in a nearly circular orbit, with a semi-major axis of 26,578 km and a period of about 12 h. The satellites continuously orient themselves to ensure that their solar panels stay pointed toward the sun and their antennas toward the earth. Each satellite carries four atomic clocks, is roughly the size of a car, and weighs about 1000 kg. The long-term frequency stability of the clocks reaches better than a few parts in $10^{-13}$ over the course of a day (cf. Scherrer 1985). The atomic clocks aboard the satellite produce the fundamental L-band frequency, 10.23 MHz.

The GPS satellites are monitored by five base stations. The main base station is in Colorado Springs, CO, and the other four are located on Ascension Island (Atlantic Ocean), Diego Garcia (Indian Ocean), Kwajalein, and Hawaii (both Pacific Ocean). All stations are equipped with precise caesium clocks and receivers to determine the broadcast ephemerides and to model the satellite clocks. Ephemerides and clock adjustments are transmitted to the satellites, which in turn use these updates in the signals that they send to GPS receivers.

Each GPS satellite transmits data on three frequencies: L1 (1575.42 MHz), L2 (1227.60 MHz), and L5 (1176.45 MHz). The L1, L2, and L5 carrier frequencies are generated by multiplying the fundamental frequency by 154, 120, and 115, respectively. Pseudorandom noise (PRN) codes, along with satellite ephemerides, ionospheric models, and satellite clock corrections are superimposed onto the carrier frequencies L1, L2, and L5. The measured transmission times of the signals that travel from the satellites to the receivers are used to compute the pseudoranges. The course/acquisition (C/A) code, sometimes called the Standard Positioning Service (SPS), is a pseudorandom noise code that is modulated onto the L1 carrier. The precision (P) code, sometimes called the Precise Positioning Service (PPS), is modulated onto the L1, L2, and L5 carriers, allowing for the removal of the effects of the ionosphere.

GPS was conceived as a ranging system from known positions of satellites in space to unknown positions on land and sea, as well as in air and space. The orbits of the GPS satellites are available by broadcast or by the International Geodetic Service (IGS) . IGS orbits are precise ephemerides after post-processing or quasi-real-time processing. All GPS receivers have an almanac programmed into their computer, which tells them where each satellite is at any given moment. The almanac is a data file containing information on orbits and clock corrections for all satellites. It is transmitted by a GPS satellite to a GPS receiver, where it facilitates rapid satellite vehicle acquisition within the GPS receivers. The GPS receivers detect, decode, and process the signals received from the satellites to create the data for code, phase, and Doppler observables. The data may be available in real time or saved for downloading. The receiver internal software is usually used to process the real-time data with the single point positioning method and to output the information to the user. Because of the limitation of the receiver software, precise positioning and navigating are usually carried out by an external computer with more powerful software. The fundamental contribution of GPS for users of the system is to inform them of their location, movements, and timing.

As the GPS technology has moved into the civilian sector, its applications have become almost limitless, and understanding GPS has become a necessity.

## 1.1.1
### GPS Modernization

GPS modernization is an ongoing effort to upgrade with new, advanced capabilities to meet growing military, civil, and commercial needs (GPS.gov 2015). A major focus of the GPS modernization program is the addition of new navigation signals to the satellite constellation. The program also involves a series of consecutive satellite acquisitions (Shaw 2011) and improvements to the GPS control segment (Bailey 2014). The specific initiatives involved in GPS modernization include the following.

*Ending Selective Availability*

The first step in the GPS modernization program occurred in May 2000, when the use of Selective Availability (SA) was ended. SA was an intentional degradation of civilian GPS accuracy implemented on a global basis from the GPS satellites. Prior to its deactivation, civil GPS readings could be off by up to 100 m. After SA was turned off, civil GPS accuracy was instantly improved by an order of magnitude, benefiting civil and commercial users worldwide.

In September 2007, the US government announced its decision to procure the future generation of GPS satellites, known as GPS III, without the SA feature. This makes the policy decision of 2000 permanent and eliminates a source of uncertainty in GPS performance that had been of concern to civil GPS users worldwide.

*New Civil Signals*

The central focus of the GPS modernization program is the addition of new navigation signals to the satellite constellation (Enge 2003). Three new signals are designed for civilian use: L2C, L5, and L1C. The legacy civil signal, called L1 C/A or C/A at L1, will continue broadcasting in the future, for a total of four civil GPS signals. The new signals are phasing in incrementally as new GPS satellites are launched to replace older ones. Most of the new signals will be of limited use until they are broadcast from 18 to 24 satellites.

L2C is the second civilian GPS signal, designed specifically to meet commercial needs. When combined with L1 C/A in a dual-frequency receiver, L2C enables ionospheric correction. Civilians with dual-frequency GPS receivers can achieve the same accuracy as the military. L2C delivers faster signal acquisition, enhanced reliability, and greater operating range. It also broadcasts at a higher effective power than the legacy L1 C/A signal, making it easier to receive under trees and even indoors.

L5 is the third civilian GPS signal, broadcast in a radio band reserved exclusively for aviation safety services. With protected spectrum, higher power, greater bandwidth features, L5 is designed to support safety-of-life transportation and other high-performance applications. It will provide users worldwide with the most advanced civilian GPS signal. In combination with L1 C/A, L5 will be used to improve accuracy through ionospheric correction and robustness via signal redundancy. When used in combination with L1 C/A and L2C, L5 will provide a highly robust service.

L1C is the fourth civilian GPS signal, designed to enable interoperability between GPS and international satellite navigation systems. The design will improve mobile GPS reception in cities and other challenging environments. Japan's Quasi-Zenith Satellite System (QZSS), the Indian Regional Navigation Satellite System (IRNSS), and China's BeiDou system also adopt L1C-like signals for international interoperability.

*New GPS Satellites*

The GPS constellation is a mix of new and legacy satellites. And the GPS modernization program involves a series of consecutive satellite acquisitions, including GPS IIR(M), GPS IIF, and GPS III.

The IIR(M) series of satellites is an upgraded version of the IIR series, completing the backbone of today's GPS constellation. The new civil and military GPS signal known as L2C is added to this generation of spacecraft. It has a 7.5-year design lifespan. It was launched in 2005–2009, and there are seven healthy IIR(M) satellites in the GPS constellation.

The IIF series expands on the capabilities of the IIR(M) series with the addition of a third civil signal on the L5 frequency for safety-of-life transportation applications. Compared to previous generations, GPS IIF satellites have a longer life expectancy and a higher accuracy requirement. Each spacecraft uses advanced atomic clocks. The IIF series will improve the accuracy, signal strength, and quality of GPS. It has a 12-year design lifespan. It was launched in 2010, and there are ten operational IIF satellites in the GPS constellation.

The III series is the most currently developed and the newest block of GPS satellites, adding a fourth civil signal on L1 (L1C). GPS III will provide more powerful signals in addition to enhanced signal reliability, accuracy, and integrity, all of which will support position, navigation, and timing services. It has a 15-year design lifespan and is planned to begin launching in 2016.

*The Control Segment Upgrades*

As part of the GPS modernization program, the GPS control segment has been continuously upgraded, including the Legacy Accuracy Improvement Initiative (L-AII); Architecture Evolution Plan (AEP); Launch and early orbit, Anomaly resolution, and Disposal Operations (LADO); Next Generation Operational Control System (OCX), and Launch Checkout Capability (LCC).

The L-AII, completed in 2008, expanded the number of monitoring sites in the operational control segment from 6 to 16, which is a 10–15 % improvement in the accuracy of the information broadcast from the GPS constellation. Ten operational GPS monitoring sites were added to help define the earth reference frame used by GPS.

In September 2007, the original legacy master control station was upgraded to an entirely new one. The AEP improved the flexibility and responsiveness of GPS operations and paved the way forward for the next generation of GPS space and control capabilities. AEP improves both monitoring stations and ground antennas, substantially improving the sustainability and accuracy of GPS. AEP is capable of managing all satellites in the constellation, including the new Block IIF satellites. It also features an alternate master control station, a fully operational backup for the master control station.

The GPS master control station can command and control a constellation of up to 32 satellites. The LADO system serves three primary functions. The first is telemetry, tracking, and control. The second is the planning and execution of satellite movements during LADO. The third function is LADO simulation of different telemetry tasks for GPS payloads and subsystems. The LADO system has been upgraded several times since 2007. In October 2010, a new version adding GPS Block IIF capability was accepted, following testing during the launch of the first GPS IIF satellite.

The OCX was developed in 2008, and will add many new capabilities to the GPS control segment, including the ability to fully control the modernized civil signals (L2C, L5, and L1C). It will be delivered in increments. OCX Block 1 will replace the existing command and control segment and support the mission operations of the initial GPS III satellites. This version will introduce the full capabilities of the L2C navigation signal. OCX Block 1 is scheduled to enter service in 2016. OCX Block 2 will support, monitor, and control additional navigation signals, including L1C and L5. OCX Block 3 will support new capabilities added to future versions of GPS III. Any increments beyond OCX Block 3 will be phased to support future satellite generations.

The LCC is a command and control centre that will check out all GPS III satellites. The LCC will be fully integrated with OCX, which will allow the operation of a single OCX-centric system that can sustain the GPS constellation from launch to disposal. The LCC component of OCX will be delivered prior to OCX Block 1 in order to support the launch and check out of the first GPS III satellite, scheduled for launch in 2015. The LCC will ensure a timely launch so constellation availability remains optimal and not impacted by the late discovery of problems.

## 1.2
## A Brief Message About GLONASS

GLONASS is the GNSS managed by the Russian Space Forces and is operated by the Coordination Scientific Information Centre (KNITs) of the Russian Defence Ministry. The system is comparable to the US GPS, and the two systems share the same principles of data transmission and positioning methods. The first GLONASS satellite was launched into orbit in 1982. The system consists of 21 satellites in three orbital planes, with three in-orbit spares. The ascending nodes of the three orbital planes are separated by 120°, and the satellites within the same orbital plane are equally spaced 45° apart. The difference in arguments of latitude for satellites in equivalent slots in two different orbital planes is 15°. Each satellite operates in nearly circular orbit, with a semi-major axis of 25,510 km. Each orbital plane has an inclination angle of 64.8°, and each satellite completes an orbit in approximately 11 h 16 min.

Caesium clocks are used on board the GLONASS satellites. The stability of the clocks reaches better than a few parts in $10^{-13}$ over a day. The satellites transmit coded signals in two frequencies located on two frequency bands, 1602–1615.5 MHz and 1246–1256.5 MHz, with a frequency interval of 0.5625 and 0.4375 MHz, respectively. The antipodal satellites, which are separated by 180° in the same orbital plane in argument of latitude, transmit on the same frequency. The signals can be received by users anywhere on the earth's surface to identify their position and velocity in real time based on ranging measurements. Coordinate and time systems used in GLONASS are different from those of the US GPS, and GLONASS satellites are distinguished by slightly different carrier frequencies rather than PRN codes. The ground control stations of GLONASS are maintained only in the territory of the former Soviet Union, for historical reasons. This lack of global coverage is not optimal for the monitoring of a GNSS.

GLONASS and GPS are not entirely compatible; however, they are generally interoperable. Combining the GLONASS and GPS resources will benefit the GNSS user community not only in increased accuracy, but also in higher system integrity on a worldwide basis.

### 1.2.1
### The Development of GLONASS

GLONASS is once again approaching full operation (Urlichich et al. 2011). Twenty-four satellites are currently in service, providing continuous global coverage (Mirgorodskaya 2013; Testoyedov 2015). These are modernized GLONASS or GLONASS-M satellites, transmitting the legacy frequency division multiple access (FDMA) navigation signals in the L1 and L2 frequency bands.

The structure of the navigation signals transmitted by the satellites determines the accuracy of the pseudorange measurements and affects a user's position accuracy. Evolution of the GLONASS navigation signals is a top priority for overall system development. A new version of the satellites, GLONASS-K, will broadcast a code division multiple access (CDMA) signal in the L3 band for the first time in the system's history. In addition to the change in signal parameters, new navigation information will be transmitted to users through this signal and will also become available in the L1 and L2 bands (Urlichich et al. 2010, 2011). The evolution of GNSS augmentation is another important aspect in the development of GLONASS. The Russian satellite-based augmentation system (SBAS) and the System for Differential Correction and Monitoring (SDCM) are in the deployment phase. Thus, interoperability and compatibility with other SBASs become important.

*Navigation Signals*

The main aspect for GLONASS development is an extension of the ensemble of navigation signals (Revnivykh 2007). This extension means that new CDMA signals in the L1, L2, and L3 bands are being added to the existing FDMA signals. The GLONASS satellites will continue to broadcast the legacy signals until the last receiver stops working.

The first phase in the implementation of CDMA technology on GLONASS-K satellites includes a new signal in the L3 band on a carrier frequency of 1202.025 MHz. The ranging code chipping rate for the CDMA signal is 10.23 megachips per second with a period of 1 ms. It is modulated onto the carrier using quadrature phase-shift keying (QPSK), with an in-phase data channel and a quadrature pilot channel.

*GLONASS Augmentation Development*

SDCM has been under development since 2002. The main elements of the system, including the network of reference stations in Russia and abroad, the central processing facility (CPF), and the SDCM information distribution channel, have been designed.

Ground Stations. The SDCM uses 14 monitoring stations in Russia and two in Antarctica. Eight more monitoring stations are being added in Russia and several more outside Russia (Revnivykh 2010).

Central Processing. Raw measurements (GLONASS and GPS L1 and L2 pseudorange and carrier phase measurements) from the ground stations come to the SDCM CPF. The CPF calculates the precise satellite ephemerides and clocks, controls integrity, and generates the SBAS messages. The format of these messages is compliant with the international standards also used by the Wide Area Augmentation System (WAAS), the European Geostationary Navigation Overlay Service (EGNOS), and the Japanese Multifunctional Transport Satellite (MTSAT) Satellite Augmentation System (MSAS).

Format Limitations. The current SBAS format has limited capacity for broadcasting corrections for GLONASS and GPS satellites combined. There is space for only 51 satellites, insufficient for the current number of satellites in orbit, and studies are looking into the efficiency of SDCM data broadcasting in an attempt to resolve this contradiction. The three main options involve using a dynamic satellite mask, using two CDMA signals, or providing an additional SBAS message.

Distribution. The main advantage of SBAS is its universal space channel to users. The SDCM orbit constellation will consist of three geostationary satellites from the multifunctional space relay system Luch, which will be used to relay communications between low earth-orbiting spacecraft and ground facilities in Russia. The satellites will also include transponders for relaying SDCM signals from CPF to users.

The development of GLONASS is entering a new historical phase. New CDMA navigation signals and deployment of a national SBAS system will provide not only a new quality of navigation service, but the basis for a regional precise navigation system with an accuracy of a few decimetres for users in Russia and neighbouring countries.

## 1.3
## Basic Information on Galileo

Galileo is a GNSS created by the European Union (EU) and the European Space Agency (ESA) to provide a highly accurate, guaranteed global positioning service under civilian control (cf., e.g., ESA homepage). While it is designed as an independent navigation system, Galileo will nonetheless be interoperable with the other two global satellite navigation systems, GPS and GLONASS. A user will be able to position with the same receiver from any of the satellites in any combination. Galileo will guarantee availability of service with higher accuracy.

The first Galileo satellite, $2.7 \times 1.2 \times 1.1$ m in size and weighing 650 kg, was launched in December 2005. The Galileo constellation consists of 30 Medium Earth orbit (MEO) satellites in three orbital planes with nine equally spaced operational satellites in each plane plus one inactive spare satellite. The ascending nodes of the orbital planes are equally spaced by 120°. The orbital planes are inclined 56°. Each Galileo satellite is in a nearly circular orbit, with a semi-major axis of 29,600 km (cf. ESA homepage) and a period of about 14 h. The Galileo satellite rotates about its earth-pointing axis such that the flat surface of the solar array always faces the sun to collect maximum solar energy. The deployed solar array spans 13 m. The antennas always point toward the earth. Once the fully deployed Galileo system is achieved, the Galileo navigation signals will provide good coverage even at latitudes up to 75°N. The large number of satellites together with the carefully optimised constellation design, plus the availability of the three active spare satellites, will ensure that the loss of one satellite has no discernible effect on the user.

The Galileo satellite has four clocks, two of each type (passive maser and rubidium, stabilities: 0.45 and 1.8 ns over 12 h, respectively). At any time, only one of each type is operational. The operating maser clock produces the reference frequency from which the navigation signal is generated. If the maser clock were to fail, the operating rubidium clock takes over instantaneously, and the two reserve clocks start up. The second maser clock takes the place of the rubidium clock after a few days when it is fully operational. The rubidium clock then goes on stand-by or reserve again. In this way, the Galileo satellite is guaranteed to generate a navigation signal at all times.

Galileo will provide ten navigation signals in right-handed circular polarization (RHCP) in the frequency ranges 1164–1215 MHz (E5a and E5b), 1215–1300 MHz (E6), and 1559–1592 MHz (E2-L1-E1) (cf. Hein et al. 2004). The interoperability and compatibility of Galileo and GPS is realized by having two common centre frequencies in E5a/L5 and L1 as well as adequate geodetic coordinate and time reference frames.

## 1.3.1
### The Development of Galileo

On 21 October 2011, the first two of four satellites designed to validate the Galileo concept both in space and on land became operational. Two more followed on 12 October 2012. This in-orbit validation (IOV) phase is now followed by additional satellite launches to reach initial operational capability (IOC) by mid-decade (Blanchard 2012). Galileo services will come with quality and integrity guarantees, marking the key difference of this first complete civil positioning system from the military systems that have come before. A range of services will be extended as the system is built up from IOC to reach full operational capability (FOC) by 2020.

Two Galileo Control Centres (GCCs) have been implemented in Europe to provide control of the satellites and to perform navigation mission management. The data provided by a global network of Galileo Sensor Stations (GSSs) will be sent to the GCCs through a redundant communications network. The GCCs will use the data from the sensor stations to compute the integrity information and to synchronise the time signal of all satellites with the ground station clocks. The exchange of data between the control centres and the satellites will be performed through up-link stations.

As an additional feature, Galileo provides a global search and rescue (SAR) function, based on the operational Cospas-Sarsat system (Bosco 2011). To do so, satellites are equipped with a transponder, which is able to transfer the distress signals from the user transmitters to regional rescue coordination centres, which will then initiate the rescue operation. At the same time, the system will send a response signal to the user, informing them that their situation has been detected and that help is on the way. This latter feature is new and is considered a major upgrade compared to the existing system, which does not provide feedback to the user (cf. ESA homepage 2015).

## 1.4
## Introduction of BeiDou

The BeiDou Navigation Satellite System (BDS), also known as BeiDou-2, is China's second-generation satellite navigation system that will be capable of providing positioning, navigation, and timing services to users on a continuous worldwide basis (ESA Navipedia 2014).

BDS consists of three major components: the space constellation, the ground control segment and the user segment. The space constellation consists of five GEO satellites, 27 MEO satellites, and three IGSO satellites. The GEO satellites are positioned at 58.75°E, 80°E, 110.5°E, 140°E, and 160°E, respectively. The MEO satellites operate in orbit at an altitude of 21,500 km and an inclination of 55° and are evenly distributed in three orbital planes. The IGSO satellites operate in orbit at an altitude of 36,000 km and an inclination of 55° and are evenly distributed in three IGSO planes. The tracks of sub-satellite points for those IGSO satellites coincide, with the intersection point at a longitude of 118°E, and a phase difference of 120°. The ground control segment consists of a Master Control Station (MCS), Time Synchronization/Upload Stations (TS/US), and Monitor Stations (MS). The main tasks of MCS are collecting observational data from each MS, processing data, generating satellite navigation messages, uploading navigation messages, monitoring satellite payload, performing mission planning and scheduling, and conducting system operation and control. The main tasks of TS/US are uploading navigation messages, exchanging data with MCS, and carrying out time synchronization and measurement under the general coordination of MCS. The main tasks of MS are the continuous tracking and monitoring of navigation satellites, receiving navigation signals, and providing observational data to the MCS for generating navigation messages. Three signals are designed for use: B1, B2, and B3. The user segment encompasses various BeiDou user terminals, including those compatible with other navigation satellite systems, to meet various application requirements from different fields and industries.

The BeiDou Time (BDT) system is used as the time reference for BDS, and the China Geodetic Coordinate System 2000 (CGCS2000) is used as the coordinate framework of BDS. Upon full system completion, BDS can provide positioning, velocity measurement, and timing services to users worldwide. It can also provide wide area differential services with accuracy of better than 1 m, as well as short message services with a capacity of 120 Chinese characters per message.

## 1.4.1
### The Development of BeiDou

BDS is steadily accelerating construction based on a "three-step" development strategy as follows:

Step I  BeiDou Navigation Satellite Demonstration System. The BeiDou Navigation Satellite Demonstration System consists of three major components: the space constellation, the ground control segment, and the user segment. The space constellation includes three GEO satellites, positioned at 80°E, 110.5°E, and 140°E, respectively, above the equator. The ground control segment consists of the ground control centre and a number of calibration stations. The ground control centre completes satellite orbit determination, ionospheric correction, user location determination, and user short message information exchanging and processing. The calibration stations mainly provide the distance measurement and six correction parameters to the ground control centre. The user segment includes hand-held type, vehicle type, command type, and other types of terminals, which are capable of sending positioning requests and receiving location information.

Step II  BDS regional services. In 2004, China initiated the construction of the BeiDou Navigation Satellite System. By the end of 2012, BDS consisted of 14 operational satellites in orbit, including five GEO satellites, five IGSO satellites, and four MEO satellites, and possessed FOC for China and the surrounding areas.

Step III  BDS global services. From 2014, additional satellites were launched, while regional service performances are advanced and expanded to the worldwide scope. Approximately 40 BeiDou navigation satellites in total will have been launched by about 2020, and the system with global coverage will be fully established.

Currently, BDS is under continuous and stable operation. As of 25 October 2012, 16 BeiDou navigation satellites had been launched to form the constellation, and they had entered into operation by the end of 2012. It possesses FOC and provides continuous passive positioning, navigation, and timing services to China and surrounding areas (China Satellite Navigation Office 2013).

Along with the construction of BDS and the development of service capabilities, BDS has been widely applied in many fields including transportation, marine fisheries, disaster forecasting, weather forecasting, forest fire prevention, time synchronization for telecommunication systems, power distribution, and disaster relief and reduction.

**Table 1.1** Frequencies of each GNSS constellation

| GNSS system | Frequency band | Frequency (MHz) |
|---|---|---|
| GPS | L1/L2/L5 | 1575.42/1227.60/1176.45 |
| GLONASS | G1/G2/G3 | 1602 + n*9/16<br>1246 + n*716<br>1202.025<br>n = −7 ~ +12 |
| Galileo | E1/E5a/E5b/E5 (E5a + E5b)/E6 | 1575.42/1176.45/1207.140/1191.795/1278.75 |
| BDS | B1/B2/B3 | 1561.098/1207.14/1268.52 |

## 1.5
## A Combined Global Navigation Satellite System

With the development of the Galileo and BeiDou systems, the GPS and GLONASS systems now face direct competition. Without a doubt, this has a positive influence on the modernisation of the GPS system and the further development of the GLONASS system. Multiple navigation systems operating independently help increase awareness and accuracy of real-time positioning and navigation. The GNSS of the future will inevitably comprise a combined system featuring an aggregation of the GPS, GLONASS, Galileo, and BeiDou systems. A constellation of hundreds of satellites among the four systems greatly increases the visibility of the satellites, especially in critical areas such as urban canyons. Many studies on multi-GNSS combinations have been conducted in recent years (Wang et al. 2001; Cai and Gao 2013; Li et al. 2015). It is expected that multi-GNSS combinations will significantly increase the number of observed satellites, optimize spatial geometry and the dilution of precision, and improve convergence, accuracy, continuity, and reliability. However, a minimum requirement for fusion of multi-GNSS data is the calibration of inter-system biases (ISB). Research related to ISB estimation and its applications, as well as ISB modelling, has been brought to the forefront recently (cf. e.g., Jiang et al. 2016).

Because GPS, GLONASS, Galileo, and BeiDou are independent systems, their time and coordinate systems differ. The four time systems are all based on UTC, and the four coordinate systems are all Cartesian systems; therefore, their relationships can be determined, and any system can be transformed from one to another. The origins of the GPS and GLONASS coordinates are meters apart from each other. The origins of GPS and Galileo coordinates have differences of a few centimetres. The GPS and BeiDou coordinates have the same origin. Several carrier frequencies are used in each system for the removal of the effects of the ionosphere. The frequency differences within the GLONASS system and between the GPS, GLONASS, Galileo, and BeiDou systems are generally not a serious problem if the carrier phase observables are considered distance surveys by multiplying the wavelength. Table 1.1 summarizes the frequencies each GNSS constellation used.

In the present edition of this book, the theory and algorithms of GPSs will be discussed in more general terms in order to take into account the differences among the GPS, GLONASS, Galileo, and BeiDou systems.

## References

Bailey Brian K (2014) GPS Modernization Update. June 2014.
Bauer M (1994) Vermessung und Ortung mit Satelliten. Wichmann Verlag, Karslruhe
Blanchard D (2012) Galileo Programme Status Update. ION GNSS 2012, November 20, pp553-587
Bosco M (2011) The European GNSS Programmes EGNOS and Galileo International Challenges Ahead. November 23, 2011.
Cai C, Gao Y (2013) Modeling and assessment of combined GPS/GLONASS precise point positioning. GPS Solutions 17(2), 223-236.
China Satellite Navigation Office (2013) Report on the Development of BeiDou (COMPASS) Navigation Satellite System (V2.2). December 2013.
Enge P (2003) GPS Modernization: Capabilities of New Civil Signals. Australian International Aerospace Congress, Brisbane, 29 July-1 August 2003.
ESA (2015) http://www.esa.int/Our_Activities/Navigation/The_future_-_Galileo/What_is_Galileo.
ESA Navipedia (2014) http://www.navipedia.net/index.php/BeiDou_General_Introduction.
GPS.gov, National Coordination Office for Space-Based Positioning, Navigation and Timing (2015) GPS Modernization, http://www.gps.gov/systems/gps/modernization/
Hein GW, Irsigler M, Avila-Rodriguez JA, Pany T (2004) Performance of Galileo L1 Signal Candidates. Proc. ENC-GNSS 2004, Rotterdam, The Netherlands, May 2004.
Hofmann-Wellenhof B, Lichtenegger H, Collins J (1997, 2001) GPS theory and practice. Springer-Press, Wien
Jiang N, Xu Y, Xu T, Xu G, Sun Z, Schuh H (2016) GPS/BDS short-term ISB modelling and prediction. GPS Solutions, DOI: 10.1007/s10291-015-0513-x
King RW, Masters EG, Rizos C, Stolz A, Collins J (1987) Surveying with Global Positioning System. Dümmler-Verlag, Bonn
Kleusberg A, Teunissen PJG (eds) (1996) GPS for geodesy. Springer-Verlag, Berlin
Leick A (1995) GPS satellite surveying. John Wiley & Sons Ltd., New York
Li X, Zhang X, Ren X, Fritsche M, Wickert J, Schuh H (2015) Precise positioning with current multi-constellation Global Navigation Satellite Systems: GPS, GLONASS, Galileo and BeiDou. Scientific reports 5, 8328.
Liu DJ, Shi YM, Guo JJ (1996) Principle of GPS and its data processing. TongJi University Press, Shanghai, (in Chinese)
Mirgorodskaya T (2013) GLONASS Government Policy, Status and Modernization Plans. IGNSS 2013, Gold Coast, Queensland, Australia, July 16, 2013.
Parkinson BW, Spilker JJ (eds) (1996) Global Positioning System: Theory and applications, Vol. I, II. American Institute of Aeronautics and Astronautics, Progress in Astronautics and Aeronautics, Vol. 163
Remondi B (1984) Using the Global Positioning System (GPS) phase observable for relative geodesy: Modelling, processing, and results. University of Texas at Austin, Center for Space Research
Revnivykh S (2007) GLONASS Status, Development and Application. International Committee on Global Navigation Satellite Systems (ICG), Bangalore, India, September 4-7, 2007.
Revnivykh S (2010) GLONASS Status and Progress. ION GNSS 2010, Portland, Oregon, September 21-24, 2010.

Scherrer R (1985) The WM GPS primer. WM Satellite Survey Company, Wild, Herrbrugg, Switzerland

Seeber G (1993) Satelliten-Geodaesie. Walter de Gruyter 1989

Shaw M (2011) GPS Modernization: On the Road to the Future GPS IIR / IIR-M and GPS III. UN/ UAE/ US Workshop On GNSS Applications, Dubai, 2011.

Strang G, Borre K (1997) Linear algebra, geodesy, and GPS. Wellesley-Cambridge Press

Testoyedov N (2015) Space Navigation in Russia: History of Development. United Nations / Russian Federation Workshop on the Applications of Global Navigation Satellite Systems, Krasnoyarsk, May 18-22, 2015.

Urlichich Y, Subbotin V, Stupak G, et al. (2010) GLONASS Developing Strategy. ION GNSS 2010, the 23$^{rd}$ International Technical Meeting of the Institute of Navigation, Portland, Oregon, September 21-24, 2010.

Urlichich Y, Subbotin V, Stupak G, et al. (2011) Innovation: GLONASS Developing strategies for the Future, GPS World, April, pp42-49

Wang G, Chen Z, Chen W, Xu G (1988) The principle of GPS precise positioning system. Surveying Press, Peking, ISBN 7-5030-0141-0/P.58, 345 p, (in Chinese)

Wang J, Rizos C, Stewart MP, Leick A (2001) GPS and GLONASS integration-Modeling and Ambiguity Resolution Issues. GPS Solutions 5(1), 55-64.

Xu QF (1994) GPS navigation and precise positioning. Army Press, Peking, ISBN 7-5065-0855-9/P.4, (in Chinese)

Xu G (2003a) A diagonalization algorithm and its application in ambiguity search. J. GPS 2(1): 35-41

Xu G (2003b) GPS – Theory, Algorithms and Applications, Springer Heidelberg, ISBN 3-540-67812-3, 315 pages, in English

Xu G (2007) GPS – Theory, Algorithms and Applications, second edition, Springer Heidelberg, ISBN 978-3-540-72714-9, 350 pages, in English

Xu G (2008) Orbits, Springer Heidelberg, ISBN 978-3-540-78521-7, 230 pages, in English

Xu G (2010) (Ed.): Sciences of Geodesy - I, Advances and Future Directions, Springer Heidelberg, chapter topics (authors): Aerogravimetry (R Forsberg), Superconducting Gravimetry (J Neumeyer), Absolute and Relative Gravimetry (L Timmen), Deformation and Tectonics (L Bastos et al.), Analytic Orbit Theory (G Xu), InSAR (Y Xia), Marine Geodesy (J Reinking), Kalman Filtering (Y Yang), Equivalence of GPS Algorithms (G Xu et al.), Earth Rotation (F Seitz, H Schuh), Satellite Laser Ranging (L Combrinck), in English, 507 pages

Xu G (2012) (Ed.): Sciences of Geodesy - II, Advances and Future Directions, Springer Heidelberg, chapter topics (authors): General Relativity and Space Geodesy (L Comblinck), Global Terrestrial Reference Systems and their Realizations (D Angermann et al), Ocean Tide Loading (M Bos, HG Scherneck), Photogrammetry (P Redweik), Regularization and Adjustment (Y Shen, G Xu), Regional Gravity Field Modelling (H Denker), VLBI (H Schuh, J Boehm), in English, 400 pages

Xu G, Xu J (2013) Orbits – 2$^{nd}$ Order Singularity-free Solutions, second edition, Springer Heidelberg, ISBN 978-3-642-32792-6, 426 pages, in English

# Chapter 2
# Coordinate and Time Systems

GPS satellites orbit the earth over time. GPS surveys are conducted mostly on land. To describe the GPS observation (distance) as a function of GPS orbit (satellite position) and the measuring position (station location), suitable coordinate and time systems must be defined.

## 2.1 Geocentric Earth-Fixed Coordinate Systems

The Earth-Centred Earth-Fixed (ECEF) coordinate system is useful for describing the location of a station on the earth's surface. This system is a right-handed Cartesian system $(x, y, z)$. Its origin and the earth's centre of mass coincide, while its $z$-axis and the mean rotational axis of the earth coincide; the $x$-axis points to the mean Greenwich meridian, while the $y$-axis is directed to complete a right-handed system (cf. Fig. 2.1). In other words, the $z$-axis points to a mean pole of the earth's rotation. The mean pole, defined by international convention, is called the Conventional International Origin (CIO). Then the $xy$-plane is called mean equatorial plane, and the $xz$-plane is called mean zero-meridian.

The ECEF coordinate system is also known as the Conventional Terrestrial System (CTS). The mean rotational axis and mean zero-meridian used here are necessary. The true rotational axis of the earth changes its direction with respect to the earth's body all the time. If such a pole were used to define a coordinate system, the coordinates of the station would also be continuously changing. Because the

**Fig. 2.1** Earth-centred earth-fixed coordinates

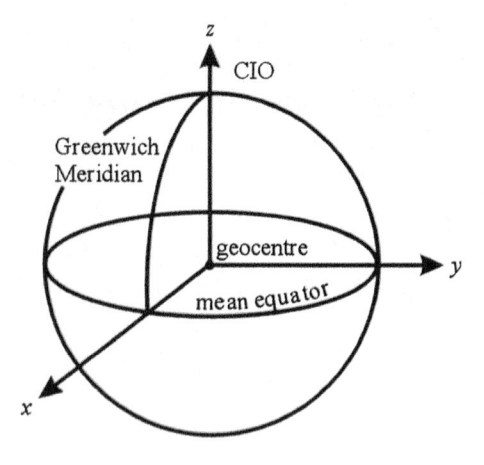

**Fig. 2.2** Cartesian and spherical coordinates

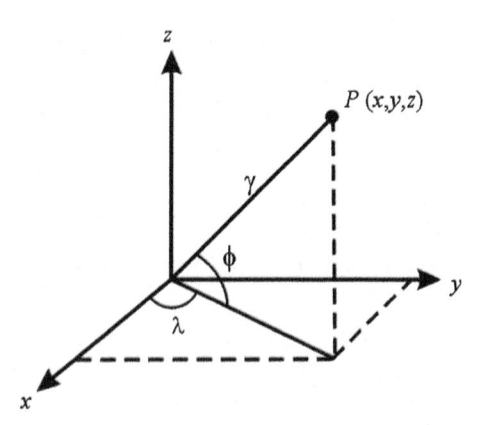

survey is made in our true world, the polar motion obviously must be taken into account, and this will be discussed later.

Of course, the ECEF coordinate system can be represented by a spherical coordinate system $(r, \phi, \lambda)$, where $r$ is the radius of the point $(x, y, z)$, and $\phi$ and $\lambda$ are the geocentric latitude and longitude, respectively (cf. Fig. 2.2). $\lambda$ is counted eastward from the zero-meridian. The relationship between $(x, y, z)$ and $(r, \phi, \lambda)$ is obvious:

$$\begin{pmatrix} x \\ y \\ z \end{pmatrix} = \begin{pmatrix} r\cos\phi\cos\lambda \\ r\cos\phi\sin\lambda \\ r\sin\phi \end{pmatrix}, \quad \text{or} \quad \begin{cases} r = \sqrt{x^2+y^2+z^2} \\ \tan\lambda = y/x \\ \tan\phi = z/\sqrt{x^2+y^2} \end{cases}. \quad (2.1)$$

An ellipsoidal coordinate system $(\varphi, \lambda, h)$ may also be defined based on the ECEF coordinates; however, geometrically, two additional parameters are needed to define the shape of the ellipsoid (cf. Fig. 2.3). $\varphi, \lambda$, and $h$ are geodetic latitude, longitude, and height, respectively. The ellipsoidal surface is a rotational ellipse.

**Fig. 2.3** Ellipsoidal coordinate system

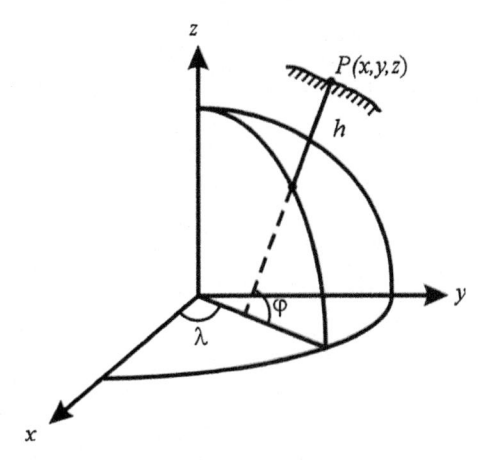

The ellipsoidal system is also called the geodetic coordinate system. Geocentric longitude and geodetic longitude are identical. The two geometric parameters can be the semi-major radius (denote by $a$) and the semi-minor radius (denote by $b$) of the rotating ellipse, or the semi-major radius and the flattening (denote by $f$) of the ellipsoid. They are equivalent sets of parameters. The relationship between ($x$, $y$, $z$) and ($\varphi$, $\lambda$, $h$) is (cf., e.g., Torge 1991):

$$\begin{pmatrix} x \\ y \\ z \end{pmatrix} = \begin{pmatrix} (N+h)\cos\varphi \cos\lambda \\ (N+h)\cos\varphi \sin\lambda \\ (N(1-e^2)+h)\sin\varphi \end{pmatrix}, \tag{2.2}$$

or

$$\begin{cases} \tan\varphi = \frac{z}{\sqrt{x^2+y^2}} \left(1 - e^2 \frac{N}{N+h}\right)^{-1} \\ \tan\lambda = \frac{y}{x} \\ h = \frac{\sqrt{x^2+y^2}}{\cos\varphi} - N \end{cases}, \tag{2.3}$$

where

$$N = \frac{a}{\sqrt{1 - e^2 \sin^2\varphi}}. \tag{2.4}$$

$N$ is the radius of curvature in the prime vertical, and $e$ is the first eccentricities. The geometric meaning of $N$ is shown in Fig. 2.4. In Eq. 2.3, $\varphi$ and $h$ must be solved by iteration; however, the iteration process converges quickly, since $h \ll N$. The flattening and the first eccentricities are defined as:

**Fig. 2.4** Radius of curvature in the prime vertical

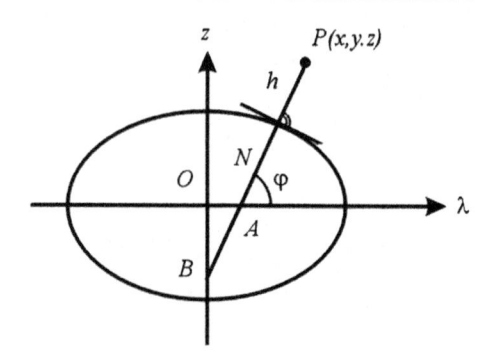

$$f = \frac{a-b}{a}, \quad \text{and} \quad e = \frac{\sqrt{a^2 - b^2}}{a}. \tag{2.5}$$

In cases where $\varphi = \pm 90°$ or $h$ is very large, the iteration formulas of Eq. 2.3 are unstable. Alternatively, using (cf. Lelgemann 2002)

$$\operatorname{ctan} \varphi = \frac{\sqrt{x^2 + y^2}}{z + \Delta z},$$

$$\Delta z = e^2 N \sin \varphi = \frac{a e^2 \sin \varphi}{\sqrt{1 - e^2 \sin^2 \varphi}},$$

may lead to a stably iterated result of $\varphi$. $\Delta z$ and $e^2 N$ are the lengths of $\overline{OB}$ and $\overline{AB}$ (cf. Fig. 2.4), respectively. $h$ can be obtained by using $\Delta z$, i.e.

$$h = \sqrt{x^2 + y^2 + (z + \Delta z)^2} - N.$$

The two geometric parameters used in the World Geodetic System 1984 (WGS-84) are ($a = 6{,}378{,}137$ m, $f = 1/298.2572236$). In International Terrestrial Reference Frame 1996 (ITRF-96), the two parameters are ($a = 6{,}378{,}136.49$ m, $f = 1/298.25645$). ITRF uses the International Earth Rotation Service (IERS) Conventions (cf. McCarthy 1996). In PZ-90 (Parameters of the Earth Year 1990) coordinate system of GLONASS, the two parameters are ($a = 6{,}378{,}136$ m, $f = 1/298.2578393$).

The relation between the geocentric and geodetic latitude $\phi$ and $\varphi$ may be given by (cf. Eqs. 2.1 and 2.3):

$$\tan \phi = \left(1 - e^2 \frac{N}{N+h}\right) \tan \varphi. \tag{2.6}$$

## 2.2
## Coordinate System Transformations

Any Cartesian coordinate system can be transformed to another Cartesian coordinate system through three successive rotations if their origins are the same and if they are both right-handed or both left-handed systems. These three rotation matrices are:

$$R_1(\alpha) = \begin{pmatrix} 1 & 0 & 0 \\ 0 & \cos\alpha & \sin\alpha \\ 0 & -\sin\alpha & \cos\alpha \end{pmatrix},$$

$$R_2(\alpha) = \begin{pmatrix} \cos\alpha & 0 & -\sin\alpha \\ 0 & 1 & 0 \\ \sin\alpha & 0 & \cos\alpha \end{pmatrix}, \quad (2.7)$$

$$R_3(\alpha) = \begin{pmatrix} \cos\alpha & \sin\alpha & 0 \\ -\sin\alpha & \cos\alpha & 0 \\ 0 & 0 & 1 \end{pmatrix},$$

where $\alpha$ is the rotating angle, which has a positive sign for a counter-clockwise rotation as viewed from the positive axis to the origin. $R_1$, $R_2$, and $R_3$ are called the rotating matrix around the $x$-, $y$-, and $z$-axis, respectively. For any rotation matrix $R$, there are $R^{-1}(\alpha) = R^T(\alpha)$ and $R^{-1}(\alpha) = R(-\alpha)$; that is, the rotation matrix is an orthogonal one, where $R^{-1}$ and $R^T$ are the inverse and transpose of the matrix $R$.

For two Cartesian coordinate systems with different origins and different length units, the general transformation can be given in vector (matrix) form as

$$X_n = X_0 + \mu R X_{old}, \text{ or,} \quad (2.8)$$

$$\begin{pmatrix} x_n \\ y_n \\ z_n \end{pmatrix} = \begin{pmatrix} x_0 \\ y_0 \\ z_0 \end{pmatrix} + \mu R \begin{pmatrix} x_{old} \\ y_{old} \\ z_{old} \end{pmatrix},$$

where $\mu$ is the scale factor (or the ratio of the two length units), and $R$ is a transformation matrix that can be formed by three suitably successive rotations. $X_n$ and $X_{old}$ denote the new and old coordinates, respectively; $X_0$ denotes the translation vector and is the coordinate vector of the origin of the old coordinate system in the new one.

If rotational angle $\alpha$ is very small, then one has $\sin\alpha \approx \alpha$ and $\cos\alpha \approx 0$. In such a case, the rotation matrix can be simplified. If the three rotational angles $\alpha_1$, $\alpha_2$, $\alpha_3$ in $R$ of Eq. 2.8 are very small, then $R$ can be written as (cf., e.g., Lelgemann and Xu 1991):

$$R = \begin{pmatrix} 1 & \alpha_3 & -\alpha_2 \\ -\alpha_3 & 1 & \alpha_1 \\ \alpha_2 & -\alpha_1 & 1 \end{pmatrix}, \tag{2.9}$$

where $\alpha_1$, $\alpha_2$, $\alpha_3$ are small rotating angles around the x-, y-, and z-axis, respectively. Using the simplified $R$, the transformation 2.8 is called the Helmert transformation.

As an example, the transformation from WGS-84 to ITRF-90 is given by McCarthy (1996):

$$\begin{pmatrix} x_{\text{ITRF}-90} \\ y_{\text{ITRF}-90} \\ z_{\text{ITRF}-90} \end{pmatrix} = \begin{pmatrix} 0.060 \\ -0.517 \\ -0.223 \end{pmatrix} + \mu \begin{pmatrix} 1 & -0.0070'' & -0.0003'' \\ 0.0070'' & 1 & -0.0183'' \\ 0.0003'' & 0.0183'' & 1 \end{pmatrix} \begin{pmatrix} x_{\text{WGS}-84} \\ y_{\text{WGS}-84} \\ z_{\text{WGS}-84} \end{pmatrix},$$

where $\mu = 0.999999989$, the translation vector has the unit of meter.

The transformations between the coordinate systems of GPS, GLONASS, and Galileo can be generally represented by Eq. 2.8 with the scale factor $\mu = 1$ (i.e. the length units used in the three systems are the same). A formula of velocity transformations between different coordinate systems can be obtained by differentiating the Eq. 2.8 with respect to the time.

## 2.3
## Local Coordinate System

The local left-handed Cartesian coordinate system $(x', y', z')$ can be defined by placing the origin to the local point $P_1(x_1, y_1, z_1)$, whose $z'$-axis points to the vertical, $x'$-axis is directed to the north, and $y'$ points to the east (cf. Fig. 2.5). The $x'$ $y'$-plane is called the horizontal plane; the vertical is defined perpendicular to the ellipsoid. Such a coordinate system is also called a local horizontal coordinate system. For any point $P_2$, whose coordinates in the global and local coordinate system are $(x_2, y_2, z_2)$ and $(x', y, z')$, respectively, one has relations of

**Fig. 2.5** Astronomical coordinate system

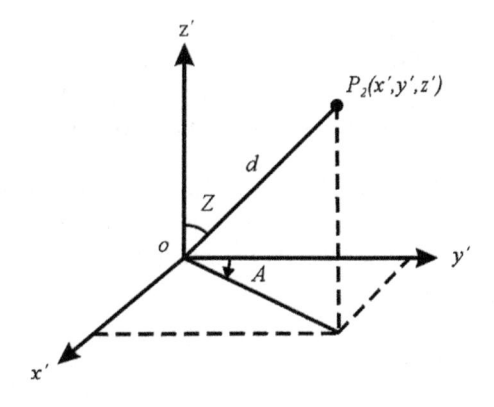

## 2.3 · Local Coordinate System

$$\begin{pmatrix} x' \\ y' \\ z' \end{pmatrix} = d \begin{pmatrix} \cos A \sin Z \\ \sin A \sin Z \\ \cos Z \end{pmatrix}, \quad \text{and} \quad \begin{pmatrix} d = \sqrt{x'^2 + y'^2 + z'^2} \\ \tan A = y'/x' \\ \cos Z = z'/d \end{pmatrix}, \quad (2.10)$$

where $A$ is the azimuth, $Z$ is the zenith distance, and $d$ is the radius of the $P_2$ in the local system. $A$ is measured from the north clockwise; $Z$ is the angle between the vertical and the radius $d$.

The local coordinate system $(x', y', z')$ can indeed be obtained by two successive rotations of the global coordinate system $(x, y, z)$ by $R_2(90 - \varphi)R_3(\lambda)$ and then by changing the $x$-axis to a right-handed system. In other words, the global system must be rotated around the $z$-axis with angle $\lambda$, then around the $y$-axis with angle $90 - \varphi$, and then change the sign of the $x$-axis. The total transformation matrix $R$ is then

$$R = \begin{pmatrix} -\sin\varphi \cos\lambda & -\sin\varphi \sin\lambda & \cos\varphi \\ -\sin\lambda & \cos\lambda & 0 \\ \cos\varphi \cos\lambda & \cos\varphi \sin\lambda & \sin\varphi \end{pmatrix}, \quad (2.11)$$

and there are:

$$X_{\text{local}} = R X_{\text{global}} \quad \text{and} \quad X_{\text{global}} = R^T X_{\text{local}}, \quad (2.12)$$

where $X_{\text{local}}$ and $X_{\text{global}}$ are the same vector represented in local and global coordinate systems. $(\varphi, \lambda)$ are the geodetic latitude and longitude of the local point.

If the vertical direction is defined as the plump line of the gravitational field at the local point, then such a local coordinate system is called an astronomic horizontal system (its $x'$-axis points to the north, left-handed system). The plump line of gravity $g$ and the vertical line of the ellipsoid at point $p$ generally do not coincide; however, the difference is very small, and is omitted in GPS in practice.

Combining Eqs. 2.10 and 2.12, the zenith angle and azimuth of a point $P_2$ (satellite) related to the station $P_1$ can be directly computed by using the global coordinates of the two points by

$$\cos Z = \frac{z'}{d} \quad \text{and} \quad \tan A = \frac{y'}{x'}, \quad (2.13)$$

where

$$d = \sqrt{(x_2 - x_1)^2 + (y_2 - y_1)^2 + (z_2 - z_1)^2},$$
$$x' = -(x_2 - x_1)\sin\varphi \cos\lambda - (y_2 - y_1)\sin\varphi \sin\lambda + (z_2 - z_1)\cos\varphi,$$
$$y' = -(x_2 - x_1)\sin\lambda + (y_2 - y_1)\cos\lambda \quad \text{and}$$
$$z' = (x_2 - x_1)\cos\varphi \cos\lambda + (y_2 - y_1)\cos\varphi \sin\lambda + (z_2 - z_1)\sin\varphi.$$

## 2.4
## Earth-Centred Inertial Coordinate System

To describe the motion of the GPS satellites, an inertial coordinate system must be defined. The motion of the satellites follows Newtonian mechanics, and Newtonian mechanics is valid and expressed in an inertial coordinate system. The Conventional Celestial Reference Frame (CRF) is suitable for our purpose. The $xy$-plane of the CRF is the plane of the earth's equator; the coordinates are celestial longitude, measured eastward along the equator from the vernal equinox, and celestial latitude. The vernal equinox is a crossover point of the ecliptic and the equator. Thus, the right-handed earth-centred inertial (ECI) system uses the earth centre as the origin, the CIO as the $z$-axis, and its $x$-axis is directed to the equinox of J2000.0 (Julian date for 1 January 2000 at 12 h). Such a coordinate system is also called equatorial coordinates of date. Because of the motion (acceleration) of the earth's centre, ECI is indeed a quasi-inertial system, and the general relativistic effects must be taken into account. The system moves around the sun, however, without rotating with respect to the CIO. This system is also called the earth-centred space-fixed (ECSF) coordinate system.

An excellent figure has been given by Torge (1991) to illustrate the motion of the earth's pole with respect to the ecliptic pole (cf. Fig. 2.6). The earth's flattening, combined with the obliquity of the ecliptic, results in a slow turning of the equator on the ecliptic due to the differential gravitational effect of the moon and the sun. The slow circular motion with a period of about 26,000 years is called precession, and the other quicker motion with periods from 14 days to 18.6 years is called nutation. Taking the precession and nutation into account, the earth's mean pole (related to the mean equator) is transformed to the earth's true pole (related to the true equator). The $x$-axis of the ECI is pointed to the vernal equinox of date.

**Fig. 2.6** Precession and nutation

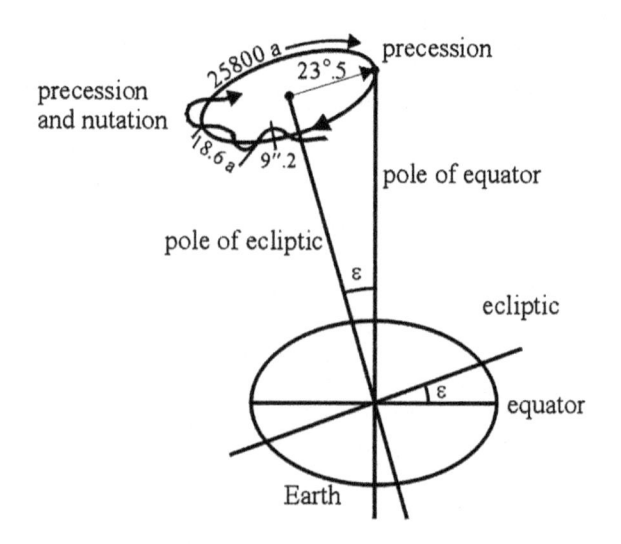

The angle of the earth's rotation from the equinox of date to the Greenwich meridian is called Greenwich Apparent Sidereal Time (GAST). Taking GAST into account (called the earth's rotation), the ECI of date is transformed to the true equatorial coordinate system. The difference between the true equatorial system and the ECEF system is the polar motion. Thus, we have transformed the ECI system in a geometric way to the ECEF system. Such a transformation process can be written as

$$X_{\text{ECEF}} = R_M R_S R_N R_P X_{\text{ECI}}, \tag{2.14}$$

where $R_P$ is the precession matrix, $R_N$ is the nutation matrix, $R_S$ is the earth rotation matrix, $R_M$ is the polar motion matrix, $X$ is the coordinate vector, and indices ECEF and ECI denote the related coordinate systems.

## Precession

The precession matrix consists of three successive rotation matrices (cf., e.g., Hofman-Wellenhof et al. 1997; Leick 1995; McCarthy 1996), i.e.

$$R_P = R_3(-z) R_2(\theta) R_3(-\zeta)$$
$$= \begin{pmatrix} \cos z \cos \theta \cos \zeta - \sin z \sin \zeta & -\cos z \cos \theta \sin \zeta - \sin z \cos \zeta & -\cos z \sin \theta \\ \sin z \cos \theta \cos \zeta + \cos z \sin \zeta & -\sin z \cos \theta \sin \zeta + \cos z \cos \zeta & -\sin z \sin \theta \\ \sin \theta \cos \zeta & -\sin \theta \sin \zeta & \cos \theta \end{pmatrix}, \tag{2.15}$$

where $z$, $\theta$, $\zeta$ are precession parameters and

$$z = 2306.''2181\,T + 1.''09468\,T^2 + 0.''018203\,T^3$$
$$\theta = 2004.''3109\,T - 0.''42665\,T^2 - 0.''041833\,T^3 \tag{2.16}$$
$$\zeta = 2306.''2181\,T + 0.''30188\,T^2 + 0.''017998\,T^3$$

where $T$ is the measuring time in Julian centuries (36,525 days) counted from J2000.0 (cf. Sect. 2.6 time systems).

## Nutation

The nutation matrix consists of three successive rotation matrices (cf., e.g., Hoffman-Wellenhof et al. 1997; Leick 1995; McCarthy 1996), i.e.

$$R_N = R_1(-\varepsilon - \Delta\varepsilon)R_3(-\Delta\psi)R_1(\varepsilon)$$

$$= \begin{pmatrix} \cos\Delta\psi & -\sin\Delta\psi\cos\varepsilon & -\sin\Delta\psi\sin\varepsilon \\ \sin\Delta\psi\cos\varepsilon_t & \cos\Delta\psi\cos\varepsilon_t\cos\varepsilon + \sin\varepsilon_t\sin\varepsilon & \cos\Delta\psi\cos\varepsilon_t\sin\varepsilon - \sin\varepsilon_t\cos\varepsilon \\ \sin\Delta\psi\sin\varepsilon_t & \cos\Delta\psi\sin\varepsilon_t\cos\varepsilon - \cos\varepsilon_t\sin\varepsilon & \cos\Delta\psi\sin\varepsilon_t\sin\varepsilon + \cos\varepsilon_t\cos\varepsilon \end{pmatrix}$$

$$\approx \begin{pmatrix} 1 & -\Delta\psi\cos\varepsilon & -\Delta\psi\sin\varepsilon \\ \Delta\psi\cos\varepsilon_t & 1 & -\Delta\varepsilon \\ \Delta\psi\sin\varepsilon_t & \Delta\varepsilon & 1 \end{pmatrix},$$

(2.17)

where $\varepsilon$ is the mean obliquity of the ecliptic angle of date, $\Delta\psi$ and $\Delta\varepsilon$ are nutation angles in longitude and obliquity, $\varepsilon_t = \varepsilon + \Delta\varepsilon$, and

$$\varepsilon = 84381.''448 - 46.''8150T - 0.''00059T^2 + 0.''001813T^3. \qquad (2.18)$$

The approximation is made by letting $\cos\Delta\psi = 1$ and $\sin\Delta\psi = \Delta\psi$ for very small $\Delta\psi$. For precise purposes, the exact rotation matrix shall be used. The nutation parameters $\Delta\psi$ and $\Delta\varepsilon$ can be computed by using the International Astronomical Union (IAU) theory or IERS theory:

$$\Delta\Psi = \sum_{i=1}^{106}(A_i + A'_i T)\sin\beta,$$

$$\Delta\varepsilon = \sum_{i=1}^{106}(B_i + B'_i T)\cos\beta$$

or

$$\Delta\Psi = \sum_{i=1}^{263}(A_i + A'_i T)\sin\beta + A''_i\cos\beta,$$

$$\Delta\varepsilon = \sum_{i=1}^{263}(B_i + B'_i T)\cos\beta + B'_i\cos\beta,$$

where argument

$$\beta = N_{1i}l + N_{2i}l' + N_{3i}F + N_{4i}D + N_{5i}\Omega,$$

where $l$ is the mean anomaly of the moon, $l'$ is the mean anomaly of the Sun, $F = L - \Omega$, $D$ is the mean elongation of the moon from the sun, $\Omega$ is the mean longitude of the ascending node of the moon, and $L$ is the mean longitude of the moon. The formulas of $l$, $l'$, $F$, $D$, and $\Omega$, are given in Sect. 11.2.8. The coefficient values of $N_{1i}$, $N_{2i}$, $N_{3i}$, $N_{4i}$, $N_{5i}$, $A_i$, $B_i$, $A'$, $B'_i$, $A_i''$, and $B_i''$ can be found in, e.g., McCarthy (1996). The updated formulas and tables can be found in updated IERS conventions. For convenience, the coefficients of the IAU 1980 nutation model are given in Appendix 1.

### Earth Rotation

The earth rotation matrix can be represented as

$$R_S = R_3(\text{GAST}), \qquad (2.19)$$

where GAST is Greenwich Apparent Sidereal Time and

$$\text{GAST} = \text{GMST} + \Delta\Psi \cos\varepsilon + 0.''00264 \sin\Omega + 0.''000063 \sin 2\Omega, \qquad (2.20)$$

where GMST is Greenwich Mean Sidereal Time. $\Omega$ is the mean longitude of the ascending node of the Moon; the second term on the right-hand side is the nutation of the equinox. Furthermore,

$$\text{GMST} = \text{GMST}_0 + \alpha \text{UT1}, \qquad (2.21)$$

$$\text{GMST}_0 = 6 \times 3600.''0 + 41 \times 60.''0 + 50.''54841$$
$$+ 8640184.''812866 T_0 + 0.''093104 T_0^2 - 6.''2 \times 10^{-6} T_0^3,$$

$$\alpha = 1.00273790093\,50795 + 5.9006 \times 10^{-11} T_0 - 5.9 \times 10^{-15} T_0^2,$$

where GMST$_0$ is Greenwich Mean Sidereal Time at midnight on the day of interest. $\alpha$ is the rate of change. UT1 is the polar motion corrected Universal Time (cf. Sect. 2.6). $T_0$ is the measuring time in Julian centuries (36,525 days) counted from J2000.0 to $0^h$UT1 of the measuring day. By computing GMST, UT1 is used (cf. Sect. 2.6).

### Polar Motion

As shown in Fig. 2.7, the polar motion is defined as the angles between the pole of date and the CIO pole. The polar motion coordinate system is defined by $xy$-plane coordinates, whose $x$-axis is pointed to the south and is coincided to the mean Greenwich meridian, and whose $y$-axis is pointed to the west. $x_p$ and $y_p$ are the angles of the pole of date, so the rotation matrix of polar motion can be represented as

$$R_M = R_2(-x_p)R_1(-y_p) = \begin{pmatrix} \cos x_p & \sin x_p \sin y_p & \sin x_p \cos y_p \\ 0 & \cos y_p & -\sin y_p \\ -\sin x_p & \cos x_p \sin y_p & \cos x_p \cos y_p \end{pmatrix}$$
$$\approx \begin{pmatrix} 1 & 0 & x_p \\ 0 & 1 & -y_p \\ -x_p & y_p & 1 \end{pmatrix}. \qquad (2.22)$$

**Fig. 2.7** Polar motion

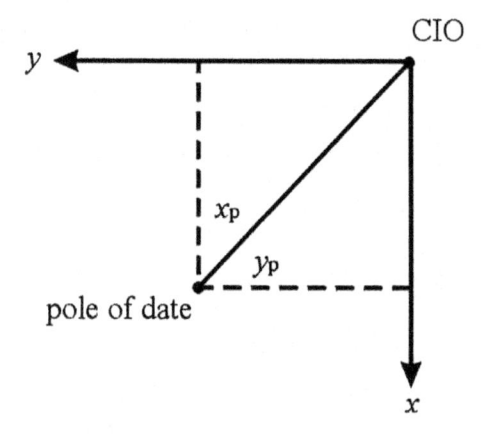

The IERS determined $x_p$ and $y_p$ can be obtained from the home pages of IERS.

## 2.5 IAU 2000 Framework

At its 2000 General Assembly, the International Astronomical Union (IAU) adopted a set of resolutions that provide a consistent framework for defining barycentric and geocentric celestial reference systems (Petit 2002). The consequence of the resolution is that coordinate transformation from celestial reference system (CRS, i.e. the ECI system) to the terrestrial reference system (TRS, i.e. the ECEF system) has the form

$$X_{ECEF} = R_M R_S R_{NP} X_{ECI}, \qquad (2.23)$$

where $R_{NP}$ is the precession-nutation matrix, $R_S$ is the earth rotation matrix, $R_M$ is the polar motion matrix, $X$ is the coordinate vector, and indices ECEF and ECI denote the related coordinate systems. The rotation matrices are functions of time $T$, which is defined (see McCarthy and Petit 2003) by

$$T = (TT - 2000 \text{ January } 1d \text{ } 12h \text{ } TT) \text{ in days}/36{,}525, \qquad (2.24)$$

where TT is the Terrestrial Time (for details see Sect. 2.7) and

$$R_M = R_2(-x_p)R_1(-y_p)R_3(s'),$$

$$R_S = R_3(\vartheta), \text{ and} \qquad (2.25)$$

$$R_{NP} = R_3(-s)R_3(-E)R_2(d)R_3(E),$$

where $x_p$ and $y_p$ are the angles of the pole of date (or polar coordinates of the Celestial Intermediate Pole (CIP) in TRS), and $s'$ is a function of $x_p$ and $y_p$:

$$s' = \frac{1}{2} \int_{T0}^{T} (x_p \dot{y}_p - \dot{x}_p y_p) dt \text{ or}$$

approximately (see McCarthy and Capitaine 2002)

$$s' = (-47 \mu as) T, \qquad (2.26)$$

where $T$ is time in Julian century counted from J2000.0 and

$$\vartheta = 2\pi (0.7790572732640 + 1.00273781191135448 \, T_u), \qquad (2.27)$$

where $T_u$ = (Julian UT1 date - 2,451,545.0) and UT1 = UTC + (UT1-UTC). (UT1-UTC) is published by the IERS.

$E$ and $d$ are such that the coordinates of the CIP in the CRS are

$$\begin{aligned} X &= \sin d \cos E \\ Y &= \sin d \sin E \\ Z &= \cos d \end{aligned} \qquad (2.28)$$

Equivalently, $R_{NP}$ can be given by

$$R_{NP} = R_3(-s) \cdot \begin{pmatrix} 1 - aX^2 & -aXY & X \\ -aXY & 1 - aY^2 & Y \\ -X & -Y & 1 - a(X^2 + Y^2) \end{pmatrix}^{-1} \qquad (2.29)$$

where

$$a = \frac{1}{1 + \cos d} \approx \frac{1}{2} + \frac{1}{8}(X^2 + Y^2). \qquad (2.30)$$

The developments of $X$ and $Y$ can be found on the website of the IERS Conventions and have the following form (in mas: microarcseconds) (Capitaine 2002):

$$\begin{aligned} X = &-16616.99'' + 2004191742.88''T - 427219.05''T^2 \\ &- 198620.54''T^3 - 46.05''T^4 + 5.98''T^5 \\ &+ \sum_i \left[ (a_{s,0})_i \sin \beta + (a_{c,0})_i \cos \beta \right] \\ &+ \sum_i \left[ (a_{s,1})_i T \sin \beta + (a_{c,1})_i T \cos \beta \right] \\ &+ \sum_i \left[ (a_{s,2})_i T^2 \sin \beta + (a_{c,2})_i T^2 \cos \beta \right] + \cdots \end{aligned} \qquad (2.31)$$

$$\begin{aligned}
Y = &-6950.78'' - 25381.99''T - 22407250.99''T^2 \\
&+ 1842.28''T^3 - 1113.06''T^4 + 0.99''T^5 \\
&+ \sum_i \left[ (b_{s,0})_i \sin\beta + (b_{c,0})_i \cos\beta \right] \\
&+ \sum_i \left[ (b_{s,1})_i T \sin\beta + (b_{c,1})_i T \cos\beta \right] \\
&+ \sum_i \left[ (b_{s,2})_i T^2 \sin\beta + (b_{c,2})_i T^2 \cos\beta \right] + \cdots
\end{aligned} \qquad (2.32)$$

In Eq. 2.29, $s$ is the accumulated rotation between the reference epoch and the date $T$ of CEO on the true equator due to the celestial motion of CIP and can be expressed as

$$s(T) = -\frac{1}{2}[X(T)Y(T) - X(T_0)Y(T_0)] + \int_{T_0}^{T} \dot{X}Y \, dt - \left( \sigma_0 N_0 - \sum_0 N_0 \right)$$

where $\sigma_0$ and $\sum_0$ are the positions of CEO at J2000.0 and the $x$-origin of CRS, respectively, and $N_0$ is the ascending node at J2000.0 in the equator of CRS. In above equation, term $s(T) + \frac{1}{2}[X(T)Y(T)]$ can be expressed as (in mas)

$$\begin{aligned}
s + XY/2 = &\; 94.0 + 3808.35T - 119.94T^2 \\
&- 72574.09T^3 + 27.70T^4 + 15.61T^5 \\
&+ \sum_i \left[ (c_{s,0})_i \sin\beta + (c_{c,0})_i \cos\beta \right] \\
&+ \sum_i \left[ (c_{s,1})_i T \sin\beta + (c_{c,1})_i T \cos\beta \right] \\
&+ \sum_i \left[ (c_{s,2})_i T^2 \sin\beta + (c_{c,2})_i T^2 \cos\beta \right] + \cdots
\end{aligned} \qquad (2.33)$$

In Eqs. 2.31–2.33, coefficients $(a_{s,j})_i, (a_{c,j})_i, (b_{s,j})_i, (b_{c,j})_i$ and $(c_{s,j})_i, (c_{c,j})_i$ can be extracted from Table 5.2a–c (available at ftp://tai.bipm.org/iers/conv2003/chapter5/). $\beta$ is the combination of the fundamental arguments of nutation theory:

$$\beta = \sum_{j=1}^{14} N_j F_j \qquad (2.34)$$

The first five $F_j$ are the Delaunay variables $l, l', F, D, \Omega$ (given in Sect. 11.2.8); the amplitudes of sines and cosines $\beta$ can be derived from the amplitudes of the precession and nutation series (see McCarthy and Petit 2003); $F_6$ to $F_{13}$ are the mean longitudes of the planets (Mercury to Neptune), including the earth; $F_{14}$ is the general precession in longitude. They are given in radians and $T$ in Julian centuries of TDB (see Sect. 2.7). The coefficients $N_j$ are functions of index $i$ and can be found in the IERS website:

$$\begin{aligned}
F_6 &= l_{\text{Me}} = 4.402608842 + 2608.7903141574T \\
F_7 &= l_{\text{Ve}} = 3.176146697 + 1021.3285546211T \\
F_8 &= l_E = 1.753470314 + 628.3075849991T \\
F_9 &= l_{\text{Ma}} = 6.203480913 + 334.0612426700T \\
F_{10} &= l_{\text{Ju}} = 0.599546497 + 52.9690962641T \\
F_{11} &= l_{\text{Sa}} = 0.874016757 + 21.3299104960T \\
F_{12} &= l_{\text{Ur}} = 5.481293872 + 7.4781598567T \\
F_{13} &= l_{\text{Ne}} = 5.311886287 + 3.8133035638T \\
F_{14} &= P_a = 0.024381750T + 0.00000538691T^2
\end{aligned} \quad (2.35)$$

Using the new paradigm, the complete procedure of transforming GCRS to ITRS, which is compatible with the IAU2000 precession-nutation, is based on the expressions of 2.31–2.33.

An equivalent way to realise the transformation between TRS and CRS under the definition of IAU 2000 can be implemented in a classical way by adding IAU2000 corrections to the corresponding rotating angles. This is done by using the transformation formula 2.14, where the three precession rotating angles (see McCarthy and Petit 2003) are

$$\begin{aligned}
z &= -2.5976176'' + 2306.0803226''T + 1.0947790''T \\
&\quad + 0.0182273''T + 0.0000470''T^4 - 0.0000003''T^5 \\
\theta &= 2004.1917476''T - 0.4269353''T - 0.0418251''T \\
&\quad - 0.0000601''T^4 - 0.0000001''T^5 \\
\zeta &= 2.5976176'' + 2306.0809506''T + 0.3019015''T \\
&\quad + 0.0179663''T - 0.0000327''T^4 - 0.0000002''T^5
\end{aligned} \quad (2.36)$$

The IAU 2000 nutation model is given by two series for nutation in longitude $\Delta\psi$ and obliquity $\Delta\varepsilon$, referred to the mean equator and equinox of date, with $T$ measured in Julian centuries from epoch J2000.0:

$$\Delta\psi = \sum_{i=1}^{N}(A_i + A'_i T)\cos\beta + (A''_i + A'''_i T)\cos\beta, \quad (2.37)$$

$$\Delta\varepsilon = \sum_{i=1}^{N}(B_i + B'_i T)\cos\beta + (B''_i + B'''_i T)\cos\beta,$$

where argument $\beta$ can be found on the IERS website. For these two formulas, rate and bias corrections are necessary because of the new definition of the Celestial Intermediate Pole and the Celestial and Terrestrial Ephemeris Origin:

$$d\Delta\psi = (-0.0166170 \pm 0.0000100)'' + (-0.29965 \pm 0.00040)''T,$$

$$d\Delta\varepsilon = (-0.0068192 \pm 0.0000100)'' + (-0.02524 \pm 0.00010)''T. \quad (2.38)$$

The earth's rotation angle (i.e. the apparent Greenwich Sidereal Time GST or GAST) can be computed by adding a correction $EO$ to the GMST in Eq. 2.27 (in mas):

$$\begin{aligned} EO = {} & 14,506 + 4,612,157,399.66T + 1,396,677.21T^2 - 93.44T^3 + 18.82T^4 \\ & + \Delta\psi \cos\varepsilon + \sum_i \left[ (d_{s,0})_i \sin\beta + (d_{c,0})_i \cos\beta \right] \\ & + \sum_i \left[ (d_{s,1})_i T \sin\beta + (d_{c,1})_i T \cos\beta \right] + \cdots \end{aligned}$$

$$(2.39)$$

where coefficients $(d_{s,j})_i, (d_{c,j})_i$ can be extracted from Table 5.4 (available at ftp://tai.bipm.org/iers/conv2003/chapter5/). $\Delta\psi$ is defined in Eq. 2.37 and $\varepsilon$ is defined in Eq. 2.18.

Similarly, the rotation matrix of polar motion shall be represented as the first formula of 2.25 and 2.26.

## 2.6
## Geocentric Ecliptic Inertial Coordinate System

As discussed above, ECI used the CIO pole in space as the $z$-axis (through consideration of the polar motion, nutation, and precession). If the ecliptic pole is used as the $z$-axis, then an ecliptic coordinate system is defined, and it may be called the Earth-Centred Ecliptic Inertial (ECEI) coordinate system. ECEI places the origin at the mass centre of the earth, its $z$-axis is directed to ecliptic pole (or the $xy$-plane is the mean ecliptic), and its $x$-axis points to the vernal equinox of date. The coordinate transformation between the ECI and ECEI systems can be represented as

$$X_{ECEI} = R_1(-\varepsilon) X_{ECI}, \quad (2.40)$$

where $\varepsilon$ is the ecliptic angle (mean obliquity) of the ecliptic plane related to the equatorial plane. The formula for $\varepsilon$ is given in Sect. 2.4. Usually, coordinates of the sun and the moon as well as planets are given in the ECEI system.

## 2.7
## Time Systems

Three time systems are used in satellite surveying. They are sidereal time, dynamic time, and atomic time (cf., e.g., Hofman-Wellenhof et al. 1997; Leick 1995; McCarthy 1996; King et al. 1987).

Sidereal time is a measure of the earth's rotation and is defined as the hour angle of the vernal equinox. If the measure is counted from the Greenwich meridian, the sidereal time is called Greenwich Sidereal Time. Universal Time (UT) is the Greenwich hour angle of the apparent sun, which orbits uniformly in the equatorial plane. Because the angular velocity of the earth's rotation is not a constant, sidereal time is not a uniformly scaled time. The oscillation of UT is also partly caused by the polar motion of the earth. The universal time corrected for the polar motion is denoted by UT1.

Dynamic time is a uniformly scaled time series used to describe the motion of bodies in a gravitational field. Barycentric Dynamic Time (TDB) is applied in an inertial coordinate system (its origin is located at the centre-of-mass, or barycentre). Terrestrial Dynamic Time (TDT) is used in a quasi-inertial coordinate system (such as ECI). Because of the motion of the earth around the sun (or in the sun's gravitational field), TDT will vary with respect to TDB. However, both the satellite and the earth are subject to almost the same gravitational perturbations. TDT may be used for describing the satellite motion without taking into account the influence of the gravitational field of the sun. TDT is also called Terrestrial Time (TT).

Atomic time is a system based on the output of atomic clocks, such as that used in the International Atomic Time (TAI) scale. It is uniformly scaled and is used in the ECEF coordinate system. In practice, TDT is realised by TAI with a constant offset (32.184 s). To take into account the slowing of the earth's rotation with respect to the sun, Coordinated Universal Time (UTC) was introduced in order to maintain the synchronisation of TAI to the solar day (by inserting leap seconds). GPS Time (GPST) is also an atomic time scale.

The relationships between different time systems are given as follows:

$$\begin{aligned} \text{TAI} &= \text{GPST} + 19.0 \, \text{s} \\ \text{TAI} &= \text{TDT} - 32.184 \, \text{s} \\ \text{TAI} &= \text{UTC} + n \, \text{s} \\ \text{UT1} &= \text{UTC} + \text{dUT1} \end{aligned} \qquad (2.41)$$

where dUT1 can be obtained by IERS, (dUT1 < 0.7 s, cf. Zhu et al. 1996), (dUT1 is also broadcasted with the navigation data), $n$ is the number of leap seconds of date and is inserted into UTC on 1 January and 1 July for the years. The actual $n$ can be found in the IERS report.

Time argument $T$ (Julian centuries) is used in the formulas given in Sect. 2.4. For convenience, $T$ is denoted by TJD, and TJD can be computed from the civil date (Year, Month, Day, and Hour) as follows:

$$JD = INT(365.25Y) + INT(30.6001\,(M+1)) + Day + Hour/24 + 1{,}720{,}981.5$$

and

$$TJD = JD/36525, \qquad (2.42)$$

where

$$Y = \text{Year}-1, \quad M = \text{Month}+12, \quad \text{if} \quad \text{Month} \leq 2,$$
$$Y = \text{Year}, \qquad M = \text{Month}, \qquad \text{if} \quad \text{Month} > 2,$$

where JD is the Julian date (JD), Hour is the time of UT, and INT denotes the integer part of a real number. The JD counted from JD2000.0 is then JD2000 = JD − JD2000.0, where JD2000.0 is the JD for 1 January 2000 at 12 h, and has a value of 2,451,545.0 days. One Julian century is 36,525 days.

Inversely, the civil date (Year, Month, Day and Hour) can be computed from the JD as follows:

$$b = INT(JD+0.5) + 1537,$$
$$c = INT\left(\frac{b-122.1}{365.25}\right),$$
$$d = INT(365.25c),$$
$$e = INT\left(\frac{b-d}{30.6001}\right),$$
$$Hour = JD + 0.5 - INT(JD+0.5),$$
$$Day = b - d - INT(30.6001\,e),$$
$$Month = e - 1 - 12\,INT\left(\frac{e}{14}\right), \text{ and}$$
$$Year = c - 4715 - INT\left(\frac{7+Month}{10}\right), \qquad (2.43)$$

where $b$, $c$, $d$, and $e$ are auxiliary numbers.

Because the GPS standard epoch is defined as JD = 2,444,244.5 (6 January 1980, 0 h), GPS week and the day of week (denoted by Week and $N$) can be computed by

$$N = \text{modulo}(INT(JD+1.5), 7) \text{ and}$$

$$Week = INT\left(\frac{JD - 2{,}444{,}244.5}{7}\right), \qquad (2.44)$$

where $N$ is the day of week ($N$ = 0 for Monday, $N$ = 1 for Tuesday, and so on).

For saving digits and counting the date from midnight instead of noon, the modified Julian date (MJD) is defined as

$$MJD = (JD-2,400,000.5). \quad (2.45)$$

GLONASS time (GLOT) is defined by Moscow time $UTC_{SU}$, which equals UTC plus three hours (corresponding to the offset of Moscow time to Greenwich time), theoretically. GLOT is permanently monitored and adjusted by the GLONASS Central Synchroniser (cf. Roßbach 2000). UTC and GLOT then has a simple relation

$$UTC = GLOT + t_c - 3h, \quad (2.46)$$

where $\tau_c$ is the system time correction with respect to $UTC_{SU}$, which is broadcasted by the GLONASS ephemerides and is less than one microsecond. Therefore, there is approximately

$$GPST = GLOT + m - 3h, \quad (2.47)$$

where $m$ refers to the number of "leap seconds" between GPS and GLONASS (UTC) time and is given in the GLONASS ephemerides. $m$ is indeed the leap seconds since GPS standard epoch (6 January 1980, 0 h).

Galileo system time (GST) will be maintained by a number of UTC laboratory clocks. GST and GPST are time systems of various UTC laboratories. After the offset of GST and GPST is made available to the user, the interoperability will be ensured. GST is apart from small differences (tens of nanoseconds), nearly identical to GPS time. The Galileo week starts at midnight Saturday/Sunday at the same second as the GPS week; The GST week as transmitted by the satellites is a 12-bit value with a roll-over after week 4095. The GST week started at zero at the first roll-over of the broadcast GPS week after 1023, i.e. on Sunday, 22 August 1999, at 00:00:00 GPS time.

The BDS Time (BDT) system is a continuous timekeeping system, with length of seconds being an SI second. BDT zero time started at 00:00:00 UTC on 1 January 2006 (GPS week 1356); therefore, BDT is 14 s behind GPS time. BDT is synchronized with UTC within 100 ns (modulo 1 s). The BDT week starts at midnight Saturday/Sunday. The BDT week is transmitted by the satellites as a 13-bit number. It has a roll-over after week 8191.

Apart from the small errors in the realizations of the different time systems, the relations between systems are:

$$GLOT = UTC = GPST - \Delta tLS \quad (2.48)$$

$$GST = GPST = UTC + \Delta tLS \quad (2.49)$$

$$BDT = UTC + \Delta tLS_{BDS} \quad (2.50)$$

where $\Delta tLS$ is the delta time between GPST and UTC due to leap seconds, such that (2005: $\Delta tLS$ = 13; 2006: $\Delta tLS$ = 14; 2008: $\Delta tLS$ = 15; 2012: $\Delta tLS$ = 16 and 2015: $\Delta tLS$ = 17). $\Delta tLS_{BDS}$ is the delta time between BDT and UTC due to leap seconds, such that (2006: $\Delta tLS_{BDS}$ = 0; 2008: $\Delta tLS_{BDS}$ = 1; 2012: $\Delta tLS_{BDS}$ = 2 and 2015: $\Delta tLS_{BDS}$ = 3).

# References

Captinaine N (2002) Comparison of "Old" and "New" Concepts: The Celestial Intermediate Pole and Earth Orientation Parameters. In: Proceedings of the IERS Workshop on the Implementation of the New IAU Resolutions, Paris, France, April 18-19, 2002, (Capitaine N et al, ed). IERS Technical Note No. 29.
Hofmann-Wellenhof B, Lichtenegger H, Collins J (1997, 2001) GPS theory and practice. Springer-Press, Wien.
King RW, Masters EG, Rizos C, Stolz A, Collins J (1987) Surveying with Global Positioning System. Dümmler-Verlag, Bonn.
Leick A (1995) GPS satellite surveying. John Wiley & Sons Ltd., New York.
Lelgemann D (2002) Lecture notes of geodesy, Technical University Berlin.
Lelgemann D, Xu G (1991) Zur Helmert-Transformation von terrestrischen und GPS-Netzen. ZfV (1).
McCarthy DD (1996) International Earth Rotation Service. IERS conventions, Paris, 95 pp. IERS Technical Note No. 21.
McCarthy DD, Capitaine N (2002) Practical Consequences of Resolution B1.6 "IAU2000 Precession-Nutation Model", Resolution B1.7 "Definition of Celestial Intermediate Pole", and Resolution B1.8 "Definition and Use of Celestial and Terrestrial Ephemeris Origin". In: Proceedings of the IERS Workshop on the Implementation of the New IAU Resolutions, Paris, France, April 18-19, 2002 (Capitaine N et al, ed.) IERS Technical Note No. 29.
McCarthy DD, Petit G (Ed) (2003) International Earth Rotation Service. IERS conventions (2003), IERS Technical Note No. 32.
Petit G (2002) Comparison of "Old" and "New" Cocepts: Coordinate Times and Time Transformations, in Proceedings of the IERS Workshop on the Implementation of the New IAU Resolutions, Paris, France, April 18-19, 2002 (Capitaine N et al, ed), IERS Technical Note No. 29.
Torge W (1991) Geodesy. Walter de Gruyter, Berlin.
Zhu SY, Reigber Ch, Massmann FH (1996) The German PAF for ERS, ERS standards used at D-PAF. D-PAF/GFZ ERS-D-STD-31101.

# Chapter 3
# Satellite Orbits

The principle of the GPS system is to measure the signal transmission paths from the satellites to the receivers. Therefore, the satellite orbits are very important topics in GPS theory. In this chapter, the basic orbits theory is briefly described. For the GPS applications in orbits correction and orbits determination, the advanced orbits perturbation theory will be discussed in Chap. 11.

## 3.1 Keplerian Motion

The simplified satellite orbiting is called Keplerian motion, and the problem is called the two-body problem. The satellite is supposed to move in a central force field. The equation of satellite motion is described by Newton's second law of motion by

$$\vec{f} = m \cdot a = m \cdot \ddot{\vec{r}}, \tag{3.1}$$

where $\vec{f}$ is the attraction force, $m$ is the mass of the satellite, $a$, or alternatively, $\ddot{\vec{r}}$ is the acceleration of the motion (second-order differentiation of vector $\vec{r}$ with respect to the time), and according to Newton's law,

$$\vec{f} = -\frac{GMm}{r^2}\frac{\vec{r}}{r}, \tag{3.2}$$

where $G$ is the universal gravitational constant, $M$ is the mass of the earth, $r$ is the distance between the mass centre of the earth and the mass centre of the satellite. The equation of satellite motion is then

$$\ddot{\vec{r}} = -\frac{\mu}{r^2}\frac{\vec{r}}{r}, \tag{3.3}$$

where $\mu$ (=$GM$) is called the earth's gravitational constant.

Equation 3.3 of satellite motion is valid only in an inertial coordinate system, so the ECSF coordinate system discussed in Chap. 2 will be used for describing the orbit of the satellite. The vector form of the equation of motion can be rewritten through three $x$, $y$, and $z$ components ($\vec{r} = (x, y, z)$) as

$$\begin{aligned} \ddot{x} &= -\frac{\mu}{r^3}x \\ \ddot{y} &= -\frac{\mu}{r^3}y \\ \ddot{z} &= -\frac{\mu}{r^3}z. \end{aligned} \tag{3.4}$$

Multiplying $y$, $z$ to the first equation of 3.4, and $x$, $z$ to the second, $x$, $y$ to the third, and then forming differences of them, one gets

$$\begin{aligned} y\ddot{z} - z\ddot{y} &= 0 \\ z\ddot{x} - x\ddot{z} &= 0 \\ x\ddot{y} - y\ddot{x} &= 0, \end{aligned} \tag{3.5}$$

or in vector form:

$$\vec{r} \times \ddot{\vec{r}} = 0. \tag{3.6}$$

Equations 3.5 and 3.6 are equivalent to

$$\begin{aligned} \frac{d(y\dot{z} - z\dot{y})}{dt} &= 0 \\ \frac{d(z\dot{x} - x\dot{z})}{dt} &= 0 \\ \frac{d(x\dot{y} - y\dot{x})}{dt} &= 0, \end{aligned} \tag{3.7}$$

$$\frac{d(\vec{r} \times \dot{\vec{r}})}{dt} = 0. \tag{3.8}$$

Integrating Eqs. 3.7 and 3.8 lead to

$$\begin{aligned} y\dot{z} - z\dot{y} &= A \\ z\dot{x} - x\dot{z} &= B \\ x\dot{y} - y\dot{x} &= C, \end{aligned} \tag{3.9}$$

$$\vec{r} \times \dot{\vec{r}} = \vec{h} = \begin{pmatrix} A \\ B \\ C \end{pmatrix}, \qquad (3.10)$$

where $A$, $B$, $C$ are integration constants; they form the integration constant vector $\vec{h}$. That is:

$$h = \sqrt{A^2 + B^2 + C^2} = |\vec{r} \times \dot{\vec{r}}|. \qquad (3.11)$$

The constant $h$ is two times of the area that the radius vector sweeps during a unit time. This is indeed Kepler's second law. Then $h/2$ is called the area velocity of the radius of the satellite.

Multiplying $x$, $y$, and $z$ to the three equations of 3.9 and adding them together, one has

$$Ax + By + Cz = 0. \qquad (3.12)$$

That is, the satellite motion fulfils the equation of a plane, and the origin of the coordinate system is in the plane. In other words, the satellite moves in a plane in the central force field of the earth. The plane is called the orbital plane of the satellite.

The angle between the orbital plane and the equatorial plane is called inclination of the satellite (denoted by $i$, cf. Fig. 3.1). Alternatively, the inclination $i$ is the angle between the vector $\vec{z} = (0, 0, 1)$ and $\vec{h} = (A, B, C)$ i.e.

$$\cos i = \frac{\vec{z} \cdot \vec{h}}{|\vec{z}| \cdot |\vec{h}|} = \frac{C}{h}. \qquad (3.13)$$

**Fig. 3.1** Orbital plane

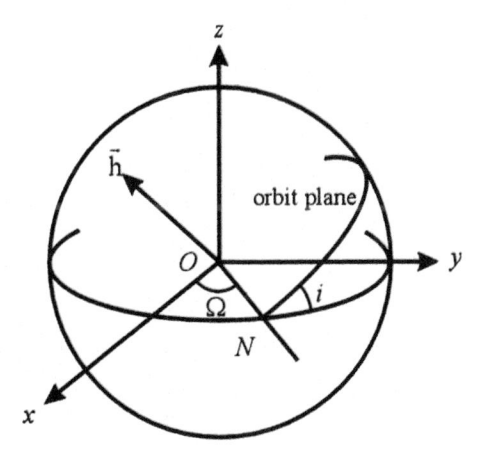

The orbital plane cuts the equator at two points. They are called ascending node $N$ and descending node. (See the next section for details). Vector $\vec{s}$ denotes the vector from the earth's centre pointed to the ascending point. The angle between the ascending node and the $x$-axis (vernal equinox) is called the right ascension of the ascending node (denoted by $\Omega$). Thus,

$$\vec{s} = \vec{z} \times \vec{h},$$

and

$$\cos \Omega = \frac{\vec{s} \cdot \vec{x}}{|\vec{s}| \cdot |\vec{x}|} = \frac{-B}{\sqrt{A^2 + B^2}},$$

$$\sin \Omega = \frac{\vec{s} \cdot \vec{y}}{|\vec{s}| \cdot |\vec{y}|} = \frac{A}{\sqrt{A^2 + B^2}}. \tag{3.14}$$

Parameters $i$ and $\Omega$ uniquely defined the place of orbital plane and therefore are called orbital plane parameters. $\Omega$, $i$, and $h$ are then selected as integration constants, which have significant geometric meanings of the satellite orbits.

## 3.1.1
### Satellite Motion in the Orbital Plane

In the orbital plane, a two-dimensional rectangular coordinate system is given in Fig. 3.2. The coordinates can be represented in polar coordinate $r$ and $\vartheta$ as

$$\begin{aligned} p &= r \cos \vartheta \\ q &= r \sin \vartheta. \end{aligned} \tag{3.15}$$

**Fig. 3.2** Polar coordinates in the orbital plane

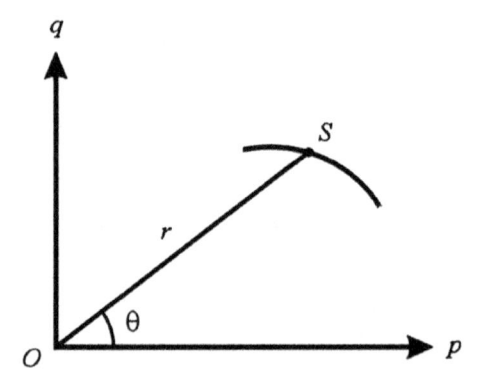

The equation of motion in *pq*-coordinates is similar to the Eq. 3.4 as

$$\ddot{p} = -\frac{\mu}{r^3}p$$
$$\ddot{q} = -\frac{\mu}{r^3}q. \tag{3.16}$$

From Eq. 3.15, one has

$$\dot{p} = \dot{r}\cos\vartheta - r\dot{\vartheta}\sin\vartheta$$
$$\dot{q} = \dot{r}\sin\vartheta + r\dot{\vartheta}\cos\vartheta$$
$$\ddot{p} = (\ddot{r} - r\dot{\vartheta}^2)\cos\vartheta - (r\ddot{\vartheta} + 2\dot{r}\dot{\vartheta})\sin\vartheta \tag{3.17}$$
$$\ddot{q} = (\ddot{r} - r\dot{\vartheta}^2)\sin\vartheta + (r\ddot{\vartheta} + 2\dot{r}\dot{\vartheta})\cos\vartheta.$$

Substituting Eqs. 3.17 and 3.15 into Eq. 3.16, one gets

$$(\ddot{r} - r\dot{\vartheta}^2)\cos\vartheta - (r\ddot{\vartheta} + 2\dot{r}\dot{\vartheta})\sin\vartheta = -\frac{\mu}{r^2}\cos\vartheta$$
$$(\ddot{r} - r\dot{\vartheta}^2)\sin\vartheta + (r\ddot{\vartheta} + 2\dot{r}\dot{\vartheta})\cos\vartheta = -\frac{\mu}{r^2}\sin\vartheta. \tag{3.18}$$

The point from which the polar angle $\vartheta$ is measured is arbitrary. So setting $\vartheta$ as zero, the equation of motion is then

$$\ddot{r} - r\dot{\vartheta}^2 = -\frac{\mu}{r^2}$$
$$r\ddot{\vartheta} + 2\dot{r}\dot{\vartheta} = 0. \tag{3.19}$$

Multiplying *r* to the second equation of 3.19, it turns out to be

$$\frac{d(r^2\dot{\vartheta})}{dt} = 0. \tag{3.20}$$

Because $r\dot{\vartheta}$ is the tangential velocity, $r^2\dot{\vartheta}$ is the two times of the area velocity of the radius of the satellite. Integrating Eq. 3.20 and comparing it with the discussion in Sect. 3.1, one has

$$r^2\dot{\vartheta} = h. \tag{3.21}$$

*h*/2 is the area velocity of the radius of the satellite.

For solving the first differential equation of 3.19, the equation must be transformed into a differential equation of $r$ with respect to variable $f$. Let

$$u = \frac{1}{r}, \qquad (3.22)$$

then from Eq. 3.21, one gets

$$\frac{d\vartheta}{dt} = hu^2 \qquad (3.23)$$

and

$$\frac{dr}{dt} = \frac{dr}{d\vartheta}\frac{d\vartheta}{dt} = \frac{d}{d\vartheta}\left(\frac{1}{u}\right)hu^2 = -h\frac{du}{d\vartheta}$$

$$\frac{d^2r}{dt^2} = -h\frac{d^2u}{d\vartheta^2}\frac{d\vartheta}{dt} = -h^2 u^2 \frac{d^2u}{d\vartheta^2}. \qquad (3.24)$$

Substituting Eqs. 3.22 and 3.24 into the first equation of 3.19, the equation of motion is then

$$\frac{d^2u}{d\vartheta^2} + u = \frac{\mu}{h^2}, \qquad (3.25)$$

and its solution is

$$u = d_1 \cos\vartheta + d_2 \sin\vartheta + \frac{\mu}{h^2},$$

where $d_1$ and $d_2$ are constants of integration. The above equation may be simplified as

$$u = \frac{\mu}{h^2}(1 + e\cos(\vartheta - \omega)), \qquad (3.26)$$

where

$$d_1 = \frac{\mu}{h^2}e\cos\omega, \quad d_2 = \frac{\mu}{h^2}e\sin\omega.$$

Thus the moving equation of satellite in the orbital plane is

$$r = \frac{h^2/\mu}{1 + e\cos(\vartheta - \omega)}. \qquad (3.27)$$

Comparing Eq. 3.27 with a standard polar equation of conic:

$$r = \frac{a(1-e^2)}{1-e\cos\phi}, \qquad (3.28)$$

orbit Eq. 3.27 is obviously a polar equation of conic section with the origin at one of the foci. Where parameter $e$ is the eccentricity, for $e = 0$, $e < 1$, $e = 1$, $e > 1$, the conic is a circle, an ellipse, a parabola, and a hyperbola, respectively. For the satellite orbiting around the earth, generally, $e < 1$. Thus, the satellite orbit is an ellipse, and this is indeed the Kepler's first law. Parameter $a$ is the semimajor axis of the ellipse, and

$$\frac{h^2}{\mu} = a(1-e^2). \qquad (3.29)$$

It is obvious that parameter $a$ has greater significance in a geometric sense than $h$, so the use of $a$ is preferred. Parameters $a$ and $e$ define the size and shape of the ellipse and are called ellipse parameters. The ellipse cuts the equator at the ascending and descending nodes. Polar angle $\varphi$ is counted from the apogee of the ellipse. This can be seen by letting $\varphi = 0$; thus $r = a(1+e)$. $\varphi$ has a 180° difference with the angle $\vartheta - \omega$. Letting $f = \vartheta - \omega$, where $f$ is called the true anomaly of the satellite counted from the perigee, then the orbit Eq. 3.27 can be written as

$$r = \frac{a(1-e^2)}{1+e\cos f}. \qquad (3.30)$$

In the case of $f = 0$, i.e. the satellite is in the point of perigee, $\omega = \vartheta$, $\vartheta$ is the polar angle of the perigee counted from the $p$-axis. Supposing the $p$-axis is an axis in the equatorial plane and points to the ascending node $N$, then $\omega$ is the angle of perigee counted from the ascending node (cf. Fig. 3.3) and is called the argument of perigee. The argument of perigee defines the axis direction of the ellipse related to the equatorial plane.

**Fig. 3.3** Ellipse of the satellite motion

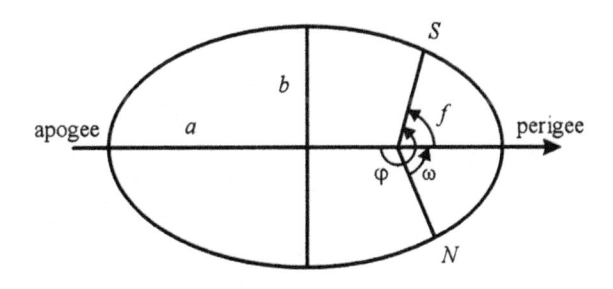

## 3.1.2
## Keplerian Equation

Thus far, five integration constants have been derived. They are inclination angle $i$, right ascension of ascending node $\Omega$, semimajor axis $a$, eccentricity $e$ of the ellipse, and argument of perigee $\omega$. Parameters $i$ and $\Omega$ decide the place of the orbital plane, $a$ and $e$ decide the size and shape of the ellipse and $\omega$ decides the direction of the ellipse (cf. Fig. 3.4). To describe the satellite position in the ellipse, velocity of the motion must be discussed.

The period $T$ of the satellite motion is the area of ellipse divided by area velocity:

$$T = \frac{\pi a b}{\frac{h}{2}} = \frac{2\pi a b}{\sqrt{\mu a(1-e^2)}} = 2\pi a^{3/2} \mu^{-1/2}. \quad (3.31)$$

The average angular velocity $n$ is then

$$n = \frac{2\pi}{T} = a^{-3/2} \mu^{1/2}. \quad (3.32)$$

Equation 3.32 is the Kepler's third law. It is obvious that it is easier to describe the angular motion of the satellite under the average angular velocity $n$ in the geometric centre of the ellipse (than in the geocentre). To simplify the problem, an angle called the eccentric anomaly is defined (denoted by $E$, cf. Fig. 3.5). $S'$ is the vertical projection of the satellite $S$ on the circle with a radius of $a$ (semimajor axis of the ellipse). The distance between the geometric centre $O'$ of the ellipse and the geocentre $O$ is $ae$. Thus,

$$\begin{aligned} x &= r\cos f = a\cos E - ae \\ y &= r\sin f = b\sin E = a\sqrt{1-e^2}\sin E, \end{aligned} \quad (3.33)$$

**Fig. 3.4** Orbital geometry

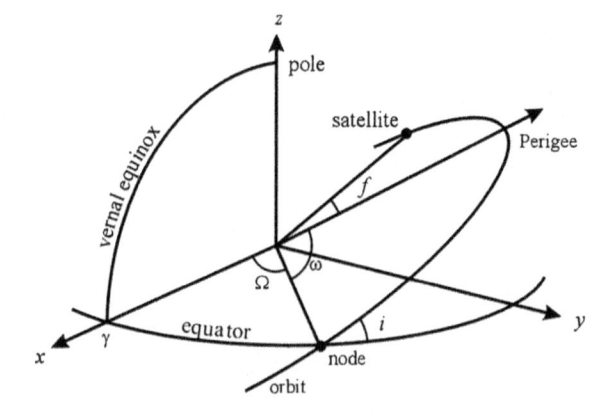

**Fig. 3.5** Mean anomaly of satellite

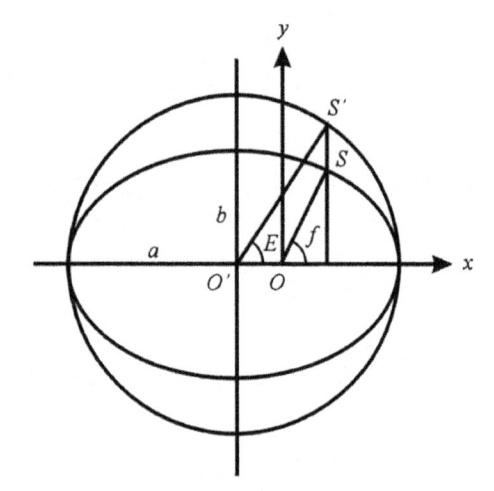

where the second equation can be obtained by substituting the first into the standard ellipse equation $((x + ae)^2/a^2 + y^2/b^2 = 1)$, where $b$ is the semiminor axis of the ellipse. The orbit equation can then be represented by variable $E$ as

$$r = a(1 - e \cos E). \tag{3.34}$$

The relation between true and eccentric anomalies can be derived by using Eqs. 3.33 and 3.34:

$$\tan\frac{f}{2} = \frac{\sin f}{1 + \cos f} = \frac{\sin E}{1 + \cos E}\frac{\sqrt{1-e^2}}{1-e} = \frac{\sqrt{1+e}}{\sqrt{1-e}}\tan\frac{E}{2}. \tag{3.35}$$

If the $xyz$-coordinates are rotated such that the $xy$-plane coincides with the orbital plane, then the area velocity formulas of Eqs. 3.9 and 3.10 have only one component along the $z$-axis, i.e.

$$x\dot{y} - y\dot{x} = h = \sqrt{\mu a(1-e^2)}. \tag{3.36}$$

From Eq. 3.33, one has

$$\begin{aligned}\dot{x} &= -a \sin E \frac{dE}{dt} \\ \dot{y} &= a\sqrt{1-e^2} \cos E \frac{dE}{dt}\end{aligned} \tag{3.37}$$

Substituting Eqs. 3.33 and 3.37 into Eq. 3.36 and taking Eq. 3.32 into account, a relation between $E$ and $t$ is obtained

$$(1 - e\cos E)dE = \sqrt{\mu}a^{-3/2}dt = ndt. \qquad (3.38)$$

Suppose at the time $t_p$ satellite is at the point perigee, i.e. $E(t_p) = 0$, and at any time $t$, $E(t) = E$, then integration of Eq. 3.38 from 0 to $E$, namely from $t_p$ to $t$ is

$$E - e\sin E = M, \qquad (3.39)$$

where

$$M = n(t - t_p). \qquad (3.40)$$

Equation 3.39 is the Keplerian equation. $E$ is given as a function of $M$, namely $t$. Because of Eq. 3.34, the Keplerian equation indirectly assigns $r$ as a function of $t$. $M$ is called the mean anomaly. $M$ describes the satellite as orbiting the earth with a mean angular velocity $n$. $t_p$ is called the perigee passage and is the sixth integration constant of the equation of satellite motion in a centre-force field.

Knowing $M$ to compute $E$, the Keplerian Eq. 3.39 may be solved iteratively. Because of the small $e$, the convergence can be achieved very quickly.

Three anomalies (true anomaly $f$, eccentric anomaly $E$ and mean anomaly $M$) are equivalent through the relations of Eqs. 3.35 and 3.39. They are functions of time $t$ (including the perigee passage $t_p$), and they describe the position changes of the satellite with the time in the ECSF coordinates.

## 3.1.3
### State Vector of the Satellite

Consider the orbital right-handed coordinate system: if the $xy$-plane is the orbital plane, the $x$-axis points to the perigee, the $z$-axis is in the direction of vector $\vec{h}$, and the origin is in the geocentre, the position vector $\vec{q}$ of the satellite is then (cf. Eq. 3.33)

$$\vec{q} = \begin{pmatrix} a(\cos E - e) \\ a\sqrt{1-e^2}\sin E \\ 0 \end{pmatrix} = \begin{pmatrix} r\cos f \\ r\sin f \\ 0 \end{pmatrix}. \qquad (3.41)$$

Differentiating Eq. 3.41 with respect to time $t$ and taking Eq. 3.38 into account, the velocity vector of the satellite is then

$$\dot{\vec{q}} = \begin{pmatrix} -\sin E \\ \sqrt{1-e^2}\cos E \\ 0 \end{pmatrix} \frac{na}{1 - e\cos E} = \begin{pmatrix} -\sin f \\ e + \cos f \\ 0 \end{pmatrix} \frac{na}{\sqrt{1-e^2}}. \qquad (3.42)$$

The second part of the above equation can be derived from the relation between $E$ and $f$. The state vector of the satellite in the orbital coordinate system can be rotated to the ECSF coordinate system by three successive rotations. First, a clockwise rotation around the 3rd-axis from the perigee to the node is given by (cf. Fig. 3.4)

$$R_3(-\omega).$$

Next, a clockwise rotation around the 1st-axis with the angle of inclination $i$ is given by

$$R_1(-i).$$

Finally, a clockwise rotation around the 3rd-axis from the node to the vernal equinox is given by

$$R_3(-\Omega).$$

So the state vector of the satellite in the ECSF coordinate system is then

$$\begin{pmatrix} \vec{r} \\ \dot{\vec{r}} \end{pmatrix} = R_3(-\Omega) R_1(-i) R_3(-\omega) \begin{pmatrix} \vec{q} \\ \dot{\vec{q}} \end{pmatrix}, \qquad (3.43)$$

where

$$\vec{r} = \begin{pmatrix} x \\ y \\ z \end{pmatrix}, \quad \dot{\vec{r}} = \begin{pmatrix} \dot{x} \\ \dot{y} \\ \dot{z} \end{pmatrix}.$$

For the six given Keplerian elements ($\Omega$, $i$, $\omega$, $a$, $e$, $M_0$) of $t_0$, where $M_0 = n(t_0 - t_p)$, the satellite state vector of time $t$ can be computed, e.g., as follows:

1. Using Eq. 3.32 to compute the mean angular velocity $n$;
2. Using Eqs. 3.40, 3.39, 3.33, and 3.30 to compute the three anomalies $M$, $E$, $f$, and $r$;
3. Using Eqs. 3.41 and 3.42 to compute the state vector $\vec{q}$ and $\dot{\vec{q}}$ in orbital coordinates;
4. Using Eq. 3.43 to rotate state vector $\vec{q}$ and $\dot{\vec{q}}$ to the ECSF coordinates.

Keplerian elements can be given in practice at any time. For example, with $t_0$, where only $f$ is a function of $t_0$, other parameters are constants. In this case, the related $E$ and $M$ can be computed by Eqs. 3.35 and 3.39; thus $t_p$ can be computed by Eq. 3.40.

From Eq. 3.42, one has

$$v^2 = \frac{a^2 n^2}{(1 - e\cos E)^2}[\sin^2 E + (1 - e^2)\cos^2 E] = \frac{a^2 n^2 (1 + e\cos E)}{1 - e\cos E}. \quad (3.44)$$

Taking Eqs. 3.32 and 3.34 into account leads to

$$v^2 = \frac{\mu(1 + e\cos E)}{r} = \frac{\mu(2 - r/a)}{r} = \mu\left(\frac{2}{r} - \frac{1}{a}\right), \quad (3.45)$$

where $v^2/2$ is the kinetic energy scaled by mass, $\mu/r$ is the potential energy, and $a$ is the semimajor axis of the ellipse. This is the total energy conservative law of mechanics.

Rotate the vector $\vec{q}$ and $\dot{\vec{q}}$ in Eqs. 3.41 and 3.42 by $R_3(-\omega)$ and denote by $\vec{p}$ and $\dot{\vec{p}}$, i.e.

$$\vec{p} = \begin{pmatrix} p_1 \\ p_2 \\ p_3 \end{pmatrix} = R_3(-\omega)\begin{pmatrix} r\cos f \\ r\sin f \\ 0 \end{pmatrix} = \begin{pmatrix} r\cos(\omega + f) \\ r\sin(\omega + f) \\ 0 \end{pmatrix}, \quad (3.46)$$

and

$$\dot{\vec{p}} = \begin{pmatrix} \dot{p}_1 \\ \dot{p}_2 \\ \dot{p}_3 \end{pmatrix} = R_3(-\omega)\begin{pmatrix} -\sin f \\ e + \cos f \\ 0 \end{pmatrix}\frac{na}{\sqrt{1 - e^2}}$$
$$= \begin{pmatrix} -\sin(\omega + f) - e\sin\omega \\ \cos(\omega + f) + e\cos\omega \\ 0 \end{pmatrix}\frac{na}{\sqrt{1 - e^2}}. \quad (3.47)$$

The reverse problem of Eq. 3.43, i.e. for given rectangular satellite state vector $(\vec{r}, \dot{\vec{r}})^T$ to compute the Keplerian elements, can be carried out as follows. $\omega + f$ is called argument of latitude and denoted by $u$.

1. Using the given state vector to compute the modulus $r$ and $v$ ($r = |\vec{r}|$, $v = |\dot{\vec{r}}|$);
2. Using Eqs. 3.10 and 3.11 to compute vector $\vec{h}$ and its modulus $h$;
3. Using Eqs. 3.13 and 3.14 to compute inclination $i$ and the right ascension of ascending node $\Omega$;
4. Using Eqs. 3.45, 3.29, and 3.32 to compute semimajor axis $a$, eccentricity $e$ and average angular velocity $n$;
5. Rotating $\vec{r}$ by $\vec{p} = R_1(i)R_3(\Omega)\vec{r}$ and then using Eq. 3.46 to compute $\omega + f$;
6. Rotating $\dot{\vec{r}}$ by $\dot{\vec{p}} = R_1(i)R_3(\Omega)\dot{\vec{r}}$ and then using Eq. 3.47 to compute $\omega$ and $f$;
7. Using Eqs. 3.33, 3.39, and 3.40 to compute $E$, $M$ and $t_p$.

To transform the GPS state vector from the ECSF coordinate system to other coordinate systems, the formulas discussed in Chap. 2 can be used.

## 3.2
## Disturbed Satellite Motion

Keplerian motion of a satellite is a motion under the assumption that the satellite is only attracted by the central force of the earth. This is, of course, an approximation. For a satellite problem, the earth cannot be considered a mass point or a homogenous sphere. The earth's total force of attraction can be considered the central force plus the non-central force. The latter is called the earth's disturbing force, which has an order of $10^{-4}$ compared with the central force. The other attraction forces, which are simply called disturbing forces, are the attraction forces of the sun and the moon, the earth and ocean tide, and surface forces such as solar radiation pressure and atmospheric drag. The satellite motion can then be considered a nominal motion (e.g., Keplerian motion) plus a disturbed motion.

If we further use the Keplerian elements to describe the disturbed motion of the satellite, all elements should be functions of time. Keplerian elements ($\Omega(t)$, $i(t)$, $\omega(t)$, $a(t)$, $e(t)$, $M(t)$) can be represented by $\sigma_j(t)$, $j = 1, \ldots, 6$, thus the polynomial approximations are

$$\sigma_j(t) = \sigma_j(t_0) + \frac{d\sigma_j(t)}{dt}\bigg|_{t=t_0}(t - t_0) + \ldots \quad j = 1, \ldots, 6. \quad (3.48)$$

In other words, the disturbed orbit can be further represented by Keplerian elements; however, all elements are time variables. If the initial elements and their changing rates are given, the instantaneous elements can be obtained. This principle is used in the broadcast ephemerides.

Detailed disturbing theory and orbit correction as well as orbit determination will be discussed in Chap. 11 later.

## 3.3
## GPS Broadcast Ephemerides

GPS broadcast ephemerides are forecasted, predicted or extrapolated satellite orbits data, which are transmitted from the satellite to the receiver in the navigation message. Because of the nature of the extrapolation, broadcast ephemerides do not have enough high qualities for precise applications. The predicted orbits are curve fitted to a set of relatively simple disturbed Keplerian elements and transmitted to the users.

The broadcast messages are

| | |
|---|---|
| SV-id | satellite number; |
| $t_c$ | reference epoch of the satellite clock; |
| $a_0, a_1, a_2$ | polynomial coefficients of the clock error; |
| $t_{oe}$ | reference epoch of the ephemerides; |
| $\sqrt{a}$ | square root of the semimajor axis of the orbital ellipse; |
| $e$ | numerical eccentricity of the ellipse; |
| $M_0$ | mean anomaly at the reference epoch $t_e$; |
| $\omega_0$ | argument of perigee; |
| $i_0$ | inclination of the orbital plane; |
| $\Omega_0$ | longitude of the ascending node at the weekly epoch; |
| $\Delta n$ | mean motion difference; |
| $idot$ | rate of inclination angle; |
| $\dot{\Omega}$ | rate of node's right ascension; |
| $C_{uc}, C_{us}$ | correction coefficients (of argument of latitude); |
| $C_{rc}, C_{rs}$ | correction coefficients (of geocentric distance); |
| $C_{ic}, C_{is}$ | correction coefficients (of inclination). |

The satellite position at epoch $t$ can be computed as follows:

$$\begin{aligned}
M &= M_0 + (\sqrt{\frac{\mu}{a^3}} + \Delta n)(t - t_{oe}), \\
\Omega &= \Omega_0 + \dot{\Omega}(t - t_{oe}), \\
\omega &= \omega_0 + C_{uc}\cos(2u_0) + C_{us}\sin(2u_0), \\
r &= r_0 + C_{rc}\cos(2u_0) + C_{rs}\sin(2u_0), \quad \text{and} \\
i &= i_0 + C_{ic}\cos(2u_0) + C_{is}\sin(2u_0) + idot(t - t_{oe}),
\end{aligned} \qquad (3.49)$$

where

$$\begin{aligned}
E &= M + e\sin E, \\
r_0 &= a(1 - e\cos E), \\
f &= 2\tan^{-1}\left(\frac{\sqrt{1+e}}{\sqrt{1-e}}\tan\frac{E}{2}\right), \quad \text{and} \\
u_0 &= \omega_0 + f.
\end{aligned} \qquad (3.50)$$

$\mu$ is the earth's gravitational constant (which can be read from the IERS Conventions, cf. table of constants). The satellite position in the orbital plane coordinate system (the 1st-axis points to the ascending node, the 3rd-axis is vertical to the orbital plane, and the 2nd-axis completes a right-handed system) is then

$$\begin{pmatrix} x' \\ y' \\ z' \end{pmatrix} = \begin{pmatrix} r \cos u \\ r \sin u \\ 0 \end{pmatrix},$$

where $u = \omega + f$. The position vector can be rotated to the ECSF coordinate system by $R_3(-\Omega)R_1(-i)$ and then rotated to the ECEF coordinate system by $R_3(\Theta)$, where $\Theta$ is Greenwich Sidereal Time and

$$\Theta = \omega_e(t - t_{oe}) + \omega_e t_{oe}, \tag{3.51}$$

where $\omega_e$ is the angular velocity of the earth (can be read from the IERS Conventions, cf. table of constants). The satellite position vector in the ECEF coordinate system is then

$$\begin{pmatrix} x \\ y \\ z \end{pmatrix}_{ECEF} = R_3(-\Omega + \Theta)R_1(-i) \begin{pmatrix} r \cos u \\ r \sin u \\ 0 \end{pmatrix}. \tag{3.52}$$

The first equation of 3.50 is the Keplerian equation, which may be solved iteratively. It is notable that the time $t$ above should be the signal transmission time. $(t - t_{oe})$ should be the actual total time difference of the two time epochs and must account for the beginning and end of week crossovers (cf. Spilker 1996). That is, if the difference is greater (or less) than 302,400 s, subtract (or add) 604,800 s. The satellite clock error can be computed by (denoting $k$ as the satellite's id)

$$\delta t_k = a_0 + a_1(t - t_c) + a_2(t - t_c)^2. \tag{3.53}$$

Unit seconds are used for the time variable; the computed clock error has units of $10^{-6}$ s.

## 3.4
## IGS Precise Ephemerides

GPS satellite precise orbits are available through the International GPS Service (IGS) in the form of post-processed results. Such orbits data are called IGS precise ephemerides. They can be downloaded for free from several internet homepages (e.g., www.gfz-potsdam.de).

IGS data are given in the ECEF coordinate system. For all possible satellites, the position vectors are given in $x$, $y$, $z$ three components (units: km), and the related clock errors are also given (units: $10^{-6}$ s). The data are given in a suitable time interval (15 min).

To obtain the ephemerides of any interested epoch, a Lagrange polynomial is used to fit the given data and then to interpolate the data at the needed epoch. The general Lagrange polynomial is (e.g., Wang et al. 1979):

$$y(t) = \sum_{j=0}^{m} L_j(t) \cdot y(t_j), \tag{3.54}$$

where

$$L_j(t) = \prod_{k=0}^{m} \frac{(t - t_k)}{(t_j - t_k)}, \quad k \neq j, \tag{3.55}$$

where symbol $\Pi$ is a multiplying operator from $k = 0$ to $k = m$, $m$ is the order of the polynomial, $y(t_j)$ are given data at the time $t_j$, $L_j(t)$ is called the base function of order $m$, and $t$ is the time on which data will be interpolated. Generally speaking, $t$ should be placed around the middle of the time duration $(t_0, t_m)$ if possible. Therefore, $m$ is usually selected as an odd number. For IGS orbit interpolation, a standard $m$ is selected as 7 or 9 from experience.

For the equal distance Lagrange interpolation there is

$$t_k = t_0 + k\Delta t$$
$$t - t_k = t - t_0 - k\Delta t$$
$$t_j - t_k = (j - k)\Delta t,$$

then

$$L_j(t) = \prod_{k=0}^{m} \frac{(t - t_0 - k\Delta t)}{(j - k)\Delta t}, \quad k \neq j, \tag{3.56}$$

where $\Delta t$ is the data interval.

In order to deal with the broadcast ephemerides in a manner similar to IGS precise ephemerides, the broadcast orbit may be first computed and then transformed to IGS-like data for use.

The forecasted IGS ephemerides are now also available to download for free.

## 3.5
## GLONASS Ephemerides

GLONASS broadcast ephemerides are forecasted, predicted, or extrapolated satellite orbit data, which are transmitted from the satellite to the receiver in the navigation message. The broadcast messages include the following: satellite number, reference epoch of the ephemerides, relative frequency offset, satellite clock offset, satellite position, satellite velocity, satellite acceleration, time system correction with respect to $UTC_{SU}$, the time difference between GLONASS time, and GPS time.

The satellite position and velocity at desired epoch $t$ can be interpolated by using the Lagrange polynomial discussed in Sect. 3.4, or alternatively, by a five-order polynomial discussed in Sect. 5.4.2 where the position, velocity, and acceleration data are used.

The precise GLONASS ephemerides are similarly available. The data has nearly the same format as that of GPS and includes the message of the time differences of the GLONASS time and GPS time.

## 3.6
## Galileo Ephemerides

The Galileo Open Service allows access to two navigation message types: F/NAV (freely accessible navigation) and I/NAV (integrity navigation). The content of the two messages differs in various items; however, in general it is very similar to the content of the GPS navigation message. But there are items in the navigation message that depend on the origin of the message (F/NAV or I/NAV): The SV clock parameters actually define the satellite clock for the dual-frequency ionosphere-free linear combination. F/NAV reports the clock parameters valid for the E5a-E1 combination; the I/NAV reports the parameters for the E5b-E1 combination.

## 3.7
## BDS Ephemerides

The BDS open service broadcast navigation message is similar in content to the GPS navigation message. The header section and the first data record (epoch, satellite clock information) are equal to the GPS navigation file. The following six records are similar to GPS. Details can be referred to IGS RINEX format (2015).

## References

IGS, RTCM-SC104 (2015) RINEX- The Receiver Independent Exchange Format (Version 3.03). ftp://igs.org/pub/data/format/rinex303.pdf

Spilker JJ (1996) GPS navigation data. In: Parkinson BW, Spilker JJ (eds) Global Positioning System: Theory and applications, Vol. I, Chapter 4

Wang LX, Fang ZD, Zhang MY, Lin GB, Gu LK, Zhong TD, Yang XA, She DP, Luo ZH, Xiao BQ, Chai H, Lin DX (1979) Mathematic handbook. Educational Press, Peking, ISBN 13012-0165

# Chapter 4
# GPS Observables

The basic GPS observables are code pseudoranges and carrier phases as well as Doppler measurements. The principle of the GPS measurements and their mathematical expressions are described.

## 4.1 Code Pseudoranges

The pseudorange is a measure of the distance between the satellite and the receiver's antenna. The distance is measured by measuring the GPS signal transmission time from the satellite to the GPS receiver's antenna. Therefore, this refers to the distance between the satellite at the time of GPS signal emission and the GPS antenna at the time of GPS signal reception. The transmission time is measured through maximum correlation analysis of the receiver code and the GPS signal. The receiver code is derived from the clock used in the GPS receiver. The GPS signal, of course, is generated by the clock used in the GPS satellite. The measured pseudorange is different from the geometric distance between the satellite and the receiver's antenna because of both the errors of the clocks and the influence of the signal transmission mediums. It is also notable that the path of the signal transmission differs slightly from the geometric path. The transmission medium not only delays transmission of the signal, but also bends its transmission path.

The GPS signal emission time of the satellite is denoted by $t_e$, and the GPS signal reception time of the receiver is denoted by $t_r$. In case of vacuum medium and error-free situation, the measured pseudorange is equal to the geometric distance and can be presented by

$$R_r^s(t_r, t_e) = (t_r - t_e)c, \qquad (4.1)$$

where $c$ denotes the speed of light, and subscript r and superscript s denote the receiver and satellite, respectively. On the left-hand side, $t_r$ denotes the epoch at which the pseudorange is measured.

$t_e$ and $t_r$ are considered true emission time and reception time of the GPS signal. Taking both the satellite and receiver clock errors into account, the pseudorange can be represented as

$$R_r^s(t_r, t_e) = (t_r - t_e)c - (\delta t_r - \delta t_s)c, \tag{4.2}$$

where $\delta t_r$ and $\delta t_s$ denote the clock errors of the receiver and satellite, respectively. The GPS satellite clock error term $\delta t_s$ is indeed known through GPS satellite orbit determination. The clock errors are usually modelled by polynomials of time. The constant term represents the bias and the linear term the drift of the clocks. These coefficients are transmitted along with the navigation message to the users. More precisely, the satellite clock error corrections can also be obtained from all IGS data centres (cf., e.g., www.gfz-potsdam.de). They are determined along with the precise IGS orbits and have higher resolution in time.

The geometric distance of the first term on the right-hand side of Eq. 4.2 is given by

$$\rho_r^s(t_r, t_e) = \sqrt{(x_s - x_r)^2 + (y_s - y_r)^2 + (z_s - z_r)^2}, \tag{4.3}$$

where the satellite coordinate vector $(x_s, y_s, z_s)$ is a vector function of the time $t_e$, and the receiver coordinate $(x_r, y_r, z_r)$ is a function of the time $t_r$. Therefore, the geometric distance is indeed a function of two time variables. Furthermore, the emission time $t_e$ is unknown in practice. Denoting the transmission time as $\Delta t$, there is

$$\Delta t = t_r - t_e. \tag{4.4}$$

For illustrating the transmission time computation, the geometric distance can be generally written as

$$\rho_r^s(t_r, t_e) = \rho_r^s(t_r, t_r - \Delta t). \tag{4.5}$$

The transmission time of the signal travelling from the GPS satellite to the receiver is about 0.07 s. The geometric distance function on the right-hand side of Eq. 4.5 can be expanded into a Taylor series at the reception time $t_r$ with respect to the transmission time by

$$\rho_r^s(t_r, t_e) = \rho_r^s(t_r) + \frac{d\rho_r^s(t_r)}{dt} \Delta t, \tag{4.6}$$

where $d\rho_r^s(t_r)/dt$ denotes the time derivation of the radial distance between satellite and receiver. The second term on the right-hand side of Eq. 4.6 is called the transmission time correction. It is notable that the coordinates of GPS antennas are usually given in the ECEF coordinate system. During the signal transmission, the receiver rotates with the earth; therefore, in computing the distance of Eq. 4.3, the so-called earth rotation correction must be considered.

Taking the ionospheric effects, tropospheric effects, the earth's tide and loading tide effects, multipath and relativistic effects, as well as remaining errors into account, the pseudorange model Eq. 4.2 can be completed by

$$R_r^s(t_r,t_e) = \rho_r^s(t_r,t_e) - (\delta t_r - \delta t_s)c + \delta_{ion} + \delta_{tro} + \delta_{tide} + \delta_{mul} + \delta_{rel} + \varepsilon. \quad (4.7)$$

where the measured pseudorange is on the left-hand side, it equals to the geometric distance between the satellite at the emission time and the antenna at the reception time plus or minus several corrections. The clock error corrections are scaled by the velocity of light $c$. $\delta_{ion}$ and $\delta_{tro}$ denote the ionospheric and tropospheric effects of the station r. $\delta_{tide}$ denotes the earth's tide and ocean loading tide effects, $\delta_{mul}$ denotes the multipath effects, and $\delta_{rel}$ denotes the relativistic effects. The remaining errors are denoted by $\varepsilon$. For convenience, unit meter is used for all terms and instrumental biases are omitted here.

The height of the GPS satellite is about 20,200 km; thus, the GPS signal transmission time is about 0.07 s. The earth rotates during the signal transition. The angular velocity of the earth's rotation is about 15 arcsecs$^{-1}$. The related earth rotation correction is about 1 arcsec (cf. Goad 1996a, b). The effects of such a correction depend on the latitude of the station. At the equator, 1 arcsec rotation is equivalent to about 31 m position displacement. The clock errors can be very big. There are examples where the negative pseudoranges are observed in practice.

The above-discussed pseudorange model is generally valid for both C/A code and P code. The precision of the pseudorange measurements depends on the electronic abilities. Generally speaking, it is no problem nowadays to measure with precision up to 1 % of the chip length. Therefore, the C/A code has a precision of about 3 m, and the P code 30 cm. The mentioned corrections will be discussed later in detail.

## 4.2
## Carrier Phases

The carrier phase is a measure of the phase of the received satellite signal relative to the receiver-generated carrier phase at the reception time. The measurement is made by shifting the receiver-generated phase to track the received phase. The number of full carrier waves between the receiver and the satellite cannot be accounted for at the initial signal acquisition. Therefore, measuring the carrier phase is to measure the fractional phase and to keep track of changes in the cycles. The carrier phase observable is indeed an accumulated carrier phase observation. The fractional carrier phase can be measured by electronics with precision better than 1 % of the wavelength, which corresponds to millimetre precision. This is also the reason why the phase measurement is more precise than that of the code. A full carrier wave is called a cycle. The ambiguous integer number of cycles in the carrier phase measurement is called ambiguity. The initial measuring has correct fractional phase and an arbitrary integer counter setting at the start epoch. Such an arbitrary initial setting will be adjusted to the correct one by modelling with ambiguity parameters.

In the case of a vacuum medium and an error-free situation, the measured phase can be presented by

$$\Phi_r^s(t_r) = \Phi_r(t_r) - \Phi^s(t_r) + N_r^s, \tag{4.8}$$

where subscript r and superscript s denote the receiver and satellite, respectively. $t_r$ denotes the GPS signal reception time of the receiver. $\Phi_r$ denotes the phase of receiver's oscillator. $\Phi^s$ denotes the received signal phase of the satellite. $N_r^s$ is the ambiguity related to receiver r and satellite s.

There is an interesting property of the signal phase transmission, i.e. the received phase of the satellite signal at the reception time is exactly the same as the phase of the emitted satellite signal at the emission time (Remondi 1984; Leick 1995), i.e.

$$\Phi^s(t_r) = \Phi_e^s(t_r - \Delta t), \tag{4.9}$$

where $\Phi_e^s$ denotes the satellite emitted phase and $\Delta t$ is the GPS signal transmission time. This can be represented by

$$\Delta t = \frac{\rho_r^s(t_r, t_e)}{c}, \tag{4.10}$$

where $\rho_r^s(t_r, t_e)$ is geometric distance between the satellite at the emission time $t_e$, and the GPS antenna at the reception time $t_r$, c is the speed of light. Then Eq. 4.8 can be written as:

$$\Phi_r^s(t_r) = \Phi_r(t_r) - \Phi_e^s(t_r - \Delta t) + N_r^s. \tag{4.11}$$

Suppose the initial time is zero and the received satellite signal and the reference carrier of the receiver have the nominal frequency $f$. Then one has

$$\Phi_r(t_r) = f t_r \text{ and} \tag{4.12}$$

$$\Phi_e^s(t_r - \Delta t) = f(t_r - \Delta t). \tag{4.13}$$

Substituting Eqs. 4.10, 4.12, and 4.13 into Eq. 4.11 gives

$$\Phi_r^s(t_r) = \frac{\rho_r^s(t_r, t_e) f}{c} + N_r^s. \tag{4.14}$$

Taking both the satellite and receiver clock errors into account, the carrier phase can be represented as

$$\Phi_r^s(t_r) = \frac{\rho_r^s(t_r, t_e) f}{c} - f(\delta t_r - \delta t_s) + N_r^s, \qquad (4.15)$$

where $\delta t_r$ and $\delta t_s$ denote the clock errors of the receiver and satellite, respectively. The frequency $f$ and wavelength $\lambda$ have the relation of

$$c = f\lambda. \qquad (4.16)$$

Taking the ionospheric effects, tropospheric effects, the earth's tide and loading tide effects, multipath and relativistic effects as well as remaining errors into account, the carrier phase model Eq. 4.15 can be completed by

$$\begin{aligned}\Phi_r^s(t_r) &= \frac{\rho_r^s(t_r, t_e)}{\lambda} - f(\delta t_r - \delta t_s) + N_r^s \\ &\quad - \frac{\delta_{\text{ion}}}{\lambda} + \frac{\delta_{\text{tro}}}{\lambda} + \frac{\delta_{\text{tide}}}{\lambda} + \frac{\delta_{\text{mul}}}{\lambda} + \frac{\delta_{\text{rel}}}{\lambda} + \frac{\varepsilon}{\lambda}\end{aligned} \qquad (4.17)$$

or

$$\begin{aligned}\lambda \Phi_r^s(t_r) &= \rho_r^s(t_r, t_e) - (\delta t_r - \delta t_s)c + \lambda N_r^s \\ &\quad - \delta_{\text{ion}} + \delta_{\text{tro}} + \delta_{\text{tide}} + \delta_{\text{mul}} + \delta_{\text{rel}} + \varepsilon,\end{aligned} \qquad (4.18)$$

where the measured phase on the left-hand side with a factor of $\lambda$ equals the geometric distance between the satellite at the emission time and the antenna at the reception time plus or minus several corrections. The clock error corrections are scaled by the speed of light $c$. $\delta_{\text{ion}}$ and $\delta_{\text{tro}}$ denote the ionospheric and tropospheric effects of the station r. $\delta_{\text{tide}}$ denotes the earth tide and ocean loading tide effects. The multipath and relativistic effects as well as remaining errors are denoted by $\delta_{\text{mul}}$, $\delta_{\text{rel}}$, $\varepsilon$ respectively. Equation 4.18 is convenient to use, because all terms have units of length (meter). It is notable that the sign of the ionospheric term is negative, whereas in the pseudorange model it is positive (see Sect. 4.1). This will be discussed later in Sect. 5.1 in detail.

During GPS signal tracking, the phase and the integer account are continuously modelled and frequently measured. In this way, the changing oscillator frequency is accounted for. Every time the phase is measured, the coefficients in the tracking loop model are updated (Remondi 1984) to ensure sufficient precision of measurement.

## 4.3
## Doppler Measurements

The Doppler effect is a phenomenon of frequency shift of the electromagnetic signal caused by the relative motion of the emitter and receiver. Supposing the emitted signal has the nominal frequency $f$, the radial velocity of the satellite related to the receiver is

**Fig. 4.1** Doppler effects

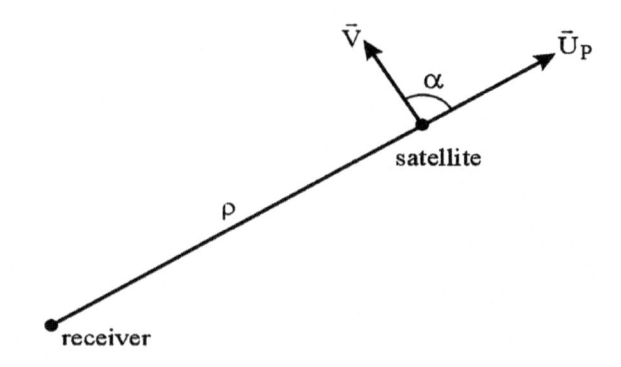

$$V_\rho = \vec{V} \cdot \vec{U}_\rho = |\vec{V}| \cos \alpha, \qquad (4.19)$$

where $\vec{V}$ is the velocity vector of the satellite related to the receiver, $V = |\vec{V}|$, $\vec{U}_\rho$ is the identity vector in the direction from the receiver to the satellite, $\alpha$ is the projection angle of the vector $\vec{V}$ to $\vec{U}_\rho$ (see Fig. 4.1), index $\rho$ is the distance from the receiver to satellite. Then the received signal has a frequency of

$$f_r = f\left(1 + \frac{V_\rho}{c}\right)^{-1} \approx f\left(1 - \frac{V_\rho}{c}\right), \qquad (4.20)$$

where $c$ is the speed of light. The Doppler frequency shift is then

$$f_d = f - f_r \approx f\frac{V_\rho}{c} = \frac{V_\rho}{\lambda} = \frac{d\rho}{\lambda dt}, \qquad (4.21)$$

where $\lambda = (f/c)$ is the wavelength.

The Doppler count (or integrated Doppler) $D$ is the historical observable of the TRANSIT satellite and is the integration of the frequency shift over a time interval (ca. 1 min). If the time interval is selected small enough, the Doppler count is the same as the instantaneous frequency shift, or

$$D = \frac{d\rho}{\lambda dt}. \qquad (4.22)$$

The approximate predicted Doppler frequency shift is required to get the satellite signal acquired. Prediction of $D$ is a part of the GPS signal tracking process. The predicted $D$ is used to predict the phase change first, and then the phase change is compared with the measured value to get the precise value of the Doppler frequency shift. The accumulated integer account of cycles is obtained through a polynomial fitting of a series of predicted phase changes and measured values (Remondi 1984). Therefore, the Doppler frequency shift is a by-product of the carrier phase measurements. However, the Doppler frequency shift is an independent observable and a measure of the instantaneous range rate.

Note that in an error-free environment, $d\rho/(\lambda dt)$ is the same as $d\Phi/dt$ and $\Phi$ is the phase measurement discussed in Sect. 4.2. Then the model of Eq. 4.22 can be obtained by differentiating the Eq. 4.17 with respect to the time $t$:

$$D = \frac{d\rho_r^s(t_r, t_e)}{\lambda dt} - f\frac{d\beta}{dt} + \delta_f + \varepsilon, \qquad (4.23)$$

where $\beta$ is the term of clock error $(\delta t_r - \delta t_s)$, $\delta_f$ is the frequency correction of the relativistic effects and $\varepsilon$ is error. Effects with low frequency properties such as ionosphere, troposphere, tide, and multipath effects are cancelled out.

## References

Goad C (1996a) Single-site GPS models. In: Kleusberg A, Teunissen PJG (eds) GPS for geodesy. Springer-Verlag, Berlin

Goad C (1996b) Short distance GPS models. In: Kleusberg A, Teunissen PJG (eds) GPS for geodesy. Springer-Verlag, Berlin

Leick A (1995) GPS satellite surveying. John Wiley & Sons Ltd., New York

Remondi B (1984) Using the Global Positioning System (GPS) phase observable for relative geodesy: Modelling, processing, and results. University of Texas at Austin, Center for Space Research

# Chapter 5
# Physical Influences of GPS Surveying

This chapter covers all physical influences of GPS observations, including ionospheric effects, tropospheric effects, relativistic effects, earth tide and ocean loading tide effects, clock errors, antenna mass centre and phase centre corrections, multipath effects, anti-spoofing and historical selective availability, and instrumental biases. Theories, models, and algorithms are discussed in detail.

## 5.1
## Ionospheric Effects

The ionospheric effect is an important error source in GPS measurements. The amount of ionospheric delay or advance of the GPS signal can vary from a few metres to more than 20 m within 1 day. It is generally difficult to model the ionospheric effects because of complicated physical interactions between the geomagnetic field and solar activity. However, the ionosphere is a dispersive medium, i.e. the ionospheric effect is frequency-dependent. This property is used to design the GPS system with several working frequencies such that ionospheric effects can be measured or corrected.

### 5.1.1
### Code Delay and Phase Advance

The phase velocity $v_p$ of an electromagnetic wave with one frequency propagating in space can be represented by

$$v_p = \lambda f, \tag{5.1}$$

where $\lambda$ is the wavelength and $f$ is the frequency; index p denotes phase. This formula is valid for both GPS L1 and L2 phase signals.

A modulated signal will propagate in space with a velocity that is called group velocity. Group velocity is different from phase velocity. The relationship between group velocity and phase velocity was described more than 100 years ago by Rayleigh (Seeber 1993):

$$v_g = v_p - \lambda \frac{dv_p}{d\lambda}, \tag{5.2}$$

where $dv_p/d\lambda$ is the differentiation of $v_p$ with respect to wave length $\lambda$; index g denotes group. This group velocity is valid for GPS code measurements.

From Eq. 5.1, one has the total differentiation

$$\frac{d\lambda}{\lambda} = -\frac{df}{f}, \tag{5.3}$$

and Eq. 5.2 can be rewritten as

$$v_g = v_p + f \frac{dv_p}{df}. \tag{5.4}$$

If the electromagnetic wave is transmitted in vacuum space, the phase velocity and the group velocity are the same and are equal to the speed of light in a vacuum. In such a case, the medium is called non-dispersive; otherwise, the medium is called dispersive. Two factors $n_p$ and $n_g$ are introduced so that both

$$v_g n_g = c \quad \text{and} \tag{5.5}$$

$$v_p n_p = c \tag{5.6}$$

are valid. These two factors $n_p$ and $n_g$ are called refractive indices. The refractive index characterises how the medium delays or advances the signal propagating velocity from the speed of light in a vacuum.

Differentiation of $v_p$ with respect to frequency $f$ can then be obtained from Eq. 5.6 by

$$\frac{dv_p}{df} = -\frac{c}{n_p^2}\left(\frac{dn_p}{df}\right). \tag{5.7}$$

Substituting the above three formulas into Eq. 5.4 yields

$$\frac{c}{n_g} = \frac{1}{n_p^2}\left(cn_p - fc\frac{dn_p}{df}\right) \quad \text{or}$$

$$n_g = \frac{n_p^2}{n_p - f\frac{dn_p}{df}}. \tag{5.8}$$

Using the mathematical expansion

$$(1-x)^{-1} = 1 + x - x^2 - \cdots \quad |x| < 1, \tag{5.9}$$

Equation 5.8 can be approximated to the first order by

$$n_g = n_p + f\left(\frac{dn_p}{df}\right). \tag{5.10}$$

The phase refractive index can be represented by

$$n_p = 1 + \frac{a_1}{f^2} + \frac{a_2}{f^3} + \cdots, \tag{5.11}$$

where coefficients $a_1$ and $a_2$ depend on the electron density $N_e$ and can be determined. Substituting Eq. 5.11 into Eq. 5.10 yields

$$n_g = 1 - \frac{a_1}{f^2} - \frac{2a_2}{f^3}. \tag{5.12}$$

The change of the length of the signal transmitting path in the medium with refractivity $n$ is

$$\Delta r = \int (n-1) ds. \tag{5.13}$$

Integration is done along the signal transmitting path. Therefore, the ionospheric effects on the phase and code signal transmission can be represented as

$$\begin{aligned} \delta_p &= \int (n_p - 1) ds = \int \left(\frac{a_1}{f^2} + \frac{a_2}{f^3}\right) ds \quad \text{and} \\ \delta_g &= \int (n_g - 1) ds = \int \left(-\frac{a_1}{f^2} - \frac{2a_2}{f^3}\right) ds. \end{aligned} \tag{5.14}$$

Omitting the second term on the right-hand side, one gets

$$\delta_p = -\delta_g = \int \left(\frac{a_1}{f^2}\right) ds. \tag{5.15}$$

Thus the ionospheric effects on the phase and code measurements have opposite signs and have approximately the same value. The coefficient $a_1$ was estimated by cf. Seeber (1993)

$$a_1 = -40.3 N_e, \tag{5.16}$$

where $N_e$ is the electron density.

The total electron content (TEC) in the zenith direction can be defined as

$$\text{TEC} = \int_{\text{zenith}} N_e \, ds, \qquad (5.17)$$

which can be computed from special models. To combine the TEC in the zenith and in the signal transmitting path, a so-called slant factor or mapping function must be introduced, which will be discussed in detail in Sect. 5.1.4.

The electron density always has a positive value; therefore, $\delta_g$ has a positive value and $\delta_p$ is negative. In other words, the ionosphere delays the code signal transmission and advances the phase signal transmission.

## 5.1.2
## Elimination of Ionospheric Effects

### Dual-Frequency Combination

The ionospheric effects on the phase (cf. Sect. 5.1.1) is rewritten as

$$\delta_p = \frac{A_1}{f^2}, \quad \text{where} \quad A_1 = \int a_1 \, ds. \qquad (5.18)$$

For the dual-frequency GPS phase observations, the ionospheric effects can be written as

$$\delta_p(f_1) = \frac{A_1}{f_1^2} \quad \text{and} \qquad (5.19)$$

$$\delta_p(f_2) = \frac{A_1}{f_2^2}. \qquad (5.20)$$

It is clear that the following combination leads to an elimination of the ionospheric effects:

$$f_1^2 \delta_p(f_1) - f_2^2 \delta_p(f_2) = 0. \qquad (5.21)$$

In other words, through linear combination of the GPS phase observations, the ionospheric effects can be eliminated. The above discussion is valid for both the code and carrier phase measurements of dual-frequencies, i.e. there is

$$f_1^2 \delta_g(f_1) - f_2^2 \delta_g(f_2) = 0. \qquad (5.22)$$

It should be pointed out that such ionosphere-free combination is indeed a first-order approximation because of the omission of the terms of the second-order ionospheric effects of Eqs. 5.14 in 5.15. Furthermore, the combinations of Eqs. 5.21 and 5.22 have to be standardised by dividing $f_1^2 - f_2^2$, so that the combined code and phase observations also have the sense that they are code and phase observables at a special frequency. The standard (first-order) ionosphere-free phase and code combinations can be represented then as

$$\frac{f_1^2 \delta_p(f_1) - f_2^2 \delta_p(f_2)}{f_1^2 - f_2^2} = 0 \quad \text{and} \quad (5.23)$$

$$\frac{f_1^2 \delta_g(f_1) - f_2^2 \delta_g(f_2)}{f_1^2 - f_2^2} = 0. \quad (5.24)$$

Formally the combined observations are observed at frequency

$$f = \frac{f_1^2 f_1 - f_2^2 f_2}{f_1^2 - f_2^2}, \quad (5.25)$$

which has a wavelength of $\lambda = c/f$, where $c$ is the speed of light in a vacuum.

### Triple-Frequency Combination

As mentioned above, a dual-frequency combination can only eliminate the first-order ionospheric effects. It is clear that a triple-frequency combination can eliminate the ionospheric effects up to the second order.

The ionospheric effects on the phase (cf. Sect. 5.1.1) are rewritten as

$$\delta_p = \frac{A_1}{f^2} + \frac{A_2}{f^3}, \quad \text{where} \quad A_1 = \int a_1 ds, \quad A_2 = \int a_2 ds. \quad (5.26)$$

For the triple-frequency GPS phase observations, the ionospheric effects can be written as

$$\delta_p(f_1) = \frac{A_1}{f_1^2} + \frac{A_2}{f_1^3}, \quad (5.27)$$

$$\delta_p(f_2) = \frac{A_1}{f_2^2} + \frac{A_2}{f_2^3} \quad \text{and} \quad (5.28)$$

$$\delta_p(f_5) = \frac{A_1}{f_5^2} + \frac{A_2}{f_5^3}. \quad (5.29)$$

The first-order ionosphere-free combinations can be formed as

$$f_1^2 \delta_p(f_1) - f_2^2 \delta_p(f_2) = \frac{A_2}{f_1} - \frac{A_2}{f_2} \quad \text{and} \tag{5.30}$$

$$f_1^2 \delta_p(f_1) - f_5^2 \delta_p(f_5) = \frac{A_2}{f_1} - \frac{A_2}{f_5}, \quad \text{or} \tag{5.31}$$

$$\frac{f_1^2 \delta_p(f_1) - f_2^2 \delta_p(f_2)}{\frac{1}{f_1} - \frac{1}{f_2}} = A_2 \quad \text{and} \tag{5.32}$$

$$\frac{f_1^2 \delta_p(f_1) - f_5^2 \delta_p(f_5)}{\frac{1}{f_1} - \frac{1}{f_5}} = A_2. \tag{5.33}$$

Then the second-order ionosphere-free combination can be formed by

$$\frac{(f_1^2 \delta_p(f_1) - f_2^2 \delta_p(f_2))(f_1 f_2)}{(f_2 - f_1)} - \frac{(f_1^2 \delta_p(f_1) - f_5^2 \delta_p(f_5))(f_1 f_5)}{(f_5 - f_1)} = 0 \tag{5.34}$$

or

$$B_1 \delta_p(f_1) + B_2 \delta_p(f_2) + B_5 \delta_p(f_5) = 0, \tag{5.35}$$

where

$$B_1 = f_1^3 f_1 \frac{(f_5 - f_2)}{(f_2 - f_1)(f_5 - f_1)}, \tag{5.36}$$

$$B_2 = -\frac{f_2^3 f_1}{f_2 - f_1} \quad \text{and} \tag{5.37}$$

$$B_5 = \frac{f_5^3 f_1}{f_5 - f_1}. \tag{5.38}$$

A standardisation of the combination in Eq. 5.35 can be made through

$$\frac{B_1 \delta_p(f_1) + B_2 \delta_p(f_2) + B_5 \delta_p(f_5)}{B_1 + B_2 + B_5} = 0 \quad \text{or} \tag{5.39}$$

$$C_1 \delta_p(f_1) + C_2 \delta_p(f_2) + C_5 \delta_p(f_5) = 0, \tag{5.40}$$

where

$$C_1 = \frac{f_1^3(f_5 - f_2)}{C_4}, \qquad (5.41)$$

$$C_2 = -\frac{f_2^3(f_5 - f_1)}{C_4}, \qquad (5.42)$$

$$C_5 = \frac{f_5^3(f_2 - f_1)}{C_4} \quad \text{and} \qquad (5.43)$$

$$C_4 = f_1^3(f_5 - f_2) - f_2^3(f_5 - f_1) + f_5^3(f_2 - f_1). \qquad (5.44)$$

The above discussion is also valid for the code measurements of triple-frequencies, i.e. there is

$$C_1 \delta_g(f_1) + C_2 \delta_g(f_2) + C_5 \delta_g(f_5) = 0. \qquad (5.45)$$

*Phase-Code Combination*

Recalling the discussion in Sect. 5.1.1 and limiting ourselves to the first-order approximation, the ionospheric effects on the phase and code measurements have opposite signs and have approximately the same value, i.e.

$$\delta_p = -\delta_g = A_1/f^2, \quad \text{where} \quad A_1 = \int a_1 \, ds. \qquad (5.46)$$

Therefore, a straightforward method to eliminate the ionospheric effects is then to combine the phase and code observables at the same frequency $f$ together, i.e.

$$\delta_p(f) + \delta_g(f) = 0. \qquad (5.47)$$

It is notable that such a combination has lower precision than that of the carrier phase and code measurements, respectively.

## 5.1.3
## Ionospheric Models

*The Broadcast Ionospheric Model*

The GPS broadcast message includes the parameters of a predicted ionospheric model (Klobuchar 1996; Leick 1995). Using the model parameters, the ionospheric effects can be computed and corrected.

The input parameters of the broadcast ionospheric model are the eight model coefficients of $\alpha_i, \beta_i, i = 1, 2, 3, 4$, geodetic latitude $\varphi$ and longitude $\lambda$ of the GPS antenna, GPS observing time $T$ in seconds, as well as the azimuth $A$ and elevation $E$ of the observed satellite. All four angular arguments $\varphi, \lambda, A$, and $E$ have units of semicircles (SC), and 1 SC equals 180 degrees. The formulas are given below:

$$F = 1 + 16(0.53 - E)^3, \tag{5.48}$$

$$\Psi = \frac{0.0137}{E + 0.11} - 0.022, \tag{5.49}$$

$$\varphi_i = \varphi + \Psi \cos A, \tag{5.50}$$

$$\varphi_i = \frac{0.416 \varphi_i}{|\varphi_i|}, \quad \text{if} \quad |\varphi_i| > 0.416, \tag{5.51}$$

$$\lambda_i = \lambda + \psi \frac{\sin A}{\cos \varphi_i}, \tag{5.52}$$

$$\phi = \varphi_i + 0.064 \cos(\lambda_i - 1.167), \tag{5.53}$$

$$t = \lambda_i 43{,}200 + T, \tag{5.54}$$

$$t = t - 86{,}400, \quad \text{if} \quad t \geq 86{,}400, \tag{5.55}$$

$$t = t + 86{,}400, \quad \text{if} \quad t < 0, \tag{5.56}$$

$$P = \sum_{i=1}^{4} \beta_i \phi^i, \tag{5.57}$$

$$P = 72{,}000, \quad \text{if} \quad P < 72{,}000, \tag{5.58}$$

$$x = \frac{2\pi(t - 50{,}400)}{P}, \tag{5.59}$$

$$Q = \sum_{i=1}^{4} \alpha_i \phi^i, \tag{5.60}$$

$$Q = 0, \quad \text{if} \quad Q < 0, \tag{5.61}$$

$$\delta_g(f_1) = cF5 \times 10^{-9}, \quad \text{if} \quad |x| > 1.57 \quad \text{and} \tag{5.62}$$

$$\delta_g(f_1) = cF \left( 5 \times 10^{-9} + Q \left( 1 - \frac{x^2}{2} + \frac{x^4}{24} \right) \right), \quad \text{if} \quad |x| < 1.57. \tag{5.63}$$

$\varphi_i$ and $\lambda_i$ are the geodetic latitude and longitude of the sub-ionospheric point. The ionospheric point is defined as the point on the sight of the satellite, which has the average ionospheric height (350 km), and the sub-ionospheric point is the projection point of the ionospheric point onto the earth's surface, which has a height of 50 km, $\phi$ being the geomagnetic latitude of the sub-ionospheric point. $\psi$ is the earth's central angle between the GPS station and the ionospheric point. $F$ is the slant factor or mapping function that maps the ionospheric effects of the zenith direction onto the signal transmitting path. The local time at the sub-ionospheric point is denoted by $t$. $P$ and $Q$ are the period and amplitude in seconds. The phase is denoted by $x$. $c$ is the speed of light. Frequency of L1 is denoted by $f_1$.

The ionospheric group delay on the L2 frequency can be computed by

$$\delta_g(f_2) = \frac{f_1^2}{f_2^2} \delta_g(f_1). \qquad (5.64)$$

The phase advance has only an opposite sign if the phase has been scaled to have units of length. Dividing the length with the wavelength can transform units of length to units of cycle.

Figure 5.1 shows the ionospheric effects of the broadcasted ionospheric model of 9 September 2001. The ionospheric parameters are

$$(\alpha_i) = (\,3073 \quad 1490 \quad -11{,}920 \quad -11{,}920\,) \times 10^{-11} \quad \text{and}$$
$$(\beta_i) = (\,1372 \quad 1638 \quad -1966 \quad 3932\,) \times 10^{+2}.$$

Station coordinates are selected as ($\varphi = 45°$, $\lambda = 0°$). Computation has been carried out for a whole day of 24 h in GPS time. The continuous line shows the ionospheric effects on a satellite that repeats its orbit every 4 h and changes its elevation and azimuth regularly from (5–85–5°) and (30–150°), respectively. The broken line shows the (zenith) ionospheric effects of a space fixed satellite in the zenith direction (elevation = 90°, azimuth = 180°). It shows the strong dependency of the

**Fig. 5.1** Broadcasted ionospheric model

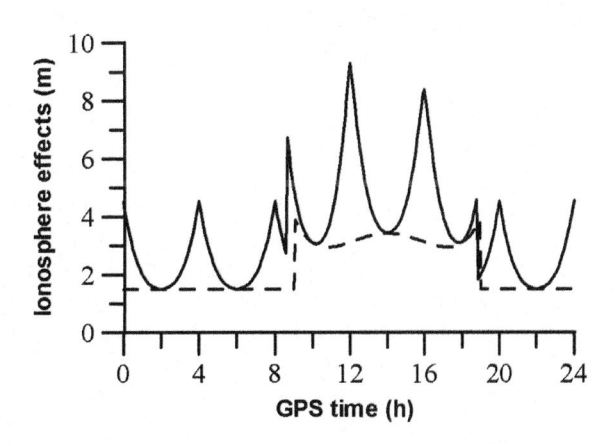

ionospheric effects on the time and zenith angle of the satellite. In the zenith direction, the ionospheric effects remain constant (1.5 m) before 9:00 and after 19:00. Strong changes happen at the time of sunrise and sunset and the ionospheric noon (14:00). Depending on the elevation of the satellite, the ionospheric effects may be amplified up to three times.

The broadcast ionospheric model can remove the ionospheric delay more than 50 % (Langley 1998a, b).

### *Dual-Frequency Ionosphere Measuring Model*

In pseudorange measurement only the ionospheric effects depend on the working frequency. Therefore, a simple difference of the pseudoranges of the dual-frequencies can eliminate all other effects except the ionospheric effects and subsequently can be used for determining the ionospheric delay:

$$R_1 - R_2 = \delta_g(f_1) - \delta_g(f_2) = \left(1 - \frac{f_1^2}{f_2^2}\right)\delta_g(f_1) \quad \text{or} \tag{5.65}$$

$$\delta_g(f_1) = \frac{R_1 - R_2}{1 - \frac{f_1^2}{f_2^2}}, \tag{5.66}$$

where $R_1$ and $R_2$ are the L1 and L2 pseudoranges, and $f_1$ and $f_2$ are the frequencies of the L1 and L2 carriers. Here the random measurement error and un-modelled bias are omitted.

Similarly, the ionospheric effects can be determined by dual-frequency phase observables. Recall the phase observable model discussed in Sect. 4.2 where a simple differential combination of both phase pseudoranges can be formed as

$$\begin{aligned}\lambda_1 \Phi_1 - \lambda_2 \Phi_2 &= \delta_p(f_1) - \delta_p(f_2) + \lambda_1 N_1 - \lambda_2 N_2 \\ &= (1 - \frac{f_1^2}{f_2^2})\delta_p(f_1) + \lambda_1 N_1 - \lambda_2 N_2,\end{aligned} \tag{5.67}$$

or

$$\delta_p(f_1) = \frac{\lambda_1 \Phi_1 - \lambda_2 \Phi_2 - \lambda_1 N_1 + \lambda_2 N_2}{1 - \frac{f_1^2}{f_2^2}}, \tag{5.68}$$

where $\Phi_1$ and $\Phi_2$ are the L1 and L2 phase pseudoranges (in units of cycles), and $N_1$ and $N_2$ are the ambiguities of the L1 and L2 carriers. The random measurement error and un-modelled bias are omitted here. As long as the phase measurements are continuous (no cycle slips), the $\lambda_1 N_1 - \lambda_2 N_2$ remains a constant. Through a long-term statistic comparison of Eqs. 5.66 and 5.68, the constant $\lambda_1 N_1 - \lambda_2 N_2$ can be approximately determined. The variation of the ionospheric effects can be determined very well by using this method.

## 5.1.4
## Mapping Functions

As mentioned in Sect. 5.1.1, in order to combine the TEC in the zenith direction and in the signal transmitting path, the slant factor or mapping function $F$ is needed such that

$$\text{TEC}_\rho = \text{TEC}_z F, \tag{5.69}$$

where indices $\rho$ and $z$ denote the path and zenith directions, respectively.

Generally, the ionosphere begins at a height of 50 km and ends at a height of about 750 km. It is therefore assumed that the ionosphere has an average height of 350 km (see Fig. 5.2). The sight line of the satellite crosses over the shell at the so-called ionospheric point. The projection of the ionospheric point at a height of 50 km is called the sub-ionospheric point. The point at the sight line of the satellite at a height of 50 km is called the sub-ionospheric point in sight. The point at the sight line of the satellite at a height of 750 km is called the sup-ionospheric point in sight. These four points are denoted by $P_{ip}$, $P_{sip}$, $P_{sips}$, and $P_{supip}$ respectively.

### Projection Mapping Function

Based on a single layer model, a homogeneous distribution of the free electrons is assumed (Fig. 5.2). This is equivalent to assuming all free electrons are concentrated in a shell of infinitesimal thickness at a height of 350 km. In such a case, the mapping function may be written as

$$F = \frac{1}{\cos z_{ip}}, \tag{5.70}$$

where $z_{ip}$ is the satellite zenith angle at the ionospheric point. Using the sinus theorem, the relationship between the $z_{ip}$ and zenith distance ($z$) of the satellite viewed from the receiver can be obtained by

$$\sin z_{ip} = \frac{r}{r+350} \sin z, \tag{5.71}$$

**Fig. 5.2** Single-layer ionospheric model

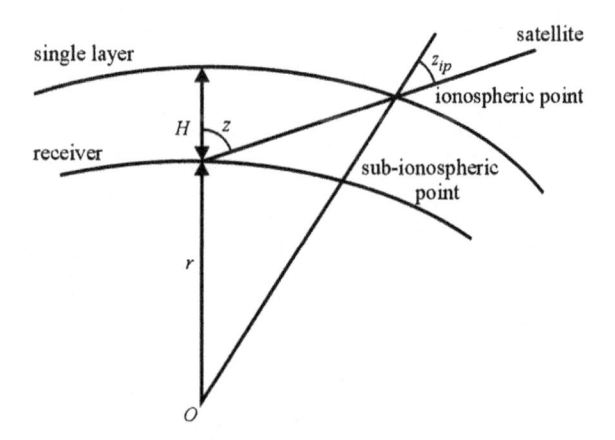

where $r$ is the mean radius of the earth in km. This mapping function is called a single layer mapping function or projection mapping function. It is notable that Eq. 5.71 is exactly valid only for the spherical zenith angles.

### Geometric Mapping Function

If a height-dependent homogeneous distribution of the free electrons is assumed, then the mapping function is a geometric one and is equivalent to

$$d\rho = dHF, \tag{5.72}$$

where $d\rho$ and $dH$ are the ionospheric path delay and zenith delay respectively.

The zenith angle of the satellite at the sub-ionospheric point (in sight) $P_{sips}$ is denoted by $z_{sips}$ (Fig. 5.3). It can be computed by using the sinus theorem

$$\sin z_{sips} = \frac{r}{r+50} \sin z, \tag{5.73}$$

where $z$ is the zenith angle of the satellite viewed from the receiver and $r$ is the mean radius of the earth in km. In the geometry, the spherical zenith angles are used here.

The difference between the spherical zenith and geodetic zenith depends on the latitude of the station and the azimuth of the satellite. The maximum difference is the difference between geodetic latitude and geocentric latitude of the computing point; this is about $(e^2/2)\sin(2\varphi)$ (Torge 1991), where $e^2$ is the first numerical eccentricity (<0.0067) and $\varphi$ is the geodetic latitude of the computing point. Therefore, the small angle difference can be omitted. Of course, for correctness the spherical zenith angle should be used here. It is the angle of the sight line to satellite with respect to the earth centred radius vector of the station. Using the cosines theorem, one has

$$(r+50+H)^2 = (r+50)^2 + \rho^2 - 2(r+50)\rho \cos(180 - z_{sips}) \tag{5.74}$$

**Fig. 5.3** Spherical ionospheric model

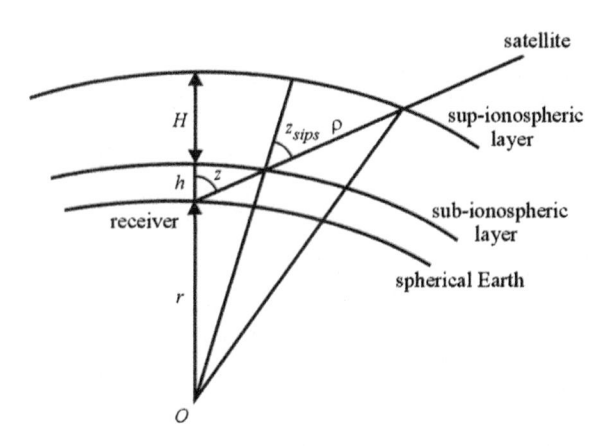

or

$$\rho^2 + 2(r+50)\cos(z_{\text{sips}})\rho + (r+50)^2 - (r+50+H)^2 = 0, \quad (5.75)$$

where $\rho$ and $H$ are the lengths of lines from the sup-ionospheric point to the sub-ionospheric point in sight and sub-ionospheric point, respectively. The second-order equation can be solved by

$$\rho = -(r+50)\cos(z_{\text{sips}}) \pm \sqrt{(r+50)^2\cos^2(z_{\text{sips}}) - (r+50)^2 + (r+50+H)^2} \quad (5.76)$$

or

$$\rho = -(r+50)\cos(z_{\text{sips}}) \pm \sqrt{(r+50+H)^2 - (r+50)^2\sin^2(z_{\text{sips}})}. \quad (5.77)$$

Because $\rho > 0$, Eq. 5.75 has a unique solution,

$$\rho = -(r+50)\cos(z_{\text{sips}}) + \sqrt{(r+50+H)^2 - (r+50)^2\sin^2(z_{\text{sips}})}. \quad (5.78)$$

Comparing Eqs. 5.72 with 5.78, one gets the geometric mapping function

$$F = -\frac{r+50}{H}\cos(z_{\text{sips}}) + \frac{\sqrt{(r+50+H)^2 - (r+50)^2\sin^2(z_{\text{sips}})}}{H} \quad (5.79)$$

or approximately

$$F = -9.183\cos(z_{\text{sips}}) + 10.183\sqrt{1 - 0.81\sin^2(z_{\text{sips}})}, \quad (5.80)$$

where $r = 6378$ km and $H = 700$ km are used.

Above, the derived geometric mapping function of Eq. 5.79 is a spherical approximation if $r$ is considered a constant. This mapping function could be called Xu's geometric mapping function (Xu 2003).

### *Ellipsoidal Mapping Function*

Taking the dependency of the radius $r$ on the latitude $\varphi$ into account, an ellipsoidal mapping function can be derived. According to Torge (1991),

$$r^2 = a^2\cos^2\beta + b^2\sin^2\beta \quad \text{and}$$
$$\tan\beta = \frac{b}{a}\tan\varphi, \quad (5.81)$$

where $r$ is the radius of the rotational ellipsoid, $a$ and $b$ are the semi-major axis and semi-minor axis of the ellipsoid, and $\beta$ is an angle that has the relation with geodetic latitude $\varphi$. Using the triangle formula

$$2\cos^2\beta - 1 = 1 - 2\sin^2\beta = \frac{1 - \tan^2\beta}{1 + \tan^2\beta}, \tag{5.82}$$

Equation 5.81 can be rewritten as

$$r^2 = \frac{a^2}{2}\left(\frac{1 - \tan^2\beta}{1 + \tan^2\beta} + 1\right) + \frac{b^2}{2}\left(1 - \frac{1 - \tan^2\beta}{1 + \tan^2\beta}\right), \tag{5.83}$$

or

$$r^2 = \frac{a^2}{2}\left(\frac{a^2 - b^2\tan^2\varphi}{a^2 + b^2\tan^2\varphi} + 1\right) + \frac{b^2}{2}\left(1 - \frac{a^2 - b^2\tan^2\varphi}{a^2 + b^2\tan^2\varphi}\right). \tag{5.84}$$

In an ellipsoid case, Eqs. 5.74 and 5.79 turn out to be

$$(r_s + 50 + H)^2 = (r_i + 50)^2 + \rho^2 - 2(r_i + 50)\rho\cos(180 - z_{sips}) \quad \text{and}$$
$$F = -\frac{r_i + 50}{H}\cos(z_{sips}) + \sqrt{(r_s + 50 + H)^2 - (r_i + 50)^2 \sin^2(z_{sips})}/H, \tag{5.85}$$

where $r_s$ and $r_i$ denote the geocentric radius of the sub-ionospheric point and sub-ionospheric point in sight, respectively. They can be obtained by substituting the geodetic latitudes $\varphi_s$ and $\varphi_i$ of the related two positions into Eq. 5.84. The ellipsoidal mapping function is then Eq. 5.85. This mapping function could be called Xu's ellipsoidal mapping function (Xu 2003).

The mapping functions are needed if the ionospheric effects have to be determined. In Eq. 5.72, $d\rho$ may be considered to be an ionospheric path delay observed by GPS, and $dH$ may be considered to be an ionospheric model, which is independent of the path zenith such as that given in Eq. 5.27. The determined parameters then have the physical meaning of the TEC in the zenith direction.

## 5.1.5
### Introduction of Commonly Used Ionospheric Models

Ionospheric delay models can generally be grouped into two types: (1) empirical models, which are based on data to represent the characteristic variation patterns seen in long data records (e.g., Klobuchar model, IRI model, NeQuick model, NTCM-GL model); and (2) mathematical function models, which are fitted by mathematical functions based on the actual measured ionospheric delay of a certain region over a period time, to better meet the positioning demand of the users in that region (e.g., POLY model, TSF model, SH function model, GIM).

*Klobuchar Model*

The GPS broadcast ephemeris uses Klobuchar model (1987) to correct the ionospheric delay of single frequency receivers (cf. Sect. 5.1.3). Klobuchar model has advantages of simple structure and convenient calculation, which is suitable for the ionospheric delay correction of real-time single-frequency receiver positioning. Klobuchar model takes the amplitude and periodic changes of daily scale of the ionosphere into account when setting the parameters, which can reflect the variation characteristics of the ionosphere and ensure the reliability of ionospheric forecasts from large scale. The downside is that it has a limited precision on the ionospheric delay correction, which is only suitable for the mid-latitude area. Because of the high-frequency activity of the ionosphere in high- and low-latitude regions, this model cannot reflect the true state of the ionosphere. Experience shows that the Klobuchar model can correct 50–60 % of the ionospheric delay generally.

Since July 2000, the Centre for Orbit Determination in Europe (CODE) began providing post-processing Klobuchar coefficients to make the eight broadcast coefficients consistent with the global ionosphere maps (GIM) products. Results show that the consistency of the post-processing Klobuchar model is better than the broadcast Klobuchar model. In the meanwhile, CODE also provides forecast Klobuchar model coefficients. Petrie et al. (2011) proved that the effect of the forecast Klobuchar model is not as significant as the post-processing model.

*IRI Model*

The International Reference Ionosphere (IRI) is the standard model for ionospheric densities and temperatures developed and updated by a joint working group of the Committee on Space Research (COSPAR) and the International Union of Radio Science (URSI) (Bilitza 2001). IRI is an empirical standard model of the ionosphere based on all available data sources. Several steadily improved editions of the model have been released. The latest version is IRI-2012 (Bilitza et al. 2014). For a given location, time, and date, IRI describes monthly averages of electron density, electron temperature, ion temperature, ion composition, and several additional parameters in the altitude range from 60 to 2000 km, and also the electron content. The National Space Science Data Center (NSSDC) provides access to on-line IRI computations through its World Wide Web (WWW) models interface at http://nssdc.gsfc.nasa.gov/model/ionospheric/iri.html. IRI provides monthly averages in the non-auroral ionosphere for magnetically quiet conditions. For high-precision observations, the instantaneous change in ionosphere should be considered.

As an empirical model, IRI has the advantage that it does not depend on an evolving theoretical understanding of the processes that shape the ionospheric plasma. A disadvantage of the empirical model is the strong dependence on the underlying database. Regions and time periods not well covered by the database will result in diminished reliability of the model in these areas (Bilitza et al. 2014).

### NeQuick Model

The NeQuick is an empirical model based on the model introduced by Di Giovanni and Radicella in 1990. The NeQuick algorithm is used by the IRI model as a default option for the upper ionosphere computation. The NeQuick is able to calculate electron density at any given location in the ionosphere. Therefore, it can provide TEC and electron density profile between any two given points (Nava et al. 2008).

### NTCM-GL Model

A new global ionospheric model, the Global Neustrelitz TEC Model (NTCM-GL), was developed at the Institute of Communications and Navigation at the German Aerospace Centre (DLR) in Neustrelitz, Germany. The model can provide values of vertical TEC (VTEC) at any given time and location. The core of the model consists of 12 coefficients, which can be autonomously used for full solar cycle. The model does not use any integration of electron density profile; therefore, it is very simple and fast (Jakowski et al. 2011). The model analytically describes daily variation, seasonal variation, equatorial altitude anomaly, and solar flux dependency as harmonic functions. All the formulas of the model algorithm can be found in Jakowski et al. (2011).

### Polynomial (POLY) Model

The POLY model is based on the difference of sun angle and latitude difference between the ionospheric pierce point (IPP) and the regional centre, which can be expressed as (Komjathy 1997)

$$\text{VTEC} = \sum_{i=0}^{n} \sum_{k=0}^{m} E_{ik} (\varphi - \varphi_0)^i (S - S_0)^k, \tag{5.86}$$

where $E_{ik}$ are the unknown model coefficients, $\varphi$ and $S$ are the geographic latitude and sun angle at the IPP, respectively, and $\varphi_0$ and $S_0$ are the geographic latitude and sun angle at the regional centre, respectively.

The POLY model has a simple structure and takes into account ionospheric changes related to latitude and sun angle. This model is used widely in regional ionospheric modelling analysis because it can obtain better results than other models over a certain period and within a certain range.

### Trigonometric Series Function (TSF) Model

Georgiadiou proposed using a trigonometric series model to build a regional ionospheric model, which further improved the simulation capabilities of local daily ionospheric variations. Based on a trigonometric series, a generalised trigonometric series with variable parameters in geomagnetic coordinates can be formed as (Mannucci et.al. 1998)

$$\text{VTEC} = A_1 + \sum_{i=1}^{N_2}\{A_{i+1}\varphi_m^i\} + \sum_{i=1}^{N_3}\{A_{i+N_2+1}h^i\} + \sum_{i=1,j=1}^{N_i,N_j}\{A_{i+N_2+N_3+1}\varphi_m^i h^j\}$$
$$+ \sum_{i=1}^{N_4}\{A_{2+N_2+N_3+N_i-1}\cos(ih) + A_{2i+N_2+N_3+N_i}\sin(ih)\},$$
(5.87)

the geomagnetic latitude at the IPP, and $h$ represents variables relative to local time. $h = 2\pi(t - 14)/T$, $T = 24$ h, where $t$ is the local time at IPP in units of hours. $\varphi_m = \varphi + 0.064\cos(\lambda - 1.617)$, where $\varphi$ and $\lambda$ are the geographical latitude and longitude, respectively, in units of radians.

### *Spherical Harmonic (SH) Function Model*

The spherical harmonic function has been widely used in global ionospheric models. A global spherical harmonic model with 15 × 15 order coefficients has been released by CODE. In global modelling, the zero-order term is the global ionosphere average TEC values. In regional modelling, although the spherical harmonic coefficients do not have orthogonality, a lower-order spherical harmonic function model can still be used to study the ionospheric region. The function model can be expressed as follows (Schaer 1999)

$$\text{VTEC} = \sum_{n=0}^{n_{\max}}\sum_{m=0}^{n} P_{nm}(\sin\beta)(A_{nm}\cos(ms) + B_{nm}\sin(ms)), \quad (5.88)$$

where $s$ is the longitudinal difference between the pierce point and sun direct spot, $P_{nm}(\sin\beta)$ is the Legendre function, and $A_{nm}$ and $B_{nm}$ are unknown model coefficients.

### *Grid Ionospheric Model (GIM)*

The grid ionospheric model is mainly applied in the Wide Area Augmentation System (WAAS). This model is based on the approximation that the ionosphere is considered to be a thin layer, an ionospheric spherical shell, to a height of 350 km above the earth's surface (Otsuka et al. 2002). The grid ionospheric model values are the vertical ionospheric delays or vertical total electron content at the specific ionospheric grid points (IGPs, the intersection points of the selected longitude and latitude lines) covering the area. The electron content of the station is usually interpolated according to the four grid points. Therefore, in order to obtain the electron content with high accuracy, space and time interpolations are needed.

The effect of the empirical ionospheric model is generally not as good as the mathematical model. This is because there are many factors that can affect the ionosphere, and these factors may contain strong randomness. Moreover, the interaction of the various factors, the change law, and internal mechanism that cause the anomalous variation of the ionospheric delay are not entirely clear, thus making it more difficult to infer only via empirical data. For the mathematical fitting function model, it can be established without the demand of deep understanding of the internal mechanism of the ionosphere. A number of small-scale changes over time are already reflected in the model. Furthermore, with the increased number of global navigation system (GNSS) satellites and their dense and uniform distribution, the regional ionospheric model can be well fitted and established within 2 to 4 h. Therefore, more desirable results can be achieved through the mathematical function model.

## 5.2
## Tropospheric Effects

The troposphere is the lower part of the atmosphere over the earth's surface. Unlike the ionosphere, the troposphere is a non-dispersive medium at GPS carrier frequencies. That is, the tropospheric effects on the GPS signal transmission are independent of the working frequency. The electromagnetic signals are affected by the neutral atoms and molecules in the troposphere. These effects are called tropospheric delay, or tropospheric refraction. Indeed, the word "tropospheric" is not used here in the exact sense; historically, tropospheric effects are simply considered to be the effects of the atmosphere below the ionosphere. The amount of tropospheric delay in the zenith direction is about 2 m, and it increases with the increase in the zenith angle of the sight line to the satellite. In the case of a lower satellite elevation by a few degrees, the tropospheric delay of the GPS signal can be a few metres or more. Therefore, the tropospheric effect is an important error source in precise GPS applications.

Generally speaking, the tropospheric delay depends on temperature, pressure, and humidity, as well as the location of the GPS antenna. Similar to the ionospheric path delay, the tropospheric path delay can be written as

$$\delta = \int (n-1) \mathrm{d}s, \qquad (5.89)$$

where $n$ is the refractive index of the troposphere, the integration is taken along the signal transmitting path, which could be simplified as the geometric path. Scaling of the refractive index anomaly $(n-1)$ is usually made by

$$N = 10^6 (n-1), \qquad (5.90)$$

where $N$ is called tropospheric refractivity. $N$ can be separated into wet (about 10 %) and dry (about 90 %) parts:

$$N = N_w + N_d, \tag{5.91}$$

where indices w and d denote the wet and dry. They are caused by water vapour and dry atmosphere, respectively. Therefore Eq. 5.89 becomes

$$\delta = \delta_w + \delta_d = 10^{-6} \int N ds, \tag{5.92}$$

where

$$\delta_w = 10^{-6} \int N_w ds \quad \text{and} \tag{5.93}$$

$$\delta_d = 10^{-6} \int N_d ds. \tag{5.94}$$

If the integrations are made along the zenith direction, then the related mapping functions should be defined by

$$\delta_w = \delta_{wz} F_w, \tag{5.95}$$

$$\delta_d = \delta_{dz} F_d \quad \text{and} \tag{5.96}$$

$$\delta = \delta_z F, \tag{5.97}$$

where index $z$ denotes the tropospheric delays in the zenith direction, and $F_w$ and $F_d$ are mapping functions related to the wet and dry components. Analogous to the discussions made in Sect. 5.1.4, mapping functions are needed to determine the related delay models in the zenith direction. All empirical tropospheric path delay models have their own mapping functions.

## 5.2.1
## Tropospheric Models

### Modified Saastamoinen Model

The modified Saastamoinen tropospheric model (Saastamoinen 1972, 1973) for calculating the tropospheric path delay can be outlined as

$$\delta = \frac{0.002277}{\cos z} \left[ P + \left( \frac{1255}{T} + 0.05 \right) e - B \tan^2 z \right] + \delta R, \tag{5.98}$$

where $z$ is the zenith angle of the satellite, $T$ is the temperature at the station (in units of Kelvin [K]), $P$ is the atmospheric pressure (in units of millibars [mb]), $e$ is the partial pressure of water vapour (in mb). $B$ and $\delta R$ are the corrections terms that

**Table 5.1** Function of $B(H)$

| Height (km) | 0.0 | 0.5 | 1.0 | 1.5 | 2.0 | 2.5 | 3.0 | 4.0 | 5.0 |
|---|---|---|---|---|---|---|---|---|---|
| $B$ (mbar) | 1.156 | 1.079 | 1.006 | 0.938 | 0.874 | 0.813 | 0.757 | 0.654 | 0.563 |

depend on $H$ and $z$, respectively. $H$ is the height of the station. $\delta$ is the tropospheric path delay (in metres), and cf., e.g., Wang et al. (1988)

$$e = R_h \exp(-37.2465 + 0.213166\,T - 0.000256908\,T^2), \qquad (5.99)$$

where $R_h$ is the relative humidity (in %) and exp() is the exponential function. $B$ and $\delta R$ can be interpolated from Tables 5.1 and 5.2, respectively.

To transform the unit of the temperature $T$ from K (Kelvin) to °C (Celsius), one may use

$$T(\text{K}) = T(\text{Celsius}) + 273.15. \qquad (5.100)$$

In the model, either measured values of pressure, temperature, and humidity or the values derived from a standard atmospheric model may be used. The height-dependent values of pressure, temperature, and humidity may be obtained by the equations

$$P = P_0[1 - 0.0000266(H - H_0)]^{5.225}, \qquad (5.101)$$

$$T = T_0 - 0.0065(H - H_0), \quad \text{and} \qquad (5.102)$$

$$R_h = R_{h0} \exp[-0.0006396(H - H_0)], \qquad (5.103)$$

where $P_0$, $T_0$, and $R_{h0}$ are called standard pressure, temperature, and humidity at the reference height $H_0$. It is clear that the values are dependent on the geographic position of the station and time as well as the weather. Without the values of $P_0$, $T_0$, and $R_{h0}$, a direct correction using the model is not possible. In such a case, the tropospheric effects are usually estimated through the factor parameters of the mapping function, which will be discussed later. Additionally, the following values will be used as standard input:

$$H_0 = 0\,\text{m}, \qquad (5.104)$$

$$P_0 = 1013.25\,\text{mbar}, \qquad (5.105)$$

$$T_0 = 18\,°\text{C}, \quad \text{and} \qquad (5.106)$$

$$R_{h0} = 50\,\%. \qquad (5.107)$$

The original Saastamoinen tropospheric model has a constant value of $B$ and $\delta R = 0$ in the modified model of Eq. 5.98. Three types of mapping functions are

**Table 5.2** Function of $\delta R(H, z)$

| z(degrees)\Height(km) | 0.0 | 0.5 | 1.0 | 1.5 | 2.0 | 3.0 | 4.0 | 5.0 |
|---|---|---|---|---|---|---|---|---|
| 60.00 | 0.003 | 0.003 | 0.002 | 0.002 | 0.002 | 0.002 | 0.001 | 0.001 |
| 66.00 | 0.006 | 0.006 | 0.005 | 0.005 | 0.004 | 0.003 | 0.003 | 0.002 |
| 70.00 | 0.012 | 0.011 | 0.010 | 0.009 | 0.008 | 0.006 | 0.005 | 0.004 |
| 73.00 | 0.020 | 0.018 | 0.017 | 0.015 | 0.013 | 0.011 | 0.009 | 0.007 |
| 75.00 | 0.031 | 0.028 | 0.025 | 0.023 | 0.021 | 0.017 | 0.014 | 0.011 |
| 76.00 | 0.039 | 0.035 | 0.032 | 0.029 | 0.026 | 0.021 | 0.017 | 0.014 |
| 77.00 | 0.050 | 0.045 | 0.041 | 0.037 | 0.033 | 0.027 | 0.022 | 0.018 |
| 78.00 | 0.065 | 0.059 | 0.054 | 0.049 | 0.044 | 0.036 | 0.030 | 0.024 |
| 78.50 | 0.075 | 0.068 | 0.062 | 0.056 | 0.051 | 0.042 | 0.034 | 0.028 |
| 79.00 | 0.087 | 0.079 | 0.072 | 0.065 | 0.059 | 0.049 | 0.040 | 0.033 |
| 79.50 | 0.102 | 0.093 | 0.085 | 0.077 | 0.070 | 0.058 | 0.047 | 0.039 |
| 79.75 | 0.111 | 0.101 | 0.092 | 0.083 | 0.076 | 0.063 | 0.052 | 0.043 |
| 80.00 | 0.121 | 0.110 | 0.100 | 0.091 | 0.083 | 0.068 | 0.056 | 0.047 |

used in the modified model. The first is obviously $1/\cos z$; this is the mapping function of the flat earth model or single layer model. The second one is $\tan^2 z/\cos z$ (cf. Eq. 5.98). The third one is implicit, represented by the numerical Table 5.2.

*Modified Hopfield Model*

The modified Hopfield model (Hopfield 1969, 1970, 1972) for calculating the tropospheric path delay can be summarised as

$$\delta = \delta_d + \delta_w \quad \text{and} \tag{5.108}$$

$$\delta_i = 10^{-6} N_i \sum_{k=1}^{9} \frac{f_{k,i}}{k} r_i^k, \quad i = d, w. \tag{5.109}$$

Subscript $i$ is used to identify the dry and wet components of the tropospheric delay, and

$$\begin{aligned}
r_i &= \sqrt{(R_E + h_i)^2 - R_E^2 \sin^2 z} - R_E \cos z, \\
f_{1,i} &= 1, \quad f_{2,i} = 4a_i, \\
f_{3,i} &= 6a_i^2 + 4b_i, \quad f_{4,i} = 4a_i(a_i^2 + 3b_i), \\
f_{5,i} &= a_i^4 + 12a_i^2 b_i + 6b_i^2, \quad f_{6,i} = 4a_i b_i(a_i^2 + 3b_i), \\
f_{7,i} &= b_i^2(6a_i^2 + 4b_i), \quad f_{8,i} = 4a_i b_i^3, \\
f_{9,i} &= b_i^4,
\end{aligned} \tag{5.110}$$

$$a_i = -\frac{\cos z}{h_i}, \quad b_i = -\frac{\sin^2 z}{2h_i R_E},$$
$$h_d = 40{,}136 + 148.72(T - 273.15)\,(\text{m}),$$
$$h_w = 11{,}000\,(\text{m}),$$
$$N_d = \frac{77.64 P}{T}\,(\text{K mb}^{-1}),\qquad\qquad (5.111)$$
$$N_w = -\frac{12.96\,e}{T} + \frac{371800\,e}{T^2},\quad \text{and}$$
$$R_E = 6378137\,\text{m},$$

where $z$ is the zenith angle of the satellite, $T$ is the temperature at the station (in units of Kelvin [K]), $P$ is the atmospheric pressure (in units of millibars [mb]), and $e$ is the partial pressure of water vapour (in mb, cf. Eq. 5.99). $R_E$ is the earth's radius. $\delta$ is the tropospheric path delay (in metres).

In the model, either measured values of pressure, temperature, and humidity or the values derived from a standard atmospheric model may be used. The height-dependent values of pressure, temperature, and humidity may be obtained using Eqs. 5.101–5.107. As mentioned before, the tropospheric effects are usually estimated through suitable parameterisation, which will be discussed in the next section.

A graph of the modified Hopfield model with standard input parameters is given in Fig. 5.4.

There are still many other models for computing the tropospheric delay, such as the Davis model (Davis and Herring 1984), original Hopfield and Saastamoinen models (cf., e.g., Hofmann-Wellenhof et al. 1997), Niellis model, and Yionoulis

**Fig. 5.4** Modified hopfield tropospheric model (troposphere delay of GPS signal)

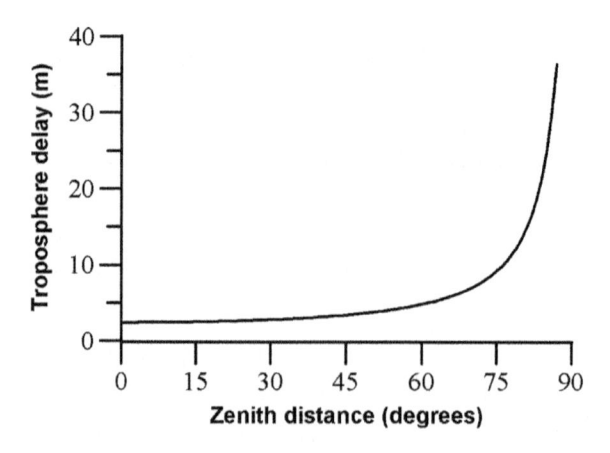

model (Zhu 2001). The differences between these models are generally very small for a zenith distance less than 75°.

## 5.2.2
## Mapping Functions and Parameterisation

In Sect. 5.1.4, the ionospheric mapping functions are discussed under the assumptions of symmetry of the sphere and rotating ellipsoid shapes of the ionosphere. For similar assumptions of the troposphere shapes, all mapping functions discussed in Sect. 5.1.4 can be directly used here for troposphere by changing the related values.

*Projection Mapping Function*

Because the troposphere has a maximum height of 50 km, the zenith angle of the satellite at the observation point may simply be used in the single layer mapping function:

$$F = \frac{1}{\cos z}. \tag{5.112}$$

*Geometric Mapping Function*

It can be similarly derived as in Sect. 5.1.4 by

$$F = -\frac{r}{H}\cos z + \frac{\sqrt{(r+H)^2 - r^2 \sin^2 z}}{H}, \tag{5.113}$$

where $r = 6378$ km, $H = 50$ km, and $z$ is the spherical zenith distance of the satellite viewed from the station. This mapping function can be called Xu's geometric mapping function (Xu 2003).

Because of the complexity of the troposphere, the so-called co-mapping function is needed. If a height-dependent homogeneous distribution of the troposphere is assumed, then the co-mapping function is a geometric one and is defined as

$$d\rho = dSF_c, \tag{5.114}$$

where $d\rho$ and $dS$ are the tropospheric path delay and the delay mapped to the line from station to the sub-tropospheric point. In the case of zenith angle $z$ equals zero, the $dS$ is zero and co-mapping function is undefined. Index c in $F_c$ is used to denote the co-mapping function. It is clear that the projection co-mapping function is

$$F_c = \frac{1}{\sin z}. \tag{5.115}$$

**Fig. 5.5** Spherical troposphere model

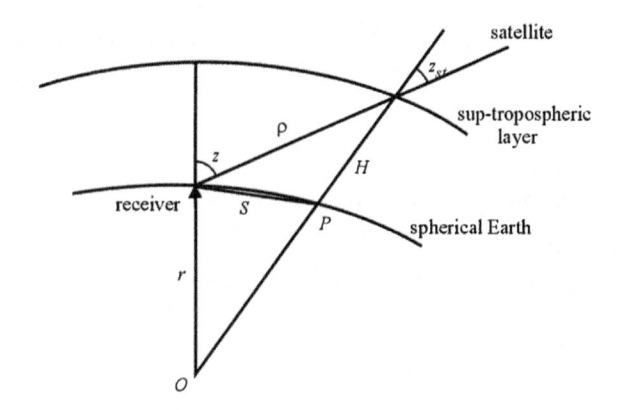

### Geometric Co-mapping Function

The zenith angle of the satellite at the sup-tropospheric point is denoted by $z_{st}$ (Fig. 5.5). It can be computed by using the sinus theorem

$$\sin z = \frac{r}{r+50} \sin z_{st}, \tag{5.116}$$

where $z$ is the zenith angle of the satellite viewed from the receiver and $r$ is the mean radius of the earth in km. Using the cosines theorem, one has

$$S^2 = H^2 + \rho^2 - 2H\rho \cos z_{st} \tag{5.117}$$

or

$$S = \sqrt{H^2 + \rho^2 - 2H\rho \cos z_{st}}, \tag{5.118}$$

where $\rho$ and $S$ are the lengths of lines from the station to the sup-tropospheric point and sub-tropospheric point, respectively. Then the geometric co-mapping function is

$$F_c = \frac{\rho}{\sqrt{H^2 + \rho^2 - 2H\rho \cos z_{st}}}, \tag{5.119}$$

where

$$\rho = -r \cos z + \sqrt{(r+H)^2 - r^2 \sin^2 z}, \tag{5.120}$$

where $r = 6378$ km, $H = 50$ km. Above, the derived geometric co-mapping function of Eq. 5.119 is a spherical approximation if $r$ is considered a constant and can be called Xu's geometric co-mapping function (Xu 2003).

The mapping and co-mapping functions are necessary for two purposes: one for the determination of a related tropospheric model, and the other for the determination of the tropospheric path delay effects on the GPS observations. Recall the definitions of $d\rho = dH \cdot F$ (cf. Eq. 5.72). Here, $dH$ may be considered a tropospheric model that is independent of the zenith distance of the signal transmitting path, whereas $d\rho$ represents observed tropospheric delays of path direction. With the observed $d\rho$ and known mapping function $F$, the parameters of the model $dH$ can be determined. The model $dH$ is generally a function of temperature, pressure, and humidity as seen in the models discussed in Sect. 5.2.1. To correct for the tropospheric effects on GPS observations, one needs tropospheric models. However, the input parameters of the models are usually not measured together with the GPS measurements. The standard method of dealing with such a problem includes two steps. First, the standard temperature, pressure, and humidity values for everywhere and any time will be used as the input of the tropospheric model to compute the path delay $d\rho$. Then, the computed $d\rho$ should be amplified with a functional factor $g$, and the $g$ must be determined by GPS data processing. The formulation of $g$ is called parameterisation of the tropospheric path delay effects. Two factorisation methods are given here:

$$g_\rho d\rho, \quad g_z \frac{d\rho}{F} + g_a \frac{d\rho}{F_c} \qquad (5.121)$$

Physically, $g_\rho$, $g_z$, and $g_a$ are factors in the path direction and zenith direction, and in the azimuth component, respectively. Mapping function $F$ and co-mapping function $F_c$ are used to map the computed $d\rho$ to the desired directions.

A step function or a first-order polynomial function

$$g = g(t) = g_i, \quad \text{if} \quad t_{i-1} \leq t < t_i, \quad i = 1,\ldots,n \qquad (5.122)$$

or

$$g = g(t) = g_{i-1} + (g_i - g_{i-1})\frac{(t - t_{i-1})}{\Delta t}, \quad \text{if} \quad t_{i-1} \leq t < t_i, \quad i = 1,\ldots,n \qquad (5.123)$$

may be used as path factor $g_\rho$. Where in Eqs. 5.122 and 5.123 $\Delta t = (t_e - t_0)/n$, $t_0$ and $t_e$ are the beginning time and ending time of the GPS surveying, $n$ is an integer that may be selected with a reasonable value, $t_i = t_0 + (i - 1)\Delta t$, $g_i$ are constant unknowns which shall be determined.

The azimuth dependency may be assumed as

$$g_a = g_1 \cos a + g_2 \sin a, \qquad (5.124)$$

where $a$ is the azimuth of the satellite at the station, and $g_1$ and $g_2$ in turn may be a step function or a first-order polynomial function given in Eqs. 5.122 and 5.123.

## 5.2.3
### Introduction of Commonly Used Tropospheric Models

The tropospheric delay can be represented as the product of the tropospheric refraction in zenith direction and a mapping function related to the elevation angle (cf. Eqs. 5.95 and 5.96). With the approximate position and the meteorological data, the zenith delay can be easily computed by the Saastamoinen model (cf. Sect. 5.2.1). Generally, the meteorological data can be obtained from actual observations or derived using a standard atmospheric value at sea level and the height of the station (Berg 1948). Meteorological data can also be determined by empirical models called GPT (Global Pressure and Temperature model, Boehm et al. 2007) or the later GPT2 model (Lagler et al. 2013). According to the computation formula of the Saastamoinen model, 1 mbar pressure change at sea level can cause a change of about 2.3 mm in a priori zenith hydrostatic delay, thus it is essential to use as accurate meteorological data as possible (Tregoning and Herring 2006). The commonly used Saastamoinen model and Hopfield model have been introduced and analysed in the Sect. 5.2.1. Therefore, the recently developed meteorological models and mapping functions are focused on in this section.

Many mapping functions were proposed in the past, such as NMF (Niell Mapping Function, Niell 1996), VMF1 (Vienna Mapping Function 1, Boehm et al. 2006b), GMF (Global Mapping Function, Boehm et al. 2006a), which were commonly researched in the recent years. By comparison, a general form of the mapping functions can be outlined as (Herring 1992; Kouba 2009)

$$MF = \frac{1 + \frac{a}{1+\frac{b}{1+c}}}{\sin\varepsilon + \frac{a}{\sin\varepsilon + \frac{b}{\sin\varepsilon + c}}}, \tag{5.125}$$

where $MF$ is the mapping function, $\varepsilon$ is the elevation angle, $a$, $b$, and $c$ are empirical coefficients with different values in various mapping functions. $(a_h, b_h, c_h)$ and $(a_w, b_w, c_w)$ are used for the hydrostatic and wet components, respectively.

The accuracy of the mapping function would definitely affect the precision of the slant delay, thus affecting the positioning precision. And when the elevation cut-off angle is lower, the impact is more significant. Applying a rule of thumb (MacMillan and Ma 1994; Boehm et al. 2006b), an error in the wet mapping function of 0.01 or in the hydrostatic mapping function of 0.001 would cause an error of 4 mm in the station height under a 5° elevation cut-off angle.

### NMF Model

The NMF model was derived by Niell (1996) based on data from 26 globally distributed balloons stations. It is widely used because of its global validity and its independence from surface meteorological observations. NMF uses 15° latitude grid tables and models the seasonal amplitudes for each of the three coefficients of Eq. (5.125). The parameter $a$ can be given as

$$a(\varphi, t) = a_{\text{avg}}(\varphi) + a_{\text{amp}}(\varphi) \cos(2\pi(t - 28)/365.25), \qquad (5.126)$$

where $\varphi$ is the latitude of the station, $t$ is the day of year, parameters $a_{\text{avg}}$ and $a_{\text{amp}}$ can be obtained by the linear interpolation between the nearest latitudes. For parameters $b$ and $c$, the same procedure can be applied.

The parametrisation of NMF is based on ray traces of radiosonde profiles spanning the latitudes 43°S to 75°N. It was extended globally by assuming longitudinal homogeneity and symmetry between the southern and the northern hemisphere. The polar regions with latitudes higher than 75° on the northern and southern hemisphere are both approximated by the 75°N latitude values, thus the precision of NMF in polar regions is poor.

### VMF1 Model

VMF1 model was derived by Boehm et al. (2006b). The coefficients $a_h$ and $a_w$ were derived from a rigorous ray tracing through pressure levels of the European Centre for Medium-Range Weather Forecasts (ECMWF) operational analysis data. These coefficients are provided by TU Vienna as site-specific or global grid (2° × 2.5°) time series with 6-hourly temporal spacing. The coefficients $b_h$ and $b_w$ were derived from 1 year of ECMWF data in a least squares fit. Whereas $b_h$ is constant, $c_h$ depends on the day of year and the latitude. $b_w$ and $c_w$ were taken from the Niell mapping function at 45° latitude, since the coefficient $a_w$ is sufficient to model the dependence of the wet mapping function on latitude. VMF1 is currently the mapping function providing globally the most accurate and reliable geodetic results. The related program is available at the official website of TU Vienna http://ggosatm.hg.tuwien.ac.at/DELAY/SOURCE.

### GMF/GPT Model

The GMF is an empirical mapping function (input arguments are only the day of year and the site location) that is consistent with VMF1 (Boehm et al. 2006a). Expressions for the coefficients $a_h$ and $a_w$ (mean values and annual signal) were derived from 3 years of ECMWF data and are provided as a spherical harmonic expansion of degree and order 9. The coefficients $b$ and $c$ are taken from the VMF1.

The GPT is an empirical model proposed by Boehm et al. (2007), which can be downloaded from http://www.hg.tuwien.ac.at/~ecmwf1. It is based on spherical harmonics up to degree and order 9 and provides pressure and temperature at any site in the vicinity of the earth's surface. The GPT model can be used for geodetic applications such as the determination of a priori hydrostatic zenith delays, reference pressure values for atmospheric loading, or thermal deformation of very-long-baseline interferometry (VLBI) radio telescopes. Input parameters of GPT are the station coordinates and the day of the year, thus also allowing one to model the annual variations of the parameters. As an improvement compared to previous models, it reproduces the large pressure anomaly over Antarctica, which can cause station height errors in the analysis of space-geodetic data of up to 1 cm if not considered properly in troposphere modelling. The pressure biases can be reduced substantially using GPT instead of the very simple approaches applied to various global navigation satellite system (GNSS) software packages. GPT also provides an appropriate model for the annual variability of global temperature.

Compared to VMF1, GMF/GPT greatly simplifies the estimation process, since no external data are required. It provides consistency with VMF1 with only a slight degradation in precision.

## GPT2 Model

The GPT2 model was proposed by TU Vienna at the end of 2012 (Lagler et al. 2013). Monthly mean profiles provided by ECMWF from 2001 to 2010 were analysed. After the calculation of mean, annual, and semi-annual terms of meteorological data based on global grid points, the results were expressed globally in the form of a 5° grid at mean earth surface heights. In addition, hydrostatic and wet mapping function coefficients were also able to be obtained. GPT2 largely eradicates the weakness in GMF/GPT, specifically their limited spatial and temporal variability. A comprehensive overview of the improvements of GPT2 in relation to GMF/GPT is given in Table 5.3.

The benefits and the improved performance of GPT2 with respect to the previous models GPT/GMF have been validated by Lagler et al. (2013). GPT2 yields a 40 % reduction in annual and semi-annual amplitude differences in station heights with

**Table 5.3** Improvements of GPT2 with respect to GMF/GPT

|  | GMF/GPT | GPT2 |
|---|---|---|
| NWM data | Monthly mean profiles from ERA-40 (23 pressure levels): 1999–2002 | Monthly mean profiles from ERA-Interim (37 levels): 2001–2010 |
| Representation | Spherical harmonics up to degree and order 9 at mean sea level | 5° grid at mean ETOPO5-based heights |
| Temporal variability | Mean and annual terms | Mean, annual, and semi-annual terms |
| Phase | Fixed to January 28 | Estimated |
| Temperature reduction | Constant lapse rate: 6.5 °C/km assumed | Mean, annual, and semi-annual terms of temperature lapse rate estimated at every grid point |
| Pressure reduction | Exponential based on standard atmosphere | Exponential based on virtual temperature at each point |
| Output parameters | Pressure ($p$), temperature ($T$), mapping function coefficients ($a_h, a_w$) | $p$, $T$, lapse rate ($dT$), water vapour pressure ($e$), $a_h$, $a_w$ |

respect to a solution based on instantaneous local pressure values and VMF1, as shown with a series of global VLBI solutions. Thus, it is recommended the replacement of the old GMF/GPT with GPT2 as empirical model in the analysis of radio space geodetic observations.

## 5.2.4
### Tropospheric Model for Airborne Kinematic Positioning

Despite the commonly used global tropospheric delay models, data from GPS ground networks have been widely used to construct precise regional tropospheric models. Thus, the tropospheric delay of a kinematic station within the region can be obtained through interpolation of the regional tropospheric model. Commonly used interpolation methods include the direct interpolation method (DIM) (Wanninger 1997) and the remove–restore method (RRM) (Zhang and Lachapelle 2001). In DIM, the tropospheric delays at a kinematic station are directly interpolated from those at reference stations. In RRM, firstly, the delays from the standard tropospheric model are removed, then the remaining values are used for interpolation. Finally, the delays from standard tropospheric model are restored. DIM is simple and applied for the flat areas while the property of RRM is suitable for the undulating areas. However, both methods are only appropriate for a kinematic station on the ground (Xu 2000). If the kinematic station is several kilometres higher than the ground reference stations, the precision of tropospheric delay will decrease rapidly (Collins and Langley 1997; Mendes and Langley 1998). Therefore, the projection extension method (PEM), proposed by Wang et al. (2010, 2011) to establish an airborne tropospheric model for high-altitude kinematic stations, is introduced in this section.

The basic concept of PEM is that the kinematic station in the air is first projected onto the average elevation of the surveyed area; the tropospheric delays at projected points are then interpolated from those of reference stations; and finally, the delays at projected points are extended upward to the airborne platform.

Suppose that the tropospheric delays $\text{dtrop}(i, t)$ at an arbitrary reference station can be expressed as

$$\text{dtrop}(i, t) = \alpha(i, t) \times \text{saas\_dtrop}(i), \qquad (5.127)$$

where $\text{saas\_dtrop}(i)$ are the tropospheric delays from a standard tropospheric model such as the Saastamoinen model, $\alpha(i, t)$ is the unknown factor, $i$ is the reference station ID, and $t$ is the time. The unknown factor $\alpha(i, t)$ can be expressed as

$$\alpha(i, t) = a_1 + a_2 \times t + \cdots + a_k \times t^{k-1} + \cdots + a_n \times t^{n-1}, \qquad (5.128)$$

where $a_k(k = 1, 2, \ldots, n)$ are the polynomial coefficients, which can be determined by a least squares fitting. The order of the polynomial can be determined based on the statistical significance (Xu and Yang 2001).

The geodetic coordinates of reference stations $B(i)$, $L(i)$, $H(i)$, and kinematic station $\bar{B}(t)$, $\bar{L}(t)$, $\bar{H}(t)$ are converted into the Gauss-Kruger coordinates $x(i)$, $y(i)$, $h(i)$, and $\bar{x}(t)$, $\bar{y}(t)$, $\bar{h}(t)$, respectively. In the projection conversion, the central meridian is the mean longitude of the reference network. The projection plane is the average height of all reference stations.

The unknown factor $\bar{\alpha}(t)$ of the projected points is obtained by interpolating the $\alpha(i, t)$ of reference stations based on the inverse distance weighted method as follows:

$$\bar{\alpha}(t) = \sum_{i=1}^{n} \alpha(i, t) \times P(i, t), \tag{5.129}$$

$$P(i, t) = \frac{r(i, t)^{-2}}{\sum_{i=1}^{n} r(i, t)^{-2}}, \tag{5.130}$$

$$r(i, t)^2 = (x(i) - \bar{x}(t))^2 + (y(i) - \bar{y}(t))^2 + (h(i) - \bar{h}(t))^2, \tag{5.131}$$

where $P(i, t)$ can be considered a kind of weight of $\alpha(i, t)$; $r(i, t)$ is the distance between the reference stations and projected points.

The standard model tropospheric delays at projected points are extended upward to the kinematic station in the air based on the following formula (Syndergaard 1999):

$$\Delta h = \text{rover\_}h(t) - \bar{h}(t), \tag{5.132}$$

$$\text{press}(t) = \overline{\text{press}} \times (1 - 0.0000226 \times \Delta h)^{5.225}, \tag{5.133}$$

$$\text{temp}(t) = \overline{\text{temp}} - 0.0065 \times \Delta h, \tag{5.134}$$

$$\text{rh}(t) = \overline{\text{rh}} \times e^{(-0.0006396 \times \Delta h)}, \tag{5.135}$$

$$\text{rover\_saas}(t) = \text{SAAS}(\text{press}(t), \text{temp}(t), \text{rh}(t)), \tag{5.136}$$

where $\Delta h$ is the elevation difference between kinematic station in the air and projected point. $\text{rover\_}h(t)$ is the elevation of the kinematic station; $\overline{\text{press}}$, $\overline{\text{temp}}$, $\overline{\text{rh}}$ are the pressure, temperature, and relative humidity of the projected point; $\text{press}(t)$, $\text{temp}(t)$, $\text{rh}(t)$ are the pressure, temperature, and relative humidity of the kinematic station; $\text{SAAS}(\text{press}(t), \text{temp}(t), \text{rh}(t))$ is the standard model tropospheric delay when given pressure and temperature and humidity. The units of height, pressure, and temperature are metres, millibars, and Kelvin, respectively.

Therefore, the tropospheric delays of the kinematic station rover_drop(t) can be obtained by

$$\text{rover\_drop}(t) = \bar{\alpha}(t) \times \text{rover\_saas}(t), \tag{5.137}$$

The numerical example for validation of this method can be referred to Wang et al. (2011).

## 5.2.5
## Water Vapour Research with Ground-Based GPS Measurement

Water vapour is mainly distributed at the bottom of the troposphere, constituting approximately 99 % of total atmospheric water vapour. It is one of the most active and variable parts of the atmospheric components and one of the most difficult meteorological parameters to characterise (Rocken et al. 1993). Water vapour plays a key role in a range of spatial and temporal scales of atmospheric processes, and its distribution is directly related to the distribution of clouds and precipitation. Research on the distribution of water vapour in the atmosphere will be of great assistance in weather forecasting and climate prediction.

Traditional water vapour measurement methods include radiosonde, water vapour radiometer, and satellite remote sensing, which may have disadvantages of heavy workload, high equipment costs, or low spatial and temporal resolution. Therefore, using GPS to detect water vapour has become an interesting area of research to meet the increasing demands of meteorological development.

Precipitable water vapour (PWV) has been highlighted in GPS meteorological calculation. By means of specific methods we can estimate zenith wet delays (ZWD) based on GPS observation data, and the relationship between ZWD and PWV can be expressed as (Bevis et al. 1992):

$$\text{PWV} = \prod \cdot \text{ZWD}, \tag{5.138}$$

where $\prod$ is a water vapour conversion factor. It can be expressed as

$$\prod = \frac{10^6}{\rho_W R_V \left[(k_3/T_m) + k_2'\right]}, \tag{5.139}$$

where $\rho_W$ is the density of water, $R_V$ is the specific gas constant for water vapour, $k_2'$ and $k_3$ are the atmospheric refractivity constants (Davis et al. 1985; Bevis et al. 1994), $T_m$ is the key variable to calculate the conversion factor $\prod$.

$T_m$ is an important parameter in ground-based GPS meteorology, and is usually expressed as a linear function of surface temperature $T_s$. The linear relationship between $T_m$ and $T_s$ has been explored by many scientists. Bevis et al. (1992) came up with the equation $T_m = 70.2 + 0.72T_s$ based on analysis of 8718 radiosonde profiles, which is suitable for mid-latitude region. However, the formula has a limited application because of the lack of surface temperature measurement at some GPS stations. Based on the analysis of 23 years of radiosonde data from 53 stations, Ross and Rosenfeld (1997) noted that the correlation between $T_m$ and $T_s$ generally weakens in the equatorial region and becomes weaker in summer than in winter. Gu et al. (2005) researched the variations of $T_m$ and computed a regional $T_m$. Li et al. (2006) verified the applicability of Bevis $T_m - T_s$ relationship in specific areas. Mao (2006) systematically studied the method of using GPS to detect water vapour. Wang et al. (2007a) derived precipitable water (PW) using the zenith path delay (ZPD) obtained from the existing ground-based GPS measurements on a global scale, and applied that to produce a near-global 2-hourly PW data set. Ding (2009) introduced the fundamental of the GPS meteorology and related calculation methods in detail. Several regionally applicable linear models were established by scientists, such as Li et al. (1999), Liu et al. (2000), Gu (2004), Wang et al. (2007b), and Lv et al. (2008). Considering seasonal variations, Yao et al. (2012) established the globally applicable GTm-I model independent of surface temperature, using radiosonde data of 135 stations in 2005–2009. To address the abnormalities of the model in some areas because of the lack of radiosonde data at sea, Yao et al. (2013) employed the GPT model and the Bevis $T_m - T_s$ relationship to provide simulated $T_m$ at sea and recomputed GTm model coefficients to improve the model's accuracy. Yeh et al. (2014) used a water vapour radiometer (WVR) to verify the PWV measured by GPS. Choy et al. (2015) compared the PWV derived from GPS with traditional radiosonde measurement and VLBI techniques, which showed good agreement. Details can be found in the respective references.

## 5.3
## Relativistic Effects

### 5.3.1
### Special Relativity and General Relativity

Einstein's special relativity is based on two postulates. The first is called the principle of relativity: "No inertial system is preferred. The equations expressing the laws of physics have the same form in all inertial systems." The second is called the principle of the constancy of the speed of light: "The speed of light is a universal constant independent of the state of motion of the source. Any light ray moves in the inertial system of coordinates with constant velocity $c$, whether the ray is emitted by a stationary or by a moving source." Of course, the speed of light refers to velocity in a vacuum (Ashby and Spilker 1996).

Consider two inertial coordinate systems S' and S in Fig. 5.6, where the $x'$-axis and $x$-axis coincide. Two origins are placed at points A and B, respectively. Origin A of system S' moves with a constant velocity $v$ along the $x$-axis toward B. The distance between A and B viewed in system S is $\Delta x$. The mirror surface is

**Fig. 5.6** Light transmission viewed in two inertial frames

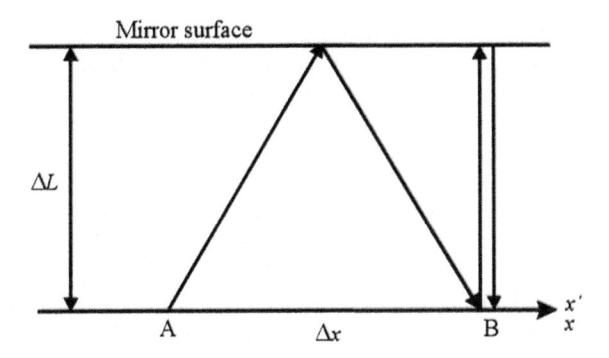

parallel to the x-axis and is faced to the x-axis. The perpendicular distance of the mirror to the x-axis is $\Delta L$. Suppose a light flash is emitted from A and the reflected light is received at B by the moving system S'. Then the transmitting time of the light measured in system S is

$$\Delta t = \frac{2\sqrt{(\Delta L)^2 + (\Delta x/2)^2}}{c}. \tag{5.140}$$

According to our assumption, one gets $\Delta x = v\Delta t$. Substituting this into Eq. 5.140, $\Delta t$ can be obtained by

$$\Delta t = \frac{2\Delta L/c}{\sqrt{1 - (v/c)^2}}. \tag{5.141}$$

Because of Einstein's postulates, the speed of light, $c$, is the same in two systems. Therefore, $2\Delta L/c$ is the light flash transmitting time viewed in the moving system S', i.e. $\Delta t' = 2\Delta L/c$, and

$$\Delta t = \frac{\Delta t'}{\sqrt{1 - (v/c)^2}}. \tag{5.142}$$

This indicates that the time interval viewed in the system at rest is shorter than the time interval viewed in the system, which is moving with velocity $v$.

Again, because of the constant $c$ in the two systems, one may denote $\Delta s = c\Delta t$ and $\Delta s' = c\Delta t'$, where $\Delta s$ and $\Delta s'$ are the lengths of the light transmitting paths viewed in the two systems. Multiplying $c$ by Eq. 5.142 gets

$$\Delta s = \frac{\Delta s'}{\sqrt{1 - (v/c)^2}}. \tag{5.143}$$

This indicates that the length viewed in the moving system is lengthened.

Consider the relation $c = f\lambda$, where $c$ is constant in both systems, $\lambda$ is wavelength, and $f$ is the related frequency. Because the wavelength $\lambda$ viewed in two systems is different, denoted by $\lambda = \Delta s$ and $\lambda' = \Delta s'$, the relationship of the frequencies $f$ and $f'$, which are viewed in two systems, can be obtained by dividing $c$ into Eq. 5.143:

$$f = f'\sqrt{1 - (v/c)^2}. \tag{5.144}$$

This indicates that the frequency $f'$ viewed in the moving system is reduced to $f$ when it is viewed by a resting system.

Using mathematical expansions

$$\frac{1}{\sqrt{1 - (v/c)^2}} = 1 + \frac{1}{2}\left(\frac{v}{c}\right)^2 \ldots, \tag{5.145}$$

$$\sqrt{1 - (v/c)^2} = 1 - \frac{1}{2}\left(\frac{v}{c}\right)^2 \ldots, \tag{5.146}$$

for Eqs. 5.142, 5.143 and 5.144, we have

$$\frac{\Delta t - \Delta t'}{\Delta t'} = \frac{\Delta s - \Delta s'}{\Delta s'} = -\frac{f - f'}{f'} = \frac{1}{2}\left(\frac{v}{c}\right)^2. \tag{5.147}$$

This is the formula of the special relativity effects caused by a constant motion of a moving inertial coordinate system viewed from a resting inertial coordinate system.

Einstein's general relativity incorporates gravitation by virtue of the principle of equivalence. The mathematics of general relativity is extremely complex. However, for treatment of the relativistic effects on GPS, only a simplified and small fraction of the theory is required. Note that the right-hand side of Eq. 5.147 is indeed the point-mass (or unit mass) kinetic energy ($v^2/2$) scaled by the speed of light $c$ (exactly $1/c^2$). That is, the special relativity effects may be interpreted as the effects caused by kinetic energy due to motion. The analogous effects may also be caused by potential energy $\Delta U$ due to the presence of the gravitation field $U$. Then

$$\frac{\Delta t - \Delta t'}{\Delta t'} = \frac{\Delta s - \Delta s'}{\Delta s'} = -\frac{f - f'}{f'} = \frac{\Delta U}{c^2} \tag{5.148}$$

represents the relativistic relations in the case of the presence of a gravitational field $U$. Thus, the total relativistic effects may be formulated as

$$\frac{\Delta t - \Delta t'}{\Delta t'} = \frac{\Delta s - \Delta s'}{\Delta s'} = -\frac{f - f'}{f'} = \frac{1}{2}\left(\frac{v}{c}\right)^2 + \frac{\Delta U}{c^2}. \tag{5.149}$$

The presence of a gravitational field indicates an acceleration of the frame $S'$ with respect to the system S at rest.

The special relativity effects of rotation may be similarly discussed. Details can be found, for example, in Ashby and Spilker (1996).

## 5.3.2
## Relativistic Effects on GPS

The inertial coordinate system at rest, its origin located at the centre of the earth, is taken as reference to view all GPS related activities. Because of the large motion velocities and near circular orbits of the GPS satellite, the non-negligible gravitational potential difference between the satellite and the users, as well as the rotation of the earth, the relativistic effects have to be taken into account. For convenience, we may imagine that the whole GPS process is viewed in an inertial reference at a point where the gravitational potential is the same as that of the geoid of the earth. Taking the earth's rotational effects into account, the view point is equivalent to the point of the GPS user on the geoid of the rotating earth.

### Frequency Effects

The fundamental frequency $f_0$ of the GPS system is selected as 10.23 MHz. All clocks on the GPS satellites and GPS receivers operate based on this frequency. If all the GPS satellites are working simply on the frequency $f' = f_0$, then we will view a frequency $f$ at our reference point, and $f$ is not the same as $f_0$ due to relativistic effects. In order to view the fundamental frequency $f = f_0$, the desired working frequency $f'$ of the GPS satellites can be computed using Eq. 5.149 by

$$-\frac{f_0 - f'}{f'} = \frac{1}{2}\left(\frac{v}{c}\right)^2 + \frac{\Delta U}{c^2}, \qquad (5.150)$$

where $v$ is the velocity of the satellite and $\Delta U$ is the difference of the earth's gravitational potential between the satellite and the geoid. The difference between the setting frequency $f'$ of the satellite clock and the fundamental frequency $f_0$ is referred to as the offset in the satellite clock frequency. Such an offset of relativistic effects has been implemented in the satellite clock settings, and therefore, users do not need to consider this effect. The offset can be computed by using the mean velocity of the satellite and $\Delta U = \mu/(R_E + H) - \mu/R_E$, where $\mu$ is the gravitational constant of the earth, $R_E$ is the earth's radius (ca. 6370 km), and $H$ is the height of the satellite above the earth (ca. 20,200 km). The offset is approximately 0.00457 Hz; in other words, the satellite clock frequency is set to $f_0 - 4.57 \times 10^{-9}$ MHz.

For the receiver fixed on the earth's surface, the frequency of the clock in the receiver is also affected by the relativistic effects. The effects can be represented analogously by Eq. 5.150, where $\Delta U = 0$ and $v$ is the velocity of the receiver due to the rotation of the earth. Such effects are corrected by the software of the receiver.

## Path Range Effects

The general relativity effects of the signal transmitted from the GPS satellite to the receiver can be represented by the Holdridge (1967) model:

$$\Delta \rho_{rel} = \frac{2\mu}{c^2} \ln \frac{\rho^j + \rho_i + \rho_i^j}{\rho^j + \rho_i - \rho_i^j}, \tag{5.151}$$

where $\rho^j$ and $\rho_i$ are the geocentric distances of the satellite $j$ and station $i$, respectively, $\rho_i^j$ is the distance between the satellite and the observing station, and $\Delta \rho_{rel}$ has units of metres and a maximum value of about 2 cm. It is notable that by computing the distance $\rho_i^j$, the effect of the rotation of the earth during signal transmission must be taken into account (if it is done in the earth's fixed system).

## Earth's Rotational Effects

All corrections related to the rotation of the earth are called Sagnac corrections. The geocentric vector of the GPS satellite is denoted by $\vec{r}_s$, the geocentric vector of the receiver by $\vec{r}_r$, and the velocity vector of the receiver by $\vec{v}_r$. These are the vectors during GPS signal emission. Suppose the transmitting time between the signal emission from satellite and signal reception of receiver is $\Delta t$. During the time of GPS signal transmission, the receiver has moved to position $\vec{r}_r + \vec{v}_r \Delta t$. Observing from the non-rotating frame, the distance of the signal transmission can be represented by

$$c \Delta t = |\vec{r}_r + \vec{v}_r \Delta t - \vec{r}_s|. \tag{5.152}$$

Therefore, the transmitting path correction due to the rotation of the earth can be presented as

$$\Delta \rho = |\vec{r}_r + \vec{v}_r \Delta t - \vec{r}_s| - |\vec{r}_r - \vec{r}_s|. \tag{5.153}$$

This can be simplified as (Ashby and Spilker 1996)

$$\Delta \rho = \frac{(\vec{r}_r - \vec{r}_s) \cdot \vec{v}_r}{c}. \tag{5.154}$$

The correction can reach up to 30 m and must be taken into account.

If the signal transmitting time $\Delta t$ has been solved through iteration of Eq. 5.152, then the Sagnac correction will automatically be taken into account.

This term of correction is also valid for the kinematic GPS receivers that are not fixed on the earth's surface. The velocity vector in Eq. 5.154 is

$$\vec{v}_r = \vec{\omega}_e \times \vec{r}_r + \vec{v}_k, \tag{5.155}$$

where the first term on the right-hand side is the velocity vector of the receiver due to the earth's rotation, and the second term $\vec{v}_k$ is the kinematic velocity vector of the receiver related to the earth's surface. A kinematic motion of 100 km h$^{-1}$ related to the earth's surface can cause additional Sagnac effects up to 2 m.

The Sagnac correction also must be taken into account for low-earth orbit (LEO) satellites (e.g., TOPEX, CHAMP, and GRACE), which are equipped with onboard GPS receivers for satellite-satellite tracking (SST).

### Relativistic Effects due to Orbit Eccentricity

The theoretical formula of the clock correction of the satellite can be written as (Ashby and Spilker 1996)

$$\Delta t_e = \frac{2}{c^2} \sqrt{\mu a}\, e \sin E + \text{const.}, \tag{5.156}$$

where $a$ is the semi-major axis of the satellite orbit, $e$ is the eccentricity of the orbit, $E$ is the eccentric anomaly of the orbit, $\mu$ is the gravitational constant of the earth, and $\Delta t_e$ is the clock correction due to the eccentricity of the orbit. The second term on the right-hand side is a constant that cannot be separated from the clock offset. This total correction has already been taken into account in the GPS orbits determination and is broadcasted in the navigation message by the parameters of the clock error polynomial. Therefore, this term of correction only needs to be considered in the satellite orbit determination.

Using the relation of $e \sin E = (x v_x + y v_y + z v_z)/\sqrt{(\mu a)}$ (c.f. Kaula 1966), the Eq. 5.156 can be presented by the position $(x, y, z)$ and velocity $(v_x, v_y, v_z)$ of the satellite.

### General Relativity Acceleration of the Satellite

The International Earth Rotation and Reference Systems Service (IERS) standard correction for the acceleration of the earth satellite is (McCarthy 1996)

$$\Delta \vec{a} = \frac{\mu}{c^2 r^3} \left\{ \left[ 4\frac{\mu}{r} - v^2 \right] \vec{r} + 4(\vec{r} \cdot \vec{v}) \vec{v} \right\}, \tag{5.157}$$

where $c$ is the speed of light, $\mu$ is the gravitational constant of the earth, $r, \vec{v}$ and $\vec{a}$ are the geocentric satellite position, velocity, and acceleration vectors, respectively.

## 5.4
## Earth Tide and Ocean Loading Tide Corrections

## 5.4.1
### Earth Tide Displacements of GPS Stations

The earth tide is a phenomenon of the deformation of the elastic body of the earth caused by the gravitational attracting force of the moon and the sun. Such a deformation depends not only on the changing of the force, but also on the physical structure and motion of the earth (Melchior 1978).

Generally, the sun-moon-earth system may be separated into two two-body systems to discuss the effects of the sun and the moon on the earth, respectively. For

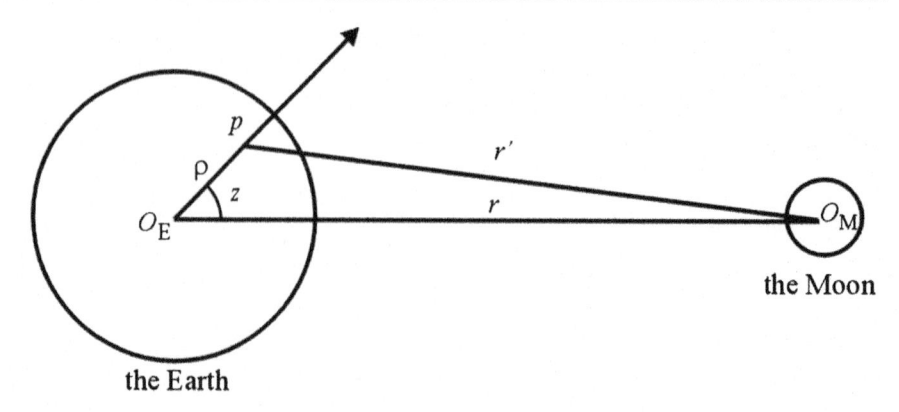

**Fig. 5.7** The earth-moon system

the moon-earth system, the mass centre can be found out according to the definition. It lies on a straight line between the centres of the earth and the moon, and has a distance to the centre of the earth of about 0.73 $R_E$, where $R_E$ is the radius of the earth (Fig. 5.7). For the point-mass $p$ (with unit mass) on the earth, the tidal potential generated by the moon can be derived as

$$W_p = \mu_m \left( \frac{1}{r'} - \frac{1}{r} - \frac{\rho}{r^2} \cos z \right), \qquad (5.158)$$

where $r$ is the geocentric distance of the moon, $\rho$ is the geocentric distance of point $p$, $\mu_m$ is the gravitational constant of the moon, $z$ is the geocentric zenith angle of the moon, and $r'$ is the distance between the point $p$ and the centre of the moon. The $1/r'$ in Eq. 5.158 can be developed by Legendre polynomials, and then

$$W_p = \mu_m \sum_{n=2}^{\infty} \frac{\rho^n}{r^{n+1}} P_n(\cos z), \qquad (5.159)$$

where $P_n(\cos z)$ is the conventional Legendre polynomials of $n$ degree. Applying the well-known formula of spherical astronomy (cf., e.g., Lambeck 1988),

$$\cos z = \sin \varphi \sin \delta + \cos \varphi \cos \delta \cos H, \qquad (5.160)$$

to Eq. 5.159 and using the addition theorem (cf., e.g., Lambeck 1988), Laplace's formula of the tidal potential can be obtained by

$$W_p = \mu_m \sum_{n=2}^{\infty} \frac{\rho^n}{r^{n+1}} [P_n(\sin \varphi) P_n(\sin \delta) + 2 \sum_{k=1}^{n} \frac{(n-k)!}{(n+k)!} P_{nk}(\sin \varphi) P_{nk}(\sin \delta) \cos kH],$$

$$(5.161)$$

where $\varphi$ is the latitude of computing point $p$, $\delta$, and $H$ are the declination and local hour angle of the moon, and $P_{nk}(x)$ is the associated Legendre polynomials of degree $n$ and order $k$. Laplace's formula shows the significant geometric and periodic characters of the tidal potential. Similar discussions can be made for the earth-sun system, and the related tidal potential can be obtained by substituting the gravitational constant of the sun $\mu_s$ and geocentric distance of the sun into Eq. 5.161. The total tidal potential is the summation of both potentials generated by the moon and the sun. The truncating order of the summation can be selected due to the precision requirement and the truncating errors can be estimated by considering $\mu_m$, $\mu_s$ and the ratio $R_E/r$ of the moon and the sun.

The tidal displacements resulting from the tidal potential are then

$$\Delta S_r = h \frac{W_p}{g} = \sum_{n=2}^{\infty} h_n \frac{W_p(n)}{g}, \tag{5.162}$$

$$\Delta S_\varphi = l \frac{\partial W_p}{g \partial \varphi} = \sum_{n=2}^{\infty} l_n \frac{\partial W_p(n)}{g \partial \varphi} \quad \text{and} \tag{5.163}$$

$$\Delta S_\lambda = l \frac{\partial W_p}{g \cos \varphi \partial \lambda} = \sum_{n=2}^{\infty} l_n \frac{\partial W_p(n)}{g \cos \varphi \partial \lambda}, \tag{5.164}$$

where $\Delta S_r$, $\Delta S_\varphi$, and $\Delta S_\lambda$ are the tidal displacements in the radial, north, and east directions, respectively; $h$ and $l$ are the Love and Shida numbers (in more detailed words, $h_n$ and $l_n$ are the Love and Shida numbers of degree $n$); $W_p(n)$ is the tidal potential of degree $n$, $g \approx \mu/R_E^2$; $\mu$ is the gravitational constant of the earth; and $R_E$ is the radius of the earth.

It is notable that the tidal potential includes a permanent (i.e. time-independent) part. This part of the tide is now included in the geoid definition, which was already accepted by the International Association of Geodesy (IAG) in 1983 (Poutanen et al. 1996). Therefore, such a term must be dealt with carefully. Examples to move or to keep the permanent tidal term from the above formulas may be found in the IERS standard (McCarthy 1996).

## 5.4.2
### Simplified Model of Earth Tide Displacements

The vector displacement of the station due to degree 2 of the tidal potential is (McCarthy 1996; Zhu et al. 1996)

$$\Delta \vec{\rho} = \sum_{j=1}^{2} \frac{\mu_j R_E^4}{\mu r_j^3} \left\{ h_2 \hat{\rho} \left[ \frac{3}{2} (\hat{r}_j \cdot \hat{\rho})^2 - \frac{1}{2} \right] + 3 l_2 (\hat{r}_j \cdot \hat{\rho}) [\hat{r}_j - (\hat{r}_j \cdot \hat{\rho}) \hat{\rho}] \right\}, \tag{5.165}$$

where $\mu$ is the gravitational constant of the earth, $R_E$ is the equatorial radius of the earth, $j = 1, 2$ are indices for the moon and sun, respectively, $\hat{r}_j$ and $\hat{\rho}$ are the geocentric identity vectors of the moon (or sun) and station, $r_j$ and $\rho$ are the magnitude of the related geocentric vectors, and $h_2$ and $l_2$ are nominal degree 2 Love and Shida numbers (for an elastic earth model, the nominal values are 0.6078 and 0.0847). In Eq. 5.165, the terms factorised by $h_2$ and $l_2$ are the radial and transverse components of the tidal displacement. Taking the latitude dependence into account, $h_2$ and $l_2$ shall have the forms of

$$h_2 = 0.6078 - 0.0006 \frac{3\sin^2\varphi - 1}{2} \quad \text{and} \tag{5.166}$$
$$l_2 = 0.0847 + 0.0002 \frac{3\sin^2\varphi - 1}{2}.$$

The vector displacement of the station due to degree 3 of the tidal potential is (McCarthy 1996)

$$\Delta\vec{\rho} = \frac{\mu_1 R_E^5}{\mu r_1^4} h_3 \hat{\rho} \left[ \frac{5}{2}(\hat{r}_1 \cdot \hat{\rho})^3 - \frac{3}{2}(\hat{r}_1 \cdot \hat{\rho}) \right], \tag{5.167}$$

where $h_3 = 0.292$. Here only the radial component of the moon is considered.

As discussed in Sect. 5.4.1, there is a permanent part of the tidal deformation included in the degree 2 tidal potential. The projections of the permanent displacement into the radial and north directions are

$$-0.0603\left(3\sin^2\varphi - 1\right) \quad \text{and} \quad -0.0252\sin 2\varphi.$$

This must be removed from the computation of Eq. 5.165 according to the IERS standard. Generally, the tidal displacements computed using the model given above have accuracy on a millimetre level.

In GPS applications, the computing time is usually the GPS time, and the coordinates of the stations are given in the CTS system. However, the ephemerides of the sun and moon are given in the CIS coordinate system with time TDT. Therefore, time and coordinates have to be transformed to a unique time-coordinate system. The details of the time-coordinate system can be found in Chap. 2.

The ephemerides of the sun and the moon are usually computed or forecasted every half-day (12 h). The ephemerides of the sun and the moon at a required epoch are interpolated from the data of the two adjacent epochs ($t_1$, $t_2$) by using a fifth-order polynomial:

$$f(t) = a + b(t - t_1) + c(t - t_1)^2 + d(t - t_1)^3 + e(t - t_1)^4 + f(t - t_1)^5.$$

For data at two epochs, e.g.,

$$t_1 : x_1, y_1, z_1, \dot{x}_1, \dot{y}_1, \dot{z}_1, \ddot{x}_1, \ddot{y}_1, \ddot{z}_1 \text{ and}$$
$$t_2 : x_2, y_2, z_2, \dot{x}_2, \dot{y}_2, \dot{z}_2, \ddot{x}_2, \ddot{y}_2, \ddot{z}_2,$$

where $\dot{x}$ and $\ddot{x}$ are the velocity and acceleration components related to $x$. Considering the formulas of $f(t)$, $df(t)/dt$, $d^2f(t)/dt^2$, and letting $t = t_1$, one gets $a = x_1$, $b = \dot{x}_1$ and $c = \ddot{x}_1/2$. Letting $t = t_2$, coefficients of $d, e, f$ can be derived theoretically, e.g., in the case of $t_2 - t_1 = 0.5$:

$$d = 80(x_2 - x_1) - 16\dot{x}_2 - 24\dot{x}_1 + \ddot{x}_2 - 3\ddot{x}_1,$$
$$e = -240(x_2 - x_1) + 56\dot{x}_2 + 64\dot{x}_1 - 4\ddot{x}_2 + 6\ddot{x}_1$$
$$f = 192(x_2 - x_1) - 48\dot{x}_2 - 48\dot{x}_1 + 4\ddot{x}_2 - 4\ddot{x}_1.$$

For $y$ and $z$ components, the formulas are similar. This interpolating algorithm is accurate enough to use the given half-day ephemerides of the sun and moon to get the data at the required epoch. The computation of the ephemerides of the sun and moon will be discussed in Sect. 11.2.8.

## 5.4.3
### Numerical Examples of Earth Tide Effects

The displacement caused by earth tide effects can reach as high as 60 cm in certain parts of the world (Melchior 1978; Poutanen et al. 1996); in Greenland, effects of 30 cm are observed (Xu and Knudsen 2000). The effects of earth tides on GPS positioning is a well-known correction term, and must be taken into account in many cases as soon as this effect is greater than the accuracy required for the GPS results. The tidal parameters (Love and Shida numbers) can also be determined through global GPS observations.

Only the GPS positioning, which is carried out on the air without fixed reference on the earth, is free from the earth tide effects. For the GPS relative positioning of a small regional area, the tidal effects may be neglected because of the small differences in the tidal displacements. In the relative airborne kinematic GPS positioning, the airborne antennas are free from earth tide effects. However, the static references fixed on the earth are not free from tidal effects. In this case, the tidal displacements are independent of the size of the applied area or lengths of the baselines and have to be taken into account.

Three examples are given to illustrate the tidal displacements (Xu and Knudsen 2000). The IERS standards are used as the principle of the earth tide effect computation (McCarthy 1996).

Three stations in Greenland were selected for computation of a whole day (GPS time used) of tidal displacements on 31 December 1998. Coordinates were quite

**Fig. 5.8** Earth tide displacements at three stations in Greenland (vertical component; date: 31 Dec. 1998)

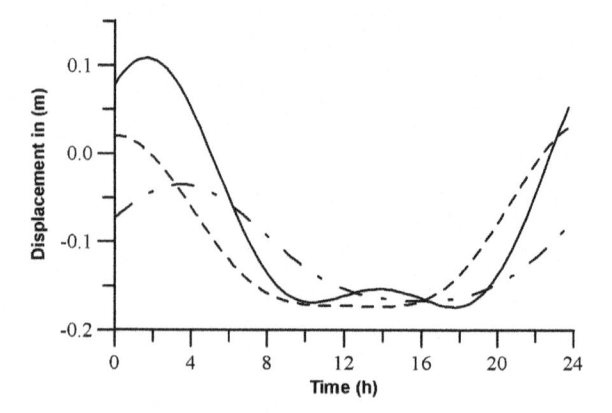

roughly selected for Narsarsuaq (60°, 315°), Scoresbysund (70°, 339°), and Thule (77°, 290°). Heights were selected as 50 m. Results of the vertical components are illustrated as 2-D graphics with the first axis time in hours and the second axis displacement in meters. Solid, dotted, and dashed lines represent the results of the first, second, and third stations, respectively (Fig. 5.8, units: meters). The tidal displacements of the three stations in Greenland have a maximum difference of about 15 cm. The size of the triangle of the three stations is about 2000 km. The change in the tidal displacements in the vertical component is about 27 cm. That change could happen within a duration of 4–5 h.

Two grid data points of $0.2° \times 0.3°$ for Denmark ($54.0° \leq \varphi \leq 57.8°$, $8° \leq \lambda \leq 12.9°$) and $1° \times 1°$ for Greenland ($59.5° \leq \varphi \leq 84°$, $285° \leq \lambda \leq 350°$) with real topography height are used to compute vertical tidal displacements at time 1:00 and 1:45, respectively. The displacements are illustrated with contour lines (units: meters) plotted in Figs. 5.9 and 5.10, with the first axis longitude in degrees and the second axis latitude in degrees. In Fig. 5.9 there is a tidal difference of 15 mm, which shows that within an 80-km distance or area in Denmark the tidal difference can reach up to 5 mm. Figure 5.10 shows that there is a vertical tidal difference of about 17 cm in Greenland.

These computations indicate that in differential GPS kinematic airborne (not touching the earth) applications, a lack of the earth tide correction could cause an accuracy of less than 30 cm in Denmark and Greenland. For the earth-touched kinematic and static differential GPS applications, a lack of earth tide correction would cause an accuracy of less than 2 cm in Denmark and 15 cm in Greenland.

It is worth mentioning that the average value of the earth tide effects on a GPS station during 24 h is generally not zero (it may be up to a few cm). This is because the earth tide is an effect that includes many periodical components. This indicates that the earth tide effects cannot be eliminated through daily averaging.

**Fig. 5.9** Earth tide displacement (in metres) in Denmark (31 Dec. 1998; GPS time 1:00; height component)

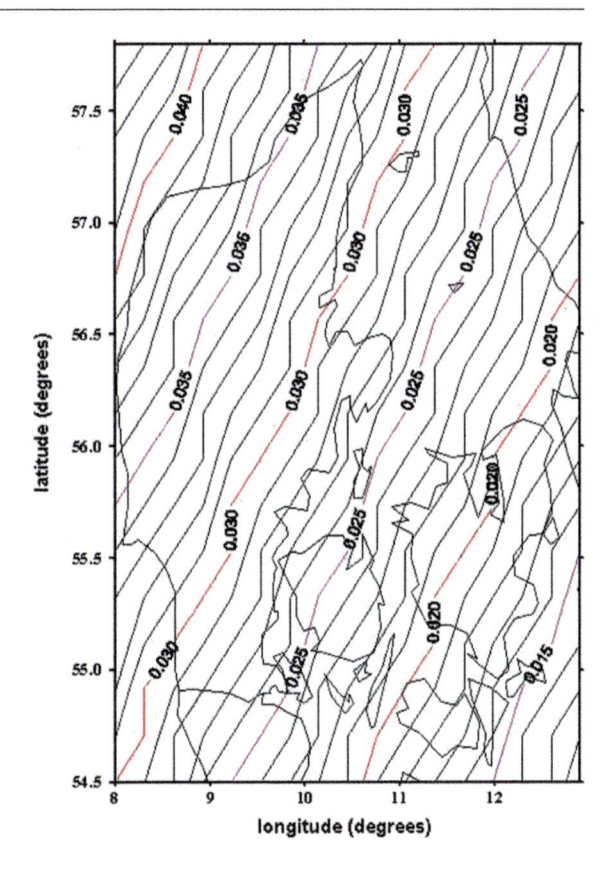

## 5.4.4
## Ocean Loading Tide Displacement

The ocean tide is a time varying load on the earth's surface. The displacement of the earth's surface due to loading is called the ocean tide loading effect. Similar to the earth tide, the loading Love numbers are introduced to describe the relationships of the loading potential and the loading displacement as

$$\Delta S_\mathrm{r} = h' \frac{W_p}{g} = \sum_{n=0}^{\infty} h'_n \frac{W_\mathrm{p}(n)}{g}, \qquad (5.168)$$

$$\Delta S_\varphi = l' \frac{\partial W_\mathrm{p}}{g \partial \varphi} = \sum_{n=0}^{\infty} l'_n \frac{\partial W_\mathrm{p}(n)}{g \partial \varphi} \quad \text{and} \qquad (5.169)$$

$$\Delta S_\lambda = l' \frac{\partial W_\mathrm{p}}{g \cos \varphi \partial \lambda} = \sum_{n=0}^{\infty} l'_n \frac{\partial W_\mathrm{p}(n)}{g \cos \varphi \partial \lambda}, \qquad (5.170)$$

**Fig. 5.10** Earth tide displacement (in metres) in Greenland (31 Dec. 1998; GPS time 1:45; height component)

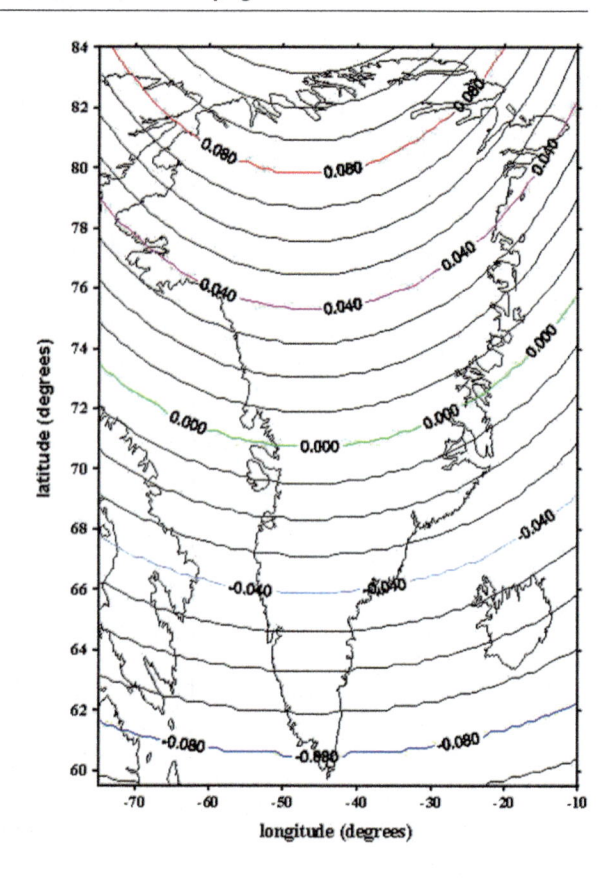

where $\Delta S_r$, $\Delta S_\varphi$, and $\Delta S_\lambda$ are the loading tide displacements in the radial, north, and east directions, respectively, $h'$ and $l'$ are the loading Love numbers (in more detail, $h'_n$ and $l'_n$ are the loading Love numbers of degree $n$), $W_p(n)$ is the loading potential of degree $n$, $g \approx \mu/R_E^2$, $\mu$ is the gravitational constant of the earth, and $R_E$ is the radius of the earth. It is notable that the zero degree loading Love number and 1 degree loading displacement exist and in the case of $n \to \infty$, $h'_n \to h_\infty$, $nl'_n \to l_\infty$. Loading Love numbers can be obtained from a theoretical model.

The loading displacement shall fulfil the elastic balance equation under the boundary condition of loading. This is called the Boussinesq boundary value problem. The response of the spherical earth under the loading of a point-mass (or unit mass) is called the Green's function. In other words, the Green's function is the solution of the partial differential equation of the Boussinesq problem under a certain spherical boundary condition of a point-mass loading. For a related boundary condition, the related Green's function can be derived. Farrell derived the following loading displacement Green's functions (Farrel 1972):

**Fig. 5.11** Ocean loading

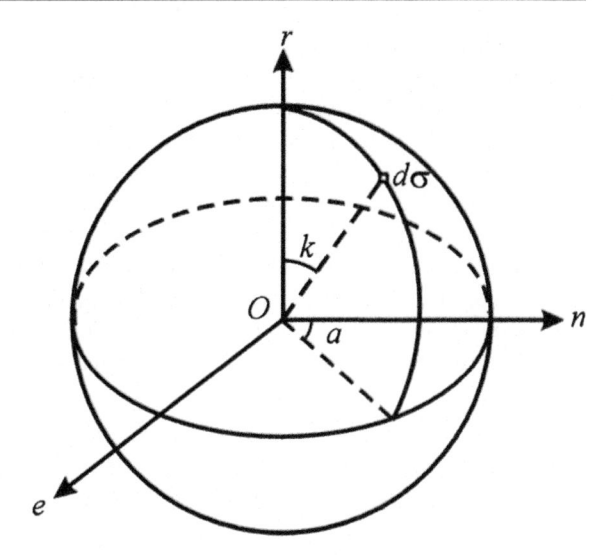

$$u(k) = \frac{Rh'_\infty}{2M_e \sin(k/2)} + \frac{R}{M_e} \sum_{n=0}^{N} (h'_n - h'_\infty) P_n(\cos k) \quad \text{and} \quad (5.171)$$

$$v(k) = \frac{-R\cos(k/2)[1 + 2\sin(k/2)]}{2M_e \sin(k/2)[1 + \sin(k/2)]} + \frac{R}{M_e} \sum_{n=1}^{N} \frac{(nl'_n - l'_\infty)}{n} \frac{\partial P_n(\cos k)}{\partial k}, \quad (5.172)$$

where $R$ is the radius of the earth, $M_e$ is the mass of the earth, $k$ is the geocentric zenith distance of the loading point (related to the computing point, see Fig. 5.11), $P_n(\cos k)$ is the Legendre function, and $u(k)$ and $v(k)$ are the radial and tangential loading displacement Green's functions, respectively.

According to the definition of the Green's function, the loading displacements of the ocean tide of the whole earth can be obtained by multiplying the tidal mass to the Green's function and integrating that over the whole ocean

$$u_r = \oiint_{\text{ocean}} \delta H u(k) \, d\sigma, \quad (5.173)$$

$$u_\varphi = \oiint_{\text{ocean}} \delta H v(k) \cos a \, d\sigma, \quad \text{and} \quad (5.174)$$

$$u_\lambda = \oiint_{\text{ocean}} \delta H v(k) \sin a \, d\sigma, \quad (5.175)$$

where $a$ is the azimuth of the integrating surface element $d\sigma$, $\delta$ is the density of the oceanic water ($\delta \approx 1.03$), $H$ is the height of ocean tide, and $u_r$, $u_\varphi$, and $u_\lambda$ are the loading displacements of radial, north, and east components, respectively. It is obvious that an ocean tide model is needed here.

One of the most commonly used models is the Schwiderski global ocean tide model, which has a resolution of $1° \times 1°$ and represents the tidal amplitude and phase (Schwiderski 1978, 1979, 1980, 1981a, b, c). The accuracy of the loading tide modelling depends on the accuracy of the loading response and ocean tide model. Because of the irregularity of coastlines, and because the loading response is more dependent on the local variable properties of the lithosphere (Farrel 1972), modelling of the loading effects cannot be done very accurately.

Let

$$H = \sum_{i=1}^{I} H_i, \quad \oiint_{ocean} F d\sigma = \sum_{n=1}^{N} F(n) d\sigma_n, \quad \sum_{n=1}^{N} d\sigma_n = \text{total surface of the ocean,}$$

where $H_i$ is the ocean tide constituent with angular velocity $\omega_i$, $I$ is the truncating wave number, $F$ is any to be integrated function, $F(n)$ and $d\sigma_n$ are the functional value and the size of the $n$th surface element, and $N$ is the total elements number. If one changes the order sequence of the summations, then Eqs. 5.173–5.175 will be

$$u_r = \sum_{i=1}^{I} \sum_{n=1}^{N} \delta H_i u(k) d\sigma_n, \tag{5.176}$$

$$u_\varphi = \sum_{i=1}^{I} \sum_{n=1}^{N} \delta H_i v(k) \cos a \, d\sigma_n, \quad \text{and} \tag{5.177}$$

$$u_\lambda = \sum_{i=1}^{I} \sum_{n=1}^{N} \delta H_i v(k) \sin a \, d\sigma_n. \tag{5.178}$$

In other words, the loading displacements can be represented by summations of displacements of the different wave of frequencies, and the amplitude and phase of the related wave are dependent on the computing position.

## 5.4.5
### *Computation of the Ocean Loading Tide Displacement*

Computation of the loading displacement depends on which ocean tide model is used. Because of the strong dependence of the loading on the coast near tide, besides a global ocean tide model a modified model near the coastlines is added quite often to raise the precision of computation. Taking advantage of the fact that the amplitude and phase of the related wave depends only on the computing position, the computation can be greatly simplified. Generally, only 11 tidal constituents are taken into account. These are the semi-diurnal waves $M_2$, $S_2$, $K_2$, and $N_2$, the diurnal waves $O_1$, $K_1$, $P_1$, and $Q_1$, and the long-period waves $M_f$, $M_m$, and $M_{sa}$. The loading displacement vector in IERS standard (McCarthy 1996) is

$$\Delta \rho_j = \sum_{i=1}^{11} f_i \cdot \text{amp}_j(i) \cdot \cos[\arg(i,t) - \text{phase}_j(i)] \quad \text{and} \tag{5.179}$$

$$\arg(i,t) = \omega_i t + \chi_i + u_i, \tag{5.180}$$

where $j = 1, 2, 3$ represents the displacement in radial, west, and south directions, respectively, $\text{amp}_j(i)$ and $\text{phase}_j(i)$ are the amplitude and phase of the $i$th wave related to the computing station of $j$th component, $\arg(i, t)$ is the argument of $i$th wave at the computing time $t$, $\omega_i$ is the angular velocity of the $i$th wave, $\chi_i$ is the astronomical argument at 0 h, and $f_i$ and $u_i$ depend on the longitude of the lunar node. $\omega_i$, $f_i$, and $u_i$ can be found in Table 26 of Doodson (1928). The $\text{amp}_j(i)$ and $\text{phase}_j(i)$ are computed for a list of stations by Scherneck (McCarthy 1996). The coefficients and software are available by Scherneck.

## 5.4.6
## Numerical Examples of Loading Tide Effects

Loading tide effects could reach up to 10 cm at some special coast regions (cf., e.g., Andersen 1994; Khan 1999). The loading displacements affect mostly only the GPS stations near the coast. The displacements at most continental stations are less than 1 cm. Loading correction has not been commonly considered in GPS data processing because the computation is more complicated and modelling is less accurate. However, for precise applications, loading effects have to be taken into account. Using software, e.g., designed by Scherneck, the loading amplitude and phase of the significant waves can be obtained for static stations. The coefficients can be computed beforehand and used for even real time applications. Kinematic GPS receivers on the air are free of loading effects. Car-borne kinematic GPS applications are generally limited within regional areas, and the relative effects are generally very small. For exactness, the loading effects can be interpolated from that of the surrounding static stations.

An example is given to illustrate the loading tide displacements of a vertical component (Fig. 5.12). The AG95 model is used as the principle of the loading effects computation (Andersen 1994). Loading displacements of two stations are computed for a whole day (GPS time used) on the 18 March 1999. Coordinates are quite roughly selected for station Brst (48.3805°, 355.5034°) and International GNSS Service (IGS) station Wtzr (49.1442°, 12.8789°). Results of the height components are illustrated as 2-D graphics with the first axis as time in hours and the second axis displacement in metres. Continuous and broken lines represent

**Fig. 5.12** Ocean loading tide effects of two GPS stations (height components)

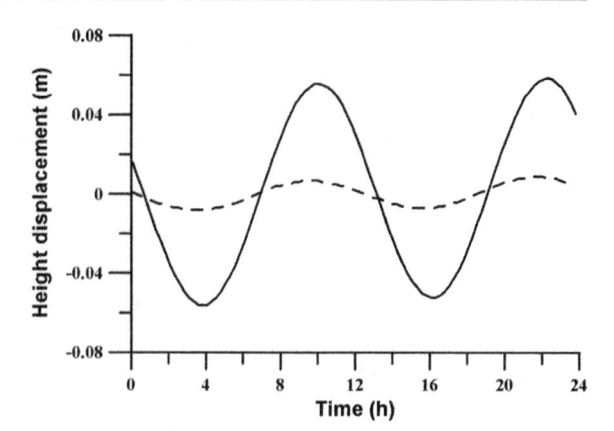

the results of the first and second stations, respectively. The loading displacements of the two stations have a maximum difference of about 6 cm.

This computation indicates that a lack of loading tide correction could cause an accuracy of a few centimetres for some applications. Even for differential applications, a lack of loading tide correction could cause an accuracy of less than 6 cm in the given example.

It is worth mentioning that the average value of the loading effects on a GPS station during 24 h is generally very small. This indicates that the loading effects can be eliminated through daily averaging. In a static case, either loading correction or no loading correction may cause a difference of the standard deviation of the length of a baseline up to 0.5 cm (Khan 1999).

As an analogue to the tropospheric model parameterisation, a loading parameter may be introduced for a static station near the coast due to less accuracy of the loading modelling. The parameter (e.g., a factor of the total loading vector) may be determined through GPS data processing.

## 5.5
## Clock Errors

As discussed in Chap. 4 in the models of GPS observables, the clocks on the satellites and receivers play a very important role in precise GPS surveying. The influences of the clock errors on the GPS may be grouped into three types. One is factored with the speed of light, $c$. Another is factored with the speed of satellites. And the third is factored with the working frequency.

The influence of the first type of clock error is obvious. For code measurements, one measures the transmitting time of the signal and multiplies the transmitting time with $c$ to obtain the transmitting path length. A clock error of $\delta t$ will cause a path length error of $c\delta t$. Similarly, a clock error of $\delta t$ will cause a phase error of $c\delta t/\lambda$. Because of the factor $c$, a small clock error may cause a very large code and phase error. Therefore, high quality clocks have to be used on the satellites and receivers. Meanwhile, clock errors must be carefully modelled. A simple model may be expressed as

$$\delta t = b + dt + at^2, \quad t_1 \leq t \leq t_2, \tag{5.181}$$

where $b$ is the bias, $d$ is the drift, and $a$ is the acceleration of the related clock. Time interval $(t_1, t_2)$ is the valid period of clock error polynomial. The length of the interval depends directly on the stability of the clock. Such a model describes that the clock has a small drift and acceleration, and the drift and acceleration as well as bias are stable. The interval may be estimated by using the drift and acceleration accordingly.

In the case of SA (selective availability, for details cf. Sect. 5.7), the frequency of the clock on the satellite is manipulated artificially. In other words, the scale of the clock on the satellite is no longer a constant; i.e. the clock is no longer stable. Therefore, in such a case, the model of Eq. 5.181 is not good enough for use. An alternative model of the satellite clock error in the case of SA is

$$\delta t = b_i, \quad t = t_i. \tag{5.182}$$

Thus the clock bias must be modelled for every measuring epoch. The clock error parameters must be determined or equivalently eliminated every epoch.

The influence of the second type of clock error is more or less implicit. Recalling the code and phase models discussed in Chap. 4, there is a geometric distance between the satellite at the signal emission time and the receiver at the signal reception time. The position and velocity of the satellite are functions of time. Therefore, a clock error causes a computing error of the position of the satellite by $\vec{v}_s \delta t$, where $\vec{v}_s$ is the velocity vector of the satellite. These errors pass through the distance function and cause errors of the computed distance. Such an influence is implicitly presented in all the GPS observation models and cannot be eliminated through forming differences. However, the influence of the clock error is factorised by the velocity of the satellite (about 3 km s$^{-1}$), so an estimation of $\delta t$ up to an accuracy of $10^{-6}$ would be enough to ensure the required accuracy of the computed satellite position. Such estimation is usually made through the single point positioning of every station at every epoch (details, cf. the section of single point positioning in Sect. 9.5.2). Of course, we must also take the relativistic effects into account.

As discussed above, the clock error causes a phase error of $c\delta t/\lambda$; this is equivalent to a frequency error of $f\delta t$. Obvious, this correction must be taken into account in Doppler data processing.

Synchronisation of the clocks on the satellites and receivers is a basic prerequisite of a meaningful GPS measurement. Clock modelling automatically leads to the synchronisation of all clocks.

A recent study showed that clock error parameters were linearly correlated with the ambiguity parameters (for details, see Sect. 9.1).

## 5.5.1
### Introduction of Commonly Used Clock Error Models

As we know, the measurement of the precise satellite navigation position is actually accurate time measurement; therefore, the highly precise satellite clock is the basis of the navigation system. The satellite clock error parameter must be precisely known when using satellite navigation systems in order to achieve rapid positioning and time synchronisation; thus the estimation and prediction of the satellite clock error are essential (Huang 2012).

In researching a satellite clock error, a precise clock error model must first be constructed. In this section, we introduce the commonly used polynomial model, spectral analysis (SA), and grey models (GM), and the autoregressive–moving-average (ARMA) model.

*Polynomial model*

The polynomial model (Kosaka 1987) is the most widely used for the prediction and fitting of clock errors, and encompasses linear, quadratic, and high-order polynomial models, which can be expressed as

$$x_i = a_0 + a_1 t_i + a_2 t_i^2 + \cdots + a_m t_i^m + e_i (0 \leq i \leq n), \tag{5.183}$$

where $x_i$ is the clock error at time $t_i$, $a_0, a_1, a_2, \ldots, a_m$ denotes a number of $m$ clock error parameters, $m$ is the order of the polynomial, and $e_i$ denotes the model error.

The polynomial model is simple and practical, with clear physical meaning, and has high precision for short-term prediction and fitting. As such, it has been extensively applied in GNSS real-time navigation positioning. The linear and quadratic models are used in the real-time satellite clock error prediction of broadcast ephemeris in GPS and GLONASS, respectively. The determination of the order of the polynomial model mainly depends on the frequency stability and frequency drift characteristics of the atomic clock. Generally, the linear polynomial is suitable for the prediction and fitting of caesium atomic clock since the short-term frequency stability of caesium atomic clock is somewhat poor and the frequency drift is not obvious. The quadratic polynomial is suitable for rubidium atomic clock since its good short-term frequency stability.

*Spectral Analysis (SA) Model*

Clock error series typically comprise not only linear and quadratic variation, but also periodic terms. Thus, SA (Percival 2006) can be used to find the obvious periodic term in the clock error series to establish a more accurate clock error model. When taking into account the linear trend term, the SA model for the clock error can be expressed as

$$x_i = a_0 + b_0 t_i + \sum_{k=1}^{p} A_k \sin(2\pi f_k t_i + \varphi_k) + e_i (0 \leq i \leq n), \tag{5.184}$$

where $a_0$ and $b_0$ are the coefficients of the linear trend term, $p$ is the number of the obvious periodic terms, $f_k$ is the frequency of the corresponding periodic term, $A_k$ and $\varphi_k$ are amplitude and phase of the corresponding periodic term, respectively, $e_i$ denotes the residual error of $x_i$. $p$, and $f_k$ can be determined by the SA.

## Grey Model (GM)

The grey system is an information processing method first proposed by Deng (1987). It works largely on system analysis that has produced poor, incomplete, or uncertain messages. The GM for the clock error can be expressed as

$$x^{(0)}(k) = (1 - e^a)\left[x^{(0)}(1) - \frac{u}{a}\right]e^{-a(k-1)}, \qquad (5.185)$$

where $x^{(0)}(k)$ is the $k$ element of the original clock error series, and $a$ and $u$ are model parameters. According to the principle of least squares, $a$ and $u$ can be estimated by $n(n \geq 4)$ observations.

The advantage of the GM is that it needs only a few data points (as few as four original data points) to estimate an unknown system, which can reduce data usage and improve the efficiency of model establishment.

## Autoregressive Moving Average (ARMA) model

The ARMA model is a conventional random time-sequence model invented by Box and Jenkins, known also as the B-J method (Box et al. 1994). There are three basic types of ARMA models: autoregressive (AR), moving average (MA), and autoregressive moving average (ARMA) models. The ARMA $(p, q)$ can be expressed as

$$x_t = \varphi_1 x_{t-1} + \cdots + \varphi_p x_{t-p} + a_t - \theta_1 a_{t-1} - \cdots - \theta_q a_{t-q}, \qquad (5.186)$$

where $x_t, x_{t-1},\ldots,x_{t-p}$ is the observation sequence, $\varphi_1,\ldots,\varphi_p$ are autoregressive coefficients, $\theta_1,\ldots,\theta_p$ are moving average coefficients, $p$ and $q$ are the order of autoregressive model and moving average model, respectively, $a_t, a_{t-1},\ldots,a_{t-p}$ is stationary white noise with a mean value of zero and a variance of $\sigma_a^2$.

The ARMA model is suitable only for a stationary random sequence. In practical applications, however, few observations can meet such a requirement, and thus difference processing is essential. The model based on the differential time sequence is called the autoregressive integrated moving average (ARIMA) model, which can be denoted as $\{x_t\} \sim$ ARIMA$(p, d, q)$, where $d$ is the order of difference.

## Parameter estimation methods

The coefficients of the clock error model mentioned above must be estimated through appropriate parameter estimation methods. Two methods are commonly used: the sequential least squares and the kinematic Kalman filter. Details of these two methods can be found in Sects. 7.3 and 7.7. Currently, the sequential least squares is most commonly applied in the real-time prediction of the GNSS clock error because of the stability and not divergence of the algorithm, while the kinematic Kalman filter is widely used in the post-processing of precise clock error by IGS.

## 5.5.2
## Impact of Frequency Reference of a GPS Receiver on the Positioning Accuracy

As we know, the vertical precision of the GPS is two to three times less accurate than its horizontal counterpart, mainly due to difficulties in correcting clock errors, multipath, tropospheric delay, antenna phase centre variations, and the asymmetry in the observed GPS constellation in the height direction (Leick 2004). Most GPS data analysis procedures utilise double differences to reduce clock and orbital errors. Carrier-phase ambiguities, cycle slips, and clock errors can be repaired by processing pseudorange signals and triple-differenced phases, while ionospheric delay can be corrected by modelling or dual-frequency combinations (Bock and Doerflinger 2001; Yeh et al. 2008). To correct clock errors, the IGS provides users with GPS satellite clock errors via the internet to increase positioning accuracy (Ray and Senior 2005; Dow et al. 2009). Therefore, the GPS receiver clock error is an important factor affecting positioning accuracy. In a study by Yeh et al. (2009), a positioning error of 1–2 cm was found due to improperly modelled receiver clock errors. In this section, we emphasise the research by Yeh et al. (2012), which explored the impact of the frequency offset and the frequency stability of the internal quartz oscillator or of an externally supplied rubidium oscillator on the positioning accuracy.

To assess the dependence of the positioning performance of the receiver on the frequency performance of its internal oscillator, this study sequentially used the receiver's clock steering function to change its clock between the internal quartz (the steering switched off) and external rubidium oscillators. The clock steering uses GPS observations to synchronise the internal quartz oscillator to GPS time. Furthermore, as the rubidium oscillator frequency may vary in time, a regular frequency calibration should be made via direct method (using a counter to calibrate the rubidium oscillator directly) or remote method (using GPS data to calculate the clock error).

In this study, undifferentiated GPS phase observables were employed to calculate frequency offsets and frequency stabilities of the GPS receivers based on the method developed by Dach et al. (2003). Bernese GPS software version 5.0 (Dach et al. 2007) was used to estimate receiver clock errors and then the clock errors in each epoch were employed to compute frequency offset and frequency stability by Allan deviation (Allan and Weiss 1980; Lesage and Ayi 1984). Experimental results showed, the rubidium oscillator on average outperformed quartz oscillators by 1–6 orders of magnitude in terms of frequency stability, while the offset of the quartz oscillator was smaller than that of the rubidium oscillator in this study since, clock steering was able to compensate for the phase difference of the quartz oscillator.

After confirming the accuracy of the frequency offset and frequency stability of both quartz and the rubidium oscillators, the effects of clock errors on GPS positioning accuracy were analysed. Three baselines (short, medium, and long) were selected to conduct static relative positioning. Experimental results showed that rubidium oscillators achieved higher positioning accuracy than quartz oscillators, regardless of whether the baseline was short, medium, or long. In this study, the accuracy with short-, middle-, and long-distance static relative positioning using rubidium oscillators was higher than that of quartz oscillators by 5, 11, and 15 %, respectively, indicating that clock quality is more influential for long-baseline GPS relative positioning. The results also showed that the average improvement of the vertical direction was twice as large as that of the horizontal direction, which agrees with the statement described at the beginning of the section.

In terms of the respective performance of the rubidium and quartz oscillators in frequency offset and frequency stability, the rubidium oscillator was much better than the quartz oscillators in frequency stability. However, for frequency offset, the quartz oscillator was better than the rubidium oscillator on average. Therefore, a conclusion that the GPS receiver with a rubidium oscillator has higher positioning accuracy than the GPS receivers with quartz oscillator because it is equipped with a relatively more stable internal frequency can be inferred. Notably, although the rubidium oscillator has a frequency offset, as long as this offset is steady and can be calculated and removed, its impact on positioning accuracy is not appreciable. Thus, the frequency stability of a receiver clock is far more critical than the frequency offset (Yeh et al. 2012).

## 5.6
## Multipath Effects

Multipath is the phenomenon whereby a GPS signal arrives at a receiver's antenna via more than one path. Multipath propagation affects both pseudorange and carrier phase measurements. In GPS static and kinematic precise positioning, the multipath effect is an error source that must be taken into account. Related studies have been carried out for many years to reduce or eliminate the multipath effects (cf., e.g., Braasch 1996; Langley 1998a, b; Hofmann-Wellenhof et al. 1997).

Multipath is a highly localised effect, which is dependent only on the local environment surrounding the antenna. As illustrated in Fig. 5.13, the receiver may receive both the directly transmitted signal and the reflected (indirect) signal. The indirect path is clearly dependent on the reflecting surface and the satellite position. The reflecting surface is usually static and is related to the receiver; however, the satellite moves with time. Therefore, the multipath effect is also a variable of time.

Consider the direct signal $s(t) = A\cos(\omega t + \varphi)$, where $A$ is the amplitude, $\omega$ is the angular velocity, and $\varphi$ is the phase; the indirect signal can then be represented as $f \cdot s(t + \delta t)$, where $f$ is a factor with the physical meaning of reduced energy through reflection, and $\delta t$ is the time delay. The multipath effect is indeed the influence of the indirect signal on the observations of the receiver. Because different receiver deals with the signals with a different manner, multipath error is highly dependent upon the architecture of the receivers.

**Fig. 5.13** Geometry of multipath effects

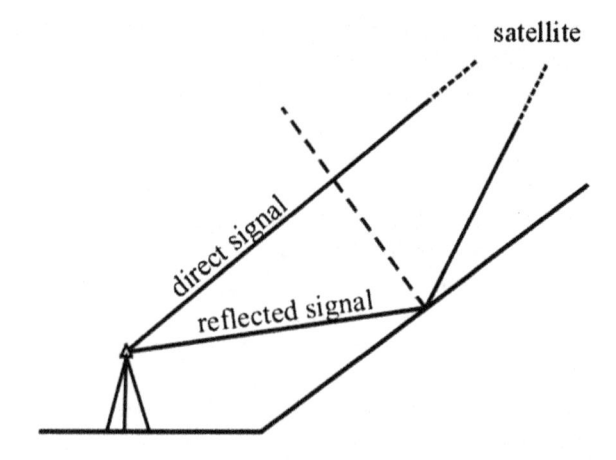

Theoretically (Braasch 1996; Langley 1998a, b), the multipath effect may reach up to 15 m for P-code measurements and 150 m for C/A-code. Because of the chip length, P-code is much less sensitive to the indirect signal. Typically, multipath error of the carrier phase is on the order of a few cm.

GPS signals are right-handed circularly polarised (RHCP); therefore, conventional GPS antennas are designed as RHCP antennas. This property helps to reject the multipath signal because the reflected signal has changed its polarisation. The pure reflected signals received by the RHCP antenna usually have only one-third of the signal-to-noise ratio compared with that of the direct signals (Knudsen et al. 1999). This may also be used to detect multipath effects. The simplest method to avoid the influence of the multipath effects is to set up the antenna far away from possible reflecting surfaces. Using only the carrier phase measurement is possibly the other method. (Code is usually used for clock error correction for the satellite coordinates computation; this would be accurate enough even if the multipath effects existed in code; for details, see the discussion in Sect. 5.5.) In the case of code positioning, a phase-smoothed code should be used. This can reduce the maximum multipath effects to a few cm.

An exact method to deal with the multipath effects is to detect the multipath using code-phase data and then reject the related phase data or set the phase data to a lower weight for phase data processing. Recalling the models of the code and phase observables discussed in Sects. 4.1 and 4.2, a code-phase difference can be formed by using Eqs. 4.7 and 4.18 as

$$R_r^s(t_r, t_e) - \lambda \Phi_r^s(t_r) = 2\delta_{\text{ion}} + \lambda N_r^s + \delta_{\text{mul}} + \varepsilon, \tag{5.187}$$

where $R_r^s(t_r, t_e)$ and $\Phi_r^s(t_r)$ are the measured pseudorange and phase, $\lambda$ is the wavelength, $t_e$ is the GPS signal emission time, and $t_r$ the signal reception time, $\delta_{\text{ion}}$ denotes the ionospheric effects of the station r, $N_r^s$ is the integer ambiguity parameter, $\delta_{\text{mul}}$ is the multipath effect of code measurements, and $\varepsilon$ is error of code measurements. The errors of phase and frequency as well as multipath in phase measurements are omitted here. Using the above formula, multipath effects in code measurements can be determined or detected. Because of the higher noise level of the code measurements, detection over a given period of time is reasonable so that the noise can be smoothed.

## 5.6.1
## GPS Altimetry, Signals Reflected from the Earth's Surface

The existence of multipath effects indicates that a GPS receiver can be used to receive the reflected GPS signal. In other words, by receiving the reflected GPS signal, GPS may be used to measure the reflecting surface topography. Early in 1993, the European Space Agency's Manuel Martin-Neira first suggested using a GPS reflected signal as a signal source for measurement. The accidental acquisition of ocean-reflected GPS signals by an airborne receiver was reported by French engineers in 1994. Katzberg and Garrison (1996) discussed how the GPS signal reflected from the ocean could be used for the determination of ionospheric effects in satellite altimetry. Komjathy et al. (1999) used the GPS signal reflected from the ocean to determine wave height, wind speed, and direction. In Denmark, in cooperation with Xu, Knudsen et al. (1999) used a downward-pointing GPS antenna to receive the reflected GPS signal to explore the possibility of using it for determining the topography of the sea surface and ice sheet as well as snow-covered land. The CHAllenging Minisatellite Payload (CHAMP) satellite has a downward-pointing GPS antenna on board for an experiment in GPS altimetry.

Profiles of footprints over the sea surface are typically measured using satellite or airborne altimetry. By using GPS altimetry, however, every footprint profile has a bandwidth, and thus GPS altimetry can be used to cover the topography of the reflecting surface. The sea and ice sheet as well as snow-covered land are good reflecting surfaces for the GPS signals (Knudsen et al. 1999).

The polarisation of the reflected signal changes after the reflection. A conventional GPS antenna is RHCP; therefore, to receive the reflected GPS signals, a left-handed circularly polarised antenna will be used (Komjathy et al. 1999; Katzberg and Garrison 1996). Such an antenna was designed and used in the experiments reported. The power of the reflected signal is then reduced insignificantly.

## 5.6.2
## Reflecting Point Positioning

The method for processing the downward-pointing antenna-measured GPS data is quite different from the known method for processing the GPS data obtained by an upward-pointing antenna. As shown in Fig. 5.14, the GPS signal is transmitted from the satellite to the downward-pointing antenna through the reflecting point R

**Fig. 5.14** Geometry of the reflecting signal

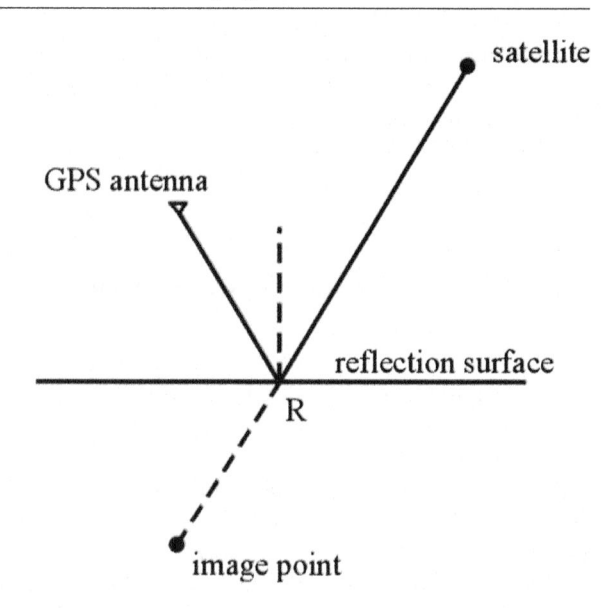

(or more exactly, a small zone surrounding R, cf. Komjathy et al. 1999) of the reflecting surface. The satellite orbit is known. The position of the downward-pointing antenna can be determined by using the data received from an upward-looking antenna. Thus the purpose of GPS altimetry is to determine the unknown point R. The vertical line of the satellite and the antenna forms a plane. Such a plane will intersect the earth's surface and form a curved line, and the reflecting point will be on the line. Because of the principle of reflection, the angle of fall in and the angle of fall out must be equal. In other words, the elevation of the antenna and that of the satellite related to the reflecting surface at the reflecting point must be the same. Therefore, the reflecting point will generally be unique if the reflecting surface is a fixed surface. Even in a static case, i.e. the GPS antenna does not move, the reflecting point R is a kinematic point, because of the movement of the satellite. Different satellites generally have different reflecting points. These points are independent if one does not take into account the a priori knowledge of the reflecting surface.

For every observed satellite of every epoch there are three new coordinate unknowns. A straightforward solution is mathematically impossible. However, suppose the reflecting surface is a geoid or a known sea surface, and the latitude and longitude of the reflecting point can be computed from the satellite position and the known antenna position. Then the remaining unknown is just one parameter of height. Suppose the reflecting surface needs to be determined only up to, say, a resolution of 2 km. The height of every point located within the 1-km radius can then be considered the same, and in this case, the GPS altimetry problem is clearly solvable.

The signal transmitting distance $d$ can be described as

$$d = \sqrt{(x_s - x_r)^2 + (y_s - y_r)^2 + (z_s - z_r)^2} + \sqrt{(x_k - x_r)^2 + (y_k - y_r)^2 + (z_k - z_r)^2}, \tag{5.188}$$

where indices s, r, and k denote satellite, reflecting point, and downward-pointing GPS antenna, respectively. Of course, the transmission time correction must also be taken into account. Cartesian coordinates $x$, $y$, and $z$ of the reflecting point can be represented by geodetic coordinates $\varphi$, $\lambda$, and $h$. $\varphi$ and $\lambda$ of the reflecting point will fulfil the following linear equation:

$$(\varphi - \varphi_k) = (\lambda - \lambda_k) \frac{\varphi_s - \varphi_k}{\lambda_s - \lambda_k}. \tag{5.189}$$

For any given $\varphi$ between values $\varphi_k$ and $\varphi_s$, a related $\lambda$ can be obtained, and the zero-height reflecting point in Cartesian coordinates can then be obtained. The zenith distances of the downward-pointing GPS antenna and the GPS satellite in relation to the reflecting point can be then computed. By using the criterion that both zenith distances shall be the same, a best set of $\varphi$ and $\lambda$ can be found. Taking the known coordinates of the zero-height reflecting point into account, there is just one parameter of height remaining as an unknown in Eq. 5.188.

By reducing the resolution rate to every two epochs—suppose that within every two epochs, the height of the reflecting point remained the same—the problem including the receiver clock error and ambiguity can be solved with sufficient redundancy.

## 5.6.3
### Image Point and Reflecting Surface Determination

An alternative method for determining the reflecting surface is proposed below.

The reflecting surface is considered a mirror, and the downward-pointing antenna is an image point behind the mirror. If the reflecting surface is a plane, then the image point positioning can be carried out using the same method as in kinematic positioning of the upward-looking antenna. The longitude and latitude of the image point can typically be obtained from the results of the upward-looking antenna; therefore, the image point positioning problem has just one coordinate-unknown height, and can be easily determined. Now one has two heights: the height of the downward-pointing antenna, and the height of the image point. The average value of these two heights is then the footprint height of the downward-pointing antenna on the reflecting surface. The longitude and latitude of the reflecting point can be determined using the method discussed in Sect. 5.6.2, and in this way, the reflecting point can be determined.

However, the reflecting surface is usually not a plane; therefore, the above-discussed image point positioning result is a kind of average height. For convenience, we define the reflecting point, which has such an average height, as a nominal reflecting point. The distances between the nominal reflecting point and the satellite and downward-pointing antenna can be computed. Comparing the computed value with the true signal transmission distance, the bias of heights of the real reflecting point and the nominal reflecting point can be determined. In this way, the reflecting surface can be calculated.

## 5.6.4
## Research Activities in GPS Altimetry

As mentioned in Sect. 5.6.1, remote sensing of the sea surface with reflected GPS signals was first introduced in 1993, when Martin-Neira (1993) proposed the concept of the PAssive Reflectometry and Interferometry System (PARIS), which uses existing GPS signals for mesoscale ocean altimetry. Further discussion about the potential use of GPS signals as ocean altimetry observables can be found in Wu et al. (1997). During tests of a GPS-based vehicle tracking system, Auber et al. (1994) reported multipath errors, especially when aircraft flew at low altitudes above the sea. Reflections from the sea surface were nearly as strong as the incident signals and caused the GPS receiver to lock to the reflected signals, yielding false positioning information. Thus, the first documented GPS reflections were measured. Since then, many theoretical and experimental studies have been carried out with different receiver designs, at different observation heights and platforms. In this section, research activities involving GPS altimetry will be briefly introduced, referring to Helm (2008).

GPS altimetry is one of the main research activities with the field of GPS reflectometry, which tries to achieve highly accurate height measurements comparable to accuracies of conventional radar altimetry (RA). Airborne campaigns have been conducted (e.g., Garrison et al. 1998; Garrison and Katzberg 2000; Rius et al. 2002) and reached a 5-cm height precision (Lowe et al. 2002b). The MEBEX balloon experiment (Cardellach et al. 2003) was conducted in August 1999 in the earth's stratosphere. The experiment confirms the feasibility of GPS reflectometry from a height of about 37 km. Lowe et al. (2002a) observed the first earth-reflected GPS signal from space by analysing SIR-C data collected on board the space shuttle during the Shuttle Radar Laboratory-2 mission in October 1994. Beyerle and Hocke (2001) found evidence of earth-reflected GPS signals in GPS/MET and CHAMP occultation data (Beyerle et al. 2002; Cardellach et al. 2004). While the CHAMP and Satelite de Aplicaciones Cientificas C (SAC-C) satellites are already equipped with earth/nadir-looking GPS antennas, work is under way to establish satellite-based GPS altimetry (Hajj and Zuffada 2003). In 2004, Gleason et al. (2005) recorded waveforms of ocean-reflected GPS signals that were experimentally detected on board the UK Disaster Monitoring Constellation (UK-DMC) satellite.

The experimental demonstration of ground-based sea surface altimetry using reflected GPS signals followed in 1997 with the BRIDGE experiment. Only the C/A code was used for correlation, leading to an altimetric accuracy of about 3 m (Martin-Neira et al. 2001). In 1998, Anderson (2000) placed GPS antennas about 10 m above the water surface and determined the water level and tides with interferometric observations of GPS signals, with an accuracy of about 12 cm. During the Crater Lake Experiment in October 1999, Treuhaft et al. (2001) were able to obtain accuracy of 2-cm using GPS L1 carrier phase data. In the pond experiment, Martin-Neira et al. (2002) reached accuracy of 1 cm. On a calm water surface inside a harbour phase altimetry was possible with the Oceanpal instrument, reaching an accuracy of about 3.1 cm (Soulat et al. 2006).

Various approaches have already been realised for using the former error source "multipath" as measurement quantity in GPS receiver design. A standard GPS receiver, either for navigation or geodetic purposes, normally does not grant access to the necessary information within the GPS signal. Access to the raw GPS signal is required to recover the altimetric information. One approach (e.g., Garrison et al. 1998; Garrison and Katzberg 2000) is to record this data stream with high rate data recorders. In a second step—in real-time or during post-processing—the data stream is analysed and processed with software, giving the receiver type also the name "software radio". Fundamentals of GPS software receivers are described in Tsui (2000). PC-based GPS civilian L1 software receivers, or so-called GPS radios, have been developed by Akos et al. (2001), Kelley et al. (2002), Beyerle (2003), Ledvina et al. (2003), Pany et al. (2004), and MacGougan et al. (2005), and have been extended to dual-frequency civilian GPS software receivers (Ledvina et al. 2004). In order to obtain adequate gain from reflected GPS signals, research has been done with multi-element digital beam-steered antenna arrays and High-gain Advanced GPS Receivers (HAGRs) (Gold et al. 2005).

## 5.7
## Anti-spoofing and Selective Availability Effects

*Anti-spoofing*

The function of anti-spoofing (AS) of the GPS system is designed for an anti-potential spoofer (or jammer). A spoofer generates a signal that mimics the GPS signal and attempts to cause the receiver to track the wrong signal. When the AS mode of operation is activated, the P code will be replaced with a secure Y code available only to authorised users, and the unauthorised receiver becomes a single L1 frequency receiver. AS has been tested frequently since 1 August 1992 and was formally activated at 00:00 UT on 31 January 1994 and now is in continuous operation on all Block II and later satellites.

The broadcasted ionospheric model (in the navigation message) may be used to overcome the problem of absence of the dual frequencies, which were originally implemented to eliminate the ionospheric effects. Of course, the method using the ionospheric model cannot be as accurate as the method using dual-frequency data, and consequently the precision is degraded. Carrier phase smoothed C/A code may be used to replace the absence of the P code.

*Selective Availability*

Selective availability (SA) is a degradation of the GPS signal with the objective of denying full position and velocity accuracy to unauthorised users by dithering the satellite clock and manipulating the ephemerides. When SA is on, the fundament frequency of the satellite clock is dithered such that the GPS measurements are affected. The broadcast ephemerides are manipulated so that the computed orbit will have slow variations. Several levels of SA effects are possible. SA is enabled on Block II and later satellites (Graas and Braasch 1996).

Authorised users may recover the un-degraded data and exploit the full system potential. To do so they must possess a key that allows them to decrypt correction data transmitted in the navigation message (Georgiadou and Daucet 1990). For high-precision users, IGS precise orbit and forecast orbit data may be used. Using known positions (or monitor stations), the range corrections can be computed. Differential GPS may also at least partially eliminate the SA effects.

SA has been switched off since May 2000.

# 5.8
# Antenna Phase Centre Offset and Variation

*Satellite Antenna Phase Centre Correction*

The geometric distance between the satellite (at signal emission time) and the receiver (at signal reception time) is, in fact, the distance between the phase centres of the two antennas. However, the orbit data, which describes the position of the satellite, usually refers to the mass centre of the satellite. Therefore, a phase centre correction (also called mass centre correction) must be applied to the satellite coordinates in precise applications.

A satellite fixed coordinate system will be set up for describing the antenna phase centre offset to the mass centre of the satellite. As shown in Fig. 5.15, the origin of the frame coincides with the mass centre of the satellite, the $z$-axis is parallel to the antenna pointing direction, the $y$-axis is parallel to the solar-panel axis, and the $x$-axis is selected to complete the right-handed frame. A solar vector is a vector from the satellite mass centre pointed to the sun. During the motion of the satellite, the $z$-axis is always pointing to the earth, and the $y$-axis (solar-panel axis) will be kept perpendicular to the solar vector. In other words, the $y$-axis is always perpendicular to the plane formed by the sun, the earth, and the satellite. The solar

**Fig. 5.15** Satellite fixed coordinate system

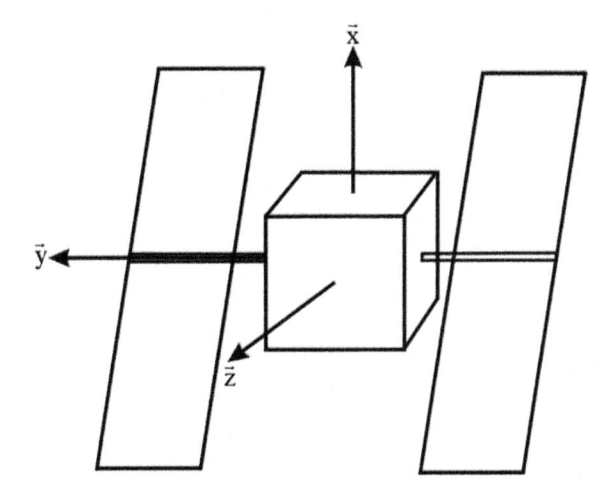

**Fig. 5.16** The sun vector in satellite fixed frame

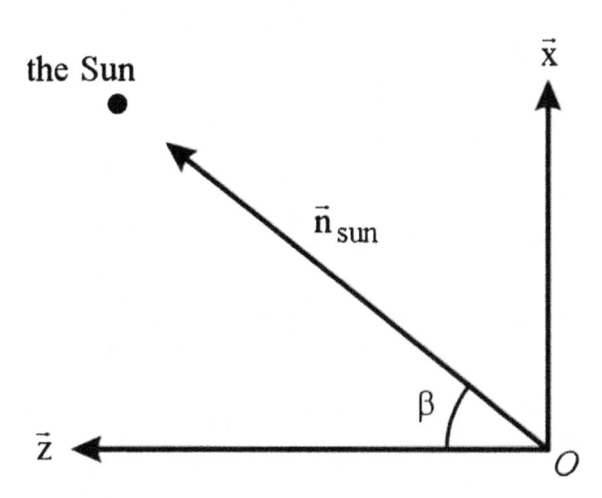

panel can be rotated around its axis to keep it perpendicular to the ray of the sun for optimal collection of the solar energy. The solar angle $\beta$ is defined as the angle between the $z$-axis and the solar identity vector $\vec{n}_{sun}$ (see Fig. 5.16). Denoting the identity vector of the satellite fixed frame as $(\vec{e}_x \vec{e}_y \vec{e}_z)$, the solar identity vector can then be represented as

$$\vec{n}_{sun} = (\sin\beta \quad 0 \quad \cos\beta). \tag{5.190}$$

$\beta$ is needed for computation of the solar radiation pressure in orbit determination.

Denoting $\vec{r}$ as the geocentric satellite vector and $\vec{r}_s$ as the geocentric solar vector (Fig. 5.17),

**Fig. 5.17** The earth-sun satellite vectors

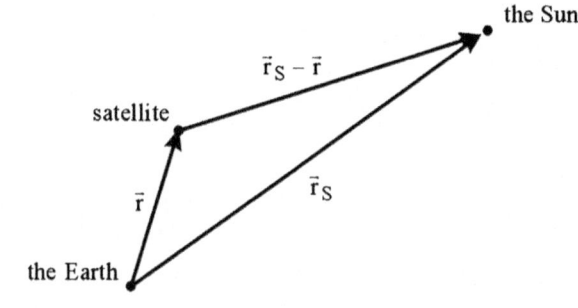

$$\vec{r} = \begin{pmatrix} X \\ Y \\ Z \end{pmatrix}, \quad \vec{r}_S = \begin{pmatrix} X_{sun} \\ Y_{sun} \\ Z_{sun} \end{pmatrix}, \tag{5.191}$$

then in a geocentric coordinate system, one has

$$\vec{e}_z = -\frac{\vec{r}}{|\vec{r}|}, \tag{5.192}$$

$$\vec{e}_y = \frac{\vec{e}_z \times \vec{n}_{sun}}{|\vec{e}_z \times \vec{n}_{sun}|}, \tag{5.193}$$

$$\vec{e}_x = \vec{e}_y \times \vec{e}_z, \tag{5.194}$$

$$\vec{n}_{sun} = \frac{\vec{r}_S - \vec{r}}{|\vec{r}_S - \vec{r}|}, \quad \text{and} \tag{5.195}$$

$$\cos \beta = \vec{n}_{sun} \cdot \vec{e}_z, \tag{5.196}$$

or

$$\vec{e}_z = \frac{-1}{r} \begin{pmatrix} X \\ Y \\ Z \end{pmatrix}, \quad r = \sqrt{X^2 + Y^2 + Z^2}, \tag{5.197}$$

$$\vec{n}_{sun} = \frac{1}{R} \begin{pmatrix} X_{sun} - X \\ Y_{sun} - Y \\ Z_{sun} - Z \end{pmatrix}, \tag{5.198}$$

$$\vec{e}_y = \frac{-1}{S} \begin{pmatrix} YZ_{\text{sun}} - Y_{\text{sun}}Z \\ ZX_{\text{sun}} - Z_{\text{sun}}X \\ XY_{\text{sun}} - X_{\text{sun}}Y \end{pmatrix}, \quad \text{and} \tag{5.199}$$

$$\vec{e}_x = \frac{1}{S \cdot r} \begin{pmatrix} (ZX_{\text{sun}} - Z_{\text{sun}}X)Z - (XY_{\text{sun}} - X_{\text{sun}}Y)Y \\ (XY_{\text{sun}} - X_{\text{sun}}Y)X - (YZ_{\text{sun}} - Y_{\text{sun}}Z)Z \\ (YZ_{\text{sun}} - Y_{\text{sun}}Z)Y - (ZX_{\text{sun}} - Z_{\text{sun}}X)X \end{pmatrix}, \tag{5.200}$$

where

$$R = \sqrt{(X_{\text{sun}} - X)^2 + (Y_{\text{sun}} - Y)^2 + (Z_{\text{sun}} - Z)^2} \quad \text{and} \tag{5.201}$$

$$S = \sqrt{(YZ_{\text{sun}} - Y_{\text{sun}}Z)^2 + (ZX_{\text{sun}} - Z_{\text{sun}}X)^2 + (XY_{\text{sun}} - X_{\text{sun}}Y)^2}. \tag{5.202}$$

Suppose the satellite antenna phase centre in the satellite fixed frame is $(x, y, z)$. The offset vector in the geocentric frame can then be obtained by substituting Eqs. 5.197, 5.199, and 5.200 into the following formula:

$$\vec{d} = x\vec{e}_x + y\vec{e}_y + z\vec{e}_z, \tag{5.203}$$

which may be added to the vector $\vec{r}$.

Beginning 5 November 2006, IGS has used different antenna phase centre vectors even within the same GPS satellite block. An example of GPS satellite antenna phase centre offsets in the satellite fixed frame are given in Table 5.4.

The dependence of the phase centre on the signal direction and frequencies is not considered for the satellite here. A mis-orientation of the $\vec{e}_y$ ($\vec{e}_x$ also) of the satellite with respect to the sun may cause errors in the geometrical phase centre correction. In the earth's shadow (for up to 55 min), the mis-orientation becomes worse. The geometrical mis-orientation may be modelled and estimated.

*Receiver Antenna Phase Centre Correction*

In the case of receiver antenna phase centre correction, the dependence of the phase centre on the signal direction and frequencies has to be taken into account. Both the phase centre offset and variation should be modelled. Generally, the phase centre corrections can be obtained through careful calibration. Receiver antenna phase centre offset is also antenna type dependent. For a GPS network, antenna phase centre corrections are usually predetermined and listed in a table for use.

**Table 5.4** GPS satellite antenna phase centre offset

| Satellites of | PRN | x | y | z |
|---|---|---|---|---|
| Block I | | 0.2100 | 0.0 | 0.8540 |
| Block II/IIA | | 0.2794 | 0.0 | 1.0259 |
| Block IIR-A | 11 | 0.0000 | 0.0 | 1.1413 |
| | 13 | 0.0000 | 0.0 | 1.3895 |
| | 14 | 0.0000 | 0.0 | 1.3454 |
| | 16 | 0.0000 | 0.0 | 1.5064 |
| | 18 | 0.0000 | 0.0 | 1.2909 |
| | 20 | 0.0000 | 0.0 | 1.3436 |
| | 21 | 0.0000 | 0.0 | 1.4054 |
| | 28 | 0.0000 | 0.0 | 1.0428 |
| Block IIR-B | 2 | 0.0000 | 0.0 | 0.7786 |
| | 19 | 0.0000 | 0.0 | 0.8496 |
| | 22 | 0.0000 | 0.0 | 0.9058 |
| | 23 | 0.0000 | 0.0 | 0.8082 |
| Block IIR-M | 4 | 0.0000 | 0.0 | 0.9656 |
| | 5 | 0.0000 | 0.0 | 0.8226 |
| | 7 | 0.0000 | 0.0 | 0.8529 |
| | 12 | 0.0000 | 0.0 | 0.8408 |
| | 15 | 0.0000 | 0.0 | 0.6811 |
| | 17 | 0.0000 | 0.0 | 0.8271 |
| | 29 | 0.0000 | 0.0 | 0.8571 |
| | 31 | 0.0000 | 0.0 | 0.9714 |
| Block IIF | 1 | 0.3940 | 0.0 | 1.5613 |
| | 3 | 0.3940 | 0.0 | 1.6000 |
| | 6 | 0.3940 | 0.0 | 1.6000 |
| | 8 | 0.3940 | 0.0 | 1.6000 |
| | 9 | 0.3940 | 0.0 | 1.6000 |
| | 10 | 0.3940 | 0.0 | 1.6000 |
| | 24 | 0.3940 | 0.0 | 1.6000 |
| | 25 | 0.3940 | 0.0 | 1.5973 |
| | 26 | 0.3940 | 0.0 | 1.6000 |
| | 27 | 0.3940 | 0.0 | 1.6000 |
| | 30 | 0.3940 | 0.0 | 1.6000 |
| | 32 | 0.3940 | 0.0 | 1.6000 |

## 5.9
## Instrumental Biases

The study of ionospheric effects using GPS observations indicates the existence of instrumental biases (cf., e.g., Yuan and Ou 1999). These biases are systematic errors, and are different from frequency to frequency and from code to phase measurements. However, they are constant for a given frequency and given observable type as well as a given instrument (receiver or GPS satellite). For code, phase, and Doppler observable of receiver $i$, satellite $j$ at working frequency $k$, instrumental biases can be modelled as

$$\begin{aligned} \delta I_c(i,k) + \delta J_c(j,k) \\ \delta I_p(i,k) + \delta J_p(j,k) \quad &\text{and} \\ \delta I_d(i,k) + \delta J_d(j,k) \end{aligned} \qquad (5.204)$$

respectively, where indices c, p, and d denote the code, phase, and Doppler observables. $\delta I$ and $\delta J$ denote the instrumental biases of the GPS receiver and GPS satellite. The separation of the instrumental biases and the ambiguities are possible because the biases of the receiver and satellite are independent of each other, whereas the ambiguity parameters are dependent on both the receiver and the satellite. However, in modelling and solving the problem, the correlation between the parameters must be carefully studied. Instrumental biases of one of the frequencies and one of the channels are linearly correlated with the clock biases. Without modelling of the instrumental biases, they may merge into the ambiguity parameters such that the integer property of the ambiguities is destroyed.

# References

Akos D, Hansson A, Normark P, Rosenlind C, Stahlberg A, Svensson F (2001) Real-time software radio architectures for GPS receivers. GPS World, pages 28–33, July 2001.

Allan D, Weiss M (1980) Accurate time and frequency transfer during common-view of a GPS satellite. In: Proceedings of 1980 IEEE frequency control symposium, Philadelphia, pp 334–356

Andersen OB (1994) M,2, and S,2, ocean tide models for the North Atlantic Ocean and adjacent seas from ERS-1 altimetry, Space at the service of our environment. In: Proceedings of the second ERS-1 symposium, Hamburg, 11–14 October 1993, Vol. 2., January 1994, Noordwijk, pp 789–794

Anderson K (2000) Determination of water level and tides using interferometric observations of GPS signals. Journal of Atmospheric and Oceanic Technology, 17: 1118-1127

Ashby N, Spilker JJ (1996) Introduction to relativistic effects on the Global Positioning System. In:Parkinson BW, Spilker JJ (eds) Global Positioning System: Theory and applications, Vol. I, Chapter 18

Auber J, Bibaut A, Rigal J (1994) Characterization of multipath on land and sea at GPS frequencies. In Proceedings of the 7th International Technical Meeting of the Satellite Division of the Institute of Navigation, pages 1155–1171, Salt Lake City, UT; US, 20–23 September 1994.

Berg, H (1948) Allgemeine Meteorologie. Ferdinand Duemmler Verlag, Bohn

Bevis M, Businger S, Herring AT, et al. (1992) GPS meteorology: remote sensing of atmospheric water vapor using the global positioning system. J Geophys Res 97(D14):15787–15801

Bevis M, Businger S, Chiswell S, et al. (1994) GPS meteorology: mapping zenith wet delays onto precipitable water. J Appl Meteorol 33:379–386

Beyerle G (2003) Opengpsrec: An open source gps receiver. http://www.geocities.com/gbeyerle/software/ download.html, July 2003.

Beyerle G, Hocke K (2001) Observation and simulation of direct and reflected GPS signals in radio occultation experiments. Geophys. Res. Lett., 28(9):1895–1898, 2001

Beyerle G, Hocke K, Wickert J, Schmidt T, Marquardt C, Reigber C (2002) GPS radio occultations with CHAMP: A radio holographic analysis of GPS signal propagation in the troposphere and surface reflections. J. Geophys. Res., 107(D24): ACL 27-1-ACL 27-14

Bilitza D (2001) International reference ionosphere 2000. Radio Science, 36(2): 261-275

Bilitza D, Altadill D, Zhang Y, et al. (2014) The international reference ionosphere 2012—a model of international collaboration. J. Space Weather Space Clim., 4, A07

Bock O, Doerflinger E (2001) Atmospheric modeling in GPS data analysis for high accuracy positioning. Phys Chem Earth 26(6–8):373–383

Boehm J, Niell A, Tregoning P, Schuh H (2006a) Global Mapping Function (GMF): A new empirical mapping function based on numerical weather model data. Geophys. Res. Lett. 33 (7), L07304

Boehm J, Werl B, Schuh H (2006b) Troposphere mapping functions for GPS and very long baseline interferometry from European Centre for Medium-Range Weather Forecasts operational analysis data. J. Geophys. Res. 111, B02406

Boehm J, Heinkelmann R, Schuh H (2007) Short Note: A global model of pressure and temperature for geodetic applications. J. Geodesy 81(10), 679-683

Box G, Jenkins G, Reinsel G (1994) Time Series Analysis, Forecasting and Control, 3$^{rd}$ Edition. Prentice Hall, Englewood Clifs

Braasch MS (1996) Multipath effects. Parkinson BW, Spilker JJ (eds) Global Positioning System: Theory and applications, Vol. I

Cardellach E, Ruffini G, Pino D, Rius A, Komjathy A, Garrison J (2003) Mediterranean ballon experiment: ocean wind speed sensing from the stratosphere, using GPS reflections. Remote Sensing of Environment, 88(3): 351-362

Cardellach E, Ao C, Torre Juarez M, Hajj G (2004) Carrier phase delay altimetry with GPS-reflection / occultation interferometry from low Earth orbiters. Geophys. Res. Lett., 31 (L10402)

Choy S, Wang CS, Yeh TK, Dawson J, Jia M, Kuleshov Y (2015) Precipitable water vapor estimates in the Australian region from ground-based GPS observations. Advances in Meteorology 2015, 956481

Collins J, Langley R (1997) Estimating the residual tropospheric delay for airborne differential GPS positioning. In: Proceedings of ION GPS-97, Kansas City, Mo., pp. 1197–1206

Dach R, Beutler G, Hugentobler U, Schaer S, Schildknecht T, Springer T, Dudle G, Prost L (2003) Time transfer using GPS carrier phase: error propagation and results. J Geodesy 77:1–14

Dach R, Hugentobler U, Fridez P, Meindl M (2007) Bernese GPS software version 5.0 user manual. Astronomical Institute University of Bern, Bern

Davis J, Herring T (1984) New atmospheric mapping function. Center of Astrophysics, Cambridge, Mass., Manuscript July 1984

Davis JL, Herring TA, Shapiro II, Rogers AEE, Elgered G (1985) Geodesy by radio interferometry: effects of atmosphericmodeling errors on estimates of baseline length. Radio Sci 20:1593–1607

Deng J (1987) The Primary Methods of Grey System Theory. Huazhong University of Science and Technology (HUST) Press. Wuhan, China

Ding JC (2009) GPS meteorology and its applications. China Meteorological Press, Beijing, pp 1–10

Doodson AT (1928) The analysis of tidal observations. Philos Tr R Soc S-A 227:223–279

Dow JM, Neilan RE, Rizos C (2009) The International GNSS service in a changing landscape of global navigation satellite systems. J Geodesy 83(3–4):191–198

Farrell WE (1972) Deformation of the Earth by surface loads. Rev Geophys Space Ge 10(3):761–797

Garrison JL, Katzberg SJ (2000) The application of reflected GPS signals to ocean remote sensing. Remote Sens Environ 73(2):175–187

Garrison JL, Katzberg SJ, Hill M (1998) Effect of sea roughness on bistatically scattered range coded signals from the global positioning system. Geophys. Res. Lett., 25(13):2257–2260

Georgiadou Y, Doucet KD (1990) The issue of selective availability. GPS World 1(5):53–56

Gleason S, Hodgart S, Sun Y, Gommenginger C, Mackin S, Adjrad M, Unwin M (2005) Detection and processing of bistatically reflected GPS signals from low earth orbit for the purpose of ocean remote sensing. IEEE Trans. Geosci. and Remote Sensing, 43(6): 1229-1241

Gold K, Brown A, Stolk K (2005) Bistatic sensing and multipath mitigation with a 109-element GPS antenna array and digital beam steering receiver. In ION National Technical Meeting, San Diego, CA, January 2005.

Graas FV, Braasch MS (1996) Selective availability. In: Parkinson BW, Spilker JJ (eds) Global Position-ing System: Theory and applications, Vol. I, Chapter 17

Gu XP (2004) Research on retrieval of GPS water vapor and method of rainfall forecast. Doctorial dissertation, China Agricultural University pp 1–15

Gu XP, Wang CY, Wu DX (2005) Research on the local algorithm for weighted atmospheric temperature used inGPSremote sensingwater vapor. Sci Metero Sin 25(1):79–83

Hajj G, Zuffada C (2003) Theoretical description of a bistatic system for ocean altimetry using the GPS signal. Radio Sci., 38(5)

Helm A (2008) Ground-based GPS altimetry with the L1 OpenGPS receiver using carrier phase-delay observations of reflected GPS signals. Scientific Technical Report STR 08/10. DOI:10.2312/GFZ.b103-08104.

Herring T (1992) Modeling atmospheric delays in the analysis of space geodetic data. In: J., M., T., S., (Eds.). Netherlands Geodetic Commision. Delft, pp. 157-164

Hofmann-Wellenhof B, Lichtenegger H, Collins J (1997, 2001) GPS theory and practice. Springer-Press, Wien

Holdridge DB (1967) An alternate expression for light time using general relativity. JPL Space Pro-gram Summary 37–48, III, pp 2–4

Hopfield HS (1969) Two-quartic tropospheric refractivity profile for correcting satellite data. J Geophys Res 74(18):4487–4499

Hopfield HS (1970) Tropospheric effect on electromagnetically measured ranges: Prediction from surface weather data. Applied Physics Laboratory, Johns Hopkins University, Baltimore, MD, July 1970

Hopfield HS (1972) Tropospheric range error parameters – further studies. Applied Physics Labora-tory, Johns Hopkins University, Baltimore, MD, June 1972

Huang G (2012) Research on Algorithms of Precise Clock Offset and Quality Evaluation of GNSS Satellite clock. Chang'an University. Xi'an, China

Jakowski N, Hoque MM, Mayer C (2011) A new global TEC model for estimating transionospheric radio wave progation errors. Journal of Geodesy, 85: 965-974

Katzberg SJ, Garrison JL (1996) Utilizing GPS to determine ionospheric delay over the ocean. NASA Technical Memorandum TM-4750, NASA Langley Research Center

Kaula WM (1966, 2001) Theory of satellite geodesy. Blaisdell Publishing Company, Dover Publications, New York

Kelley C, Barnes J, Cheng J (2002) Open Source GPS: Open source software for learning about GPS. In ION GPS 2002, pages 2524–2533, Portland, USA, September 2002.

Khan SA (1999) Ocean loading tide effects on GPS positioning. MSc. thesis, Copenhagen University

Klobuchar JA (1996) Ionospheric effects on GPS. In: Parkinson BW, Spilker JJ (eds) Global Position-ing System: Theory and applications, Vol. I, Chapter 12

Klobuchar J A (1987) Ionospheric time-delay algorithm for single-frequency GPS users. Aerospace and Electronic Systems, IEEE Transactions on, (3): 325-331

Knudsen P, Olsen H, Xu G (1999) GPS-altimetry tests – Measuring GPS signal reflected from the Earth surface. Poster on the 22st IUGG General Assembly, IAG Symposium, England

Komjathy A (1997) Global ionospheric total electron content mapping using the Global Positioning System. University of New Brunswick, Fredericton, New Brunswick, Canada

Komjathy A, Garrison J, Zavorotny V (1999) GPS: A new tool for ocean science. GPS World 10 (4):50–56

Kosaka M (1987) Evaluation method of polynomial models' prediction performance for random clock error. Journal of Guidance, Control, and Dynamics 10(6): 523-527

Kouba J (2009) Testing of global pressure/temperature (GPT) model and global mapping function (GMF) in GPS analyses. J. Geodesy 83(3-4), 199-208

Lagler K, Schindelegger M, Bohm J, Krasna H, Nilsson T (2013) GPT2: Empirical slant delay model for radio space geodetic techniques. Geophys. Res. Lett. 40(6), 1069-1073

Lambeck K (1988) Geophysical geodesy – The slow deformations of the Earth. Oxford Science Publications

Langley RB (1998a) Propagation of the GPS signals. In: Kleusberg A, Teunissen PJG (eds) GPS for geodesy. Springer-Verlag, Berlin

Langley RB (1998b) GPS receivers and the observables. In: Kleusberg A, Teunissen PJG (eds) GPS for geodesy. Springer-Verlag, Berlin

Ledvina B, Psiaki M, Powell S, Kintner P (2003) A 12-channel real-time GPS L1 software receiver. In Proc. of the Institute of Navigation National Technical Meeting, Anaheim, CA, January 22-24, 2003.

Ledvina B, Psiaki M, Sheinfeld D, Cerruti A, Powell S, Kintner P (2004) A real-time GPS civilian L1/L2 software receiver. In Proc. of the Institute of Navigation GNSS, Long Beach, CA, September 21-24 2004.

Leick A (1995) GPS satellite surveying. John Wiley & Sons Ltd., New York

Leick A (2004) GPS Satellite Surveying (3rd ed.) xxiv + 664 p. John Wiley, New York, NY

Lesage P, Ayi T (1984) Characterization of frequency stability: analysis of the modified allan variance and properties of its estimate. IEEE Trans Instrum Measure IM-33(4):332–336

Li JG, Mao JT, Li CC (1999) The approach to remote sensing of water vapor based on GPS and linear regression $T_m$ in eastern region of China. Acta Meteor Sin 57(3):283–292

Li GP, Huang GF, Liu BQ (2006) Experiment on driving precipitable water vapor form ground-basedGPS network inChengdu Plain. Geomat Inf Sci 31(12):1086–1089

Liu M, Yang Y, Stein S, Zhu S, Engeln J (2000) Crustal shortening in the Andes: Why do GPS rate differ from geological rates? Geophys Res Lett 27(18):3005–3008

Lowe S, LaBrecque J, Zuffada C, Romans L, Young L, Hajj G (2002a) First spaceborne observation of an earth reflected GPS signal. Radio Sci., 37(1)

Lowe S, Zuffada C, Chao Y, Kroger P, Young J (2002b) 5-cm-precision aircraft ocean altimetry using GPS reflections. Geophys. Res. Lett., 29(10)

Lv YP, Yin HT, Huang DF (2008) Modeling of weighted mean atmospheric temperature and application in GPS/PWV of Chengdu region. Sci Survey Mapp 33(4):103–105

MacGougan G, Normark P-L, Ståhlberg C (2005) Satellite navigation evolution. The software GNSS receiver. GPS World 16(1): 48-52

MacMillan D, Ma C (1994) Evaluation of very long baseline interferometry atmospheric modeling improvements. J. Geophys. Res. 99(B1), 637-651

Mannucci A J, Wilson B D, Yuan D N, et al. (1998) A global mapping technique for GPS—derived ionospheric total electron content measurements. Radio Science 33(3): 565–582

Mao JT (2006) Research of remote sensing of atmospheric water vapor using Global Positioning System (GPS). Doctorial dissertation, Beijing University pp 1–5

Martin-Neira M (1993) A passive reflectometry and interferometry system (PARIS): Application to ocean altimetry. ESA Journal, 17:331–355

Martin-Neira M, Caparrini M, Font-Rossello J, Lannelongue S, Serra C (2001) The paris concept: An experimental demonstration of sea surface altimetry using GPS reflected signals. IEEE Trans. Geosci. and Remote Sensing, 39:142–150

Martin-Neira M, Colmenarejo P, Ruffini G, Serra C (2002) Altimetry precision of 1 cm over a pond using the wide-lane carrier phase of gps reflected signals. Ca. J. Remote Sensing, 28(3): pp. 394–403

McCarthy DD (1996) International Earth Rotation Service. IERS conventions, Paris, 95 pp. IERS Technical Note No. 21

Melchior P (1978) The tides of the planet Earth. Pergamon Press

Mendes V, Langley R (1998) Tropospheric zenith delay prediction accuracy for airborne GPS high-precision positioning, Proceedings of ION 54th Annual Meeting, Denver, Colorado, pp. 337–348

Nava B, Coïsson P, Radicella S M (2008) A new version of the NeQuick ionosphere electron density model. Journal of Atmospheric and Solar-Terrestrial Physics, 70(15): 1856-1862

Niell AE (1996) Global mapping functions for the atmosphere delay at radio wavelengths. J. Geophys. Res. 101(B2), 3227-3246

Otsuka Y, Ogawa T, Saito A, Tsugawa T, et al. (2002) A new technique for mapping of total electron content using GPS network in Japan. Earth Planets Space 54: 63-70

Pany T, Eisfeller B, Hein G, Moon S, Sanroma D (2004) ipexSR: A PC based software GNSS receiver completely developed in Europe. In Europeen Navigation Conference GNSS 2004, Rotterdam, May 2004.

Percival D (2006) Spectral Analysis of Clock Noise: A Primer. Metrologia 43(4): S299-S310

Petrie EJ, Hernández-Pajares M, Spalla P, et al. (2011) A review of higher order ionospheric refraction effects on dual frequency GPS. Surveys in Geophysics, 32(3): 197-253

Poutanen M, Vermeer M, Maekinen J (1996) The permanent tide in GPS positioning. J Geodesy 70: 499–504

Ray J, Senior K (2005) Geodetic techniques for time and frequency comparisons using GPS phase and code measurements. Metrologia 42(3):215–232

Rius A, Aparicio J, Cardellach E, Martin-Neira M, Chapron B (2002) Sea surface state measured using GPS reflected signals. Geophys. Res. Lett., 29(23): 371-374

Rocken C, Ware R,VanHove T, Solheim F, Alber C, Johnson J, Bevis M, Businger S (1993) Sensing atmospheric water vapor with the global positioning system. Geophys Res Lett 20 (23):2631–2634

Ross RJ, Rosenfeld S (1997) Estimating mean weighted temperature of the atmosphere for Global Positioning System. J Geophys Res 102(18):21719–21730

Saastamoinen J (1972) Contribution to the theory of atmospheric refraction. B Geod 105–106

Saastamoinen J (1973) Contribution to the theory of atmospheric refraction. B Geod 107

Schaer S (1999) Mapping and predicting the earth's ionosphere using the global positioning system. Astronomical Institute, University of Berne, Switzerland

Schwiderski EW (1978) Global ocean tide, I. A detailed hydrodynamical interpolation model. Rep. NSWC/DL TR 3866, Nav. Surf. Weapons Cent. Dahlgren, Va.

Schwiderski EW (1979) Ocean tide, II. The semidiurnal principal lunar tide (M2). Rep. NSWC TR 79-414, Nav. Surf. Weapons Cent. Dahlgren, Va.

Schwiderski EW (1980) On charting global ocean tide. Rev Geophys 18:243–268

Schwiderski EW (1981a) Ocean tide, III. The semidiurnal principal solar tide (S2). Rep. NSWC TR 81-122, Nav. Surf. Weapons Cent. Dahlgren, Va.

Schwiderski EW (1981b) Ocean tide, IV. The diurnal luni-solar declination tide (K1). Rep. NSWC TR 81-142, Nav. Surf. Weapons Cent. Dahlgren, Va.

Schwiderski EW (1981c) Ocean tide, V. The diurnal principal lunar tide (O1). Rep. NSWC TR 81-144, Nav. Surf. Weapons Cent. Dahlgren, Va.

Seeber G (1993) Satelliten-Geodaesie. Walter de Gruyter 1989

Soulat F, Caparrini M, Farres E, Dunne S, Chapron B, Buck C, Ruffini G (2006) Oceanpal experimental campaigns. Proceedings of the GNSSR'06 workshop, ESTEC, Noordwijk, the Netherlands, 14-15 June 2006.

Syndergaard S (1999) Retrieval analysis and methodologies in atmospheric limb sounding using the GNSS radio occultation technique. Dissertation, Niels Bohr Institute for Astronomy, Physics and Geophysics, Faculty of Science, University of Copenhagen

Torge W (1991) Geodesy. Walter de Gruyter, Berlin

Tregoning P, Herring TA (2006) Impact of a priori zenith hydrostatic delay errors on GPS estimates of station heights and zenith total delays. Geophys. Res. Lett. 33, L23303

Treuhaft R, Lowe S, Zuffada C, Chao Y (2001) 2-cm GPS altimetry over Crater Lake. Geophys. Res. Lett., 22(23):4343–4346

Tsui J (2000) Fundamentals of Global Positioning System Receivers: A Software Approach, volume ISBN: 0-471-38154-3. John Wiley & Sons, Inc., 1 edition, May 2000.

Wang G, Chen Z, Chen W, Xu G (1988) The principle of GPS precise positioning system. Surveying Press, Peking, ISBN 7-5030-0141-0/P.58, 345 p, (in Chinese)

Wang JH, Zhang LY, Dai AG, Van Hove T, Van Baelen J (2007a) A near-global, 2-hourly data set of atmosoheric precipitable water from ground-based GPS measurements. J Geophys Res 112, D11107

Wang Y, Liu LT, Hao XG et al (2007b) The application study of the GPS meteorology network in wuhan region. Acta Geodaet Cartogr Sinica 36(2):141–145

Wang Q, Xu T, Xu G (2010) HALO_GPS (High Altitude and Long Range Airborne GPS Positioning Software) – Software User Manual, Scientific Technical Report, German Research Centre for Geosciences, 2010, ISSN: 1610-0956

Wang Q, Xu T, Xu G (2011) Adaptively Changing Reference Station Algorithm and Its Application in GPS Long Range Airborne Kinematic Relative Positioning. Acta Geodaetica et Cartographica Sinica, 40(4): 429-434

Wanninger L (1997) Real-time differential GPS error modeling in regional reference station networks. In: Proc. 1997 IAG Symposium, 86–92, Rio de Janeiro, Brazil

Wu S, Meehan T, Young L (1997) The potential use of GPS signals as ocean altimetry observables. 1997 national technical meeting, Santa Monica, California, January 1997.

Xu G (2000) A concept of precise kinematic positioning and flight-state monitoring from the AGMASCO practice. Earth Planets Space 52(10):831–836

Xu G (2003) GPS – Theory, Algorithms and Applications, Springer Heidelberg, ISBN 3-540-67812-3, 315 pages, in English

Xu G, Knudsen P (2000) Earth tide effects on kinematic/static GPS positioning in Denmark and Greenland. Phys Chem Earth 25(A4):409–414

Xu T, Yang Y. (2001) The hypothesis testing of scale parameter in coordinate transformation model. Geomatics and Information Science of Wuhan University, 26, 70–74

Yao YB, Zhu S, Yue SQ (2012) A globally applicable, seasonspecific model for estimating the weighted mean temperature of the atmosphere. J Geod 86:1125–1135

Yao YB, Zhang B, Yue SQ, Xu CQ, Peng WF (2013) Global empirical model for mapping zenith wet delays onto precipitable water. J Geod 87: 439-448

Yeh TK, Hwang C, Xu G (2008) GPS height and gravity variations due to ocean tidal loading around Taiwan. Surv Geophys 29(1):37–50

Yeh TK, Hwang C, Xu G, Wang CS, Lee CC (2009) Determination of global positioning system (GPS) receiver clock errors: impact on positioning accuracy. Meas Sci Technol 20:075105

Yeh TK, Chen CH, Xu G, Wang CS, Chen KH (2012) The impact on the positioning accuracy of the frequency reference of a GPS receiver. Surv Geophys 34: 73-87

Yeh TK, Hong JS, Wang CS, Hsiao TY, Fong CT (2014) Applying the water vapor radiometer to verify the precipitable water vapor measured by GPS. Terr. Atmos. Ocean. Sci. 25: 189-201

Yuan YB, Ou JK (1999) The effects of instrumental bias in GPS observations on determining ionospheric delays and the methods of its calibration. Acta Geod Cartogr Sinica 28(2)

Zhang J, Lachapelle G (2001) Precise estimation of residual tropospheric delays using a regional GPS network for real-time kinematic applications. Journal of Geodesy 75, 255–266

Zhu SY (2001) Private communication and the source code of the EPOS-OC software

Zhu SY, Reigber Ch, Massmann FH (1996) The German PAF for ERS, ERS standards used at D-PAF. D-PAF/GFZ ERS-D-STD-31101

# Chapter 6
# GPS Observation Equations and Equivalence Properties

In this chapter, we begin with a discussion of the general mathematical model of GPS observation and its linearisation. All partial derivatives of the observation function are given in detail. These are necessary for forming GPS observation equations. We then outline the linear transformation and covariance propagation. In the section on data combinations, we discuss all meaningful and useful data combinations, such as ionosphere-free, geometry-free, code–phase combinations, and ionospheric residuals, as well as differential Doppler and Doppler integration. In the data differentiation section, we discuss single, double, and triple differences and their related observation equations and weight propagation. The parameters in the equations are greatly reduced through difference forming; however, the covariance derivations are tedious. In the last two sections, we discuss the equivalent properties between the uncombined and combining and the undifferenced and differencing algorithms. We propose a unified GPS data processing method, which is described in detail. The method is selectively equivalent to the zero-, single-, double-, triple-, and user-defined differential methods.

## 6.1
## General Mathematical Models of GPS Observations

Recalling the discussions in Chap. 4, the GPS code pseudorange, carrier phase, and Doppler observables are formulated as (cf. Eqs. 4.7, 4.18 and 4.23)

$$R_i^k(t_r, t_e) = \rho_i^k(t_r, t_e) - (\delta t_r - \delta t_k)c + \delta_{\text{ion}} + \delta_{\text{trop}} + \delta_{\text{tide}} + \delta_{\text{rel}} + \varepsilon_c, \quad (6.1)$$

$$\lambda \Phi_i^k(t_r, t_e) = \rho_i^k(t_r, t_e) - (\delta t_r - \delta t_k)c + \lambda N_i^k - \delta_{\text{ion}} + \delta_{\text{trop}} + \delta_{\text{tide}} + \delta_{\text{rel}} + \varepsilon_p, \text{ and} \quad (6.2)$$

$$D = \frac{d\rho_i^k(t_r, t_e)}{\lambda dt} - f\frac{d(\delta t_r - \delta t_k)}{dt} + \delta_{\text{rel\_}f} + \varepsilon_d, \quad (6.3)$$

Where ionospheric effects can be approximated as (cf. Sect. 5.1.2, Eq. 5.26)

$$\delta_{\text{ion}} = \frac{A_1}{f^2} + \frac{A_2}{f^3},$$

and $R$ is the observed pseudorange, $\Phi$ is the observed phase, $D$ is Doppler measurement, $t_e$ denotes the GPS signal emission time of the satellite $k$, $t_r$ denotes the GPS signal reception time of the receiver $i$, $c$ denotes the speed of light, subscript $i$ and superscript $k$ denote the receiver and satellite, and $\delta t_r$ and $\delta t_k$ denote the clock errors of the receiver and satellite at the time $t_r$ and $t_e$, respectively. The terms $\delta_{\text{ion}}$, $\delta_{\text{trop}}$, $\delta_{\text{tide}}$, and $\delta_{\text{rel}}$ denote the ionospheric, tropospheric, tidal, and relativistic effects, respectively. Tidal effects include earth tides and ocean tidal loading. The multipath effect was discussed in Sect. 5.6 and is omitted here. $\varepsilon_c$, $\varepsilon_p$. and $\varepsilon_d$ are the remaining errors, respectively. $f$ is the frequency, wavelength is denoted by $\lambda$, $A_1$, and $A_2$ are ionospheric parameters, $N_i^k$ is the ambiguity related to receiver $i$ and satellite $k$, $\delta_{\text{rel}\_f}$ is the frequency correction of the relativistic effects, the $\rho_i^k$ is the geometric distance, and (cf. Eq. 4.6)

$$\rho_i^k(t_r, t_e) = \rho_i^k(t_r) + \frac{d\rho_i^k(t_r)}{dt}\Delta t, \tag{6.4}$$

where $\Delta t$ denotes the signal transmitting time and $\Delta t = t_r - t_e$. $d\rho_i^k(t_r)/dt$ denotes the time derivation of the radial distance between the satellite and receiver at time $t_r$. All terms in Eqs. 6.1 and 6.2 have units of length (metres).

Considering Eq. 6.4 in the ECEF coordinate system, the geometric distance is a function of station state vector $(x_i, y_i, z_i, \dot{x}_i, \dot{y}_i, \dot{z}_i i)$ (denoted by $X_i$) and satellite state vector $(x_k, y_k, z_k, \dot{x}_k, \dot{y}_k, \dot{z}_k)$ (denoted by $X_k$). GPS observation Eqs. 6.1, 6.2, and 6.3 can then be generally presented as

$$O = F(X_i, X_k, \delta t_i, \delta t_k, \delta_{\text{ion}}, \delta_{\text{trop}}, \delta_{\text{tide}}, \delta_{\text{rel}}, N_i^k, \delta_{\text{rel}\_f}), \tag{6.5}$$

where $O$ denotes observation and $F$ denotes implicit function. In other words, the GPS observable is a function of state vectors of the station and satellite, and numbers of physical effects as well as ambiguity parameters. In principle, GPS observations can be used to solve for the desired parameters of the function in Eq. 6.5. This is why GPS is now widely used for positioning and navigation (to determine the state vector of the station), orbit determination (to determine the state vector of satellite), timing (to synchronise clocks), meteorological applications (i.e. tropospheric profiling), and ionospheric occultation (i.e. ionospheric sounding). In turn, the satellite orbit is a function of the earth's gravitational field and the number of disturbing effects such as solar radiation pressure and atmospheric drag. GPS is now also used for gravity field mapping and for solar and earth system study.

It is obvious that Eq. 6.5 is non-linear. The straightforward mathematical method for solving problem 6.5 is to search for the optimal solution using various effective search algorithms. The so-called ambiguity function (AF, see Sects. 8.5

and 12.2) method is an example. Generally speaking, solving a non-linear problem is much more complicated than first linearising the problem and then solving the linearised problem.

It is worth noting that the satellite and station state vectors will be represented in the same coordinate system; otherwise, coordinate transformation, as discussed in Chap. 2, will be carried out. Because the rotations are "distance-keeping" transformations, the distances computed in two different coordinate systems must be the same. However, because of the earth's rotation, the velocities expressed in the ECI and ECEF coordinate systems are not the same. Generally, the station coordinates and both the ionospheric and tropospheric effects are given and presented in the ECEF system. A satellite state vector may be given in both the ECSF and ECEF systems, depending on the need of the specific application.

## 6.2
## Linearisation of the Observation Model

The non-linear multivariable function $F$ in Eq. 6.5 can be further generalised as

$$O = F(Y) = F(y_1, y_2, \ldots, y_n), \tag{6.6}$$

where variable vector $Y$ has $n$ elements. The linearisation is accomplished by expanding the function in a Taylor series to the first order (linear term) as

$$O = F(Y^0) + \frac{\partial F(Y)}{\partial Y}\bigg|_{Y^0} \cdot dY + \varepsilon(dY), \tag{6.7}$$

where

$$\frac{\partial F(Y)}{\partial Y} = \begin{pmatrix} \frac{\partial F}{\partial y_1} & \frac{\partial F}{\partial y_2} & \cdots & \frac{\partial F}{\partial y_n} \end{pmatrix}, \quad \text{and} \quad dY = (Y - Y^0) = \begin{pmatrix} dy_1 \\ dy_2 \\ \vdots \\ dy_n \end{pmatrix};$$

the symbol $|_{Y^0}$ means that the partial derivative $\partial F(Y)/\partial Y$ takes the value of $Y = Y^0$ and $\varepsilon$ is the truncation error, which is a function of the second-order partial derivative and $dY$. $Y^0$ is called the initial value vector. Equation 6.7 turns out then to be

$$O - C = \begin{pmatrix} \frac{\partial F}{\partial y_1} & \frac{\partial F}{\partial y_2} & \cdots & \frac{\partial F}{\partial y_n} \end{pmatrix}_{Y^0} \cdot \begin{pmatrix} dy_1 \\ dy_2 \\ \vdots \\ dy_n \end{pmatrix} + \varepsilon, \tag{6.8}$$

where $F(Y^0)$ is denoted by $C$ (or say, the computed value). So GPS observation Eq. 6.6 is linearised as a linear equation (Eq. 6.8). Denoting the observation and truncation errors as $v$ and $O$–$C$ as $l$, partial derivative $(\partial F/\partial y_j)|_{y^0} = a_j$, then Eq. 6.8 can be written as

$$l_i = \begin{pmatrix} a_{i1} & a_{i2} & \cdots & a_{in} \end{pmatrix} \cdot \begin{pmatrix} dy_1 \\ dy_2 \\ \vdots \\ dy_n \end{pmatrix} + v_i, \quad (i = 1, 2, \ldots, m), \quad (6.9)$$

where $l$ is also often called "observable" in adjustment or $O$–$C$ (observed minus computed), and $j$ and $i$ are indices of unknowns and the observations. Equation 6.9 is a linear error equation. A set of GPS observables then forms a linear error equation system:

$$\begin{pmatrix} l_1 \\ l_2 \\ \vdots \\ l_m \end{pmatrix} = \begin{pmatrix} a_{11} & a_{12} & \cdots & a_{1n} \\ a_{21} & a_{22} & \cdots & a_{2n} \\ \vdots & \vdots & \vdots & \vdots \\ a_{m1} & a_{m2} & \cdots & a_{mn} \end{pmatrix} \cdot \begin{pmatrix} dy_1 \\ dy_2 \\ \vdots \\ dy_n \end{pmatrix} + \begin{pmatrix} v_1 \\ v_2 \\ \vdots \\ v_m \end{pmatrix},$$

or in matrix form ($dY$ is denoted by $X$)

$$L = AX + V, \quad (6.10)$$

where $m$ is the observable number. A number of adjustment and filtering methods (cf. Chap. 7) can be applied for solving GPS problem 6.10. The solved parameter vector is $X$ (or $dY$). The original unknown vector $Y$ can be obtained by adding $dY$ to $Y^0$. $V$ is the residual vector. Statistically, $V$ is assumed to be a random vector, and is normally distributed with zero expectation and variance var($V$). To characterise the different qualities and correlation situations of the observables, a so-called weight matrix $P$ is introduced to Eq. 6.10. Supposing all observations are linearly independent or uncorrelated, the covariance of observable vector $L$ is

$$Q_{LL} = \mathrm{cov}(L) = \sigma^2 E \quad (6.11)$$

or

$$P = Q_{LL}^{-1} = \frac{1}{\sigma^2} E, \quad (6.12)$$

where $E$ is an identity matrix of dimension $m \times m$, superscript $^{-1}$ is an inversion operator, and cov($L$) is the covariance of $L$.

Generally, the linearisation process is considered to have been done well only when the solved unknown vector $dY$ is sufficiently small. Therefore, the initial vector $Y^0$ must be carefully given. In cases in which the initial vector is not well known or not well given, the linearisation process must be repeated. In other words, the initial vector that is not well known must be modified by the solved vector $dY$, and the linearisation process must be repeated until $dY$ converges. If $X = 0$, then $L = V$; therefore, the "observable" vector $L$ is sometimes also called a residual vector. If the initial vector $Y^0$ is well known or well given, then the residual vector $V$ can also be used as a criterion to judge the "goodness or badness" of the original observable vector. This property is used in robust Kalman filtering to adjust the weight of the observable (cf. Chap. 7).

## 6.3
## Partial Derivatives of Observation Function

### Partial Derivatives of Geometric Path Distance with Respect to the State Vector $(x_i, y_i, z_i, \dot{x}_i, \dot{y}_i, \dot{z}_i)$ of the GPS Receiver

The signal transmitting path is described by (cf. Eqs. 4.3 and 4.6 in Chap. 4)

$$\rho_i^k(t_r, t_e) = \sqrt{(x_k(t_e) - x_i)^2 + (y_k(t_e) - y_i)^2 + (z_k(t_e) - z_i)^2}, \quad \text{and} \quad (6.13)$$

$$\rho_i^k(t_r, t_e) \approx \rho_i^k(t_r, t_r) + \frac{d\rho_i^k(t_r, t_r)}{dt} \Delta t, \quad (6.14)$$

where index $k$ denotes the satellite, and the satellite coordinates are related to the signal emission time $t_e$, $i$ denotes the station, and station coordinates are related to the signal reception time $t_r$, $\Delta t = t_e - t_r$. Then one has

$$\frac{d\rho_i^k(t_r, t_r)}{dt} = \frac{1}{\rho_i^k(t_r, t_r)}$$
$$\times \left( (x_k - x_i)(\dot{x}_k - \dot{x}_i) \quad (y_k - y_i)(\dot{y}_k - \dot{y}_i) \quad (z_k - z_i)(\dot{z}_k - \dot{z}_i) \right), \quad (6.15)$$

where the satellite state vector is related to the time $t_r$, and

$$\frac{\partial \rho_i^k(t_r, t_e)}{\partial(x_i, y_i, z_i)} = \frac{-1}{\rho_i^k(t_r, t_e)} (x_k - x_i \quad y_k - y_i \quad z_k - z_i), \quad (6.16)$$

$$\frac{\partial \rho_i^k(t_r, t_e)}{\partial(\dot{x}_i, \dot{y}_i, \dot{z}_i)} = \frac{-\Delta t}{\rho_i^k(t_r, t_r)} (x_k - x_i \quad y_k - y_i \quad z_k - z_i). \quad (6.17)$$

### Partial Derivatives of Geometric Path Distance with Respect to the State Vector $(x_k, y_k, z_k, \dot{x}_k, \dot{y}_k, \dot{z}_k)$ of the GPS Satellite

Similar to above, one has

$$\frac{\partial \rho_i^k(t_r, t_e)}{\partial (x_k, y_k, z_k)} = \frac{1}{\rho_i^k(t_r, t_e)} \begin{pmatrix} x_k - x_i & y_k - y_i & z_k - z_i \end{pmatrix}, \quad (6.18)$$

$$\frac{\partial \rho_i^k(t_r, t_e)}{\partial (\dot{x}_k, \dot{y}_k, \dot{z}_k)} = \frac{\Delta t}{\rho_i^k(t_r, t_r)} \begin{pmatrix} x_k - x_i & y_k - y_i & z_k - z_i \end{pmatrix}. \quad (6.19)$$

***Partial Derivatives of the Doppler Observable with Respect to the Velocity Vector of the Station***

The time differentiation of the geometric signal path distance can be derived as

$$\frac{d\rho_i^k(t_r, t_e)}{dt} = \frac{1}{\rho_i^k(t_r, t_e)}$$
$$((x_k(t_e) - x_i)(\dot{x}_k(t_e) - \dot{x}_i) + (y_k(t_e) - y_i)(\dot{y}_k(t_e) - \dot{y}_i) + (z_k(t_e) - z_i)(\dot{z}_k(t_e) - \dot{z}_i)); \quad (6.20)$$

then one has

$$\frac{\partial (d\rho_i^k(t_r, t_e)/dt)}{\partial (\dot{x}_i, \dot{y}_i, \dot{z}_i)} = \frac{-1}{\rho_i^k(t_r, t_e)} \begin{pmatrix} x_k(t_e) - x_i & y_k(t_e) - y_i & z_k(t_e) - z_i \end{pmatrix}. \quad (6.21)$$

***Partial Derivatives of Clock Errors with Respect to the Clock Parameters***

If the clock errors are modelled by Eq. 5.163 (cf. Sect. 5.5)

$$\delta t_i = b_i + d_i t + e_i t^2, \quad \delta t_k = b_k + d_k t + e_k t^2, \quad (6.22)$$

where $i$ and $k$ are the indices of the clock error parameters of the receiver and satellite, then one has

$$\frac{\partial \delta t_i}{\partial (b_i, d_i, e_i)} = \begin{pmatrix} 1 & t & t^2 \end{pmatrix} \text{ and}$$
$$\frac{\partial \delta t_k}{\partial (b_k, d_k, e_k)} = \begin{pmatrix} 1 & t & t^2 \end{pmatrix}. \quad (6.23)$$

If the clock errors are modelled by Eq. 5.164 (cf. Sect. 5.5)

$$\delta t_i = b_i, \quad \delta t_k = b_k, \quad (6.24)$$

then

$$\frac{\partial \delta t_i}{\partial b_i} = 1, \quad \frac{\partial \delta t_k}{\partial b_k} = 1. \tag{6.25}$$

The above derivatives are valid for both the code and phase observable equations. For the Doppler observable, denote (cf. Eq. 6.3)

$$\delta_{\text{clock}} = f \frac{d(\delta t_i - \delta t_k)}{dt}, \tag{6.26}$$

then for the clock error model of Eq. 6.22 one has

$$\frac{\partial \delta_{\text{clock}}}{\partial (d_i, e_i)} = (1 \quad 2t)f \quad \text{and} \quad \frac{\partial \delta_{\text{clock}}}{\partial (d_k, e_k)} = (1 \quad 2t)f. \tag{6.27}$$

*Partial Derivatives of Tropospheric Effects with Respect to the Tropospheric Parameters*

If the tropospheric effects can be modelled by (cf. Sect. 5.2)

$$\begin{aligned} \text{I}: & \quad \delta_{\text{trop}} = f_p d\rho \quad \text{and} \\ \text{II}: & \quad \delta_{\text{trop}} = \frac{f_z d\rho}{F} + \frac{f_a d\rho}{F_c}, \end{aligned} \tag{6.28}$$

where $d\rho$ is the tropospheric effect computed by using the standard tropospheric model, $f_p$, $f_z$, $f_a$ are parameters of the tropospheric delay in path, zenith, azimuth directions, and $F$ and $F_c$ are the mapping and co-mapping functions discussed in Sect. 5.2. The derivatives with respect to the parameters $f_p$, $f_z$, $f_a$ are then

$$\begin{aligned} \text{I}: & \quad \frac{\partial \delta_{\text{trop}}}{\partial f_p} = d\rho \quad \text{and} \\ \text{II}: & \quad \frac{\partial \delta_{\text{trop}}}{\partial (f_z, f_a)} = \left( \frac{d\rho}{F} \quad \frac{d\rho}{F_c} \right). \end{aligned} \tag{6.29}$$

Furthermore, if the tropospheric parameters are defined as a step function or first-order polynomial (cf. Sect. 5.2) by

$$\begin{aligned} \text{I}: & \quad f_p = f_z = f_j \quad \text{if} \quad t_{j-1} < t \leq t_j, \quad j = 1, 2, \ldots, n \quad \text{and} \\ \text{II}: & \quad f_p = f_z = f_{j-1} + (f_j - f_{j-1})\frac{t - t_{j-1}}{\Delta t} \quad \text{if} \quad t_{j-1} < t \leq t_j, \quad j = 1, 2, \ldots, n+1, \end{aligned} \tag{6.30}$$

where $\Delta t = (t_n - t_0)/n$, $t_0$ and $t_n$ are the beginning and the ending times of the GPS survey, and $\Delta t$ is usually selected by 2–4 h. Then one has

$$\text{I}: \quad \frac{\partial f_p}{\partial f_j} = \frac{\partial f_z}{\partial f_j} = 1 \quad \text{and}$$

$$\text{II}: \quad \frac{\partial f_p}{\partial (f_{j-1}, f_j)} = \frac{\partial f_z}{\partial (f_{j-1}, f_j)} = \left(1 + \frac{-t + t_{j-1}}{\Delta t} \quad \frac{t - t_{j-1}}{\Delta t}\right). \tag{6.31}$$

The azimuth dependence may be assumed to be (cf. Eq. 5.121)

$$f_a = g_1 \cos a + g_2 \sin a, \tag{6.32}$$

where $a$ is the azimuth, and $g_1$ and $g_2$ are called azimuth-dependent parameters. Then one gets

$$\frac{\partial f_a}{\partial (g_1, g_2)} = (\cos a \quad \sin a). \tag{6.33}$$

If parameters $g_1$ and $g_2$ are also defined as step functions or first-order polynomials like Eq. 6.30, the partial derivatives can be obtained in a similar manner to Eq. 6.31.

### Partial Derivatives of the Phase Observable with Respect to the Ambiguity Parameters

Depending on the scale that one prefers, there is

$$\frac{\partial \lambda N}{\partial \lambda N} = 1 \quad \text{or} \quad \frac{\partial \lambda N}{\partial N} = \lambda. \tag{6.34}$$

### Partial Derivatives of Tidal Effects with Respect to the Tidal Parameters

If the earth tide model in Eqs. 5.147 and 5.149 is used, then the tidal effects can generally be written as

$$\delta_{\text{earth-tide}} = s_1 h_2 + s_2 l_2 + s_3 h_3, \tag{6.35}$$

where $s_1$, $s_2$, and $s_3$ are the coefficient functions, which are given in Sect. 5.4.2 in detail, and $h_2$, $h_3$, and $l_2$ are the love numbers and Shida number, respectively. Then one has

$$\frac{\partial \delta_{\text{earth-tide}}}{\partial (h_2, l_2, h_3)} = (s_1 \quad s_2 \quad s_3). \tag{6.36}$$

Ocean loading tide effects can be modelled as

$$\delta_{\text{loading-tide}} = f_{\text{load}} ( \, dx_{\text{load}} \quad dy_{\text{load}} \quad dz_{\text{load}} \, ), \tag{6.37}$$

where $f_{\text{load}}$ is the factor of the computed ocean loading effect vector ($dx_{\text{load}}$ $dy_{\text{load}}$ $dz_{\text{load}}$). Then one has

$$\frac{\partial \delta_{\text{loading-tide}}}{\partial f_{\text{load}}} = ( \, dx_{\text{load}} \quad dy_{\text{load}} \quad dz_{\text{load}} \, ). \tag{6.38}$$

## 6.4
## Linear Transformation and Covariance Propagation

For any linear equation system

$$L = AX \tag{6.39}$$

or

$$\begin{pmatrix} l_1 \\ l_2 \\ \vdots \\ l_m \end{pmatrix} = \begin{pmatrix} a_{11} & a_{12} & \cdots & a_{1n} \\ a_{21} & a_{22} & \cdots & a_{2n} \\ \vdots & \vdots & \vdots & \vdots \\ a_{m1} & a_{m2} & \cdots & a_{mn} \end{pmatrix} \begin{pmatrix} x_1 \\ x_2 \\ \vdots \\ x_n \end{pmatrix},$$

a linear transformation can be defined as a multiplying operation of matrix $T$ to Eq. 6.39, i.e.

$$TL = TAX \tag{6.40}$$

or

$$\begin{pmatrix} t_{11} & t_{12} & \cdots & t_{1m} \\ t_{21} & t_{22} & \cdots & t_{2m} \\ \vdots & \vdots & \vdots & \vdots \\ t_{k1} & t_{k2} & \cdots & t_{km} \end{pmatrix} \begin{pmatrix} l_1 \\ l_2 \\ \vdots \\ l_m \end{pmatrix} = \begin{pmatrix} t_{11} & t_{12} & \cdots & t_{1m} \\ t_{21} & t_{22} & \cdots & t_{2m} \\ \vdots & \vdots & \vdots & \vdots \\ t_{k1} & t_{k2} & \cdots & t_{km} \end{pmatrix} \begin{pmatrix} a_{11} & a_{12} & \cdots & a_{1n} \\ a_{21} & a_{22} & \cdots & a_{2n} \\ \vdots & \vdots & \vdots & \vdots \\ a_{m1} & a_{m2} & \cdots & a_{mn} \end{pmatrix} \begin{pmatrix} x_1 \\ x_2 \\ \vdots \\ x_n \end{pmatrix},$$

where $T$ is called the linear transformation matrix and has a dimension of $k \times m$. An inverse transformation of $T$ is denoted by $T^{-1}$. An invertible linear transformation does not change the property (and solutions) of the original linear equations. This may be verified by multiplying $T^{-1}$ to Eq. 6.40. A non-invertible linear transformation is called a rank-deficient (or not full rank) transformation.

The covariance matrix of $L$ is denoted by cov($L$) or $Q_{LL}$ (cf. Sect. 6.2); the covariance of the transformed $L$ (i.e. $TL$) can then be obtained by the covariance propagation theorem by (cf., e.g., Koch 1988)

$$\operatorname{cov}(TL) = T \operatorname{cov}(L) T^{\mathrm{T}} = T Q_{LL} T^{\mathrm{T}}, \qquad (6.41)$$

where superscript $^{\mathrm{T}}$ denotes the transpose of the transformation matrix.

If transformation matrix $T$ is a vector (i.e. $k = 1$) and $L$ is an inhomogeneous and independent observable vector (i.e. covariance matrix $Q_{LL}$ is a diagonal matrix with elements of $\sigma_j^2$, where $\sigma_j^2$ is the variance ($\sigma_j$ is called standard deviation) of the observable $l_j$), then Eqs. 6.40 and 6.41 can be written as

$$(t_1 \; t_2 \; \ldots \; t_m) \begin{pmatrix} l_1 \\ l_2 \\ \vdots \\ l_m \end{pmatrix} = (t_1 \; t_2 \; \ldots \; t_m) \begin{pmatrix} a_{11} & a_{12} & \ldots & a_{1n} \\ a_{21} & a_{22} & \ldots & a_{2n} \\ \vdots & \vdots & \vdots & \vdots \\ a_{m1} & a_{m2} & \ldots & a_{mn} \end{pmatrix} \begin{pmatrix} x_1 \\ x_2 \\ \vdots \\ x_n \end{pmatrix} \quad \text{and}$$

$$\operatorname{cov}(TL) = (t_1 \; t_2 \; \ldots \; t_m) \begin{pmatrix} \sigma_1^2 & 0 & \ldots & 0 \\ 0 & \sigma_2^2 & \ldots & 0 \\ \vdots & \vdots & \vdots & \vdots \\ 0 & 0 & \ldots & \sigma_m^2 \end{pmatrix} \begin{pmatrix} t_1 \\ t_2 \\ \vdots \\ t_m \end{pmatrix}.$$

$$(6.42)$$

Denoting cov($TL$) as $\sigma_{TL}^2$, one gets

$$\sigma_{TL}^2 = t_1^2 \sigma_1^2 + t_2^2 \sigma_2^2 + \ldots + t_m^2 \sigma_m^2 = \sum_{j=1}^m t_j^2 \sigma_j^2. \qquad (6.43)$$

Equation 6.43 is called the error propagation theorem.

## 6.5
## Data Combinations

Data combinations are methods of combining GPS data measured with the same receiver at the same station. Generally, the observables are the code pseudoranges, carrier phases and Doppler at working frequencies such as C/A code, $P_1$ and $P_2$ codes, L1 phase $\Phi_1$ and L2 phase $\Phi_2$, and Doppler $D_1$ and $D_2$. In the future, there will also be $P_5$ code, L5 phase $\Phi_5$ and Doppler $D_5$. According to the observation equations of the observables, a suitable combination can be advantageous for understanding and solving GPS problems.

For convenience, the code, phase, and Doppler observables are simplified and rewritten as (cf. Eqs. 6.1–6.3)

$$R_j = \rho - (\delta t_{\mathrm{r}} - \delta t_k) c + \delta_{\mathrm{ion}}(j) + \delta_{\mathrm{trop}} + \delta_{\mathrm{tide}} + \delta_{\mathrm{rel}} + \varepsilon_{\mathrm{c}}, \qquad (6.44)$$

$$\lambda_j \Phi_j = \rho - (\delta t_r - \delta t_k)c + \lambda_j N_j - \delta_{\text{ion}}(j) + \delta_{\text{trop}} + \delta_{\text{tide}} + \delta_{\text{rel}} + \varepsilon_p, \quad (6.45)$$

$$D_j = \frac{d\rho}{\lambda_j dt} - f_j \frac{d(\delta t_r - \delta t_k)}{dt} + \varepsilon_d, \quad \text{and} \quad (6.46)$$

$$\delta_{\text{ion}}(j) = \frac{A_1}{f_j^2} + \frac{A_2}{f_j^3}. \quad (6.47)$$

where $j$ is the index of frequency $f$, the means of the other symbols are the same as the notes of Eqs. 6.1–6.3. Equation 6.47 is an approximation for code.

A general code–code combination can be formed by $n_1 R_1 + n_2 R_2 + n_5 R_5$, where $n_1$, $n_2$, and $n_5$ are arbitrary constants. However, in order to make such a combination that still has the sense of a code survey, a standardised combination has to be formed by

$$R = \frac{n_1 R_1 + n_2 R_2 + n_5 R_5}{n_1 + n_2 + n_5}. \quad (6.48)$$

The newly formed code $R$ can then be interpreted as a weight-averaged code survey of $R_1$, $R_2$, and $R_5$. The mathematical model of the observable Eq. 6.44 is generally still valid for $R$. Denoting the standard deviation of code observable $R_i$ as $\sigma_{ci}(i = 1, 2, 5)$, the newly-formed code observation $R$ has the variance of

$$\sigma_c^2 = \frac{1}{(n_1 + n_2 + n_5)^2} \left( n_1^2 \sigma_{c1}^2 + n_2^2 \sigma_{c2}^2 + n_5^2 \sigma_{c5}^2 \right).$$

Because

$$\left| \frac{n_1 + n_2 + \ldots + n_m}{m} \right| \leq \sqrt{\frac{n_1^2 + n_2^2 + \ldots + n_m^2}{m}},$$

(cf., e.g., Wang et al. 1979; Bronstein and Semendjajew 1987), one has the property of

$$(n_1 + n_2 + \ldots + n_m)^2 \leq m(n_1^2 + n_2^2 + \ldots + n_m^2),$$

where $m$ is the maximum index. Therefore, in our case, one has

$$\sigma_c^2 \geq m \cdot \min\{\sigma_{c1}^2, \sigma_{c2}^2, \sigma_{c5}^2\}, \quad m = 2 \text{ or } 3$$

for combinations of two or three code observables.

A general phase–phase linear combination can be formed by

$$\Phi = n_1 \Phi_1 + n_2 \Phi_2 + n_5 \Phi_5, \quad (6.49)$$

where the combined signal has the frequency and wavelength

$$f = n_1 f_1 + n_2 f_2 + n_5 f_5 \quad \text{and} \quad \lambda = \frac{c}{f}. \tag{6.50}$$

$\lambda \Phi$ means the measured distance (with ambiguity!) and can be presented alternatively as

$$\lambda \Phi = \frac{1}{f}(n_1 f_1 \lambda_1 \Phi_1 + n_2 f_2 \lambda_2 \Phi_2 + n_5 f_5 \lambda_5 \Phi_5). \tag{6.51}$$

The mathematical model of Eq. 6.45 is generally still valid for the newly formed $\lambda \Phi$. Denoting the standard deviation of phase observable $\lambda_i \Phi$ as $\sigma_i$ ($i = 1, 2, 5$), the newly formed observation has a variance of

$$\sigma^2 = \frac{1}{f^2}\left(n_1^2 f_1^2 \sigma_1^2 + n_2^2 f_2^2 \sigma_2^2 + n_5^2 f_5^2 \sigma_5^2\right) \tag{6.52}$$

and

$$\sigma^2 \geq m \cdot \min\{\sigma_1^2, \sigma_2^2, \sigma_5^2\},$$

with $m = 2$ or $3$ for combinations of two or three phases.

That is, the data combination will degrade the quality of the original data.

Linear combinations $\Phi_W = \Phi_1 - \Phi_2$ and $\Phi_X = 2\Phi_1 - \Phi_2$ are called wide-lane and x-lane combinations with wavelengths of about 86.2 and 15.5 cm. They reduce the first-order ionospheric effects on frequency $f_2$ to 40 % and 20 %, respectively. $\Phi_N = \Phi_1 + \Phi_2$ is called a narrow-lane combination.

### 6.5.1
### Ionosphere-Free Combinations

Due to Eqs. 6.44–6.47, phase–phase and code–code ionosphere-free combinations can be formed by (cf. Sect. 5.1)

$$\lambda \Phi = \frac{f_1^2 \lambda_1 \Phi_1 - f_2^2 \lambda_2 \Phi_2}{f_1^2 - f_2^2} = \lambda(f_1 \Phi_1 - f_2 \Phi_2) \quad \text{and} \tag{6.53}$$

$$R = \frac{f_1^2 R_1 - f_2^2 R_2}{f_1^2 - f_2^2}. \tag{6.54}$$

The related observation equations can be formed from Eqs. 6.44 and 6.45 as

$$R = \rho - (\delta t_r - \delta t_k)c + \delta_{\text{trop}} + \delta_{\text{tide}} + \delta_{\text{rel}} + \varepsilon_{\text{cc}} \quad \text{and} \tag{6.55}$$

$$\lambda \Phi = \rho - (\delta t_r - \delta t_k)c + \lambda N + \delta_{\text{trop}} + \delta_{\text{tide}} + \delta_{\text{rel}} + \varepsilon_{\text{pc}}, \tag{6.56}$$

where

$$N = f_1 N_1 - f_2 N_2, \quad \lambda = \frac{c}{f_1^2 - f_2^2}, \tag{6.57}$$

$\varepsilon_{\text{cc}}$ and $\varepsilon_{\text{pc}}$ denote the residuals after the combination of code and phase, respectively.

The advantages of such ionosphere-free combinations are that the ionospheric effects have disappeared from the observation Eqs. 6.55 and 6.56 and the other terms of the equations have remained the same. However, the combined ambiguity is not an integer anymore, and the combined observables have higher standard deviations. Equations 6.55 and 6.56 are indeed first-order ionosphere-free combinations.

Second-order ionosphere-free combinations can be formed by (see Sect. 5.1.2 for details)

$$\lambda \Phi = C_1 \lambda_1 \Phi_1 + C_2 \lambda_2 \Phi_2 + C_5 \lambda_5 \Phi_5 \quad \text{and} \tag{6.58}$$

$$R = C_1 R_1 + C_2 R_2 + C_5 R_5, \tag{6.59}$$

where

$$C_1 = \frac{f_1^3(f_5 - f_2)}{C_4}, \quad C_2 = \frac{-f_2^3(f_5 - f_1)}{C_4},$$

$$C_5 = \frac{f_5^3(f_2 - f_1)}{C_4}, \quad C_4 = f_1^3(f_5 - f_2) - f_2^3(f_5 - f_1) + f_5^3(f_2 - f_1),$$

$$\lambda = \frac{c}{C_4}, \quad N = C_4(C_1 N_1 + C_2 N_2 + C_5 N_5).$$

The related observation equations are the same as Eqs. 6.55 and 6.56, with $\lambda$ and $N$ given above.

## 6.5.2
## Geometry-Free Combinations

Given Eqs. 6.44–6.46, code–code, phase–phase, and phase–code geometry-free combinations can be formed by

$$R_1 - R_2 = \delta_{\text{ion}}(1) - \delta_{\text{ion}}(2) + \Delta\varepsilon_c = \frac{A_1}{f_1^2} - \frac{A_1}{f_2^2} + \Delta\varepsilon_c, \tag{6.60}$$

$$\lambda_1\Phi_1 - \lambda_2\Phi_2 = \lambda_1 N_1 - \lambda_2 N_2 - \frac{A_1}{f_1^2} + \frac{A_1}{f_2^2} + \Delta\varepsilon_p, \tag{6.61}$$

$$\lambda_1 D_1 - \lambda_2 D_2 = \Delta\varepsilon_d, \tag{6.62}$$

$$\lambda_j \Phi_j - R_j = \lambda_j N_j - 2\delta_{\text{ion}}(j) + \Delta\varepsilon_{\text{pc}}, \quad \text{and} \quad j = 1, 2, 5, \tag{6.63}$$

where

$$\Delta\delta_{\text{ion}} = \delta_{\text{ion}}(1) - \delta_{\text{ion}}(2) = \frac{A_1}{f_1^2} - \frac{A_1}{f_2^2}. \tag{6.64}$$

For an ionospheric model of the second order, one has approximately

$$\Delta\delta_{\text{ion}} = \delta_{\text{ion}}(1) - \delta_{\text{ion}}(2) = \frac{A_1}{f_1^2} - \frac{A_1}{f_2^2} + \frac{A_2}{f_1^3} - \frac{A_2}{f_2^3}.$$

The geometry-free code–code and phase–phase combinations cancel out all other terms in the observation equations except the ionospheric term and the ambiguity parameters. Recalling the discussions of Sect. 5.1, $\delta_{\text{ion}}$ is the ionospheric path delay and can be considered a mapping of the zenith delay $\delta_{\text{ion}}^z$ or $\delta_{\text{ion}} = \delta_{\text{ion}}^z F$, where $F$ is the mapping function (cf. Sect. 5.1). So one has

$$\delta_{\text{ion}}(1) = \frac{A_1^z}{f_1^2} F = \frac{A_1}{f_1^2}, \tag{6.65}$$

where $A_1$ and $A_1^z$ have the physical meaning of total electron content at the signal path direction and the zenith direction, respectively. $A_1^z$ is then independent from the zenith angle of the satellite. If the variability of the electron content at the zenith direction is stable enough, $A_1^z$ can be modelled by a step function or a first-order polynomial with a reasonably short time interval $\Delta t$ by

$$A_1^z = g_j \quad \text{if} \quad t_{j-1} < t \leq t_j, \quad j = 1, 2, \ldots, n+1 \tag{6.66}$$

or

$$A_1^z = g_{j-1} + (g_j - g_{j-1})\frac{t - t_{j-1}}{\Delta t} \quad \text{if} \quad t_{j-1} < t \leq t_j, j = 1, 2, \ldots, n+1, \tag{6.67}$$

where $\Delta t = (t_n - t_0)/n$, and $t_0$ and $t_n$ are the beginning and ending time of the GPS survey. $\Delta t$ can be selected, for example, as 30 min. $g_j$ is the coefficient of the polynomial.

## 6.5 · Data Combinations

Geometry-free combinations of Eqs. 6.60, 6.61, and 6.63 (only for $j = 1$) can be considered a linear transformation of the original observable vector $L = (R_1 \ R_2 \ \lambda_1 \Phi_1 \ \lambda_2 \Phi_2)^T$ by

$$\begin{pmatrix} 1 & -1 & 0 & 0 \\ 0 & 0 & 1 & -1 \\ -1 & 0 & 1 & 0 \end{pmatrix} \cdot \begin{pmatrix} R_1 \\ R_2 \\ \lambda_1 \Phi_1 \\ \lambda_2 \Phi_2 \end{pmatrix} = \begin{pmatrix} 0 & 0 & g \\ \lambda_1 & -\lambda_2 & -g \\ \lambda_1 & 0 & d \end{pmatrix} \cdot \begin{pmatrix} N_1 \\ N_2 \\ A_1 \end{pmatrix} + \begin{pmatrix} \Delta \varepsilon_c \\ \Delta \varepsilon_p \\ \Delta \varepsilon_{pc} \end{pmatrix}, \tag{6.68}$$

where Eq. 6.65 is used and

$$g = \left( \frac{1}{f_1^2} - \frac{1}{f_2^2} \right), \quad d = -\frac{2}{f_1^2} \quad \text{and} \quad T = \begin{pmatrix} 1 & -1 & 0 & 0 \\ 0 & 0 & 1 & -1 \\ -1 & 0 & 1 & 0 \end{pmatrix}.$$

Equation 6.68 is called an ambiguity-ionospheric equation. For any viewed GPS satellite, Eq. 6.68 is solvable. If the variance vector of the observable vector is

$$\begin{pmatrix} \sigma_c^2 & \sigma_c^2 & \sigma_p^2 & \sigma_p^2 \end{pmatrix}^T,$$

then the covariance matrix of the original observable vector is (cf. Sect. 6.2)

$$Q_{LL} = \begin{pmatrix} \sigma_c^2 & 0 & 0 & 0 \\ 0 & \sigma_c^2 & 0 & 0 \\ 0 & 0 & \sigma_p^2 & 0 \\ 0 & 0 & 0 & \sigma_p^2 \end{pmatrix},$$

and the covariance matrix of the transformed observable vector (left side of Eq. 6.68) is (cf. Sect. 6.4)

$$\text{cov}(TL) = T Q_{LL} T^T = \begin{pmatrix} 2\sigma_c^2 & 0 & -\sigma_c^2 \\ 0 & 2\sigma_p^2 & \sigma_p^2 \\ -\sigma_c^2 & \sigma_p^2 & \sigma_c^2 + \sigma_p^2 \end{pmatrix},$$

and

$$P = (\text{cov}(TL))^{-1} = \frac{1}{2} \begin{pmatrix} h + \sigma_c^{-2} & -h & 2h \\ -h & h + \sigma_p^{-2} & -2h \\ 2h & -2h & 4h \end{pmatrix}, \quad h = \frac{1}{\sigma_c^2 + \sigma_p^2}. \tag{6.69}$$

Taking all measured data at a station into account, the ambiguity and ionospheric parameters (as a step function of the polynomial) can be solved by using Eq. 6.68 with the weight of Eq. 6.69. Taking into account the data station by station, all ambiguity and ionospheric parameters can be determined. The different weights of the code and phase measurements are considered exactly here. Because of the physical property of the ionosphere, all solved ionospheric parameters will have the same sign. Even though observation Eq. 6.68 is already a linear equation system, an initialisation is still helpful to avoid numbers from ambiguities that are too large. The broadcasting ionospheric model can be used for initialisation of the related ionospheric parameters.

A geometry-free combination of Eq. 6.62 can be used as a quality check of the Doppler data.

### 6.5.3
### Standard Phase–Code Combination

Traditionally, phase and code combinations are used to compute the wide-lane ambiguity (cf. Sjoeberg 1999; Hofmann-Wellenhof et al. 1997). The formulas can be derived as follows. Dividing $\lambda_j$ into Eq. 6.63 and forming the difference for $j = 1$ and $j = 2$, one gets

$$\Phi_w - \frac{R_1}{\lambda_1} + \frac{R_2}{\lambda_2} = N_w - \frac{2A_1}{c}\left(\frac{1}{f_1} + \frac{1}{f_2}\right), \tag{6.70}$$

where $\Phi_w = \Phi_1 - \Phi_2$, $N_w = N_1 - N_2$, and they are called wide-lane observable and ambiguity; $c$ is the velocity of light and $A_1$ is the ionospheric parameter. The error term is omitted here. Equation 6.60 can be rewritten (by omitting the error term) as

$$A_1 = (R_1 - R_2)\frac{f_1^2 f_2^2}{f_2^2 - f_1^2}, \tag{6.71}$$

and then one gets

$$\frac{A_1}{c}\left(\frac{1}{f_1} - \frac{1}{f_2}\right) = \left(\frac{R_1}{\lambda_1 f_1} - \frac{R_2}{\lambda_2 f_2}\right)\frac{f_1 f_2}{f_2 + f_1} = \frac{R_1}{\lambda_1}\frac{f_2}{(f_1 + f_2)} - \frac{R_2}{\lambda_2}\frac{f_1}{(f_1 + f_2)}. \tag{6.72}$$

Substituting Eq. 6.72 into 6.70 yields

$$N_w = \Phi_w - \frac{f_1 - f_2}{f_1 + f_2}\left(\frac{R_1}{\lambda_1} + \frac{R_2}{\lambda_2}\right). \tag{6.73}$$

Equation 6.73 is the most popular formula for computing wide-lane ambiguities using phase and code observables. The undifferenced ambiguity $N_1$ can be derived as follows. Setting $\Phi_2 = \Phi_1 - \Phi_w$, $N_2 = N_1 - N_w$ into Eq. 6.61 and omitting the error term, one has

$$\lambda_1 N_1 - \lambda_2(N_1 - N_w) = \frac{A_1}{f_1^2} - \frac{A_1}{f_2^2} + \lambda_1 \Phi_1 - \lambda_2(\Phi_1 - \Phi_w),$$

$$N_1 = \Phi_1 - (\Phi_w - N_w)\frac{f_1}{f_w} + \frac{A_1 f_1 + f_2}{c \; f_1 f_2}$$

or

$$N_1 = \Phi_1 - (\Phi_w - N_w)\frac{f_1}{f_w} - \frac{R_1 f_2}{\lambda_1 f_w} + \frac{R_2 f_1}{\lambda_2 f_w}, \tag{6.74}$$

where $f_w = f_1 - f_2$ is the wide-lane frequency.

Compared with the adjustment method derived in Sect. 6.5.2, it is obvious that the quality differences of the phase and code data are not considered using Eqs. 6.73 and 6.74 for determining the ambiguity parameters. Therefore, we suggest that the method proposed in Sect. 6.5.2 be used.

## 6.5.4
## Ionospheric Residuals

Considering the GPS observables as a time series, the geometry-free combinations of Eqs. 6.60–6.64 can be rewritten as

$$R_1(t_j) - R_2(t_j) = \Delta\delta_{\text{ion}}(t_j) + \Delta\varepsilon_c, \tag{6.75}$$

$$\lambda_1 \Phi_1(t_j) - \lambda_2 \Phi_2(t_j) = \lambda_1 N_1 - \lambda_2 N_2 - \Delta\delta_{\text{ion}}(t_j) + \Delta\varepsilon \quad \text{and} \tag{6.76}$$

$$\lambda_i \Phi_i(t_j) - R_i(t_j) = \lambda_i N_i - 2\delta_{\text{ion}}(i, t_j) + \Delta\varepsilon_{\text{pc}}, \quad i = 1, 2, 5, \tag{6.77}$$

where

$$\Delta\delta_{\text{ion}}(t_j) = \delta_{\text{ion}}(1, t_j) - \delta_{\text{ion}}(2, t_j) = \frac{A_1(t_j)}{f_1^2} - \frac{A_1(t_j)}{f_2^2}, \quad j = 1, 2, \ldots, m. \tag{6.78}$$

The differences of the above observable combinations at the two consecutive epochs $t_j$ and $t_{j-1}$ can be formed as

$$\Delta_t R_1(t_j) - \Delta_t R_2(t_j) = \Delta_t \Delta\delta_{\text{ion}}(t_j) + \Delta_t \Delta\varepsilon_c, \tag{6.79}$$

$$\lambda_1 \Delta_t \Phi_1(t_j) - \lambda_2 \Delta_t \Phi_2(t_j) = \lambda_1 \Delta_t N_1 - \lambda_2 \Delta_t N_2 - \Delta_t \Delta\delta_{\text{ion}}(t_j) + \Delta_t \Delta\varepsilon_p, \quad \text{and} \tag{6.80}$$

$$\lambda_i \Delta_t \Phi_i(t_j) - \Delta_t R_i(t_j) = \lambda_i \Delta_t N_i - 2\Delta_t \delta_{\text{ion}}(i, t_j) + \Delta_t \Delta\varepsilon_{\text{pc}}, \quad i = 1, 2, 5, \tag{6.81}$$

where $\Delta_t$ is a time difference operator, and for any time function $G(t)$, $\Delta_t G(t_j) = G(t_j)-G(t_{j-1})$ is valid.

Because the time differences of the ionospheric effects $\Delta_t \delta_{\text{ion}}$ and $\Delta_t \Delta\delta_{\text{ion}}$ are generally very small, they are called ionospheric residuals. In the case of no cycle slips, i.e. ambiguities $N_1$ and $N_2$ are constant, $\Delta N_1$ and $\Delta N_2$ equal zero. Equations 6.79–6.81 are called ionospheric residual combinations. The first combination of Eq. 6.79 can be used for a consistency check of two code measurements. Equations 6.80 and 6.81 can be used for a cycle slip check. Equation 6.81 is a phase–code combination, due to the lower accuracy of the code measurements; it can be used only to check for large cycle slips. Equation 6.80 is a phase–phase combination, and therefore it has higher sensitivity to cycle slips. However, two special cycle slips, $\Delta N_1$ and $\Delta N_2$, can lead to a very small combination of $\delta_1 \Delta_t N_1 - \delta_2 \Delta_t N_2$. Examples of such combinations can be found in Hofmann-Wellenhof et al. (1997). Thus even the ionospheric residual of Eq. 6.80 is very small; it may not guarantee that there are no cycle slips.

## 6.5.5
## Differential Doppler and Doppler Integration

### Differential Doppler

The numerical differentiation of the original observables given in Eqs. 6.44 and 6.45 at the two consecutive epochs $t_j$ and $t_{j-1}$ can be formed as

$$\frac{\Delta_t R_j}{\lambda_j \Delta t} = \frac{\Delta_t \rho}{\lambda_j \Delta t} - f_j \frac{\Delta_t(\delta t_r - \delta t_k)}{\Delta t} + \frac{\Delta_t \varepsilon_c}{\lambda_j \Delta t}, \quad j = 1, 2, \quad \text{and} \quad (6.82)$$

$$\frac{\Delta_t \Phi_j}{\Delta t} = \frac{\Delta_t \rho}{\lambda_j \Delta t} - f_j \frac{\Delta_t(\delta t_r - \delta t_k)}{\Delta t} + \frac{\Delta_t \varepsilon_p}{\lambda_j \Delta t}, \quad j = 1, 2, \quad (6.83)$$

where $\Delta_t/\Delta t$ is a numerical differentiation operator and $\Delta t = t_j - t_{j-1}$.

The left-hand side of Eq. 6.83 is called differential Doppler. Ionospheric residuals are negligible and are omitted here. The third terms of Eqs. 6.82 and 6.83 on the right-hand side are small residual errors. For convenience of comparison, the Doppler observable model of Eq. 6.46 is copied below:

$$D_j = \frac{d\rho}{\lambda_j dt} - f_j \frac{d(\delta t_r - \delta t_k)}{dt} + \varepsilon_d. \quad (6.84)$$

It is clear that Eqs. 6.83 and 6.84 are nearly the same. The only difference is that in Doppler Eq. 6.84, the observed Doppler is an instantaneous one, and its model is presented by theoretical differentiation, whereas the term on the left-hand side of Eq. 6.83 is the numerically differenced Doppler (formed by phases), and its model is presented by numerical differentiation. Doppler measurement measures the instantaneous motion of the GPS antenna, whereas differential Doppler describes a kind of average velocity of the antenna over two consecutive epochs. The velocity solution of Eq. 6.83 (denoted by $(\dot{x} \; \dot{y} \; \dot{z})^T$) can be used to predict the future kinematic position by

$$\begin{pmatrix} x_{j+1} \\ y_{j+1} \\ z_{j+1} \end{pmatrix} = \begin{pmatrix} x_j \\ y_j \\ z_j \end{pmatrix} + \begin{pmatrix} \dot{x}_j \\ \dot{y}_j \\ \dot{z}_j \end{pmatrix} \cdot \Delta t. \tag{6.85}$$

In other words, differential Doppler can be used as the system equation of a Kalman filter for kinematic positioning. The Kalman filter will be discussed in the next chapter. A Kalman filter using differential Doppler will be discussed in Sect. 9.8.

### Doppler Integration

Integrating the instantaneous Doppler Eq. 6.84, one has

$$\lambda_j \int_{t_{j-1}}^{t_j} D_j \mathrm{d}t = \Delta_t \rho - \Delta_t(\delta t_r - \delta t_k)c + \varepsilon_d.$$

Using the operator $\Delta_t$ to the undifferenced phase Eq. 6.45 and code Eq. 6.44, one gets

$$\begin{aligned} \lambda_j \Delta_t \Phi_j &= \Delta_t \rho - \Delta_t(\delta t_r - \delta t_k)c + \lambda_j \Delta_t N_j + \varepsilon_p \quad \text{and} \\ \Delta_t R_j &= \Delta_t \rho - \Delta_t(\delta t_r - \delta t_k)c + \varepsilon_c, \end{aligned} \tag{6.86}$$

where the same symbols are used for the error terms (later too). Differencing the first equation of Eq. 6.86 with the integrated Doppler leads to

$$\lambda_j \Delta_t N_j = \lambda_j \Delta_t \Phi_j - \lambda_j \int_{t_{j-1}}^{t_j} D_j \mathrm{d}t + \varepsilon_1$$

or

$$\Delta_t N_j = \Delta_t \Phi_j - \int_{t_{j-1}}^{t_j} D_j \mathrm{d}t + \varepsilon_1, \quad j = 1, 2, 5. \tag{6.87}$$

Thus, integrated Doppler can be used for cycle slip detection. This detection method is very reasonable. The phase is measured by keeping track of the partial phase and accumulating the integer count. If any loss of lock of the signal happens during this time, the integer accumulation will be wrong, i.e. a cycle slip occurs. Therefore, an external instantaneous Doppler integration can be used as an alternative method of cycle slip detection. The integration can be achieved by first fitting the Doppler with a suitable order polynomial, and then integrating that within the time interval.

### Code Smoothing

Comparing the two formulas of Eq. 6.86, one has

$$\Delta_t R_j = \lambda_j \Delta_t \Phi_j - \lambda_j \Delta_t N_j + \varepsilon_2$$

or

$$\Delta_t R_j = \lambda_j \Delta_t \Phi_j + \varepsilon_3. \tag{6.88}$$

Equation 6.88 can be used for smoothing the code survey by phase if there are no cycle slips.

### Differential Phases

The first formula of Eq. 6.86 is the numerical difference of the phases at the two consecutive epochs $t_j$ and $t_{j-1}$

$$\lambda_j \Delta_t \Phi_j = \Delta_t \rho - \Delta_t(\delta t_r - \delta t_k)c + \lambda_j \Delta_t N_j + \varepsilon_p, \quad j = 1, 2.$$

All terms on the right-hand side except the ambiguity term are of low variation. Any cycle slips will lead to a sudden jump in the time difference of the phases. Therefore, the time-differenced phase can be used as an alternative method of cycle slip detection.

## 6.6
## Data Differentiations

Data differentiations are methods of combining GPS data (of the same type) measured at different stations. For the convenience of later discussions, tidal effects and relativistic effects are considered corrected before forming the differences. The original code, phase, and Doppler observables as well as their standardised combinations can be rewritten as (cf. Eqs. 6.44–6.47)

$$R_i^k(j) = \rho_i^k - c\delta t_i + c\delta t_k + \delta_{\text{ion}}(j) + \delta_{\text{trop}} + \varepsilon_c, \tag{6.89}$$

$$\lambda_j \Phi_i^k(j) = \rho_i^k - c\delta t_i + c\delta t_k + \lambda_j N_i^k(j) - \delta_{\text{ion}}(j) + \delta_{\text{trop}} + \varepsilon_p, \tag{6.90}$$

$$\delta_{\text{ion}}(j) = \frac{A_1}{f_j^2} + \frac{A_2}{f_j^3}, \quad \text{and} \tag{6.91}$$

$$D_i^k(j) = \frac{d\rho_i^k}{\lambda_j dt} - f_j \frac{d(\delta t_i - \delta t_k)}{dt} + \varepsilon_d, \tag{6.92}$$

where $j$ ($j = 1,2,5$) is the index of frequency $f$, subscript $i$ is the index of the station number, and superscript $k$ is the ID number of the satellite.

## 6.6.1
## *Single Differences*

Single difference (SD) is the difference formed by data observed at two stations on the same satellite as

$$\text{SD}_{i1,i2}^{k}(O) = O_{i2}^{k} - O_{i1}^{k}, \tag{6.93}$$

where $O$ is the original observable, and $i1$ and $i2$ are the two ID numbers of the stations. Supposing the original observables have the same variance of $\sigma^2$, then the single-difference observable has a variance of $2\sigma^2$. Considering Eqs. 6.89–6.92, one has

$$\text{SD}_{i1,i2}^{k}(R(j)) = \rho_{i2}^{k} - \rho_{i1}^{k} - c\delta t_{i2} + c\delta t_{i1} + d\delta_{\text{ion}}(j) + d\delta_{\text{trop}} + d\varepsilon_c, \tag{6.94}$$

$$\begin{aligned}\text{SD}_{i1,i2}^{k}(\lambda_j \Phi(j)) &= \rho_{i2}^{k} - \rho_{i1}^{k} - c\delta t_{i2} + c\delta t_{i1} + \lambda_j N_{i2}^{k}(j) - \lambda_j N_{i1}^{k}(j) \\ &\quad - d\delta_{\text{ion}}(j) + d\delta_{\text{trop}} + d\varepsilon_p, \quad \text{and}\end{aligned} \tag{6.95}$$

$$\text{SD}_{i1,i2}^{k}(D(j)) = \frac{\dot{\rho}_{i2}^{k} - \dot{\rho}_{i1}^{k}}{\lambda_j} - f_j \frac{d(\delta t_{i2} - \delta t_{i1})}{dt} + d\varepsilon_d, \tag{6.96}$$

where $\dot{\rho}$ is the time differentiation of $\rho$, and $d\delta_{\text{ion}}(j)$ and $d\delta_{\text{trop}}$ are the differenced ionospheric and tropospheric effects at the two stations related to the satellite $k$, respectively.

The most important property of single differences is that the satellite clock error terms in the model are eliminated. However, it should be emphasised that the satellite clock error, which implicitly affects the computation of satellite position, must still be carefully considered. Ionospheric and tropospheric effects are reduced through difference forming, especially for those stations that are not very far apart. Because of the identical mathematical models of the station clock errors and ambiguities, not all clock and ambiguity parameters can be resolved in the single-difference equations of Eqs. 6.94–6.96.

For the original observable vector of station $i1$ and $i2$,

$$O = \begin{pmatrix} O_{i1}^{k1} & O_{i1}^{k2} & O_{i1}^{k3} & O_{i2}^{k1} & O_{i2}^{k2} & O_{i2}^{k3} \end{pmatrix}^T, \quad \text{cov}(O) = \sigma^2 E,$$

the single differences

$$\mathrm{SD}(O) = \begin{pmatrix} O^{k1}_{i1,i2} & O^{k2}_{i1,i2} & O^{k3}_{i1,i2} \end{pmatrix}^{\mathrm{T}},$$

can be formed by a linear transformation

$$\mathrm{SD}(O) = C \cdot O \quad \text{and}$$

$$C = \begin{pmatrix} -1 & 0 & 0 & 1 & 0 & 0 \\ 0 & -1 & 0 & 0 & 1 & 0 \\ 0 & 0 & -1 & 0 & 0 & 1 \end{pmatrix} = (-E \quad E). \tag{6.97}$$

Where common satellites $k1, k2, k3$ are observed, $E$ is an identity matrix whose size is that of the observed satellite number; in the above example the size is $3 \times 3$.

The covariance matrix of the single differences is then

$$\mathrm{cov}(\mathrm{SD}(O)) = C \cdot \mathrm{cov}(O) \cdot C^{\mathrm{T}} = \sigma^2 C \cdot C^{\mathrm{T}} = 2\sigma^2 E, \tag{6.98}$$

i.e. the weight matrix is

$$P = \frac{1}{2\sigma^2} E.$$

In other words, the single differences are uncorrelated observables in the case of a single baseline. $C$ in Eq. 6.97 is a general form, so $C$ is denoted by $C_s = (-E_{n \times n} \; E_{n \times n})$, and $n$ is the number of commonly viewed satellites.

Single differences can be formed for any baselines as long as the two stations have common satellites in sight. However, these should be a set of "independent" baselines. The most widely used methods involve the formation of radial or transverse baselines. Supposing the stations' ID vector is $(i1, i2, i3,\ldots, i(m-1), im)$, and the baseline between station $i1$ and $i2$ is denoted by $(i1, i2)$, then the radial baselines can be formed, for example, by $(i1, i2),(i1, i3),\ldots,(i1, im)$, and the transverse baselines by $(i1, i2), (i2, i3),\ldots,(i(m-1), im)$. Station $i1$ is called a reference station and is freely selectable. In some cases, mixed radial and transverse baselines must be formed—for example, by $(i1, i2), (i1, i3), (i3, i4),\ldots,(i3, i(m-1)), (i3, im)$. Sometimes the baselines must be formed by several groups, and thus several references must be selected. A method of forming independent and optimal baseline networks will be discussed in Sects. 9.1 and 9.2.

In the case in which three stations are used to measure the GPS data, the original observable vector of stations $i1, i2$ and $i3$ is

$$O_i = \begin{pmatrix} O_i^{k1} & \cdots & O_i^{kn} \end{pmatrix}^{\mathrm{T}}, \quad \mathrm{cov}(O_i) = \sigma^2 E_{n \times n}, \quad i = i1, i2, i3,$$

where $n$ is the commonly observed satellite number. The single differences of the baseline $(i, j)$ are

$$\mathrm{SD}_{i,j}(O) = \begin{pmatrix} O_{i,j}^{k1} & \cdots & O_{i,j}^{kn} \end{pmatrix}^T \quad i,j = i1, i2, i3, \quad i \neq j.$$

If the baselines are formed in a radial way, i.e. as $(i1, i2)$ and $(i1, i3)$, then one has

$$\begin{pmatrix} \mathrm{SD}_{i1,i2}(O) \\ \mathrm{SD}_{i1,i3}(O) \end{pmatrix} = \begin{pmatrix} -E & E & 0 \\ -E & 0 & E \end{pmatrix} \begin{pmatrix} O_{i1} \\ O_{i2} \\ O_{i3} \end{pmatrix},$$

and

$$\mathrm{cov}(\mathrm{SD}) = \sigma^2 \begin{pmatrix} -E & E & 0 \\ -E & 0 & E \end{pmatrix} \begin{pmatrix} -E & -E \\ E & 0 \\ 0 & E \end{pmatrix} = \sigma^2 \begin{pmatrix} 2E & E \\ E & 2E \end{pmatrix} \quad \text{and}$$

$$P_s = [\mathrm{cov}(\mathrm{SD})]^{-1} = \frac{1}{3\sigma^2} \begin{pmatrix} 2E & -E \\ -E & 2E \end{pmatrix}. \tag{6.99}$$

If the baselines are formed in a transverse way, i.e. as $(i1, i2)$ and $(i2, i3)$, then one has

$$\begin{pmatrix} \mathrm{SD}_{i1,i2}(O) \\ \mathrm{SD}_{i2,i3}(O) \end{pmatrix} = \begin{pmatrix} -E & E & 0 \\ 0 & -E & E \end{pmatrix} \begin{pmatrix} O_{i1} \\ O_{i2} \\ O_{i3} \end{pmatrix},$$

$$\mathrm{cov}(\mathrm{SD}) = \sigma^2 \begin{pmatrix} -E & E & 0 \\ 0 & -E & E \end{pmatrix} \begin{pmatrix} -E & 0 \\ E & -E \\ 0 & E \end{pmatrix} = \sigma^2 \begin{pmatrix} 2E & -E \\ -E & 2E \end{pmatrix} \quad \text{and}$$

$$P_s = [\mathrm{cov}(\mathrm{SD})]^{-1} = \frac{1}{3\sigma^2} \begin{pmatrix} 2E & E \\ E & 2E \end{pmatrix}.$$

It is obvious that the single differences are correlated if the number of stations is greater than two, and the correlation depends on the ways the baselines are formed. Therefore, it is not possible to derive a general covariance formula for the single differences of a network. Furthermore, the commonly viewed satellite number $n$ could be different from baseline to baseline, further complicating the formulation of the covariance matrix.

A baseline-wise processing of the GPS data of a network using single differences is equivalent to an omission of the correlation between the baselines.

## 6.6.2
## Double Differences

Double differences are formed between two single differences related to two observed satellites as

$$DD_{i1,i2}^{k1,k2}(O) = SD_{i1,i2}^{k2}(O) - SD_{i1,i2}^{k1}(O) \tag{6.100}$$

or

$$DD_{i1,i2}^{k1,k2}(O) = (O_{i2}^{k2} - O_{i1}^{k2}) - (O_{i2}^{k1} - O_{i1}^{k1}), \tag{6.101}$$

where $k1$ and $k2$ are the two id numbers of the satellites. Supposing the original observables have the same variance of $\sigma^2$, then the double-differenced observables have a variance of $4\sigma^2$. Considering Eqs. 6.89–6.92, one has

$$DD_{i1,i2}^{k1,k2}(R(j)) = \rho_{i2}^{k2} - \rho_{i1}^{k2} - \rho_{i2}^{k1} + \rho_{i1}^{k1} + dd\delta_{ion}(j) + dd\delta_{trop} + dd\varepsilon_c, \tag{6.102}$$

$$DD_{i1,i2}^{k1,k2}(\lambda_j\Phi(j)) = \rho_{i2}^{k2} - \rho_{i1}^{k2} - \rho_{i2}^{k1} + \rho_{i1}^{k1} + \lambda_j(N_{i2}^{k2}(j) - N_{i1}^{k2}(j) \\ - N_{i2}^{k1}(j) + N_{i1}^{k1}(j)) - dd\delta_{ion}(j) + dd\delta_{trop} + dd\varepsilon_p, \quad \text{and} \tag{6.103}$$

$$DD_{i1,i2}^{k1,k2}(D(j)) = \frac{\dot\rho_{i2}^{k2} - \dot\rho_{i1}^{k2} - \dot\rho_{i2}^{k1} + \dot\rho_{i1}^{k1}}{\lambda_j} + dd\varepsilon_d, \tag{6.104}$$

where $dd\delta_{ion}(j)$ and $dd\delta_{trop}$ are the differenced ionospheric and tropospheric effects at the two stations related to the two satellites, respectively. For the ionosphere-free combined observables (denoted by $j = 4$ for distinguishing), the ionospheric error terms have vanished from above equations.

The most important property of double differences is that the clock error terms in the equation (model) are completely eliminated. It should be emphasised that the clock error, which implicitly affects the computation of the position of the satellite, must still be carefully considered. Ionospheric and tropospheric effects are reduced greatly through difference forming, especially for those stations that are not far apart. Double-differenced Doppler directly describes the geometry change. Double-differenced ambiguities can be denoted by

$$N_{i1,i2}^{k1,k2}(j) = N_{i2}^{k2}(j) - N_{i1}^{k2}(j) - N_{i2}^{k1}(j) + N_{i1}^{k1}(j). \tag{6.105}$$

For convenience, the original ambiguities used in Eq. 6.103 are for the case of the reference satellite changing.

For the single-difference observable vector

$$SD(O) = \begin{pmatrix} O_{i1,i2}^{k1} & O_{i1,i2}^{k2} & O_{i1,i2}^{k3} \end{pmatrix}^T \quad \text{and} \quad \text{cov}(SD(O)) = 2\sigma^2 E, \tag{6.106}$$

the double differences

$$DD(O) = \begin{pmatrix} O^{k1,k2}_{i1,i2} & O^{k1,k3}_{i1,i2} \end{pmatrix}^T \quad (6.107)$$

can be formed by a linear transformation

$$DD(O) = C_d \cdot SD(O), \quad (6.108)$$

$$C_d = \begin{pmatrix} -1 & 1 & 0 \\ -1 & 0 & 1 \end{pmatrix} = (-I_m \quad E_{m \times m}) \quad \text{(here } m = 2\text{)}, \quad (6.109)$$

where $E$ is an identity matrix of size $m \times m$, $I$ is a 1 vector of size $m$ (all elements of the vector are 1), $m$ is the number of formed double differences, and $m = n - 1$. The covariance matrix of the double differences is then

$$\text{cov}(DD(O)) = C_d \cdot \text{cov}(SD(O)) \cdot C_d^T = 2\sigma^2 C_d \cdot C_d^T = 2\sigma^2 \begin{pmatrix} 2 & 1 \\ 1 & 2 \end{pmatrix}. \quad (6.110)$$

For single and double differences

$$SD(O) = \begin{pmatrix} O^{k1}_{i1,i2} & O^{k2}_{i1,i2} & O^{k3}_{i1,i2} & O^{k4}_{i1,i2} \end{pmatrix}^T, \quad \text{cov}(SD(O)) = 2\sigma^2 E, \quad \text{and} \quad (6.111)$$

$$DD(O) = \begin{pmatrix} O^{k1,k2}_{i1,i2} & O^{k1,k3}_{i1,i2} & O^{k1,k4}_{i1,i2} \end{pmatrix}^T, \quad (6.112)$$

the linear transformation matrix $C_d$ and the covariance matrix can be obtained by

$$C_d = \begin{pmatrix} -1 & 1 & 0 & 0 \\ -1 & 0 & 1 & 0 \\ -1 & 0 & 0 & 1 \end{pmatrix} = (-I \quad E) \quad \text{and} \quad (6.113)$$

$$\text{cov}(DD(O)) = C_d \cdot \text{cov}(SD(O)) \cdot C_d^T = 2\sigma^2 C_d \cdot C_d^T = 2\sigma^2 \begin{pmatrix} 2 & 1 & 1 \\ 1 & 2 & 1 \\ 1 & 1 & 2 \end{pmatrix}. \quad (6.114)$$

For the general case of

$$SD(O) = \begin{pmatrix} O^{k1}_{i1,i2} & O^{k2}_{i1,i2} & O^{k3}_{i1,i2} & \cdots & O^{kn}_{i1,i2} \end{pmatrix}^T, \quad \text{cov}(SD(O)) = 2\sigma^2 E, \quad \text{and}$$

$$DD(O) = \begin{pmatrix} O^{k1,k2}_{i1,i2} & O^{k1,k3}_{i1,i2} & \cdots & O^{k1,km}_{i1,i2} \end{pmatrix}^T,$$

$$(6.115)$$

it is obvious that the general transformation matrix $C_d$ and the related covariance matrix can be represented as

$$C_d = \begin{pmatrix} -I_m & E_{m \times m} \end{pmatrix} \quad \text{and} \tag{6.116}$$

$$\text{cov}(\text{DD}(O)) = C_d \text{cov}(\text{SD}(O))C_d^T = 2\sigma^2 C_d C_d^T = 2\sigma^2 (I_{m \times m} + E_{m \times m}) \tag{6.117}$$

where $I_{m \times m}$ is an $m \times m$ matrix whose elements are all 1, and the weight matrix has the form of

$$P = [\text{cov}(\text{DD}(O))]^{-1} = \frac{1}{2\sigma^2 n}(nE_{m \times m} - I_{m \times m}), \tag{6.118}$$

where $n = m + 1$. Equation 6.118 can be verified by an identity matrix test (i.e. $P \cdot \text{cov}(\text{DD}(O)) = E$).

In the case of three stations, supposing $n$ common satellites ($k1, k2,\ldots, kn$) are viewed, then the single and double differences can be written as

$$\begin{aligned} \text{SD}_{i,j}(O) &= \begin{pmatrix} O_{i,j}^{k1} & O_{i,j}^{k2} & O_{i,j}^{k3} & \cdots & O_{i,j}^{kn} \end{pmatrix}^T \quad \text{and} \\ \text{DD}_{i,j}(O) &= \begin{pmatrix} O_{i,j}^{k1,k2} & O_{i,j}^{k1,k3} & \cdots & O_{i,j}^{k1,km} \end{pmatrix}^T \quad i,j = i1, i2, i3, 4 \quad i \neq j. \end{aligned} \tag{6.119}$$

Then one has the transformation and covariance

$$\begin{pmatrix} \text{DD}_{i1,i2}(O) \\ \text{DD}_{i1,i3}(O) \end{pmatrix} = \begin{pmatrix} C_d & 0 \\ 0 & C_d \end{pmatrix} \begin{pmatrix} \text{SD}_{i1,i2}(O) \\ \text{SD}_{i1,i3}(O) \end{pmatrix} \quad \text{and}$$

$$\text{cov}(\text{DD}) = \begin{pmatrix} C_d & 0 \\ 0 & C_d \end{pmatrix} \text{cov}(\text{SD}) \begin{pmatrix} C_d & 0 \\ 0 & C_d \end{pmatrix}^T = \sigma^2 \begin{pmatrix} 2E & -E \\ -E & 2E \end{pmatrix}(C_d C_d^T).$$

Because of the dependence of the cov(SD) on the baselines forming, cov(DD) is also dependent on the baselines forming. A baseline-wise processing of a network GPS data using double differences is equivalent to an omission of the correlation between the baselines.

### 6.6.3
### Triple Differences

Triple differences are formed between two double differences related to the same stations and satellites at the two adjacent epochs as

$$\text{TD}_{i1,i2}^{k1,k2}(O(t1,t2)) = \text{DD}_{i1,i2}^{k1,k2}(O(t2)) - \text{DD}_{i1,i2}^{k1,k2}(O(t1))$$

or

$$\mathrm{TD}_{i1,i2}^{k1,k2}(O(t1,t2)) = O_{i2}^{k2}(t2) - O_{i1}^{k2}(t2) - O_{i2}^{k1}(t2) + O_{i1}^{k1}(t2) \\ - O_{i2}^{k2}(t1) + O_{i1}^{k2}(t1) + O_{i2}^{k1}(t1) - O_{i1}^{k1}(t1),$$ (6.120)

where $t1$ and $t2$ are two adjacent epochs. Supposing the original observables have the same variance of $\sigma^2$, then the triple-differenced observables have a variance of $8\sigma^2$. Considering Eqs. 6.102–6.104, one has

$$\mathrm{TD}_{i1,i2}^{k1,k2}(R(j,t1,t2)) = \rho_{i2}^{k2}(t2) - \rho_{i1}^{k2}(t2) - \rho_{i2}^{k1}(t2) + \rho_{i1}^{k1}(t2) - \rho_{i2}^{k2}(t1) \\ + \rho_{i1}^{k2}(t1) + \rho_{i2}^{k1}(t1) - \rho_{i1}^{k1}(t1) + td\varepsilon_c,$$ (6.121)

$$\mathrm{TD}_{i1,i2}^{k1,k2}(\lambda_j \Phi(j,t1,t2)) = \rho_{i2}^{k2}(t2) - \rho_{i1}^{k2}(t2) - \rho_{i2}^{k1}(t2) + \rho_{i1}^{k1}(t2) - \rho_{i2}^{k2}(t1) \\ + \rho_{i1}^{k2}(t1) + \rho_{i2}^{k1}(t1) - \rho_{i1}^{k1}(t1) + \delta N + td\varepsilon_p, \quad \text{and}$$ (6.122)

$$\mathrm{TD}_{i1,i2}^{k1,k2}(D(j,t1,t2)) = \frac{\dot{\rho}_{i2}^{k2}(t2) - \dot{\rho}_{i1}^{k2}(t2) - \dot{\rho}_{i2}^{k1}(t2) + \dot{\rho}_{i1}^{k1}(t2)}{\lambda_j} \\ - \frac{\dot{\rho}_{i2}^{k2}(t1) - \dot{\rho}_{i1}^{k2}(t1) - \dot{\rho}_{i2}^{k1}(t1) + \dot{\rho}_{i1}^{k1}(t1)}{\lambda_j} + td\varepsilon_d,$$ (6.123)

where

$$\delta N = \lambda_j (N_{i1,i2}^{k1,k2}(j,t2) - N_{i1,i2}^{k1,k2}(j,t1)).$$ (6.124)

Ionospheric and tropospheric effects are eliminated. If there are no cycle slips during the time, the term of Eq. 6.124 is zero. Therefore, triple differences of Eq. 6.122 can also be used as a check for the cycle slips. Through triple-difference forming, the systematic cycle slip turns out to be an effect like an outlier.

The most important property of triple differences is that only the geometric change is left in the models. Triple differences of Doppler describe the acceleration of the position.

For double differences

$$\mathrm{DD}(O(t)) = \begin{pmatrix} O_{i1,i2}^{k1,k2}(t) & O_{i1,i2}^{k1,k3}(t) & \cdots & O_{i1,i2}^{k1,km}(t) \end{pmatrix}^{\mathrm{T}},$$ (6.125)

one has

$$\mathrm{TD}(O(t1,t2)) = C_T \cdot \begin{pmatrix} \mathrm{DD}(O(t1)) \\ \mathrm{DD}(O(t2)) \end{pmatrix},$$ (6.126)

where

$$C_T = \begin{pmatrix} -E_{m\times m} & E_{m\times m} \end{pmatrix}. \tag{6.127}$$

Then the related covariance matrix can be represented as

$$\begin{aligned} \operatorname{cov}(\mathrm{TD}(O(t1,t2))) &= C_T \cdot \operatorname{cov}(\mathrm{DD}(O)) \cdot C_T^{\mathrm{T}} \\ &= C_T \cdot C_{d2} \operatorname{cov}(\mathrm{SD}(O)) \cdot C_{d2}^{\mathrm{T}} C_T^{\mathrm{T}} = 2\sigma^2 C_T C_{d2} C_{d2}^{\mathrm{T}} C_T^{\mathrm{T}}, \end{aligned} \tag{6.128}$$

where $C_{d2}$ is the double-difference transformation matrix of two epochs. Because double differences are independent epoch wise, $C_{d2}$ is a diagonal matrix of $C_d$, i.e.

$$C_{d2} = \begin{pmatrix} C_d & 0 \\ 0 & C_d \end{pmatrix}. \tag{6.129}$$

It is worth noting that the triple differences formed by epochs ($t1$, $t2$) are correlated to the differences formed by epochs ($t0$, $t1$) and ($t1$, $t2$). Such correlation makes a sequential processing of the triple-difference data very complicated. Sequentially using the above covariance formula indicates an omission of the correlation related to the previous epoch and the next epoch.

Taking the correlation between the baselines into account, an exact correlation description of the triple differences of a GPS network becomes very difficult.

## 6.7
## Equivalence of the Uncombined and Combining Algorithms

Uncombined and combining algorithms are standard GPS data processing methods, which can often be found in the literature (cf., e.g., Leick 2004; Hofmann-Wellenhof et al. 2001). Different combinations own different properties and are beneficial for dealing with the data and solving the problem in different cases (Hugentobler et al. 2001; Kouba and Heroux 2001; Zumberge et al. 1997). The equivalence between the undifferenced and differencing algorithms was proved and a unified equivalent data processing method proposed by Xu (2002, cf. Sect. 6.8). The question of whether the uncombined and combining algorithms are also equivalent is an interesting topic and will be addressed here in detail (cf. Xu et al. 2006a).

### 6.7.1
### Uncombined GPS Data Processing Algorithms

*Original GPS Observation Equations*

The original GPS code pseudorange and carrier phase measurements represented in Eqs. 6.44 and 6.45 (cf. Sect. 6.5) can be simplified as

$$R_j = C_\rho + \delta_{\text{ion}}(j), \tag{6.130}$$

$$\lambda_j \Phi_j = C_\rho + \lambda_j N_j - \delta_{\text{ion}}(j), \quad j = 1, 2 \tag{6.131}$$

where

$$C_\rho = \rho - (\delta t_r - \delta t_k)c + \delta_{\text{trop}} + \delta_{\text{tide}} + \delta_{\text{rel}} + \varepsilon_i, \quad i = c, p \tag{6.132}$$

$$\delta_{\text{ion}}(j) = \frac{A_1}{f_j^2} = \frac{A_1^z}{f_j^2} F = \frac{f_s^2 B_1}{f_j^2} = \frac{f_s^2 B_1^z}{f_j^2} F. \tag{6.133}$$

Where symbols have the same meanings as those of Eqs. 6.44–6.47. $j$ is the index of the frequency $f$ and wavelength $\lambda$. $A_1$ and $A_1^z$ are the ionospheric parameters in the path and zenith directions; $B_1$ and $B_1^z$ are scaled $A_1$ and $A_1^z$ with $f_s^2$ for numerical reasons. $c$ denotes the speed of light, index $c$ denotes code. $C_\rho$ is called geometry and $N_j$ is the ambiguity. For simplicity, the residuals of the codes (and phases) are denoted with the same symbol $\varepsilon_c$ (and $\varepsilon_p$) and have the same standard deviations of $\sigma_c$ (and $\sigma_p$). Equations 6.130 and 6.131 can be written in a matrix form with weight matrix $P$ as (Blewitt 1998)

$$\begin{pmatrix} R_1 \\ R_2 \\ \lambda_1 \Phi_1 \\ \lambda_2 \Phi_2 \end{pmatrix} = \begin{pmatrix} 0 & 0 & f_s^2/f_1^2 & 1 \\ 0 & 0 & f_s^2/f_2^2 & 1 \\ 1 & 0 & -f_s^2/f_1^2 & 1 \\ 0 & 1 & -f_s^2/f_2^2 & 1 \end{pmatrix} \begin{pmatrix} \lambda_1 N_1 \\ \lambda_2 N_2 \\ B_1 \\ C_\rho \end{pmatrix}, \quad P = \begin{pmatrix} \sigma_c^2 & 0 & 0 & 0 \\ 0 & \sigma_c^2 & 0 & 0 \\ 0 & 0 & \sigma_p^2 & 0 \\ 0 & 0 & 0 & \sigma_p^2 \end{pmatrix}^{-1}. \tag{6.134}$$

*Solutions of Uncombined Observation Equations*

Equation 6.134 includes the observations of one satellite viewed by one receiver at one epoch. Alternatively, Eq. 6.134 can be considered a transformation between the observations and unknowns, and the transformation is a linear and invertible one. Denoting

$$a = \frac{f_1^2}{f_1^2 - f_2^2}, \quad b = \frac{-f_2^2}{f_1^2 - f_2^2}, \quad g = \frac{1}{f_1^2} - \frac{1}{f_2^2}, \quad q = g f_s^2, \tag{6.135}$$

then one has relations of

$$1 - a = b, \quad \frac{1}{f_1^2 g} = b, \quad \frac{1}{f_2^2 g} = -a \qquad (6.136)$$

and

$$\begin{pmatrix} 0 & 0 & f_s^2/f_1^2 & 1 \\ 0 & 0 & f_s^2/f_2^2 & 1 \\ 1 & 0 & -f_s^2/f_1^2 & 1 \\ 0 & 1 & -f_s^2/f_2^2 & 1 \end{pmatrix}^{-1} \begin{pmatrix} 1-2a & -2b & 1 & 0 \\ -2a & 2a-1 & 0 & 1 \\ 1/q & -1/q & 0 & 0 \\ a & b & 0 & 0 \end{pmatrix} = T. \qquad (6.137)$$

Where $a$ and $b$ are the coefficients of the ionosphere-free combinations of the observables of L1 and L2. The solution of Eq. 6.134 has a form of (by multiplying the transformation matrix $T$ to Eq. 6.134)

$$\begin{pmatrix} \lambda_1 N_1 \\ \lambda_2 N_2 \\ B_1 \\ C_\rho \end{pmatrix} = \begin{pmatrix} 1-2a & -2b & 1 & 0 \\ -2a & 2a-1 & 0 & 1 \\ 1/q & -1/q & 0 & 0 \\ a & b & 0 & 0 \end{pmatrix} \begin{pmatrix} R_1 \\ R_2 \\ \lambda_1 \Phi_1 \\ \lambda_2 \Phi_2 \end{pmatrix}. \qquad (6.138)$$

The related covariance matrix of the above solution vector is then

$$Q = \text{cov}\begin{pmatrix} \lambda_1 N_1 \\ \lambda_2 N_2 \\ B_1 \\ C_\rho \end{pmatrix} = T \begin{pmatrix} \sigma_c^2 & 0 & 0 & 0 \\ 0 & \sigma_c^2 & 0 & 0 \\ 0 & 0 & \sigma_p^2 & 0 \\ 0 & 0 & 0 & \sigma_p^2 \end{pmatrix} T^T$$

$$= \begin{pmatrix} (1-2a)^2 + 4b^2 + \frac{\sigma_p^2}{\sigma_c^2} & 4a^2 - 4ab - 2a + 2b & \frac{1-2a+2b}{q} & a - 2a^2 - 2b^2 \\ 4a^2 - 4ab - 2a + 2b & 8a^2 - 4a + 1 + \frac{\sigma_p^2}{\sigma_c^2} & \frac{1-4a}{q} & -2a^2 + 2ab - b \\ \frac{1-2a+2b}{q} & \frac{1-4a}{q} & \frac{2}{q^2} & \frac{a-b}{q} \\ a - 2a^2 - 2b^2 & -2a^2 + 2ab - b & \frac{a-b}{q} & a^2 + b^2 \end{pmatrix} \sigma_c^2.$$

(6.139)

Equation 6.139 can be simplified by using the relation of $1 - a = b$ and neglecting the terms of $(\sigma_p/\sigma_c)^2$ (because $(\sigma_p/\sigma_c)$ is less than 0.01) as well as letting $f_s = f_1$ (so that $q = 1/b$). Taking the relationships of ratios of the frequencies into account ($f_1 = 154 f_0$ and $f_2 = 120 f_0$, $f_0$ is the fundamental frequency), one has approximately

$$\text{cov}\begin{pmatrix} \lambda_1 N_1 \\ \lambda_2 N_2 \\ B_1 \\ C_\rho \end{pmatrix} = \begin{pmatrix} 26.2971 & 33.4800 & 11.1028 & -15.1943 \\ 33.4800 & 42.6629 & 14.1943 & -19.2857 \\ 11.1028 & 14.1943 & 4.7786 & -6.3243 \\ -15.1943 & -19.2857 & -6.3243 & 8.8700 \end{pmatrix} \sigma_c^2 \quad (6.140)$$

The precision of the solutions will be further discussed in Sect. 6.7.3. The parameterisation of the GPS observation models is an important issue and can be found in Chap. 9 or (Blewitt 1998; Xu 2004) if interested.

## 6.7.2
## Combining Algorithms of GPS Data Processing

### Ionosphere-Free Combinations

Letting transformation matrix

$$T_1 = \begin{pmatrix} 1 & -1 & 0 & 0 \\ a & b & 0 & 0 \\ 0 & 0 & a & b \\ 1/2 & 0 & 1/2 & 0 \end{pmatrix}, \quad (6.141)$$

and applying the transform to the Eq. 6.134, one has

$$T_1 \begin{pmatrix} R_1 \\ R_2 \\ \lambda_1 \Phi_1 \\ \lambda_2 \Phi_2 \end{pmatrix} = \begin{pmatrix} 0 & 0 & q & 0 \\ 0 & 0 & 0 & 1 \\ a & b & 0 & 1 \\ 1/2 & 0 & 0 & 1 \end{pmatrix} \begin{pmatrix} \lambda_1 N_1 \\ \lambda_2 N_2 \\ B_1 \\ C_\rho \end{pmatrix}. \quad (6.142)$$

The ionospheric parameter in Eq. 6.142 is free in the last three equations, which are traditionally called ionosphere-free combinations. Solving the ionosphere-free equations or the whole Eq. 6.142 will lead to the same results. Equation 6.142 has a unique solution vector of

$$\begin{pmatrix} \lambda_1 N_1 \\ \lambda_2 N_2 \\ B_1 \\ C_\rho \end{pmatrix} = \begin{pmatrix} 0 & -2 & 0 & 2 \\ 0 & (2a-1)/b & 1/b & -2a/b \\ 1/q & 0 & 0 & 0 \\ 0 & 1 & 0 & 0 \end{pmatrix} T_1 \begin{pmatrix} R_1 \\ R_2 \\ \lambda_1 \Phi_1 \\ \lambda_2 \Phi_2 \end{pmatrix}, \quad (6.143)$$

or (noticing $(1 - a) = b$, cf. Eq. 6.136)

$$\begin{pmatrix} \lambda_1 N_1 \\ \lambda_2 N_2 \\ B_1 \\ C_\rho \end{pmatrix} = \begin{pmatrix} 1-2a & -2b & 1 & 0 \\ -2a & 2a-1 & 0 & 1 \\ 1/q & -1/q & 0 & 0 \\ a & b & 0 & 0 \end{pmatrix} \begin{pmatrix} R_1 \\ R_2 \\ \lambda_1 \Phi_1 \\ \lambda_2 \Phi_2 \end{pmatrix}. \quad (6.144)$$

Equations 6.144 and 6.138 are identical. Therefore, the covariance matrix of the solution vector on the left side of Eq. 6.144 is the same as that given in Eq. 6.139. This shows that the uncombined algorithms and the ionosphere-free combinations are equivalent in this case.

### Geometry-Free Combinations

Letting transformation matrix

$$T_2 = \begin{pmatrix} a & b & 0 & 0 \\ 1 & -1 & 0 & 0 \\ 0 & 0 & 1 & -1 \\ -1 & 0 & 1 & 0 \end{pmatrix}, \quad (6.145)$$

and applying the transformation to Eq. 6.134, one has

$$T_2 \begin{pmatrix} R_1 \\ R_2 \\ \lambda_1 \Phi_1 \\ \lambda_2 \Phi_2 \end{pmatrix} = \begin{pmatrix} 0 & 0 & 0 & 1 \\ 0 & 0 & q & 0 \\ 1 & -1 & -q & 0 \\ 1 & 0 & -2f_s^2/f_1^2 & 0 \end{pmatrix} \begin{pmatrix} \lambda_1 N_1 \\ \lambda_2 N_2 \\ B_1 \\ C_\rho \end{pmatrix}. \quad (6.146)$$

The geometric component in Eq. 6.146 is free in the last three equations, which are traditionally called geometry-free combinations. Solving the geometry-free equations or Eq. 6.146 will lead to the same results. Equation 6.146 has a unique solution vector of

$$\begin{pmatrix} \lambda_1 N_1 \\ \lambda_2 N_2 \\ B_1 \\ C_\rho \end{pmatrix} = \begin{pmatrix} 0 & 2/(f_1^2 g) & 0 & 1 \\ 0 & 2/(f_1^2 g) - 1 & -1 & 1 \\ 0 & 1/q & 0 & 0 \\ 1 & 0 & 0 & 0 \end{pmatrix} T_2 \begin{pmatrix} R_1 \\ R_2 \\ \lambda_1 \Phi_1 \\ \lambda_2 \Phi_2 \end{pmatrix}, \quad (6.147)$$

or (noticing $1/(f_1^2 g) = b$, cf. Equation 6.136)

$$\begin{pmatrix} \lambda_1 N_1 \\ \lambda_2 N_2 \\ B_1 \\ C_\rho \end{pmatrix} = \begin{pmatrix} 2b-1 & -2b & 1 & 0 \\ 2b-2 & 1-2b & 0 & 1 \\ 1/q & -1/q & 0 & 0 \\ a & b & 0 & 0 \end{pmatrix} \begin{pmatrix} R_1 \\ R_2 \\ \lambda_1 \Phi_1 \\ \lambda_2 \Phi_2 \end{pmatrix}. \quad (6.148)$$

Taking the relations of Eq. 6.136 (i.e. $b = 1 - a$) into account, Eqs. 6.148 and 6.138 are identical. Therefore, the covariance matrix of the solution vector on the left side of Eq. 6.148 is identical with that of Eq. 6.139. This shows that the uncombined algorithms and the geometry-free combinations are equivalent in this case.

### Ionosphere-Free and Geometry-Free Combinations

Letting transformation matrix

## 6.7 · Equivalence of the Uncombined and Combining Algorithms

$$T_3 = \begin{pmatrix} 1 & 0 & 0 & 0 \\ 0 & 1 & 0 & 0 \\ 0 & -1 & 1 & 0 \\ 0 & -1 & 0 & 1 \end{pmatrix}, \tag{6.149}$$

one then has

$$T_3 T_1 = \begin{pmatrix} 1 & 0 & 0 & 0 \\ 0 & 1 & 0 & 0 \\ 0 & -1 & 1 & 0 \\ 0 & -1 & 0 & 1 \end{pmatrix} \begin{pmatrix} 1 & -1 & 0 & 0 \\ a & b & 0 & 0 \\ 0 & 0 & a & b \\ 1/2 & 0 & 1/2 & 0 \end{pmatrix}$$

$$= \begin{pmatrix} 1 & -1 & 0 & 0 \\ a & b & 0 & 0 \\ -a & -b & a & b \\ 1/2 - a & -b & 1/2 & 0 \end{pmatrix}. \tag{6.150}$$

Applying the transformation 6.150 to Eq. 6.134 or applying the transformation 6.149 to Eq. 6.142 leads to the same results, and one has

$$T_3 T_1 \begin{pmatrix} R_1 \\ R_2 \\ \lambda_1 \Phi_1 \\ \lambda_2 \Phi_2 \end{pmatrix} = \begin{pmatrix} 0 & 0 & q & 0 \\ 0 & 0 & 0 & 1 \\ a & b & 0 & 0 \\ 1/2 & 0 & 0 & 0 \end{pmatrix} \begin{pmatrix} \lambda_1 N_1 \\ \lambda_2 N_2 \\ B_1 \\ C_\rho \end{pmatrix} \tag{6.151}$$

or

$$\begin{pmatrix} R_1 - R_2 \\ aR_1 + bR_2 \\ a\lambda_1 \Phi_1 + b\lambda_2 \Phi_2 - aR_1 - bR_2 \\ (\lambda_1 \Phi_1 + R_1)/2 - aR_1 - bR_2 \end{pmatrix} = \begin{pmatrix} 0 & 0 & q & 0 \\ 0 & 0 & 0 & 1 \\ a & b & 0 & 0 \\ 1/2 & 0 & 0 & 0 \end{pmatrix} \begin{pmatrix} \lambda_1 N_1 \\ \lambda_2 N_2 \\ B_1 \\ C_\rho \end{pmatrix}. \tag{6.152}$$

The ionosphere and geometry are both free in the last two equations, which are called ionosphere-geometry-free combinations. Solving the ionosphere-free and geometry-free equations or directly solving Eq. 6.152 will lead to the same results. Equation 6.152 has a unique solution vector of

$$\begin{pmatrix} \lambda_1 N_1 \\ \lambda_2 N_2 \\ B_1 \\ C_\rho \end{pmatrix} = \begin{pmatrix} 0 & 0 & 0 & 2 \\ 0 & 0 & 1/b & -2a/b \\ 1/q & 0 & 0 & 0 \\ 0 & 1 & 0 & 0 \end{pmatrix} T_3 T_1 \begin{pmatrix} R_1 \\ R_2 \\ \lambda_1 \Phi_1 \\ \lambda_2 \Phi_2 \end{pmatrix}, \tag{6.153}$$

or (noticing $(1 - a)/b = 1$, cf. Eq. 6.136)

$$\begin{pmatrix} \lambda_1 N_1 \\ \lambda_2 N_2 \\ B_1 \\ C_\rho \end{pmatrix} = \begin{pmatrix} 1-2a & -2b & 1 & 0 \\ -2a & 2a-1 & 0 & 1 \\ 1/q & -1/q & 0 & 0 \\ a & b & 0 & 0 \end{pmatrix} \begin{pmatrix} R_1 \\ R_2 \\ \lambda_1 \Phi_1 \\ \lambda_2 \Phi_2 \end{pmatrix}. \tag{6.154}$$

Equations 6.154 and 6.138 are identical. This shows that the uncombined algorithms and the ionosphere-geometry-free combinations are equivalent in this discussed case.

### Diagonal Combinations

Letting transformation matrix

$$T_4 = \begin{pmatrix} 1 & 0 & 0 & 0 \\ 0 & 1 & 0 & 0 \\ 0 & 0 & 1 & -2a \\ 0 & 0 & 0 & 1 \end{pmatrix}, \tag{6.155}$$

one has

$$\begin{aligned} T_4 T_3 T_1 &= \begin{pmatrix} 1 & 0 & 0 & 0 \\ 0 & 1 & 0 & 0 \\ 0 & 0 & 1 & -2a \\ 0 & 0 & 0 & 1 \end{pmatrix} \begin{pmatrix} 1 & -1 & 0 & 0 \\ a & b & 0 & 0 \\ -a & -b & a & b \\ 1/2-a & -b & 1/2 & 0 \end{pmatrix} \\ &= \begin{pmatrix} 1 & -1 & 0 & 0 \\ a & b & 0 & 0 \\ -2ab & b(2a-1) & 0 & b \\ 1/2-a & -b & 1/2 & 0 \end{pmatrix}. \end{aligned} \tag{6.156}$$

If applying the transformation 6.156 to Eq. 6.134 or applying the transformation 6.155 to Eq. 6.151, one has the same results of

$$T_4 T_3 T_1 \begin{pmatrix} R_1 \\ R_2 \\ \lambda_1 \Phi_1 \\ \lambda_2 \Phi_2 \end{pmatrix} = \begin{pmatrix} 0 & 0 & q & 0 \\ 0 & 0 & 0 & 1 \\ 0 & b & 0 & 0 \\ 1/2 & 0 & 0 & 0 \end{pmatrix} \begin{pmatrix} \lambda_1 N_1 \\ \lambda_2 N_2 \\ B_1 \\ C_\rho \end{pmatrix}. \tag{6.157}$$

In the above equation, the ionosphere and geometry as well as the ambiguities are diagonal to each other. Such combinations are called diagonal ones. The solution vector of Eq. 6.157 may be easily derived

$$\begin{pmatrix} \lambda_1 N_1 \\ \lambda_2 N_2 \\ B_1 \\ C_\rho \end{pmatrix} = \begin{pmatrix} 0 & 0 & 0 & 2 \\ 0 & 0 & 1/b & 0 \\ 1/q & 0 & 0 & 0 \\ 0 & 1 & 0 & 0 \end{pmatrix} T_4 T_3 T_1 \begin{pmatrix} R_1 \\ R_2 \\ \lambda_1 \Phi_1 \\ \lambda_2 \Phi_2 \end{pmatrix}. \tag{6.158}$$

or

$$\begin{pmatrix} \lambda_1 N_1 \\ \lambda_2 N_2 \\ B_1 \\ C_\rho \end{pmatrix} = \begin{pmatrix} 1-2a & -2b & 1 & 0 \\ -2a & 2a-1 & 0 & 1 \\ 1/q & -1/q & 0 & 0 \\ a & b & 0 & 0 \end{pmatrix} \begin{pmatrix} R_1 \\ R_2 \\ \lambda_1 \Phi_1 \\ \lambda_2 \Phi_2 \end{pmatrix}. \quad (6.159)$$

Equations 6.159 and 6.138 are identical, which shows that the uncombined algorithms and diagonal combinations are equivalent in the case discussed.

## *General Combinations*

For arbitrary combinations, once the transformation matrix is invertible, the transformed equations are equivalent to the original equations based on algebraic theory. The solution vector and the variance–covariance matrix are identical. In other words, regardless of the combinations used, neither the solutions nor the precision of the solutions obtained will differ. Various combinations lead to an easier resolution of specific related problems.

## *Wide- and Narrow-Lane Combinations*

Denoting

$$T_5 = \begin{pmatrix} 0 & 0 & 0 & 2 \\ 0 & 0 & 1/b & 0 \\ 1/q & 0 & 0 & 0 \\ 0 & 1 & 0 & 0 \end{pmatrix} \quad (6.160)$$

and letting transformation matrix

$$T_6 = \begin{pmatrix} \frac{1}{\lambda_1} & \frac{-1}{\lambda_2} & 0 & 0 \\ \frac{1}{\lambda_1} & \frac{1}{\lambda_2} & 0 & 0 \\ 0 & 0 & 1 & 0 \\ 0 & 0 & 0 & 1 \end{pmatrix}, \quad (6.161)$$

one may form the wide and narrow lanes (Petovello 2006) directly by multiplying Eq. 6.161 by Eq. 6.158 to obtain the related wide- and narrow-lane ambiguities

$$\begin{pmatrix} N_1 - N_2 \\ N_1 + N_2 \\ B_1 \\ C_\rho \end{pmatrix} = T_6 T_5 T_4 T_3 T_1 \begin{pmatrix} R_1 \\ R_2 \\ \lambda_1 \Phi_1 \\ \lambda_2 \Phi_2 \end{pmatrix}. \quad (6.162)$$

Indeed, there is $T_5 T_4 T_3 T_1 = T$. Because of the unique properties of the solutions of different combinations, any direct combinations of the solutions must be equivalent to each other. No one combination will lead to a better solution or greater precision of solutions than any other combination. From this rigorous theoretical aspect, the traditional wide-lane ambiguity fixing technique may lead to a more effective search, but not a better solution and precision of the ambiguity.

## 6.7.3
### Secondary GPS Data Processing Algorithms

*In the Case of More Satellites in View*

Up to now, the discussions have been limited for the observations of one satellite viewed by one receiver at one epoch. The original observation equation is given in Eq. 6.134. The solution vector and its covariance matrix are given in Eqs. 6.138 and 6.139, respectively. The elements of the covariance matrix depend on the coefficients of Eq. 6.134, and the coefficients of the observation equation depend on the method of parameterisation. For example, if instead of $B_1$, $B_1^z$ is used, then Eq. 6.134 becomes

$$\begin{pmatrix} R_1(k) \\ R_2(k) \\ \lambda_1 \Phi_1(k) \\ \lambda_2 \Phi_2(k) \end{pmatrix} = \begin{pmatrix} 0 & 0 & F_k f_s^2/f_1^2 & 1 \\ 0 & 0 & F_k f_s^2/f_2^2 & 1 \\ 1 & 0 & -F_k f_s^2/f_1^2 & 1 \\ 0 & 1 & -F_k f_s^2/f_2^2 & 1 \end{pmatrix} \begin{pmatrix} \lambda_1 N_1(k) \\ \lambda_2 N_2(k) \\ B_1^z \\ C_\rho(k) \end{pmatrix}, \qquad (6.163)$$

where $k$ is the index of the satellite. Ionospheric mapping function $F_k$ is dependent on the zenith distance of the satellite $k$. The solution vector of Eq. 6.163 is then similar to that of Eq. 6.138:

$$\begin{pmatrix} \lambda_1 N_1(k) \\ \lambda_2 N_2(k) \\ B_1^z \\ C_\rho(k) \end{pmatrix} = \begin{pmatrix} 1-2a & -2b & 1 & 0 \\ -2a & 2a-1 & 0 & 1 \\ 1/q_k & -1/q_k & 0 & 0 \\ a & b & 0 & 0 \end{pmatrix} \begin{pmatrix} R_1(k) \\ R_2(k) \\ \lambda_1 \Phi_1(k) \\ \lambda_2 \Phi_2(k) \end{pmatrix}, \quad Q(k), \qquad (6.164)$$

where $q_k = qF_k$ and $Q(k)$ is the covariance matrix, which can be similarly derived and given by adding the index $k$ to $q$ in $Q$ of Eq. 6.139. The terms on the right-hand side can be considered secondary "observations" of the unknowns on the left-hand side. If $K$ satellites are viewed, one has the observation equations of one receiver

## 6.7 · Equivalence of the Uncombined and Combining Algorithms

$$
\begin{pmatrix} \lambda_1 N_1(1) \\ \lambda_2 N_2(1) \\ B_1^z \\ C_\rho(1) \\ \vdots \\ \lambda_1 N_1(K) \\ \lambda_2 N_2(K) \\ B_1^z \\ C_\rho(K) \end{pmatrix} = \begin{pmatrix} 1-2a & -2b & 1 & 0 & \ldots & 0 & 0 & 0 & 0 \\ -2a & 2a-1 & 0 & 1 & \ldots & 0 & 0 & 0 & 0 \\ 1/q_1 & -1/q_1 & 0 & 0 & \ldots & 0 & 0 & 0 & 0 \\ a & b & 0 & 0 & \ldots & 0 & 0 & 0 & 0 \\ \vdots & \vdots & \vdots & \vdots & \ldots & \vdots & \vdots & \vdots & \vdots \\ 0 & 0 & 0 & 0 & \ldots & 1-2a & -2b & 1 & 0 \\ 0 & 0 & 0 & 0 & \ldots & -2a & 2a-1 & 0 & 1 \\ 0 & 0 & 0 & 0 & \ldots & 1/q_K & -1/q_K & 0 & 0 \\ 0 & 0 & 0 & 0 & \ldots & a & b & 0 & 0 \end{pmatrix} \begin{pmatrix} R_1(1) \\ R_2(1) \\ \lambda_1 \Phi_1(1) \\ \lambda_2 \Phi_2(1) \\ \vdots \\ R_1(K) \\ R_2(K) \\ \lambda_1 \Phi_1(K) \\ \lambda_2 \Phi_2(K) \end{pmatrix},
$$

(6.165)

and variance matrix

$$
Q_K = \begin{pmatrix} Q(1) & \ldots & 0 \\ \vdots & \ldots & \vdots \\ 0 & \ldots & Q(K) \end{pmatrix}.
$$

(6.166)

Multiplying a transformation matrix

$$
T(K) = \begin{pmatrix} 1 & 0 & 0 & 0 & \ldots & 0 & 0 & 0 & 0 \\ 0 & 1 & 0 & 0 & \ldots & 0 & 0 & 0 & 0 \\ 0 & 0 & 1/K & 0 & \ldots & 0 & 0 & 1/K & 0 \\ 0 & 0 & 0 & 1 & \ldots & 0 & 0 & 0 & 0 \\ \vdots & \vdots & \vdots & \vdots & \ldots & \vdots & \vdots & \vdots & \vdots \\ 0 & 0 & 0 & 0 & \ldots & 1 & 0 & 0 & 0 \\ 0 & 0 & 0 & 0 & \ldots & 0 & 1 & 0 & 0 \\ 0 & 0 & 0 & 0 & \ldots & 0 & 0 & 0 & 1 \end{pmatrix}
$$

(6.167)

to Eq. 6.165, one has the solutions of GPS observation equations of one station

$$
\begin{pmatrix} \lambda_1 N_1(1) \\ \lambda_2 N_2(1) \\ B_1^z \\ C_\rho(1) \\ \vdots \\ \lambda_1 N_1(K) \\ \lambda_2 N_2(K) \\ C_\rho(K) \end{pmatrix} = T(K) \begin{pmatrix} 1-2a & -2b & 1 & 0 & \ldots & 0 & 0 & 0 & 0 \\ -2a & 2a-1 & 0 & 1 & \ldots & 0 & 0 & 0 & 0 \\ 1/q_1 & -1/q_1 & 0 & 0 & \ldots & 0 & 0 & 0 & 0 \\ a & b & 0 & 0 & \ldots & 0 & 0 & 0 & 0 \\ \vdots & \vdots & \vdots & \vdots & \ldots & \vdots & \vdots & \vdots & \vdots \\ 0 & 0 & 0 & 0 & \ldots & 1-2a & -2b & 1 & 0 \\ 0 & 0 & 0 & 0 & \ldots & -2a & 2a-1 & 0 & 1 \\ 0 & 0 & 0 & 0 & \ldots & 1/q_K & -1/q_K & 0 & 0 \\ 0 & 0 & 0 & 0 & \ldots & a & b & 0 & 0 \end{pmatrix} \begin{pmatrix} R_1(1) \\ R_2(1) \\ \lambda_1 \Phi_1(1) \\ \lambda_2 \Phi_2(1) \\ \vdots \\ R_1(K) \\ R_2(K) \\ \lambda_1 \Phi_1(K) \\ \lambda_2 \Phi_2(K) \end{pmatrix},
$$

(6.168)

and the related

$$Q = T(K)Q_K(T(K))^\mathrm{T} \tag{6.169}$$

where mapping function is used to combine the $K$ ionospheric parameters into one. Similar discussions can be made for the cases of using more receivers. The original observation vector and the so-called secondary "observation" vector are

$$\begin{pmatrix} R_1(k) \\ R_2(k) \\ \lambda_1 \Phi_1(k) \\ \lambda_2 \Phi_2(k) \end{pmatrix}, \quad \begin{pmatrix} \lambda_1 N_1(k) \\ \lambda_2 N_2(k) \\ B_1(k) \\ C_\rho(k) \end{pmatrix}. \tag{6.170}$$

The two vectors are equivalent, as proved in Sect. 6.7.2, and they can be uniquely transformed from one to the other. Any further data processing can be considered processing based on the secondary "observations". The secondary "observations" own the equivalence property whether they are uncombined or combining. Therefore the equivalence property is valid for further data processing based on the secondary "observations".

### *GPS Data Processing Using Secondary "Observations"*

A by-product of the above equivalence discussions is that GPS data processing can be performed directly using so-called secondary observations. In addition to the two ambiguity parameters (scaled with the wavelengths), the other two secondary observations are the electron density in the observation path (scaled by square of $f_1$) and the geometry. The geometry includes the whole observation model with the exception of the ionospheric and ambiguity terms. For a time series of the secondary "observations", the electron density (or, for simplicity, "ionosphere") and the "geometry" are real-time observations, whereas the "ambiguities" are constants in case no cycle slip occurs (Langley 1998a, b). Sequential adjustment or filtering methods can be used to deal with the observation time series. It is worth noting that the secondary "observations" are correlated with one another (see the covariance matrix Eq. 6.139). However, the "ambiguities" are direct observations of the ambiguity parameters, and the "ionosphere" and "geometry" are modelled by Eqs. 6.132 and 6.133, respectively. The "ambiguity" observables are ionosphere-geometry-free. The "ionosphere" observable is geometry-free and ambiguity-free. The "geometry" observable is ionosphere-free. But although some algorithms may be more effective, the results and the precision of the solutions are equivalent regardless of the algorithm used. It should be emphasised that all the above discussions are based on the observation model 6.134. The problem concerning the parameterisation of the GPS observation model will not affect the conclusions of these discussions and will be further explored in Chap. 9.

### *Precision Analysis*

If the sequential time series of the original observations are considered time-independent, as they traditionally have been, then the secondary "observations" and their precision are also independent time series. From Eq. 6.140, the standard deviations of the L1 and L2 ambiguities are approximately $5.1281\sigma_c$ and $6.5317\sigma_c$, respectively. The standard deviation of ionosphere and geometry "observations" are about $2.1860\sigma_c$ and $2.9783\sigma_c$, respectively. Thus the precision of the "observed" ambiguities is lower than that of the others at one epoch. If the standard deviation of the P code is approximately 1 dm (phase-smoothed), then the precision of the ambiguities determined by one epoch is lower than 0.5 m. However, an average filter of m epoch data will raise the precision by a factor of sqrt(m) (square root of m). After 100 or 10,000 epochs, the ambiguities are able to be determined with precision of about 5 cm or 5 mm. "Ionospheric" effects are observed with better precision. However, due to the high dynamic of the electron movements, ionospheric effects may not be easily smoothed to improve precision. The "geometry" model is the most complicated, and discussions on static, kinematic, and dynamic applications can be found in numerous publications (cf., e.g., ION proceedings, Chap. 10).

## 6.7.4
### Summary

Here, the equivalence properties between uncombined and combining algorithms have been proved theoretically by algebraic linear transformations. The solution vector and related covariance matrix are identical regardless of the algorithms used. Different combinations can lead to a more effective and easier way of dealing with the data. So-called ionosphere-geometry-free and diagonal combinations have been derived, which have better properties than those of the traditional combinations. A data processing algorithm using the uniquely transformed secondary "observations" has been outlined and used to prove the equivalence. Because of the unique properties of solutions for different combinations, any direct combination of solutions must be equivalent to each other. No one combination will yield a better solution or one with greater precision than any other combination. In this respect, the traditional wide-lane ambiguity fixing technique may lead to a more effective search of ambiguity, but it will not lead to a better solution and precision of the ambiguity. The equivalence of the uncombined and combining algorithms can be called Xu's equivalence theory of GNSS data combinations (Xu 2003, 2007).

## 6.8
## Equivalence of Undifferenced and Differencing Algorithms

In Sect. 6.6, the single, double, and triple differences and their related observation equations were discussed. The number of unknown parameters in the equations was greatly reduced through difference forming; however, the covariance derivations are tedious, especially for a GPS network.

In this section, a unified GPS data processing method based on equivalently eliminated equations is proposed, and the equivalence between undifferenced and differencing algorithms is proved. The theoretical background of the method is also given. By selecting the eliminated unknown vector as a vector of zero, a vector of satellite clock error, a vector of all clock error, a vector of clock and ambiguity parameters, or a vector of user-defined unknowns, the respective selectively eliminated equivalent observation equations can be formed. The equations are equivalent to the zero-, single-, double-, triple-, or user-defined differencing equations. The advantage in such a technique is that the different GPS data processing methods are unified into one unique method, while the original observation vector is retained, and the weight matrix maintains the uncorrelated diagonal form. In other words, the use of this equivalent method allows one to selectively reduce the unknown number, without having to deal with the complicated correlation problem. Several special cases of single, double, and triple difference are discussed in detail to illustrate the theory. The reference-related parameters are dealt with using the a priori datum method.

### *6.8.1*
### *Introduction*

In practice, the common methods for GPS data processing are the so-called zero-difference (non-differential), single-difference, double-difference, and triple-difference methods (Bauer 1994; Hofmann-Wellenhof et al. 1997; King et al. 1987; Leick 1995; Remondi 1984; Seeber 1993; Strang and Borre 1997; Wang et al. 1988). It is well known that the observation equations of the differencing methods can be obtained by carrying out a related linear transformation to the original equations. When the weight matrix is similarly transformed according to the law of covariance propagation, all methods are equivalent theoretically. A theoretical proof of the equivalence between the non-differential and differential methods was described by Schaffrin and Grafarend (1986). A comparison of the advantages and disadvantages of the non-differential and differential methods can be found, for example, in de Jong (1998). The advantage of the differential methods is that there are fewer unknown parameters, and the whole problem to be solved thus becomes smaller. The disadvantage of the differential methods is that there is a correlation problem that appears in cases of multiple baselines of single difference and in all double as well as triple differences. The correlation problem is often complicated and difficult to deal with exactly (compared with the uncorrelated problem). The advantages and disadvantages reach a balance. If one wants to deal with a reduced problem (cancellation of many unknowns), then one has to deal with the correlation problem. As an alternative, we use the equivalent observation equation approach to unify the non-differential and differential methods while retaining all the advantages of both methods.

In the following sections, the theoretical basis of the equivalently eliminated equations is presented, based on the derivation described by Zhou (1985). Several cases are then discussed in detail to illustrate the theory. The reference-related parameters are dealt with using the a priori datum method. A summary of the selectively eliminated equivalent GPS data processing method is outlined at the end.

## 6.8.2
### Formation of Equivalent Observation Equations

For convenience of later discussion, the method for forming an equivalently eliminated equation system is outlined here. The theory is given in Sect. 7.6 in detail. In practice, there may be only one group of unknowns of interest, and it is better to eliminate the other group of unknowns (called nuisance parameters), for example, because of their size. In this case, the use of the so-called equivalently eliminated observation equation system can be very beneficial (Wang et al. 1988; Xu and Qian 1986; Zhou 1985). The nuisance parameters can be eliminated directly from the observation equations instead of from the normal equations.

The linearised observation equation system can be represented using the matrix

$$V = L - (A \quad B) \begin{pmatrix} X_1 \\ X_2 \end{pmatrix} \quad \text{and} \quad P, \qquad (6.171)$$

where $L$ is an observation vector of dimension $n$, $A$ and $B$ are coefficient matrices of dimension $n \times (s - r)$ and $n \times r$, $X_1$ and $X_2$ are unknown vectors of dimension $s - r$ and $r$, $V$ is residual error, $s$ is the total number of unknowns, and $P$ is the weight matrix of dimension $n \times n$.

The related least squares normal equation can then be formed as

$$(A \quad B)^T P (A \quad B) \begin{pmatrix} X_1 \\ X_2 \end{pmatrix} = (A \quad B)^T PL \qquad (6.172)$$

or

$$M_{11}X_1 + M_{12}X_2 = B_1 \quad \text{and} \tag{6.173}$$

$$M_{21}X_1 + M_{22}X_2 = B_2, \tag{6.174}$$

where

$$B_1 = A^T PL, \quad B_2 = B^T PL \quad \text{and}$$
$$\begin{pmatrix} A^T PA & A^T PB \\ B^T PA & B^T PB \end{pmatrix} = \begin{pmatrix} M_{11} & M_{12} \\ M_{21} & M_{22} \end{pmatrix}. \tag{6.175}$$

After eliminating the unknown vector $X_1$, the eliminated equivalent normal equation system is then

$$M_2 X_2 = R_2, \tag{6.176}$$

where

$$M_2 = -M_{21} M_{11}^{-1} M_{12} + M_{22} = B^T PB - B^T PA M_{11}^{-1} A^T PB \quad \text{and} \tag{6.177}$$

$$R_2 = B_2 - M_{21} M_{11}^{-1} B_1 \tag{6.178}$$

The related equivalent observation equation of Eq. 6.176 is then (cf. Sect. 7.6; Xu and Qian 1986; Zhou 1985)

$$U = L - (E - J)BX_2, \quad P, \tag{6.179}$$

where

$$J = AM_{11}^{-1} A^T P. \tag{6.180}$$

$E$ is an identity matrix of size $n$, $L$ and $P$ are the original observation vector and weight matrix, and $U$ is the residual vector, which has the same property as $V$ in Eq. 6.171. The advantage of using Eq. 6.179 is that the unknown vector $X_1$ has been eliminated; however, $L$ vector and $P$ matrix remain the same as the originals.

Similarly, the $X_2$-eliminated equivalent equation system is

$$U_1 = L - (E - K)AX_1 \quad \text{and} \quad P, \tag{6.181}$$

where

$$K = BM_{22}^{-1} B^T P, \quad M_{22} = B^T PB,$$

and $U_1$ is the residual vector (which has the same property as $V$).

We have separated the observation Eq. 6.171 into two equations, Eqs. 6.179 and 6.181; each equation contains only one of the unknown vectors. Each unknown vector can be solved independently and separately. Equations 6.179 and 6.181 are called equivalent observation equations of Eq. 6.171.

The equivalence property of Eqs. 6.171 and 6.179 is valid under three implicit assumptions. The first is that an identical observation vector is used, the second is that the parameterisation of $X_2$ is identical, and the third is that the $X_1$ is able to be eliminated. Otherwise, the equivalence does not hold.

## 6.8.3
## Equivalent Equations of Single Differences

In this section, equivalent equations are first formed to eliminate the satellite clock errors from the original zero-difference equations, and then the equivalence of the single differences (in two cases) related to the original zero-difference equations is proved.

Single differences cancel all satellite clock errors out of the observation equations. This can also be achieved by forming equivalent equations where satellite clock errors are eliminated. Considering Eq. 6.171 the original observation equation, and $X_1$ the vector of satellite clock errors, the equivalent equations of single differences can be formed as outlined in Sect. 6.8.2.

Suppose $n$ common satellites ($k1, k2, \ldots, kn$) are observed at stations $i1$ and $i2$. The original observation equation can then be written as

$$\begin{pmatrix} V_{i1} \\ V_{i2} \end{pmatrix} = \begin{pmatrix} L_{i1} \\ L_{i2} \end{pmatrix} - \begin{pmatrix} E & B_{i1} \\ E & B_{i2} \end{pmatrix} \cdot \begin{pmatrix} X_1 \\ X_2 \end{pmatrix} \quad \text{and} \quad P = \frac{1}{\sigma^2} \begin{pmatrix} E & 0 \\ 0 & E \end{pmatrix}, \quad (6.182)$$

where $X_1$ is the vector of satellite clock errors and $X_2$ is the vector of other unknowns. For simplicity, clock errors are scaled by the speed of light $c$ and directly used as unknowns; the $X_1$-related coefficient matrix is then an identity matrix, $E$.

Comparing Eq. 6.182 with Eq. 6.171, one has (cf. Sect. 6.8.2)

$$A = \begin{pmatrix} E \\ E \end{pmatrix}, \quad B = \begin{pmatrix} B_{i1} \\ B_{i2} \end{pmatrix}, \quad L = \begin{pmatrix} L_{i1} \\ L_{i2} \end{pmatrix} \quad \text{and} \quad V = \begin{pmatrix} V_{i1} \\ V_{i2} \end{pmatrix},$$

and

$$M_{11} = (E\ \ E)\frac{1}{\sigma^2}\begin{pmatrix}E & 0\\ 0 & E\end{pmatrix}\begin{pmatrix}E\\ E\end{pmatrix} = \frac{2}{\sigma^2}E,$$

$$J = \begin{pmatrix}E\\ E\end{pmatrix}\frac{\sigma^2}{2}E(E\ \ E)P = \frac{1}{2}\begin{pmatrix}E & E\\ E & E\end{pmatrix},$$

$$E_{2n\times 2n} - J = \frac{1}{2}\begin{pmatrix}E & -E\\ -E & E\end{pmatrix} \quad \text{and}$$

$$(E_{2n\times 2n} - J)B = \frac{1}{2}\begin{pmatrix}B_{i1} - B_{i2}\\ B_{i2} - B_{i1}\end{pmatrix}.$$

So the equivalently eliminated equation system of Eq. 6.182 is

$$\begin{pmatrix}U_{i1}\\ U_{i2}\end{pmatrix} = \begin{pmatrix}L_{i1}\\ L_{i2}\end{pmatrix} - \frac{1}{2}\begin{pmatrix}B_{i1} - B_{i2}\\ B_{i2} - B_{i1}\end{pmatrix}\cdot X_2, \quad P = \frac{1}{\sigma^2}\begin{pmatrix}E & 0\\ 0 & E\end{pmatrix}, \quad (6.183)$$

where the satellite clock error vector $X_1$ is eliminated, and the observable vector and weight matrix are unchanged.

Denoting $B_s = B_{i2} - B_{i1}$, the least squares normal equation of Eq. 6.183 can then be formed as (cf. Chap. 7) (suppose Eq. 6.183 is solvable)

$$\frac{1}{2}(-B_s^T\ \ B_s^T)\cdot P\cdot\begin{pmatrix}-B_s\\ B_s\end{pmatrix}\cdot X_2 = (-B_s^T\ \ B_s^T)\cdot P\cdot\begin{pmatrix}L_{i1}\\ L_{i2}\end{pmatrix}$$

or

$$B_s^T B_s \cdot X_2 = B_s^T(L_{i2} - L_{i1}). \tag{6.184}$$

Alternatively, a single-difference equation can be obtained by multiplying Eq. 6.182 with a transformation matrix $C_s$

$$C_s = (-E\ \ E),$$

giving

$$C_s\cdot\begin{pmatrix}V_{i1}\\ V_{i2}\end{pmatrix} = C_s\cdot\begin{pmatrix}L_{i1}\\ L_{i2}\end{pmatrix} - C_s\cdot\begin{pmatrix}E & B_{i1}\\ E & B_{i2}\end{pmatrix}\cdot\begin{pmatrix}X_1\\ X_2\end{pmatrix}$$

or

$$V_{i2} - V_{i1} = (L_{i2} - L_{i1}) - (B_{i2} - B_{i1})X_2 \tag{6.185}$$

and

$$\text{cov}(SD(O)) = C_s\sigma^2 \begin{pmatrix} E & 0 \\ 0 & E \end{pmatrix} C_s^T = 2\sigma^2 E \quad \text{and} \quad P_s = \frac{1}{2\sigma^2} E, \tag{6.186}$$

where $P_s$ is the weight matrix of single differences, and cov(SD(O)) is the covariance of the single-difference (SD) observation vector (O). Supposing Eq. 6.185 is solvable, the least squares normal equation system of Eq. 6.185 is then

$$(B_{i2} - B_{i1})^T (B_{i2} - B_{i1}) X_2 = (B_{i2} - B_{i1})^T (L_{i2} - L_{i1}). \tag{6.187}$$

It is clear that Eqs. 6.187 and 6.184 are identical. Therefore, in the case of two stations, the single-difference Eq. 6.185 is equivalent to the equivalently eliminated Eq. 6.183, and is consequently equivalent to the original zero-difference equation.

Suppose $n$ common satellites ($k1, k2, \ldots, kn$) are observed at stations $i1$, $i2$ and $i3$. The original observation equation can then be written as

$$\begin{pmatrix} V_{i1} \\ V_{i2} \\ V_{i3} \end{pmatrix} = \begin{pmatrix} L_{i1} \\ L_{i2} \\ L_{i3} \end{pmatrix} - \begin{pmatrix} E & B_{i1} \\ E & B_{i2} \\ E & B_{i3} \end{pmatrix} \cdot \begin{pmatrix} X_1 \\ X_2 \end{pmatrix} \quad \text{and} \quad P = \frac{1}{\sigma^2} \begin{pmatrix} E & 0 & 0 \\ 0 & E & 0 \\ 0 & 0 & E \end{pmatrix}. \tag{6.188}$$

Comparing Eq. 6.188 with Eq. 6.171, one has (cf. Section 6.8.2)

$$A = \begin{pmatrix} E \\ E \\ E \end{pmatrix}, \quad B = \begin{pmatrix} B_{i1} \\ B_{i2} \\ B_{i3} \end{pmatrix}, \quad L = \begin{pmatrix} L_{i1} \\ L_{i2} \\ L_{i3} \end{pmatrix} \quad \text{and} \quad V = \begin{pmatrix} V_{i1} \\ V_{i2} \\ V_{i3} \end{pmatrix},$$

and

$$M_{11} = A^T P A = \frac{3}{\sigma^2} E,$$

$$J = A \frac{\sigma^2}{3} E A^T P = \frac{1}{3} \begin{pmatrix} E & E & E \\ E & E & E \\ E & E & E \end{pmatrix},$$

$$E_{3n \times 3n} - J = \frac{1}{3} \begin{pmatrix} 2E & -E & -E \\ -E & 2E & -E \\ -E & -E & 2E \end{pmatrix}, \quad \text{and}$$

$$(E_{3n \times 3n} - J)B = \frac{1}{3} \begin{pmatrix} 2B_{i1} - B_{i2} - B_{i3} \\ -B_{i1} + 2B_{i2} - B_{i3} \\ -B_{i1} - B_{i2} + 2B_{i3} \end{pmatrix}.$$

So the equivalently eliminated equation system of Eq. 6.188 is

$$\begin{pmatrix} U_{i1} \\ U_{i2} \\ U_{i3} \end{pmatrix} = \begin{pmatrix} L_{i1} \\ L_{i2} \\ L_{i3} \end{pmatrix} - \frac{1}{3} \begin{pmatrix} 2B_{i1} - B_{i2} - B_{i3} \\ -B_{i1} + 2B_{i2} - B_{i3} \\ -B_{i1} - B_{i2} + 2B_{i3} \end{pmatrix} \cdot X_2, \quad P = \frac{1}{\sigma^2} \begin{pmatrix} E & 0 & 0 \\ 0 & E & 0 \\ 0 & 0 & E \end{pmatrix},$$

(6.189)

and the related least squares normal equation can be formed as

$$\frac{1}{3} \begin{pmatrix} 2B_{i1} - B_{i2} - B_{i3} \\ -B_{i1} + 2B_{i2} - B_{i3} \\ -B_{i1} - B_{i2} + 2B_{i3} \end{pmatrix}^T \begin{pmatrix} 2B_{i1} - B_{i2} - B_{i3} \\ -B_{i1} + 2B_{i2} - B_{i3} \\ -B_{i1} - B_{i2} + 2B_{i3} \end{pmatrix} X_2$$
$$= \begin{pmatrix} 2B_{i1} - B_{i2} - B_{i3} \\ -B_{i1} + 2B_{i2} - B_{i3} \\ -B_{i1} - B_{i2} + 2B_{i3} \end{pmatrix}^T \begin{pmatrix} L_{i1} \\ L_{i2} \\ L_{i3} \end{pmatrix}.$$

(6.190)

Alternatively, for Eq. system 6.188, single differences can be formed using transformation (cf. Sect. 6.6.1)

$$C_s = \begin{pmatrix} -E & E & 0 \\ 0 & -E & E \end{pmatrix}$$

and

$$P_s = [\text{cov}(\text{SD})]^{-1} = \frac{1}{3\sigma^2} \begin{pmatrix} 2E & E \\ E & 2E \end{pmatrix}.$$

The correlation problem appears in the case of single differences of multiple baselines. The related observation equations and the least squares normal equation can be written as

$$\begin{pmatrix} V_{i2} - V_{i1} \\ V_{i3} - V_{i2} \end{pmatrix} = \begin{pmatrix} L_{i2} - L_{i1} \\ L_{i3} - L_{i2} \end{pmatrix} - \begin{pmatrix} B_{i2} - B_{i1} \\ B_{i3} - B_{i2} \end{pmatrix} X_2, \quad P_s \quad \text{and} \qquad (6.191)$$

$$\begin{pmatrix} B_{i2} - B_{i1} \\ B_{i3} - B_{i2} \end{pmatrix}^T \begin{pmatrix} 2E & E \\ E & 2E \end{pmatrix} \begin{pmatrix} B_{i2} - B_{i1} \\ B_{i3} - B_{i2} \end{pmatrix} X_2$$
$$= \begin{pmatrix} B_{i2} - B_{i1} \\ B_{i3} - B_{i2} \end{pmatrix}^T \begin{pmatrix} 2E & E \\ E & 2E \end{pmatrix} \begin{pmatrix} L_{i2} - L_{i1} \\ L_{i3} - L_{i2} \end{pmatrix}. \qquad (6.192)$$

Equations 6.190 and 6.192 are identical. This may be proved by expanding both equations and comparing the results. Again, this shows that the equivalently eliminated equations are equivalent to the single-difference equations, but without the need to deal with the correlation problem.

## 6.8.4
### Equivalent Equations of Double Differences

Double differences cancel all clock errors out of the observation equations. This can also be achieved by forming equivalent equations where all clock errors are eliminated. Considering Eq. 6.171 the original observation equation, and $X_1$ the vector of all clock errors, the equivalent equation of double differences can be formed as outlined in Sect. 6.8.2.

In the case of two stations, supposing $n$ common satellites ($k1, k2,..., kn$) are observed at station $i1$ and $i2$, the equivalent single-difference observation equation is then Eq. 6.183. Denoting $B_{s1} = B_{i2} - B_{i1}$, the station clock error parameter as $\delta t_{i1} - \delta t_{i2}$ (cf. Eqs. 6.89–6.92), and assigning the coefficients of the first column to the station clock errors, i.e. $B_{s1} = (I_{n\times 1} \ B_s)$, Eq. 6.183 turns out to be

$$\begin{pmatrix} U_{i1} \\ U_{i2} \end{pmatrix} = \begin{pmatrix} L_{i1} \\ L_{i2} \end{pmatrix} - \frac{1}{2} \begin{pmatrix} -I_{n\times 1} & -B_s \\ I_{n\times 1} & B_s \end{pmatrix} \begin{pmatrix} X_c \\ X_3 \end{pmatrix} \quad \text{and} \quad P = \frac{1}{\sigma^2} \begin{pmatrix} E & 0 \\ 0 & E \end{pmatrix}, \tag{6.193}$$

where $X_c$ is the station clock error vector, $X_3$ is the other unknown vector, $B_s$ is the $X_3$-related coefficient matrix, $I_{n\times 1}$ is a 1 matrix (where all elements are 1), and clock errors are scaled by the speed of light.

Comparing Eq. 6.193 with Eq. 6.171, one has (cf. Sect. 6.8.2)

$$A = \frac{1}{2}\begin{pmatrix} -I_{n\times 1} \\ I_{n\times 1} \end{pmatrix}, \quad B = \frac{1}{2}\begin{pmatrix} -B_s \\ B_s \end{pmatrix}, \quad L = \begin{pmatrix} L_{i1} \\ L_{i2} \end{pmatrix} \quad \text{and} \quad V = \begin{pmatrix} U_{i1} \\ U_{i2} \end{pmatrix},$$

and

$$M_{11} = \frac{1}{4}\begin{pmatrix} -I^T_{n\times 1} & I^T_{n\times 1} \end{pmatrix} \frac{1}{\sigma^2}\begin{pmatrix} E & 0 \\ 0 & E \end{pmatrix}\begin{pmatrix} -I_{n\times 1} \\ I_{n\times 1} \end{pmatrix} = \frac{n}{2\sigma^2},$$

$$J = \begin{pmatrix} -I_{n\times 1} \\ I_{n\times 1} \end{pmatrix}\frac{\sigma^2}{2n}\begin{pmatrix} -I^T_{n\times 1} & I^T_{n\times 1} \end{pmatrix} \cdot P = \frac{1}{2n}\begin{pmatrix} I_{n\times n} & -I_{n\times n} \\ -I_{n\times n} & I_{n\times n} \end{pmatrix}, \quad \text{and}$$

$$(E_{2n\times 2n} - J)\frac{1}{2}\begin{pmatrix} -B_s \\ B_s \end{pmatrix} = \frac{1}{2}\begin{pmatrix} -E_{n\times n} + \frac{1}{n}I_{n\times n} \\ E_{n\times n} - \frac{1}{n}I_{n\times n} \end{pmatrix} B_s.$$

So the equivalently eliminated equation system of Eq. 6.193 is

$$\begin{pmatrix} U_{i1} \\ U_{i2} \end{pmatrix} = \begin{pmatrix} L_{i1} \\ L_{i2} \end{pmatrix} - \frac{1}{2}\begin{pmatrix} -E_{n\times n} + \frac{1}{n}I_{n\times n} \\ E_{n\times n} - \frac{1}{n}I_{n\times n} \end{pmatrix} B_s X_3 \quad \text{and} \quad P = \frac{1}{\sigma^2}\begin{pmatrix} E & 0 \\ 0 & E \end{pmatrix}, \tag{6.194}$$

where the receiver clock error vector $X_c$ is eliminated, observable vector and weight matrix are unchanged. The normal equation has a simple form of

$$B_s^T \left( E_{n\times n} - \frac{1}{n} I_{n\times n} \right) B_s X_3 = B_s^T \left( E_{n\times n} - \frac{1}{n} I_{n\times n} \right) (L_{i2} - L_{i1}). \quad (6.195)$$

Alternatively, the traditional single-difference observation Eqs. 6.185 and 6.186 can be rewritten as

$$V_{i2} - V_{i1} = (L_{i2} - L_{i1}) - \begin{pmatrix} I_{n\times 1} & B_s \end{pmatrix} \begin{pmatrix} X_c \\ X_3 \end{pmatrix}$$

or

$$\begin{pmatrix} V_{i2}^1 - V_{i1}^1 \\ V_{i2}^k - V_{i1}^k \end{pmatrix} = \begin{pmatrix} L_{i2}^1 - L_{i1}^1 \\ L_{i2}^k - L_{i1}^k \end{pmatrix} - \begin{pmatrix} 1 & B_s^1 \\ I_{m\times 1} & B_s^k \end{pmatrix} \begin{pmatrix} X_c \\ X_3 \end{pmatrix} \quad (6.196)$$

and

$$\mathrm{cov}(\mathrm{SD}(O)) = C_s \sigma^2 \begin{pmatrix} E & 0 \\ 0 & E \end{pmatrix} C_s^T = 2\sigma^2 E \quad \text{and} \quad P_s = \frac{1}{2\sigma^2} E,$$

where $m = n - 1$, and the superscript 1 and $k$ denote the first row and remaining rows of the matrices (or columns in case of vectors), respectively. The double-difference transformation matrix and covariance are (cf. Sect. 6.6.2, Eqs. 6.116–6.118)

$$C_d = \begin{pmatrix} -I_{m\times 1} & E_{m\times m} \end{pmatrix},$$
$$\mathrm{cov}(\mathrm{DD}(O)) = C_d \mathrm{cov}(\mathrm{SD}(O)) C_d^T = 2\sigma^2 C_d C_d^T = 2\sigma^2 (I_{m\times m} + E_{m\times m}) \quad \text{and}$$
$$P_d = [\mathrm{cov}(\mathrm{DD}(O))]^{-1} = \frac{1}{2\sigma^2 n} (n E_{m\times m} - I_{m\times m}).$$

The double-difference observation equation and related normal equation are

$$C_d \begin{pmatrix} V_{i2}^1 - V_{i1}^1 \\ V_{i2}^k - V_{i1}^k \end{pmatrix} = C_d \begin{pmatrix} L_{i2}^1 - L_{i1}^1 \\ L_{i2}^k - L_{i1}^k \end{pmatrix} - C_d \begin{pmatrix} 1 & B_s^1 \\ I_{m\times 1} & B_s^k \end{pmatrix} \begin{pmatrix} X_c \\ X_3 \end{pmatrix}$$

or

$$C_d \begin{pmatrix} V_{i2}^1 - V_{i1}^1 \\ V_{i2}^k - V_{i1}^k \end{pmatrix} = C_d \begin{pmatrix} L_{i2}^1 - L_{i1}^1 \\ L_{i2}^k - L_{i1}^k \end{pmatrix} - C_d \begin{pmatrix} B_s^1 \\ B_s^k \end{pmatrix} X_3,$$

i.e.

$$C_d (V_{i2} - V_{i1}) = C_d (L_{i2} - L_{i1}) - C_d B_s X_3 \quad (6.197)$$

and

$$B_s^T C_d^T P_d C_d B_s X_3 = B_s^T C_d^T P_d C_d (L_{i2} - L_{i1}), \tag{6.198}$$

where

$$C_d^T P_d C_d = \frac{1}{2\sigma^2 n}\begin{pmatrix}-I_{m\times 1} & E_{m\times m}\end{pmatrix}^T (n E_{m\times m} - I_{m\times m})\begin{pmatrix}-I_{m\times 1} & E_{m\times m}\end{pmatrix}, \tag{6.199}$$

$$\begin{pmatrix}-I_{m\times 1} & E_{m\times m}\end{pmatrix}^T (n E_{m\times m} - I_{m\times m}) = \begin{pmatrix}-I_{m\times 1} & n E_{m\times m} - I_{m\times m}\end{pmatrix}^T \text{ and } \tag{6.200}$$

$$\begin{pmatrix}-I_{m\times 1} & n E_{m\times m} - I_{m\times m}\end{pmatrix}^T \begin{pmatrix}-I_{m\times 1} & E_{m\times m}\end{pmatrix} = n E_{n\times n} - I_{n\times n}. \tag{6.201}$$

The above three equations can be readily proved. Substituting Eqs. 6.199–6.201 into Eqs. 6.198, 6.198 then becomes the same as Eq. 6.195, thus proving the equivalence between the double-difference equation and the directly formed equivalent Eq. 6.193.

## 6.8.5
### Equivalent Equations of Triple Differences

Triple differences cancel all clock errors and ambiguities out of the observation equations. This can also be achieved by forming equivalent equations where all clock errors and ambiguities are eliminated. Considering Eq. 6.171 the original observation equation, and $X_1$ the parameter vector of all clock errors and ambiguities, the equivalent equations of triple differences can then be formed as outlined in Sect. 6.8.2.

It is well known that traditional triple differences are correlated between adjacent epochs and between baselines. In the case of sequential (epoch-by-epoch) data processing of triple differences, the correlation problem is difficult to deal with. However, with the use of the equivalently eliminated equations, the weight matrix remains diagonal, and the original GPS observables are retained.

An alternative method for proving the equivalence between triple differences and zero-difference is proposed and derived in Xu (2016). Considering the definition of triple differences and Eq. 6.120 given in Sect. 6.6.3, the triple-difference equation can be rearranged as

$$\begin{aligned}\text{TD}_{i1,i2}^{k1,k2}(O(t1,t2)) &= \{[O_{i2}^{k2}(t2) - O_{i2}^{k2}(t1)] - [O_{i1}^{k2}(t2) - O_{i1}^{k2}(t1)]\} \\ &\quad - \{[O_{i2}^{k1}(t2) - O_{i2}^{k1}(t1)] - [O_{i1}^{k1}(t2) - O_{i1}^{k1}(t1)]\} \\ &= (D^t \cdot O_{i2}^{k2} - D^t \cdot O_{i1}^{k2}) - (D^t \cdot O_{i2}^{k1} - D^t \cdot O_{i1}^{k1}),\end{aligned} \tag{6.202}$$

where $D^t$ represents the time difference observables between time $t1$ and $t2$.

From Eq. 6.202, triple differences can be regarded first as forming the time difference of the same satellite between two adjacent epochs at the station, and then to be formed by double differences between two single differences related to two observed satellites. The time difference equation is proved to be equivalent to the zero-difference equation in Xu (2016), and the time-differenced observable between two adjacent epochs has the same property as the original one, which is still uncorrelated. Moreover, considering the equivalence between the double-difference and zero-difference equations (cf. Sect. 6.8.4), we can thus conclude that the triple-difference equation is equivalent to the zero-difference equation.

## 6.8.6
### Method of Dealing with the Reference Parameters

In differential GPS data processing, the reference-related parameters are usually considered to be known and are fixed (or not adjusted). This may be realised by the a priori datum method (for details cf. Sect. 7.8.2). Here we outline only the basic principle.

The equivalent observation Eq. system 6.179 can be rewritten as

$$U = L - \begin{pmatrix} D_1 & D_2 \end{pmatrix} \begin{pmatrix} X_{21} \\ X_{22} \end{pmatrix} \quad \text{and} \quad P, \qquad (6.203)$$

where

$$D = \begin{pmatrix} D_1 & D_2 \end{pmatrix} \quad \text{and} \quad X_2 = \begin{pmatrix} X_{21} \\ X_{22} \end{pmatrix}.$$

Suppose there are a priori constraints of (cf. e.g. Zhou et al. 1997)

$$W = \overline{X}_{22} - X_{22} \quad \text{and} \quad P_2, \qquad (6.204)$$

where $\overline{X}_{22}$ is the "directly observed" parameter sub-vector, $P_2$ is the weight matrix with respect to the parameter sub-vector $X_{22}$, and $W$ is a residual vector, which has the same property as $U$. Typically, $\overline{X}_{22}$ is "observed" independently, so $P_2$ is a diagonal matrix. If $X_{22}$ is a sub-vector of station coordinates, then the constraint of Eq. 6.204 is referred to as a datum constraint (this is also the reason that the term a priori datum is used). Here we consider $X_{22}$ a vector of reference-related parameters (such as clock errors and ambiguities of the reference satellite and reference station). Generally, the a priori weight matrix $P_2$ is given by covariance matrix $Q_W$ and

$$P_2 = Q_W^{-1}. \tag{6.205}$$

In practice, the sub-vector $\overline{X}_{22}$ is usually a zero vector; this can be achieved through careful initialisation by forming observation Eq. 6.171.

A least squares normal equation of the a priori datum problem of Eqs. 6.203 and 6.204 can be formed (cf. Sect. 7.8.2). Compared with the normal equation of Eq. 6.203, the only difference is that the a priori weight matrix $P_2$ has been added to the normal matrix. This indicates that the a priori datum problem can be dealt with simply by adding $P_2$ to the normal equation of observation Eq. 6.203.

If some diagonal components of the weight matrix $P_2$ are set to zero, the related parameters (in $X_{22}$) are then free parameters (or free datum) of the adjustment problem (without a priori constraints). Otherwise, parameters with a priori constraints are called a priori datum. Large weight values indicate strong constraint and small weight values indicate soft constraint. The strongest constraint is keeping the datum fixed. The reference-related datum (coordinates and clock errors as well as ambiguities) can be fixed by applying the strongest constraints to the related parameters, i.e. by adding the strongest constraints to the datum-related diagonal elements of the normal matrix.

## 6.8.7
### Summary of the Unified Equivalent Algorithm

For any linearised zero-difference GPS observation Eq. system 6.171

$$V = L - (A \quad B)\begin{pmatrix} X_1 \\ X_2 \end{pmatrix} \quad \text{and} \quad P, \tag{6.206}$$

the $X_1$-eliminated equivalent GPS observation equation system is then Eq. 6.179:

$$U = L - (E - J)BX_2 \quad \text{and} \quad P, \tag{6.207}$$

where

$$J = AM_{11}^{-1}A^{T}P, \quad M_{11} = A^{T}PA,$$

$E$ is an identity matrix, $L$ is original observation vector, $P$ is original weight matrix, and $U$ is residual vector, which has the same property as $V$.

Similarly, the $X_2$ eliminated equivalent equation system is Eq. 6.181

$$U_1 = L - (E - K)AX_1 \quad \text{and} \quad P, \tag{6.208}$$

where

$$K = BM_{22}^{-1}B^{T}P, \quad M_{22} = B^{T}PB,$$

and $U_1$ is the residual vector (which has the same property as $V$).

Fixing the values of sub-vector $X_{22}$ (of $X_2$) can be realised by adding the strongest constraints to the $X_{22}$-related diagonal elements of the normal matrix formed by Eq. 6.207. Alternatively, we may first apply the strongest constraints directly to the normal equation formed by Eq. 6.206. In this way, the reference-related parameters (clock errors, ambiguities, coordinates, etc.) are fixed. We may then form the equivalently eliminated observation Eq. 6.207. Thus, relative and differential GPS data processing can be realised by using Eq. 6.207 after selecting the $X_1$ to be eliminated.

The GPS data processing algorithm using Eq. 6.207 is then a selectively eliminated equivalent method. Selecting $X_1$ in Eq. 6.206 as a zero vector, the algorithm is identical to the zero-difference method. Selecting $X_1$ in Eq. 6.206 as the satellite clock error vector, the vector of all clock errors, the clock error and ambiguity vector, and any user-defined vector, the algorithm is equivalent to the single-difference, double-difference, triple-difference, and user-defined elimination methods, respectively. The eliminated unknown $X_1$ can be solved separately if desired.

The advantages of this method (compared with non-differential and differential methods) are as follows:

- Non-differential and differential GPS data processing can be dealt with in an equivalent and unified way. The data processing scenarios can be selected by a switch and used in a combinative way.
- The eliminated parameters can also be solved separately with the same algorithm.
- The weight matrix remains the original diagonal one.
- The original observations are used; no differencing is required.

It is clear that the described algorithm has all the advantages of both non-differential and differential GPS data processing methods. The equivalence theory of the undifferenced and differencing GPS data processing algorithms may be described as Xu's equivalence theory.

# References

Bauer M (1994) Vermessung und Ortung mit Satelliten. Wichmann Verlag, Karslruhe
Blewitt G (1998) GPS data processing methodology. In: Teunissen, P.J.G., and Kleusberg, A. (eds), GPS for Geodesy, Springer-Verlag, Berlin, Heidelberg, New York, 231-270.
Bronstein IN, Semendjajew KA (1987) Taschenbuch der Mathematik.B. G. Teubner Verlagsgesellschaft, Leipzig, ISBN 3-322-00259-4
Jong PJ de (1998) A processing strategy of the application of the GPS in networks. PhD theis, Nether-lands Geodetic Commision, Delft, The Netherland

# References

Hofmann-Wellenhof B, Lichtenegger H, Collins J (1997) GPS theory and practice. Springer-Press, Wien

Hofmann-Wellenhof B, Lichtenegger H, Collins J (2001) GPS theory and practice. Springer-Press, Wien

Hugentobler U, Schaer S, Fridez P (2001) Bernese GPS Software: Version 4.2. Astronomical Inst. Univ. of Berne, Switzerland.

King RW, Masters EG, Rizos C, Stolz A, Collins J (1987) Surveying with Global Positioning System. Dümmler-Verlag, Bonn

Koch KR (1988) Parameter estimation and hypothesis testing in linear models. Springer-Verlag, Berlin

Kouba J, Heroux P (2001) Precise point positioning using IGS orbit and clock products. GPS Solutions, 5(2):12-28

Langley RB (1998a) Propagation of the GPS signals. In: Kleusberg A, Teunissen PJG (eds) GPS for geodesy. Springer-Verlag, Berlin

Langley RB (1998b) GPS receivers and the observables. In: Kleusberg A, Teunissen PJG (eds) GPS for geodesy. Springer-Verlag, Berlin

Leick A (1995) GPS satellite surveying. John Wiley & Sons Ltd., New York

Leick A (2004) GPS Satellite Surveying (3rd ed.) xxiv + 664 p. John Wiley, New York, NY

Petovello MG (2006) Narrowlane: is it worth it? GPS Solutions, DOI 10.1007/s10291-006-0020-1

Remondi B (1984) Using the Global Positioning System (GPS) phase observable for relative geodesy: Modelling, processing, and results. University of Texas at Austin, Center for Space Research

Schaffrin B, Grafarend E (1986) Generating classes of equivalent linear models by nuisance parameter elimination. Manuscr Geodaet 11:262–271

Seeber G (1993) Satelliten-Geodaesie. Walter de Gruyter 1989

Sjoeberg LE (1999) Unbiased vs biased estimation of GPS phase ambiguities from dual-frequency code and phase observables. J Geodesy 73:118–124

Strang G, Borre K (1997) Linear algebra, geodesy, and GPS. Wellesley-Cambridge Press

Wang LX, Fang ZD, Zhang MY, Lin GB, Gu LK, Zhong TD, Yang XA, She DP, Luo ZH, Xiao BQ, Chai H, Lin DX (1979) Mathematic handbook. Educational Press, Peking, ISBN 13012-0165

Wang G, Chen Z, Chen W, Xu G (1988) The principle of GPS precise positioning system. Surveying Press, Peking, ISBN 7-5030-0141-0/P.58, 345 p, (in Chinese)

Xu G (2002) GPS data processing with equivalent observation equations, GPS Solutions, Vol. 6, No. 1-2, 6:28-33

Xu G (2003) GPS – Theory, Algorithms and Applications, Springer Heidelberg, ISBN 3-540-67812-3, 315 pages, in English

Xu G (2004) MFGsoft – Multi-Functional GPS/(Galileo) Software – Software User Manual, (Version of 2004), Scientific Technical Report STR04/17 of GeoForschungsZentrum (GFZ) Potsdam, ISSN 1610-0956, 70 pages, www.gfz-potsdam.de/bib/pub/str0417/0417.pdf

Xu G (2007) GPS – Theory, Algorithms and Applications, second edition, Springer Heidelberg, ISBN 978-3-540-72714-9, 350 pages, in English

Xu Y (2016) GNSS Precise Point Positioning with Application of Equivalence Principle. Dissertation, Technical University of Berlin, Berlin, Germany

Xu G, Qian Z (1986) The application of block elimination adjustment method for processing of the VLBI Data. Crustal Deformation and Earthquake, Vol. 6, No. 4, (in Chinese)

Xu G, Guo J, Yeh TK (2006a) Equivalence of the uncombined and combining GPS algorithms

Zhou J (1985) On the Jie factor. Acta Geodaetica et Geophysica 5 (in Chinese)

Zhou J, Huang Y, Yang Y, Ou J (1997) Robust least squares method. Publishing House of Huazhong University of Science and Technology, Wuhan

Zumberge JF, Heflin MB, Jefferson DC, Watkins MM, Webb FH (1997) Precise point positioning for the efficient and robust analysis of GPS data from large networks. J. Geophysical Res., 102 (B3): 5005-5017.

# Chapter 7
# Adjustment and Filtering Methods

## 7.1 Introduction

In this chapter, we outline the most useful and necessary adjustment and filtering algorithms for statistical and kinematic as well as dynamic GPS data processing. We derive the necessary estimators, and provide a detailed discussion of the relationships between the methods presented.

The adjustment algorithms discussed here include least squares adjustment, sequential application of least squares adjustment via accumulation, sequential least squares adjustment, conditional least squares adjustment, a sequential application of conditional least squares adjustment, block-wise least squares adjustment, a sequential application of block-wise least squares adjustment, a special application of block-wise least squares adjustment for code–phase combinations, an equivalent algorithm to form the eliminated observation equation system, and an algorithm to diagonalise the normal and equivalent observation equations.

The filtering algorithms discussed here include the classic Kalman filter, the sequential least squares adjustment method as a special case of Kalman filtering, the robust Kalman filter, and the adaptively robust Kalman filter.

A priori constrained adjustment and filtering are discussed for solving rank-deficient problems. After a general discussion on a priori parameter constraints, a special case of the so-called a priori datum method is provided. A quasi-stable datum method is also discussed.

A summary is presented at the end of this chapter, and applications of the GPS data processing methods discussed are outlined.

## 7.2 Least Squares Adjustment

The principle of least squares adjustment can be summarised as outlined below (Gotthardt 1978; Cui et al. 1982):

1. The linearised observation equation system can be represented by

$$V = L - AX, P \qquad (7.1)$$

where
- $L$   observation vector of dimension $m$,
- $A$   coefficient matrix of dimension $m \times n$,
- $X$   unknown parameter vector of dimension $n$,
- $V$   residual vector of dimension $m$,
- $n$   number of unknowns,
- $m$   number of observations, and
- $P$   symmetric and definite weight matrix of dimension $m \times m$.

2. The least squares criterion for solving the observation equations is well known as

$$V^T PV = \min, \qquad (7.2)$$

where
- $V^T$   the transpose of the related vector $V$.

3. To solve $X$ and compute $V$, a function $F$ is set as

$$F = V^T PV. \qquad (7.3)$$

The function $F$ reaches minimum value if the partial differentiation of $F$ with respect to $X$ equals zero, i.e.

$$\frac{\partial F}{\partial X} = 2V^T P(-A) = 0$$

or

$$A^T PV = 0, \qquad (7.4)$$

where
- $A^T$   transpose matrix of $A$.

4. Multiplying $A^T P$ with Eq. 7.1, one has

$$A^T PAX - A^T PL = -A^T PV. \qquad (7.5)$$

Setting Eq. 7.4 into 7.5, one has

$$A^T PAX - A^T PL = 0. \qquad (7.6)$$

5. For simplification, let $M = A^T PA$, $Q = M^{-1}$, where superscript $^{-1}$ is an inverse operator, and $M$ is usually called a normal matrix. The least squares solution of Eq. 7.1 is then

$$X = Q(A^T PL). \tag{7.7}$$

6. The precision of the $i$th element of the estimated parameter is

$$p[i] = m_0 \sqrt{Q[i][i]}, \tag{7.8}$$

where $i$ is the element index of a vector or a matrix, $m_0$ is the so-called standard deviation (or sigma), $p[i]$ is the $i$th element of the precision vector, $Q[i][i]$ is the $i$th diagonal element of the cofactor matrix $Q$, and

$$m_0 = \sqrt{\frac{V^T PV}{m-n}}, \quad \text{if} \quad (m > n). \tag{7.9}$$

7. For convenience of sequential computation, $V^T PV$ can be calculated by using

$$V^T PV = L^T PL - (A^T PL)^T X. \tag{7.10}$$

This can be obtained by substituting Eq. 7.1 into $V^T PV$ and considering Eq. 7.4. Thus far, we have derived the complete formulas of least squares adjustment.

## 7.2.1
## Least Squares Adjustment with Sequential Observation Groups

Suppose one has two sequential observation equation systems

$$V_1 = L_1 - A_1 X \quad \text{and} \tag{7.11}$$

$$V_2 = L_2 - A_2 X, \tag{7.12}$$

with weight matrices $P_1$ and $P_2$. These two equation systems are uncorrelated or independent and have the common unknown vector $X$. The combined problem can be represented as

$$\begin{pmatrix} V_1 \\ V_2 \end{pmatrix} = \begin{pmatrix} L_1 \\ L_2 \end{pmatrix} - \begin{pmatrix} A_1 \\ A_2 \end{pmatrix} X \quad \text{and} \quad P = \begin{pmatrix} P_1 & 0 \\ 0 & P_2 \end{pmatrix}. \tag{7.13}$$

The least squares normal equation can be formed then as

$$\begin{pmatrix} A_1^T & A_2^T \end{pmatrix} \begin{pmatrix} P_1 & 0 \\ 0 & P_2 \end{pmatrix} \begin{pmatrix} A_1 \\ A_2 \end{pmatrix} X = \begin{pmatrix} A_1^T & A_2^T \end{pmatrix} \begin{pmatrix} P_1 & 0 \\ 0 & P_2 \end{pmatrix} \begin{pmatrix} L_1 \\ L_2 \end{pmatrix}$$

or
$$(A_1^T P_1 A_1 + A_2^T P_2 A_2) X = (A_1^T P_1 L_1 + A_2^T P_2 L_2). \tag{7.14}$$

This is indeed the accumulation of the two least squares normal equations formed from Eqs. 7.11 and 7.12, respectively:

$$(A_1^T P_1 A_1) X = A_1^T P_1 L_1 \quad \text{and} \tag{7.15}$$

$$(A_2^T P_2 A_2) X = A_2^T P_2 L_2. \tag{7.16}$$

The solution is then

$$X = (A_1^T P_1 A_1 + A_2^T P_2 A_2)^{-1} (A_1^T P_1 L_1 + A_2^T P_2 L_2). \tag{7.17}$$

The precision of the $i$th element of the estimated parameter is

$$p[i] = m_0 \sqrt{Q[i][i]}, \tag{7.18}$$

where

$$m_0 = \sqrt{\frac{V^T P V}{m - n}}, \quad \text{if} \quad (m > n), \quad \text{and} \tag{7.19}$$

$$Q = (A_1^T P_1 A_1 + A_2^T P_2 A_2)^{-1}, \tag{7.20}$$

where $m$ is the number of total observations and $n$ is the number of unknowns. And $V^T P V$ can be calculated by using

$$\begin{aligned} V^T P V &= V_1^T P_1 V_1 + V_2^T P_2 V_2 \\ &= L_1^T P_1 L_1 + L_2^T P_2 L_2 - (A_1^T P_1 L_1)^T X - (A_2^T P_2 L_2)^T X. \\ &= (L_1^T P_1 L_1 + L_2^T P_2 L_2) - (A_1^T P_1 L_1 + A_2^T P_2 L_2)^T X \end{aligned} \tag{7.21}$$

Equation 7.17 indicates that the sequential least squares problem can be solved by simply accumulating the normal equations of the observation equations. The weighted squares residuals can also be computed by accumulating the individual quadratic forms of the residuals using Eq. 7.21.

For further sequential and independent observation equation systems,

$$V_1 = L_1 - A_1 X, \quad P_1, \tag{7.22}$$

$$V_2 = L_2 - A_2 X, \quad P_2, \tag{7.23}$$

$$\ldots$$

$$V_i = L_i - A_i X, \quad P_i, \tag{7.24}$$

the solution can be similarly derived as

$$X = (A_1^T P_1 A_1 + A_2^T P_2 A_2 + \cdots + A_i^T P_i A_i)^{-1} (A_1^T P_1 L_1 + A_2^T P_2 L_2 + \cdots + A_i^T P_i L_i) \tag{7.25}$$

and

$$V^T P V = (L_1^T P_1 L_1 + L_2^T P_2 L_2 + \cdots + L_i^T P_i L_i) \\ - (A_1^T P_1 L_1 + A_2^T P_2 L_2 + \cdots + A_i^T P_i L_i)^T X. \tag{7.26}$$

Obviously, if a solution is needed for every epoch, then the accumulated equation system must be solved at each epoch. The accumulations must always be made with the sequential normal equations. Of course, the solutions can be computed after a defined epoch or at the last epoch, which could be very useful if the solution to the problem is unstable at the beginning.

## 7.3
## Sequential Least Squares Adjustment

Recalling the discussions in Sect. 7.2, one has sequential observation equation systems

$$V_1 = L_1 - A_1 X, \quad P_1 \quad \text{and} \tag{7.27}$$

$$V_2 = L_2 - A_2 X, \quad P_2. \tag{7.28}$$

These two equation systems are uncorrelated. The sequential problem can then be solved by accumulating the individual normal equations as discussed in Sect. 7.2:

$$(A_1^T P_1 A_1 + A_2^T P_2 A_2) X = (A_1^T P_1 L_1 + A_2^T P_2 L_2) \quad \text{or} \tag{7.29}$$

$$X = (A_1^T P_1 A_1 + A_2^T P_2 A_2)^{-1} (A_1^T P_1 L_1 + A_2^T P_2 L_2). \tag{7.30}$$

And $V^T P V$ can be calculated by using

$$V^T P V = (L_1^T P_1 L_1 + L_2^T P_2 L_2) - (A_1^T P_1 L_1 + A_2^T P_2 L_2)^T X. \tag{7.31}$$

If Eq. 7.27 is solvable, the least squares solution can then be represented as

$$X = (A_1^T P_1 A_1)^{-1}(A_1^T P_1 L_1) \quad \text{and} \tag{7.32}$$

$$V^T P V = L_1^T P_1 L_1 - (A_1^T P_1 L_1)^T X. \tag{7.33}$$

For convenience, the estimated vector of $X$ using the first group of observations is denoted by $X_1$ and the quadratic form of the residuals by $(V^T P V)_1$ as well as $Q_1 = (A_1^T P_1 A_1)^{-1}$.

Using the formula (Cui et al. 1982; Gotthardt 1978)

$$(D + ACB)^{-1} = D^{-1} - D^{-1} A K B D^{-1}, \tag{7.34}$$

where $A$ and $B$ are any matrices, $C$ and $D$ are matrices that can be inverted and

$$K = (C^{-1} + B D^{-1} A)^{-1}, \tag{7.35}$$

the inversion of the accumulated normal matrix can be represented as $Q$:

$$\begin{aligned} Q &= (A_1^T P_1 A_1 + A_2^T P_2 A_2)^{-1} \\ &= (A_1^T P_1 A_1)^{-1} - (A_1^T P_1 A_1)^{-1} A_2^T K A_2 (A_1^T P_1 A_1)^{-1} \text{ and} \\ &= Q_1 - Q_1 A_2^T K A_2 Q_1 \\ &= (E - Q_1 A_2^T K A_2) Q_1 \end{aligned} \tag{7.36}$$

$$K = (P_2^{-1} + A_2 Q_1 A_2^T)^{-1}, \tag{7.37}$$

where $E$ is an identity matrix. The total term in the parentheses on the right-hand side of Eq. 7.36 can be interpreted as a modifying factor for $Q_1$ matrix; in other words, due to the sequential Eq. 7.28, the $Q$ matrix can be computed by multiplying a factor to the $Q_1$ matrix. Thus the sequential least squares solution of Eqs. 7.27 and 7.28 can be obtained:

$$\begin{aligned} X &= (Q_1 - Q_1 A_2^T K A_2 Q_1)(A_1^T P_1 L_1 + A_2^T P_2 L_2). \\ &= (E - Q_1 A_2^T K A_2) X_1 + Q(A_2^T P_2 L_2) \end{aligned} \tag{7.38}$$

Mathematically, the solutions to the sequential problem of Eqs. 7.27 and 7.28 will be the same regardless of whether they are solved using accumulation of the least squares, as discussed in Sect. 7.2.1, or using sequential adjustment, as discussed above. However, in practice, the accuracy of the computation is always limited by the effective digits of the computer being used. Such limitations cause inaccuracy in numerical computation, and this inaccuracy will be accumulated and propagated in further computing processes. By comparing the results obtained with the above-mentioned methods, we note that the sequential method will produce a drift in the results. This drift will increase with time and will generally become non-negligible after a long time interval.

## 7.4
## Conditional Least Squares Adjustment

The principle of least squares adjustment with condition equations can be summarised as follows (Gotthardt 1978; Cui et al. 1982):

1. The linearised observation equation system can be represented by Eq. 7.1 (cf. Sect. 7.2).
2. The corresponding condition equation system can be written as

$$CX - W = 0, \qquad (7.39)$$

where
- $C$    coefficient matrix of dimension $r \times n$,
- $W$    constant vector of dimension $r$, and
- $r$    number of conditions.

3. The least squares criterion for solving the observation equations with condition equations is well known as

$$V^T P V = \min, \qquad (7.40)$$

where $V^T$ is the transpose of the related vector $V$.

4. To solve $X$ and compute $V$, a function $F$ can be formed as

$$F = V^T P V + 2K^T(CX - W), \qquad (7.41)$$

where $K$ is a gain vector (of dimension $r$) to be determined.

The function $F$ reaches minimum value if the partial differentiation of $F$ with respect to $X$ equals zero, i.e.

$$\frac{\partial F}{\partial X} = 2V^T P(-A) + 2K^T C = 0;$$

then one has

$$-A^T P V + C^T K = 0 \qquad (7.42)$$

or

$$A^T P A X + C^T K - A^T P L = 0, \qquad (7.43)$$

where $A^T$, $C^T$ are transposed matrices of $A$ and $C$, respectively.

5. Combining Eqs. 7.43 and 7.39, one has

$$A^T PAX + C^T K - A^T PL = 0 \quad \text{and} \tag{7.44}$$

$$CX - W = 0. \tag{7.45}$$

6. For simplification, let $M = A^T PA$, $W_1 = A^T PL$, $Q = M^{-1}$, where superscript $^{-1}$ is an inverse operator. The solutions of Eqs. 7.44 and 7.45 are then

$$\begin{aligned} K &= (CQC^T)^{-1}(CQW_1 - W), \\ X &= -Q(C^T K - W_1) \end{aligned} \tag{7.46}$$

or

$$\begin{aligned} X &= (A^T PA)^{-1}(A^T PL) - (A^T PA)^{-1} C^T K. \\ &= (A^T PA)^{-1}(A^T PL - C^T K) \end{aligned} \tag{7.47}$$

7. The precision of the solutions is then

$$p[i] = m_0 \sqrt{Q_c[i][i]}, \tag{7.48}$$

where $i$ is the element index of a vector or a matrix, $\sqrt{\phantom{x}}$ is the square root operator, $m_0$ is the so-called standard deviation (or sigma), $p[i]$ is the $i$th element of the precision vector, $Q_c[i][i]$ is the $i$th diagonal element of the quadratic matrix $Q_c$, and

$$Q_c = Q - QC^T Q_2 CQ, \tag{7.49}$$

$$Q_2 = (CQC^T)^{-1} \quad \text{and} \tag{7.50}$$

$$m_0 = \sqrt{\frac{V^T PV}{m - n + r}}, \quad \text{if} \quad (m > n - r). \tag{7.51}$$

8. For convenience of sequential computation, $V^T PV$ can be calculated using

$$V^T PV = L^T PL - (A^T PL)^T X - W^T K. \tag{7.52}$$

This can be obtained by substituting Eq. 7.1 into $V^T PV$ and using the relations of Eqs. 7.39 and 7.42.

Thus far, we have derived the complete formulas of conditional least squares adjustment.

## 7.4.1
## Sequential Application of Conditional Least Squares Adjustment

Recalling the least squares adjustment discussed in Sect. 7.2, the linearised observation equation system

$$V = L - AX, \quad P \tag{7.53}$$

has the solution

$$X = (A^T PA)^{-1}(A^T PL). \tag{7.54}$$

The precision of the solutions can be obtained by

$$p[i] = m_0 \sqrt{Q[i][i]}, \tag{7.55}$$

where

$$m_0 = \sqrt{\frac{V^T PV}{m-n}}, \quad \text{if} \quad (m > n), \tag{7.56}$$

and $V^T PV$ can be calculated by using

$$V^T PV = L^T PL - (A^T PL)^T X. \tag{7.57}$$

For convenience, the least squares solution vector is denoted by $X_0$ and weighted residuals square by $(V^T PV)_0$.

Similarly, in the conditional least squares adjustment discussed in Sect. 7.4, the linearised observation equation system and conditional equations read

$$V = L - AX \quad \text{and} \tag{7.58}$$

$$CX - W = 0; \tag{7.59}$$

the solution follows

$$X = (A^T PA)^{-1}(A^T PL - C^T K), \tag{7.60}$$

where $K$ is the gain, and

$$K = (CQC^T)^{-1}(CQW_1 - W). \tag{7.61}$$

The precision vector of the solution vector can be obtained by using Eqs. 7.48–7.52. Using the notations obtained in least squares solution, one has

$$X = X_0 - QC^T K \tag{7.62}$$

and

$$V^T PV = (V^T PV)_0 + (A^T PL)^T QC^T K - W^T K. \tag{7.63}$$

Equation 7.62 indicates that the conditional least squares problem can be solved first without the conditions, and then through the gain $K$ to compute a modification term. The change of the solution is caused by the conditions. For computing the weighted squares of the residuals, Eq. 7.63 can be used (by adding two modification terms to the weighted squares of residuals of the least squares solution). This property is very important for many practical applications such as ambiguity fixing or coordinates fixing. For example, after the least squares solution and fixing the ambiguity values, one needs to compute the ambiguity fixed solution. Of course, one can put the fixed ambiguities as known parameters and go back to solve the problem once again. However, using the above formulas, one can use the fixed ambiguities as conditions to compute the gain and the modification terms to get the ambiguity fixed solution directly. Similarly, this property can be also used for solutions with some fixed station coordinates.

## 7.5
## Block-Wise Least Squares Adjustment

The principle of block-wise least squares adjustment can be summarised as follows (Gotthardt 1978; Cui et al. 1982):

1. The linearised observation equation system can be represented by Eq. 7.1 (cf. Sect. 7.2).
2. The unknown vector $X$ and observable vector $L$ are rewritten as two sub-vectors:

$$\begin{pmatrix} V_1 \\ V_2 \end{pmatrix} = \begin{pmatrix} L_1 \\ L_2 \end{pmatrix} - \begin{pmatrix} A_{11} & A_{12} \\ A_{21} & A_{22} \end{pmatrix} \begin{pmatrix} X_1 \\ X_2 \end{pmatrix} \quad \text{and} \quad P = \begin{pmatrix} P_1 & 0 \\ 0 & P_2 \end{pmatrix}. \tag{7.64}$$

The least squares normal equation can then be formed as

$$\begin{pmatrix} A_{11} & A_{12} \\ A_{21} & A_{22} \end{pmatrix}^T \begin{pmatrix} P_1 & 0 \\ 0 & P_2 \end{pmatrix} \begin{pmatrix} A_{11} & A_{12} \\ A_{21} & A_{22} \end{pmatrix} \begin{pmatrix} X_1 \\ X_2 \end{pmatrix}$$
$$= \begin{pmatrix} A_{11} & A_{12} \\ A_{21} & A_{22} \end{pmatrix}^T \begin{pmatrix} P_1 & 0 \\ 0 & P_2 \end{pmatrix} \begin{pmatrix} L_1 \\ L_2 \end{pmatrix}. \tag{7.65}$$

The normal equation can be denoted by

$$\begin{pmatrix} M_{11} & M_{12} \\ M_{21} & M_{22} \end{pmatrix} \begin{pmatrix} X_1 \\ X_2 \end{pmatrix} = \begin{pmatrix} B_1 \\ B_2 \end{pmatrix} \quad (7.66)$$

or

$$M_{11}X_1 + M_{12}X_2 = B_1 \quad \text{and} \quad (7.67)$$

$$M_{21}X_1 + M_{22}X_2 = B_2, \quad (7.68)$$

where

$$M_{11} = A_{11}^T P_1 A_{11} + A_{21}^T P_2 A_{21}, \quad (7.69)$$

$$M_{12} = M_{21}^T = A_{11}^T P_1 A_{12} + A_{21}^T P_2 A_{22}, \quad (7.70)$$

$$M_{22} = A_{12}^T P_1 A_{12} + A_{22}^T P_2 A_{22}, \quad (7.71)$$

$$B_1 = A_{11}^T P_1 L_1 + A_{21}^T P_2 L_2 \quad \text{and} \quad (7.72)$$

$$B_2 = A_{12}^T P_1 L_1 + A_{22}^T P_2 L_2. \quad (7.73)$$

3. Normal Eqs. 7.67 and 7.68 can be solved as follows: from Eq. 7.67, one has

$$X_1 = M_{11}^{-1}(B_1 - M_{12}X_2). \quad (7.74)$$

Substituting $X_1$ into Eq. 7.68, one gets a normal equation related to the second block of unknowns:

$$M_2 X_2 = R_2, \quad (7.75)$$

where

$$M_2 = M_{22} - M_{21} M_{11}^{-1} M_{12} \quad \text{and} \quad (7.76)$$

$$R_2 = B_2 - M_{21} M_{11}^{-1} B_1. \quad (7.77)$$

The solution of Eq. 7.75 is then

$$X_2 = M_2^{-1} R_2. \quad (7.78)$$

From Eqs. 7.78 and 7.74, the block-wise least squares solution of Eqs. 7.1 and 7.64 can be computed. For estimating the precision of the solved vector, one has (see discussion in Sect. 7.2):

$$p[i] = m_0 \sqrt{Q[i][i]} \qquad (7.79)$$

where

$$m_0 = \sqrt{\frac{V^T P V}{m-n}}, \quad \text{if} \quad (m > n). \qquad (7.80)$$

$Q$ is the inversion of the total normal matrix $M$. $m$ is the number of total observations, and $n$ is the number of unknowns.

Furthermore,

$$Q = \begin{pmatrix} M_{11} & M_{12} \\ M_{21} & M_{22} \end{pmatrix}^{-1} = \begin{pmatrix} Q_{11} & Q_{12} \\ Q_{21} & Q_{22} \end{pmatrix} \quad \text{is denoted,} \qquad (7.81)$$

where (Gotthardt 1978; Cui et al. 1982)

$$Q_{11} = (M_{11} - M_{12} M_{22}^{-1} M_{21})^{-1}, \qquad (7.82)$$

$$Q_{22} = (M_{22} - M_{21} M_{11}^{-1} M_{12})^{-1}, \qquad (7.83)$$

$$Q_{12} = M_{11}^{-1}(-M_{12} Q_{22}), \text{ and} \qquad (7.84)$$

$$Q_{21} = M_{22}^{-1}(-M_{21} Q_{11}). \qquad (7.85)$$

And $V^T P V$ can be calculated by using

$$V^T P V = L^T P L - (A^T P L)^T X. \qquad (7.86)$$

One finds very important applications in GPS data processing by separating the unknowns into two groups, which will be discussed in the next sub-section.

## 7.5.1
### Sequential Solution of Block-Wise Least Squares Adjustment

Suppose one has two sequential observation equation systems

$$V_{t1} = L_{t1} - A_{t1} Y_{t1} \quad \text{and} \qquad (7.87)$$

$$V_{t2} = L_{t2} - A_{t2} Y_{t2}, \qquad (7.88)$$

## 7.5 · Block-Wise Least Squares Adjustment

with weight matrices $P_{t1}$ and $P_{t2}$. The unknown vector $Y$ can be separated into two sub-vectors; one is sequence-dependent and the other is time-independent. Let us assume

$$Y_{t1} = \begin{pmatrix} X_{t1} \\ X_2 \end{pmatrix} \quad \text{and} \quad Y_{t2} = \begin{pmatrix} X_{t2} \\ X_2 \end{pmatrix}, \tag{7.89}$$

where $X_2$ is the common unknown vector, and $X_{t1}$ and $X_{t2}$ are sequential (time) independent unknowns (i.e. they are different from each other).

Equations 7.87 and 7.88 can be solved separately using the block-wise least squares method, as follows (cf. Sect. 7.5):

$$X_{t1} = (M_{11})_{t1}^{-1}(B_1 - M_{12}X_2)_{t1}, \tag{7.90}$$

$$(M_2)_{t1}X_2 = (R_2)_{t1} \quad \text{and} \tag{7.91}$$

$$X_2 = (M_2)_{t1}^{-1}(R_2)_{t1}, \tag{7.92}$$

and

$$X_{t2} = (M_{11})_{t2}^{-1}(B_1 - M_{12}X_2)_{t2}, \tag{7.93}$$

$$(M_2)_{t2}X_2 = (R_2)_{t2} \quad \text{and} \tag{7.94}$$

$$X_2 = (M_2)_{t2}^{-1}(R_2)_{t2}, \tag{7.95}$$

where indices $t1$ and $t2$ outside of the parenthesis indicate that the matrices and vectors are related to Eqs. 7.87 and 7.88, respectively.

The combined solution of Eqs. 7.87 and 7.88 then can be derived as

$$X_{t1} = (M_{11})_{t1}^{-1}((B_1)_{t1} - (M_{12})_{t1}(X_2)_{ta}), \tag{7.96}$$

$$X_{t2} = (M_{11})_{t2}^{-1}((B_1)_{t2} - (M_{12})_{t2}(X_2)_{ta}), \tag{7.97}$$

$$((M_2)_{t1} + (M_2)_{t2})(X_2)_{ta} = (R_2)_{t1} + (R_2)_{t2} \quad \text{and} \tag{7.98}$$

$$(X_2)_{ta} = ((M_2)_{t1} + (M_2)_{t2})^{-1}((R_2)_{t1} + (R_2)_{t2}), \tag{7.99}$$

where index $ta$ means that the solution is related to all equations. The normal equations related to the common unknowns are accumulated and solved for. The solved common unknowns are used for computing sequentially different unknowns.

In the case of many sequential observations, a combined solution could be difficult or even impossible because of the large number of unknowns and the requirement of the computing capacities. Therefore, a sequential solution could be a good alternative. For the sequential observation equations

$$V_{t1} = L_{t1} - A_{t1}Y_{t1}, \quad P_{t1}, \tag{7.100}$$

$$V_{ti} = L_{ti} - A_{ti}Y_{ti}, \quad P_{ti}, \tag{7.101}$$

the sequential solutions are

$$X_{t1} = (M_{11})_{t1}^{-1}(B_1 - M_{12}X_2)_{t1}, \tag{7.102}$$

$$(M_2)_{t1}X_2 = (R_2)_{t1}, \tag{7.103}$$

$$X_2 = (M_2)_{t1}^{-1}(R_2)_{t1}, \tag{7.104}$$

$$X_{ti} = (M_{11})_{ti}^{-1}((B_1)_{ti} - (M_{12})_{ti}X_2), \tag{7.105}$$

$$((M_2)_{t1} + \cdots + (M_2)_{ti})X_2 = (R_2)_{t1} + \cdots + (R_2)_{ti}, \quad \text{and} \tag{7.106}$$

$$X_2 = ((M_2)_{t1} + \cdots + (M_2)_{ti})^{-1}((R_2)_{t1} + \cdots + (R_2)_{ti}). \tag{7.107}$$

It is notable that the sequential solution of the second unknown sub-vector $X_2$ is exactly the same as the combined solution at the last step. The only difference between the combined solution and the sequential solution is that the $X_2$ used are different. In the sequential solution, only the up-to-date $X_2$ is used. Therefore, at end of the sequential solution (Eq. 7.107), the last obtained $X_2$ has to be substituted into all $X_{tj}$ computing formulas, where $j < i$. This can be done in two ways. The first way is to remember all formulas for computing $X_{tj}$, after $X_2$ is obtained from Eq. 7.107, using $X_2$ to compute $X_{tj}$. The second way is to go back to the beginning after the $X_2$ is obtained, and use $X_2$ as the known vector to solve $X_{tj}$ once again. In these ways, the combined sequential observation equations can be solved exactly in a sequential way.

## 7.5.2
## Block-Wise Least Squares for Code–Phase Combination

Recalling the block-wise observation equations discussed in Sect. 7.5, one has

$$\begin{pmatrix} V_1 \\ V_2 \end{pmatrix} = \begin{pmatrix} L_1 \\ L_2 \end{pmatrix} - \begin{pmatrix} A_{11} & A_{12} \\ A_{21} & A_{22} \end{pmatrix} \begin{pmatrix} X_1 \\ X_2 \end{pmatrix} \quad \text{and} \quad P = \begin{pmatrix} P_1 & 0 \\ 0 & P_2 \end{pmatrix}. \tag{7.108}$$

Such an observation equation can be used for solving the problem of code-phase combination. Supposing $L_1$ and $L_2$ are phase and code observation vectors, respectively, and they have the same dimensions, then $X_2$ is a sub-vector that only exists in phase observation equations. Then one has $A_{22} = 0$, and $A_{11} = A_{21}$, as well as $P_1 = w_p P_0$, $P_2 = w_c P_0$, where $P_0$ is the weight matrix, and $w_p$ and $w_c$ are weight factors of phase and code observables. In order to keep the coefficient matrices $A_{11} = A_{21}$, the observable vectors $L_1$ and $L_2$ must be carefully scaled. Equation 7.108 can be rewritten as

$$\begin{pmatrix} V_1 \\ V_2 \end{pmatrix} = \begin{pmatrix} L_1 \\ L_2 \end{pmatrix} - \begin{pmatrix} A_{11} & A_{12} \\ A_{11} & 0 \end{pmatrix} \begin{pmatrix} X_1 \\ X_2 \end{pmatrix} \quad \text{and} \quad P = \begin{pmatrix} w_p P_0 & 0 \\ 0 & w_c P_0 \end{pmatrix}. \tag{7.109}$$

The least squares normal equation can then be formed as

$$\begin{pmatrix} A_{11} & A_{12} \\ A_{11} & 0 \end{pmatrix}^T \begin{pmatrix} w_p P_0 & 0 \\ 0 & w_c P_0 \end{pmatrix} \begin{pmatrix} A_{11} & A_{12} \\ A_{11} & 0 \end{pmatrix} \begin{pmatrix} X_1 \\ X_2 \end{pmatrix}$$
$$= \begin{pmatrix} A_{11} & A_{12} \\ A_{11} & 0 \end{pmatrix}^T \begin{pmatrix} w_p P_0 & 0 \\ 0 & w_c P_0 \end{pmatrix} \begin{pmatrix} L_1 \\ L_2 \end{pmatrix}. \tag{7.110}$$

The normal equation can be denoted by

$$\begin{pmatrix} M_{11} & M_{12} \\ M_{21} & M_{22} \end{pmatrix} \begin{pmatrix} X_1 \\ X_2 \end{pmatrix} = \begin{pmatrix} B_1 \\ B_2 \end{pmatrix}, \tag{7.111}$$

where

$$M_{11} = (w_p + w_c) A_{11}^T P_0 A_{11}, \tag{7.112}$$

$$M_{12} = M_{21}^T = w_p A_{11}^T P_0 A_{12}, \tag{7.113}$$

$$M_{22} = w_p A_{12}^T P_0 A_{12}, \tag{7.114}$$

$$B_1 = A_{11}^T P_0 (w_p L_1 + w_c L_2), \quad \text{and} \tag{7.115}$$

$$B_2 = w_p A_{12}^T P_0 L_1. \tag{7.116}$$

Normal Eq. 7.111 can be solved using the general formulas derived in Sects. 7.2 and 7.5.

## 7.6
## Zhou's Theory: Equivalently Eliminated Observation Equation System

In least squares adjustment, the unknowns can be divided into two groups and then solved in a block-wise manner, as discussed in Sect. 7.5. In practice, sometimes only one group of unknowns is of interest, and it is better to eliminate the other group of unknowns (called nuisance parameters) because of its size, for example. In this case, using the so-called equivalently eliminated observation equation system could be very beneficial (Wang et al. 1988; Xu and Qian 1986; Zhou 1985). The nuisance parameters can be eliminated directly from the observation equations instead of from the normal equations.

The linearised observation equation system can be represented by

$$V = L - (A \quad B)\begin{pmatrix} X_1 \\ X_2 \end{pmatrix}, \quad P. \tag{7.117}$$

where
- $L$ observation vector of dimension $m$,
- $A, B$ coefficient matrices of dimension $m \times (n - r)$ and $m \times r$,
- $X_1, X_2$ unknown vectors of dimension $n-r$ and $r$,
- $V$ residual vector of dimension $m$,
- $n$ number of total unknowns,
- $m$ number of observations, and
- $P$ symmetric and definite weight matrix, of dimension $m \times m$.

The least squares normal equation can then be formed by

$$\begin{pmatrix} M_{11} & M_{12} \\ M_{21} & M_{22} \end{pmatrix} \begin{pmatrix} X_1 \\ X_2 \end{pmatrix} = \begin{pmatrix} B_1 \\ B_2 \end{pmatrix}, \tag{7.118}$$

where

$$\begin{pmatrix} M_{11} & M_{12} \\ M_{21} & M_{22} \end{pmatrix} = \begin{pmatrix} A^\mathrm{T} PA & A^\mathrm{T} PB \\ B^\mathrm{T} PA & B^\mathrm{T} PB \end{pmatrix}, \tag{7.119}$$

$$B_1 = A^\mathrm{T} PL, \quad B_2 = B^\mathrm{T} PL. \tag{7.120}$$

The elimination matrix

$$\begin{pmatrix} E & 0 \\ -Z & E \end{pmatrix} \quad \text{is formed}, \tag{7.121}$$

where $E$ is the identity matrix, 0 is a zero matrix, and $Z = M_{21} M_{11}^{-1}$. $M_{11}^{-1}$ is the inversion of $M_{11}$. Multiplying the elimination matrix Eq. 7.121 to the normal Eq. 7.118 one has

$$\begin{pmatrix} E & 0 \\ -Z & E \end{pmatrix} \begin{pmatrix} M_{11} & M_{12} \\ M_{21} & M_{22} \end{pmatrix} \begin{pmatrix} X_1 \\ X_2 \end{pmatrix} = \begin{pmatrix} E & 0 \\ -Z & E \end{pmatrix} \begin{pmatrix} B_1 \\ B_2 \end{pmatrix},$$

## 7.6 · Zhou's Theory: Equivalently Eliminated Observation Equation System

or

$$\begin{pmatrix} M_{11} & M_{12} \\ 0 & M_2 \end{pmatrix} \begin{pmatrix} X_1 \\ X_2 \end{pmatrix} = \begin{pmatrix} B_1 \\ R_2 \end{pmatrix} \quad (7.122)$$

where

$$M_2 = -M_{21}M_{11}^{-1}M_{12} + M_{22}$$
$$= B^{\mathrm{T}}PB - B^{\mathrm{T}}PAM_{11}^{-1}A^{\mathrm{T}}PB = B^{\mathrm{T}}P(E - AM_{11}^{-1}A^{\mathrm{T}}P)B. \quad (7.123)$$

$$R_2 = B_2 - M_{21}M_{11}^{-1}B_1 = B^{\mathrm{T}}P(E - AM_{11}^{-1}A^{\mathrm{T}}P)L. \quad (7.124)$$

If we are interested only in the unknown vector $X_2$, then only the second equation of Eq. 7.122 needs to be solved. The solution is identical to that obtained by solving all of Eq. 7.122. The above elimination process is similar to the Gauss-Jordan algorithm, which has often been used for the inversion of the normal matrix (or for solving linear equation systems). Indeed, the second equation of Eq. 7.122 is identical to Eq. 7.75 derived in the block-wise least squares adjustment (cf. Section 7.5).

Letting

$$J = AM_{11}^{-1}A^{\mathrm{T}}P, \quad (7.125)$$

one has properties of

$$J^2 = (AM_{11}^{-1}A^{\mathrm{T}}P)(AM_{11}^{-1}A^{\mathrm{T}}P) = AM_{11}^{-1}A^{\mathrm{T}}PAM_{11}^{-1}A^{\mathrm{T}}P = AM_{11}^{-1}A^{\mathrm{T}}P = J,$$
$$(E - J)(E - J) = E^2 - 2EJ + J^2 = E - 2J + J = E - J \quad \text{and}$$
$$[P(E - J)]^{\mathrm{T}} = (E - J^{\mathrm{T}})P = P - (AM_{11}^{-1}A^{\mathrm{T}}P)^{\mathrm{T}}P = P - PAM_{11}^{-1}A^{\mathrm{T}}P = P(E - J),$$

i.e. matrices $J$ and $(E-J)$ are idempotent and $(E-J)^{\mathrm{T}}P$ is symmetric, or

$$J^2 = J, \quad (E - J)^2 = E - J \quad \text{and} \quad (E - J)^{\mathrm{T}}P = P(E - J). \quad (7.126)$$

Using the above derived properties, $M_2$ in Eq. 7.123 and $R_2$ in Eq. 7.124 can be rewritten as

$$M_2 = B^{\mathrm{T}}P(E - J)B = B^{\mathrm{T}}P(E - J)(E - J)B = B^{\mathrm{T}}(E - J)^{\mathrm{T}}P(E - J)B \quad \text{and} \quad (7.127)$$

$$R_2 = B^{\mathrm{T}}P(E - J)L = B^{\mathrm{T}}(E - J)^{\mathrm{T}}PL. \quad (7.128)$$

Denoting

$$D_2 = (E - J)B, \tag{7.129}$$

then the eliminated normal equation (the second equation of Eq. 7.122) can be rewritten as

$$B^T(E-J)^T P(E-J)BX_2 = B^T(E-J)^T PL \quad \text{or} \tag{7.130}$$

$$D_2^T P D_2 X_2 = D_2^T P L. \tag{7.131}$$

This is the least squares normal equation of the following linear observation equation:

$$U_2 = L - D_2 X_2, \quad P \tag{7.132}$$

or

$$U_2 = L - (E-J)BX_2, \quad P, \tag{7.133}$$

where $L$ and $P$ are the original observation vector and weight matrix, and $U_2$ is the residual vector, which has the same property as $V$ in Eq. 7.117.

The advantage in using Eq. 7.133 is that the unknown vector $X_1$ has been eliminated; however, $L$ vector and $P$ matrix remain the same as the originals. Applications of this theory can be found in Sect. 6.8, 8.3, and 9.2. The theory was proposed by Jiangwen Zhou in 1985.

## 7.6.1
## Zhou–Xu's Theory: Diagonalised Normal Equation and the Equivalent Observation Equation

In least squares adjustment, the unknowns can be divided into two groups. One group of unknowns can be eliminated by matrix partitioning to obtain an equivalently eliminated normal equation system of the other group of unknowns. Using the elimination process twice for the two groups of unknowns respectively, the normal equation can be diagonalised. The algorithm can be outlined as follows.

A linearised observation equation and the normal equations can be represented by Eqs. 7.117 and 7.118. From the first equation of 7.118, one has

$$X_1 = M_{11}^{-1}(B_1 - M_{12}X_2). \tag{7.134}$$

## 7.6 · Zhou's Theory: Equivalently Eliminated Observation Equation System

Setting $X_1$ into the second equation of 7.118, one gets an equivalently eliminated normal equation of $X_2$:

$$M_2 X_2 = R_2, \qquad (7.135)$$

where

$$\begin{aligned} M_2 &= M_{22} - M_{21} M_{11}^{-1} M_{12}. \\ R_2 &= B_2 - M_{21} M_{11}^{-1} B_1 \end{aligned} \qquad (7.136)$$

Similarly, from the second equation of 7.118, one has

$$X_2 = M_{22}^{-1}(B_2 - M_{21} X_1). \qquad (7.137)$$

Setting $X_2$ into the first equation of 7.118, one gets an equivalently eliminated normal equation of $X_1$:

$$M_1 X_1 = R_1, \qquad (7.138)$$

where

$$\begin{aligned} M_1 &= M_{11} - M_{12} M_{22}^{-1} M_{21}. \\ R_1 &= B_1 - M_{12} M_{22}^{-1} B_2 \end{aligned} \qquad (7.139)$$

Combining Eqs. 7.138 and 7.135, one has

$$\begin{pmatrix} M_1 & 0 \\ 0 & M_2 \end{pmatrix} \begin{pmatrix} X_1 \\ X_2 \end{pmatrix} = \begin{pmatrix} R_1 \\ R_2 \end{pmatrix}, \qquad (7.140)$$

where (cf., e.g., Cui et al. 1982; Gotthardt 1978)

$$\begin{aligned} Q_{11} &= M_1^{-1}, & Q_{22} &= M_2^{-1} \\ Q_{12} &= -M_{11}^{-1}(M_{12} Q_{22}), & Q_{21} &= -M_{22}^{-1}(M_{21} Q_{11}) \end{aligned}. \qquad (7.141)$$

It is obvious that Eqs. 7.118 and 7.140 are two equivalent normal equations. The solutions of the both equations are identical. Equation 7.140 is a diagonalised normal equation related to $X_1$ and $X_2$. The process of forming Eq. 7.140 from Eq. 7.118 is called the diagonalisation process of a normal equation.

As discussed in Sect. 7.6, the equivalently eliminated observation equation of the second equation of Eq. 7.140 is Eq. 7.133. Similarly, if

$$\begin{aligned} I &= B M_{22}^{-1} B^\mathrm{T} P \text{ and} \\ D_1 &= (E - I) A, \end{aligned}$$

then the equivalently eliminated observation equation of the first normal equation of Eq. 7.140 has the form

$$U_1 = L - (E - I)AX_1, \quad P,$$

where $U_1$ is a residual vector that has the same property as $V$ in Eq. 7.117. $L$ and $P$ are the original observation vector and weight matrix.

The above equation and Eq. 7.133 can be written together as

$$\begin{pmatrix} U_1 \\ U_2 \end{pmatrix} = \begin{pmatrix} L \\ L \end{pmatrix} - \begin{pmatrix} D_1 & 0 \\ 0 & D_2 \end{pmatrix} \begin{pmatrix} X_1 \\ X_2 \end{pmatrix}, \quad \begin{pmatrix} P & 0 \\ 0 & P \end{pmatrix}. \tag{7.142}$$

Equation 7.142 is derived from the normal Eq. 7.140; therefore, it is true inversely, i.e. Equation 7.140 is the least squares normal equation of the observation Eq. 7.142. Equations 7.118 and 7.140 are normal equations of the observation Eqs. 7.117 and 7.142. Thus, Eq. 7.142 is an equivalent observation equation of Eq. 7.117. Equations 7.140 and 7.142 are called diagonalised equations of 7.118 and 7.117, respectively. This diagonalised normal equation and the equivalent observation equation could be called Zhou–Xu diagonalisation and equivalent theory (Xu 2003).

## 7.7
## Kalman Filter

### 7.7.1
### Classic Kalman Filter

The principle of the classical Kalman filter can be summarised as follows (Yang et al. 1999):

The linearised observation equation system can be represented by

$$V_i = L_i - A_i X_i, \quad P_i, \tag{7.143}$$

where
- $L$    observation vector of dimension $m$,
- $A$    coefficient matrix of dimension $m \times n$,
- $X$    unknown vector of dimension $n$,
- $V$    residual vector of dimension $m$,
- $n$    number of unknowns,
- $m$    number of observations,
- $i$    sequential index, $i = 1,2,3,\ldots$, and
- $P_i$   weight matrix of index $i$.

Suppose the system equations are known and can be presented as

$$U_i = X_i - F_{i,i-1}X_{i-1}, \quad i = 2, 3, \ldots, \tag{7.144}$$

where
F  transition matrix of dimension $n \times n$, and
U  residual vector of dimension $n$.

$U$ and $V$ are uncorrelated and have zero expectations. Using the covariance propagation law, one has from Eq. 7.144

$$Q(X_i) = F_{i,i-1}Q(X_{i-1})(F_{i,i-1})^T + Q_U. \tag{7.145}$$

The normal Eq. 7.143 can be formed as

$$M_i X_i = B_i. \tag{7.146}$$

For the initial step or epoch, i.e. $i = 1$, Eq. 7.146 has the solution under the least squares principle

$$\tilde{X}_i = Q_i B_i, \quad \text{where} \quad Q_i = M_i^{-1}, \tag{7.147}$$

and here one will assume

$$\tilde{Q}_i = Q_i, \tag{7.148}$$

where $\tilde{X}_i$ and $\tilde{Q}_i$ are called estimated values. Using the estimated values and transition matrix, one can predict the unknown values and covariance matrix of the next epoch (say $i = 2$):

$$\underline{X}_i = F_{i,i-1}\tilde{X}_{i-1} \quad \text{and} \tag{7.149}$$

$$\underline{Q}_i = F_{i,i-1}\tilde{Q}_{i-1}(F_{i,i-1})^T + Q_U, \tag{7.150}$$

where $\underline{X}_i$ and $\underline{Q}_i$ are called predicted values (vector and matrix). Then estimated values of this epoch can be calculated by

$$\tilde{X}_i = \underline{X}_i + K(L_i - A_i\underline{X}_i), \tag{7.151}$$

$$\tilde{Q}_i = (E - KA_i)\underline{Q}_i, \quad \text{and} \tag{7.152}$$

$$K = \underline{Q}_i A_i^T (A_i \underline{Q}_i A_i^T + Q_V)^{-1}, \tag{7.153}$$

where $K$ is the gain matrix.

For the next sequential step $i$, the predicted values must be computed by using Eqs. 7.149 and 7.150, and the estimated values can be computed by using Eqs. 7.151 and 7.152. This iterative process is called Kalman filtering.

In classical Kalman filtering, it is assumed that for the problem of Eq. 7.143 there exists a system transition matrix $F_{i,i-1}$ in Eq. 7.144 and the cofactor $Q_U$. Therefore, the estimated values in the Kalman filter process are dependent on $F_{i,i-1}$ and $Q_U$. The transition matrix will be based on strengthened physical models, and the cofactor will be well known or reasonably given. If the system description is accurate enough, of course Kalman filtering will lead to a more precise solution. However, if the system is not sufficiently well known, the results of Kalman filter will sometimes not converge to the true values (divergence). Furthermore, a kinematic process is generally difficult to be precisely represented by theoretical system equations. However, for a dynamic process (such as onboard GPS for satellite to satellite tracking or orbit determination) the system equation can be well formulated (by an orbital equation of motion). Another problem of Kalman filtering is the strong dependency of the given initial values. Many studies have been made in this area to overcome the above-mentioned shortages.

## 7.7.2
## Kalman Filter: A General Form of Sequential Least Squares Adjustment

The sequential least squares problem is a special case of the classic Kalman filter. If one lets

$$F_{i,i-1} = E, \quad (7.154)$$

then the system Eq. 7.144 in Sect. 7.7.1 turns out to be

$$X_i = X_{i-1}, \quad U = 0 \quad \text{and} \quad Q_U = 0. \quad (7.155)$$

The Kalman filter process is then as follows, for the initial step or epoch, i.e. $i = 1$, Eq. 7.27 in Sect. 7.3 has the solution under the least squares principle:

$$\tilde{X}_i = Q_i B_i, Q_i = M_i^{-1}, \quad (7.156)$$

with

$$\tilde{Q}_i = Q_i, \quad (7.157)$$

where $\tilde{X}_i$ and $\tilde{Q}_i$ are called estimated values. The predicted unknown values and covariance matrix of the next epoch (say $i = 2$) of Eqs. 7.149 and 7.150 in Sect. 7.7.1 are then

$$\underline{X}_i = \tilde{X}_{i-1} \text{ and} \tag{7.158}$$

$$\underline{Q}_i = \tilde{Q}_{i-1}. \tag{7.159}$$

The estimated values of Eqs. 7.151, 7.152 and 7.153 in Sect. 7.7.1 can be simplified as

$$\tilde{X}_i = \tilde{X}_{i-1} + G(L_i - A_i\tilde{X}_{i-1}), \tag{7.160}$$

$$\tilde{Q}_i = (E - GA_i)\tilde{Q}_{i-1}, \text{ and} \tag{7.161}$$

$$G = \tilde{Q}_{i-1}A_i^T(A_i\tilde{Q}_{i-1}A_i^T + Q_V)^{-1}, \tag{7.162}$$

where $G$ denotes the gain matrix. If one notices that $Q_V = (P_i)^{-1}$ and applies the formula of Bennet (Cui et al. 1982; Koch 1986), one has

$$\tilde{Q}_{i-1}A_i^T(A_i\tilde{Q}_{i-1}A_i^T + Q_V)^{-1} = \tilde{Q}_{i-1}A_i^T P_i. \tag{7.163}$$

Equation 7.160 can then be rewritten as

$$\begin{aligned}\tilde{X}_i &= (E - GA_i)\tilde{X}_{i-1} + GL_i. \\ &= (E - GA_i)\tilde{X}_{i-1} + \tilde{Q}_i A_i^T P_i L_i\end{aligned} \tag{7.164}$$

Comparing the derived Eqs. 7.161 and 7.164 with the Eqs. 7.36 and 7.38 derived in Sect. 7.3, one can easily determine that they are identical. Therefore, the sequential least squares adjustment is a special case of Kalman filtering.

### 7.7.3
### Robust Kalman Filter

The classical Kalman filter is suitable for real-time applications. The chief problem in Kalman filtering is the divergence caused by the inexact descriptions of system equations and its statistical properties, as well as the divergence caused by data with inhomogeneous precision.

Efforts have been made to modify the performance of Kalman filtering. In the classical Kalman filter, the weight matrix $P$ of the observables is static, i.e. $P$ is assumed to be a definite matrix. Taking the residuals of Kalman filtering into account, one may adjust the weight $P$ of the observables accordingly. This process is called robust Kalman filtering (Koch and Yang 1998a, b; Yang 1999).

Generally, observations are either accepted or rejected in least squares adjustment and the classical Kalman filter. In other words, the weight is either set as 1 (accepted) or zero (rejected). In the robust Kalman filter, a continuous weight between 1 and zero is introduced.

Originally, one has $P = (Q_V)^{-1}$, the adjusted $P$ is denoted by $\bar{P}$; then the Eq. 7.153 in the classical Kalman filter can be rewritten as

$$K = \underline{Q}_i A_i^T (A_i \underline{Q}_i A_i^T + \bar{P}_i^{-1})^{-1}. \tag{7.165}$$

In the case of independent observations, $P_i$ is a diagonal matrix. Taking the residuals into account, $P_i$ may be adjusted as (Huber 1964; Yang et al. 2000)

$$\bar{P}_i(k) = \begin{cases} P_i(k) \\ P_i(k) \frac{c}{|V_i(k)/\sigma_i|}, \end{cases} \text{if} \quad \begin{aligned} &|V_i(k)/\sigma_i| \leq c \\ &|V_i(k)/\sigma_i| > c \end{aligned}, \tag{7.166}$$

where $V_i(k)$ is the $k$th element of the vector $V$, $P_i(k)$ is the diagonal element of matrix $P_i$, and $c$ is a constant, which is usually chosen as 1.3–2.0 (Yang et al. 2000). $V_i$ is the residual of the observation $L_i$, $\sigma_i$ is the standard deviation of the $i$th epoch, and $P_i = 1/\sigma_i$. In this way, the weight of the observation $L_i$ is adjusted due to the related residual.

If the observations are correlated with each other, the weight matrix may be given by (Yang et al. 2000)

$$\bar{P}_{kj} = \begin{cases} P_{kj} \\ P_{kj} \frac{c}{\max\{|V_i(k)/\sigma_i|,|V_i(j)/\sigma_i|\}}, \end{cases} \text{if} \quad \begin{aligned} &|V_i(k)/\sigma_i| \leq c \text{ and } |V_i(j)/\sigma_i| \leq c \\ &|V_i(k)/\sigma_i| > c \text{ or } |V_i(j)/\sigma_i| > c \end{aligned}. \tag{7.167}$$

It is obvious that an adjusted weight matrix can better reflect the different data quality and can better fit the reality of the observations.

Usually the outlier will be rejected if the absolute value of the residual is greater than $e\sigma_i$, i.e. $|V_i| > e\sigma_i$, where $e$ is a constant, $e$ may be selected as 3–4, $\sigma_i$ is the standard deviation, and $i$ is the iterative calculation index. That is, $P_{-i} = 0$ if $|V_i/\sigma_i| \geq e$. Setting $|V_i/\sigma_i| = e$ into Eq. 7.166 one gets $\bar{P}_i = (c/e)P_i$. In other words, the weight definitions of Eqs. 7.166 and 7.167 are not continuous at point $e$. A modification (Xu 2003) of Eq. 7.166 can be made by defining

$$\bar{P}_i(k) = \begin{cases} p_i(k) \\ y_1 P_i(k) \\ y_2 P_i(k) \\ 0 \end{cases} \text{if} \quad \begin{cases} |V_i(k)/\sigma_i| \leq c \\ c < |V_i(k)/\sigma_i| \leq d \\ d < |V_i(k)/\sigma_i| \leq e \\ |V_i(k)/\sigma_i| \leq e \end{cases}, \tag{7.168}$$

where

$$y_1 = 1 - \frac{1-b}{(d-c)^2} \left( \left| \frac{V_i(k)}{\sigma_i} \right| - c \right)^2 \quad \text{and} \tag{7.169}$$

$$y_2 = \frac{b}{(e-d)^2}\left(e - \left|\frac{V_i(k)}{\sigma_i}\right|\right)^2, \qquad (7.170)$$

where $b$ is the value of $y_1$ if $|V_i(k)/\sigma_i| = d$. $c, d, e$ are constants, and $0 < c < d < e$. For simplification, if one lets $b = (e-d)/(e-c)$, then one has $1 - b = (d-c)/(e-c)$. One may let $d = (e+c)/2$ for further simplification and have

$$y_1 = 1 - \frac{2}{(e-c)^2}\left(\left|\frac{V_i(k)}{\sigma_i}\right| - c\right)^2 \quad \text{and}$$

$$y_2 = \frac{2}{(e-c)^2}\left(e - \left|\frac{V_i(k)}{\sigma_i}\right|\right)^2.$$

By selecting $c = 1$, $e = 3$, and using the above assumptions, the weight functions of Eqs. 7.166 and 7.168 are shown in Fig. 7.1 with broken and continuous lines. It is obvious that Eq. 7.168 is a more reasonable weight function, which may make the Kalman filter more robust.

Similar determinations can be made similarly for correlated cases. Denoting $|V_i(k)/\sigma_i|$ as $v(k)$, a modification of Eq. 7.167 can be rewritten as (Xu 2007)

$$\bar{P}_i(k,j) = \begin{cases} p_i(k,j) \\ z_1 P_i(k,j) \\ z_2 P_i(k,j) \\ 0 \end{cases}, \quad \text{if} \quad \begin{cases} \max\{v(k), v(j)\} \leq c \\ c < \max\{v(k), v(j)\} \leq d \\ d < \max\{v(k), v(j)\} \leq e \\ \max\{v(k), v(j)\} > e \end{cases}, \qquad (7.171)$$

where

$$z_1 = 1 - \frac{1-b}{(d-c)^2}(\max\{v(k), v(j)\} - c)^2 \quad \text{and} \qquad (7.172)$$

**Fig. 7.1** Weight functions

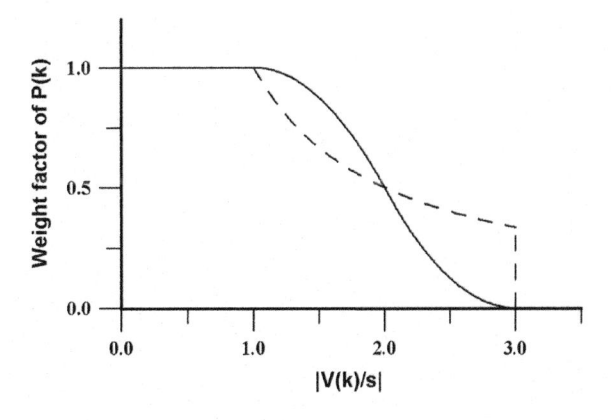

$$z_2 = \frac{b}{(e-d)^2}(e - \max\{v(k), v(j)\})^2, \qquad (7.173)$$

where $b$ is the value of $z_1$ if $\max\{v(k), v(j)\} = d$. For simplification, if one lets $b = (e-d)/(e-c)$, then one has $1-b = (d-c)/(e-c)$. Further if one lets $d = (e-c)/2$, then one has

$$z_1 = 1 - \frac{2}{(e-c)^2}(\max\{v(k), v(j)\} - c)^2 \quad \text{and}$$

$$z_2 = \frac{2}{(e-c)^2}(e - \max\{v(k), v(j)\})^2.$$

### 7.7.4
### Yang's Filter: Adaptively Robust Kalman Filtering

The reliability of the linear filtering results, however, will degrade when the noise of the kinematic model is not accurately modelled in filtering or the measurement noise at any measurement epoch is not normally distributed. In this section, we introduce a new adaptively robust filtering technique proposed by Yang et al. (2001a, b) based on the robust M (maximum likelihood type) estimation. It consists in weighting the influence of the updated parameters in accordance with the magnitude of the discrepancy between the updated parameters and the robust estimates obtained from the kinematic measurements, and in weighting individual measurements at each discrete epoch. The new procedure is different from functional model error compensation; it changes the covariance matrix or, equivalently, changes the weight matrix of the predicted parameters to cover the model errors. A general estimator for an adaptively robust filter is presented, which includes the estimators of the classical Kalman filter, adaptive Kalman filter, robust filter, sequential least squares (LS) adjustment, and robust sequential adjustment. The procedure not only resists the influence of outlying kinematic model errors, but also controls the effects of measurement outliers. In addition to the robustising properties, feasibility in implementation of the new filter is achieved through the equivalent weights of the measurements and the predicted state parameters.

Applications of the Kalman filter in dynamic or kinematic positioning have sometimes encountered difficulties, which have been referred to as divergences. These divergences can often be traced to three factors: (1) insufficient accuracy in modelling the dynamics or kinematics (functional model errors of the state equations); (2) insufficient accuracy in modelling the observations (functional model errors of observation equations); and (3) insufficient accuracy in modelling the distributions or the priori covariance matrices of the measurements and the updated parameters (stochastic model errors).

The current basic procedure for the quality control of a Kalman filter consists of the following:

- Functional model compensation for model errors by introducing uncertain parameters into the state and/or the observation equations. Any model error term can be arbitrarily introduced into the models, and the state can then be augmented (Jazwinski 1970, p. 308). A similar approach was developed by Schaffrin (1991, pp. 32–34). Here, the state vector is partitioned into $h$ groups, each affected by a common scale error, and $h \times 1$ vectors of scale parameters are then introduced into the models. This type of approach may, of course, lead to a high-dimensional state vector, which in turn greatly increases the filter computational load (Jazwinski 1970, p. 305).
- Stochastic model compensation by introducing a variance–covariance matrix of the model errors. In taking this approach to prevent divergence, one must determine which covariance matrix to add. A reasonable covariance matrix may compensate for the model errors. An ineffective covariance matrix, however, adds to the model divergence. For instance, when the model is accurate in some dynamic or kinematic periods, an unsuitable increase of the covariance matrix of model error will degrade the state estimators. Thus an effective covariance matrix for model errors can be determined only by trial and error.
- The DIA procedure—detection, identification, and adaptation (Teunissen 1990). This approach employs a recursive testing procedure to eliminate outliers. In the detection step, one looks for unspecified model errors, and in the identification step, one tries to find the cause of the model error and its most likely starting time. After a model error has been detected and identified, the bias in the state estimate caused by the model error must be eliminated as well. This model recovery from errors is called adaptation (Salzmanm 1995). The identification of the model, however, is quite difficult, especially when the measurements are not accurate enough to detect the unspecified model errors.
- The sequential least squares procedure. A rather different procedure frequently used for kinematic positioning does not use the dynamic model information at all, but determines discrete positions at the measurement epochs (Cannon et al. 1986). In this case, there is no assumption made on a dynamic model, and only the measurements at the discrete epoch are employed to estimate the state parameters. The model error, therefore, does not affect the estimates of new state parameters. This method is typically presented as a sequential least squares algorithm (Schwarz et al. 1989). The current limitation to this approach is that it wastes the good information of the state model in cases when the model accurately describes the dynamic process.
- Adaptive Kalman filtering. An innovation-based adaptive Kalman filter for an integrated INS/GPS was developed by Mohamed and Schwarz (1999), based on the maximum likelihood criterion by proper choice of the filter weight. Another adaptive Kalman filter algorithm to directly estimate the variance and covariance components for the measurements was studied by Wang et al. (1999). Both

algorithms need to collect the residuals of the measurements or the updated series to calculate the state variance–covariance matrices.
- A robust filter based on the min–max robust theory. The deviation of observation error distribution from Gaussian distribution may also seriously degrade the performance of Kalman filtering. Thus, there appears to be considerable motivation for considering filters which are robustised to perform fairly well in non-Gaussian environments. To address this problem, Masreliez and Martin (1977) applied the influence function of the min–max robust theory to replace the score function of the classical Kalman filter. The key disadvantages with this kind of robust filter are that the estimator requires symmetric distribution of the unknown contamination, and this filter does not work as well as the standard Kalman filter in Gaussian noise.
- A robust filter based on M estimation theory (Huber 1964) and Bayesian statistics. To resist the negative influence of both state model errors and measurement outliers, a robust M–M filter was developed (Yang 1991, 1997a, b; Zhou et al. 1997, p. 299). Here, measurement outliers are controlled by robust equivalent weights of the measurements, and the model errors are resisted by the equivalent weights of the updated parameters according to the divergence of the predicted parameters from the estimated parameters. In addition, a robust filter for rank-deficient observation models was developed by Koch and Yang (1998a, b), using Bayesian statistics and applying the robust M estimate.

All of the methods described above require knowledge of the dynamic model errors, with which the functional or stochastic models to compensate for the model errors and the equivalent weights for the robust filter are constructed. In practical applications, it is very difficult to predict the error distribution or the error type of the updated parameters or the dynamic model errors, and thus it is very difficult to construct functional and stochastic models. Furthermore, when a moving vehicle accelerates from zero or decelerates to a stop, the acceleration profile is discontinuous. If this discontinuity falls between two measurement epochs, the dynamics cannot be accurately modelled or predicted by state equations; in this case, one should not rely too heavily on the information predicted from the dynamic model. Thus, the filtering procedure should weaken the effects of the updated parameters. In addition, if the updated parameter vector is contaminated by model error, it is generally distorted in its entirety. Therefore, it is not necessary to consider the error influence of the individual element of the updated parameter vector as is done with the robust M–M filter. In this case, an adaptive filter is suitable for balancing the dynamic model information and the measurements.

1. ***General Estimator of Adaptively Robust Filtering***

An adaptively robust filter is constructed as (cf. Yang et al. 2001a, b)

$$\tilde{X}_i = (A_i^T \bar{P}_i A_i + \alpha P_{\underline{X}_i})^{-1}(A_i^T \bar{P}_i L_i + \alpha P_{\underline{X}_i} \underline{X}_i) \quad \text{and} \quad (7.174)$$

$$Q_{\tilde{X}_i} = (A_i^T \bar{P}_i A_i + \alpha P_{\underline{X}_i})^{-1} \sigma_0^2, \tag{7.175}$$

where $\bar{P}_i$ is the equivalent weight matrix of the observation vector, $P_{\underline{X}_i}$ is the weight matrix of the predicted vector $\underline{X}_i$, $Q_{\tilde{X}_i}$ is the covariance matrix of the estimated state vector, $\sigma_0^2$ is a scale factor, and $\alpha$ is an adaptive factor, which can be chosen as

$$\alpha = \begin{cases} 1 & |\Delta \tilde{X}_i| \leq c_0 \\ \frac{c_0}{|\Delta \tilde{X}_i|} (\frac{c_1 - |\Delta \tilde{X}_i|}{c_1 - c_0})^2 & c_0 < |\Delta \tilde{X}_i| \leq c_1 , \\ 0 & |\Delta \tilde{X}_i| > c_1 \end{cases} \tag{7.176}$$

where $c_0$ and $c_1$ are constants that are experienced, valued as $c_0 = 1.0$–$1.5$, $c_1 = 3.0$–$4.5$,

$$\Delta \tilde{X}_i = \frac{\|\hat{X}_i - \underline{\hat{X}}_i\|}{\sqrt{tr\{Q_{\hat{X}_i}\}}}, \tag{7.177}$$

and $\hat{X}_i$ is a robust estimate of state vector (state position), which is evaluated only by new measurements at epoch i, and the raw velocity observations are not included in it. $\underline{\hat{X}}_i$ is a predicted position from Eq. 7.149 in which the a priori velocity components are not included. The change in the position expressed by Eq. 7.177 can also reflect the stability of the velocity (cf. Yang et al. 2001a, b).

Expression 7.174 is a general estimator of an adaptively robust filter. In the case of $\alpha \neq 0$, Eq. 7.174 is changed, using the matrix identities (Koch 1988, p. 40), into

$$\tilde{X}_i = \underline{X}_i + Q_{\underline{X}_i} A_i^T (A_i Q_{\underline{X}_i} A_i^T + \alpha Q_V)^{-1} (L_i - A_i \underline{X}_i). \tag{7.178}$$

## 2. Special Estimators

The adaptive factor $\alpha$ changes between zero and one, which balances the contribution of the new measurements and the updated parameters to the new estimates of state parameters.

**Case 1**: If $\alpha = 0$ and $\bar{P}_i = P_i$, then

$$\tilde{X}_i = (A_i^T P_i A_i)^{-1} A_i^T P_i L_i, \tag{7.179}$$

which is an LS estimator by using only the new measurements at epoch i. This estimator is suitable in the case where the measurements are not contaminated by outliers and the updated parameters are biased to such a degree that $\Delta \tilde{X}_i$ in Eq. 7.177 is larger than $c_1$ (rejecting point), and the information of updated parameters is completely forgotten.

**Case 2**: If $\alpha = 1$ and $\bar{P}_i = P_i$, then

$$\tilde{X}_i = (A_i^T P_i A_i + P_{\underline{X}_i})^{-1}(A_i^T P_i L_i + P_{\underline{X}_i}\underline{X}_i), \tag{7.180}$$

which is a general estimator of the classical Kalman filter.

**Case 3**: If $\alpha$ is determined by Eq. 7.177 and $\bar{P}_i = P_i$, then

$$\tilde{X}_i = (A_i^T P_i A_i + \alpha P_{\underline{X}_i})^{-1}(A_i^T P_i L_i + \alpha P_{\underline{X}_i}\underline{X}_i), \tag{7.181}$$

which is an adaptive LS estimator of the Kalman filter. It balances the contribution of the updated parameters and the measurements. The only difference between Eqs. 7.174 and 7.181 is the weight matrix of $L_i$. The former uses the equivalent weights and the latter uses the original weights of $L_i$.

**Case 4**: If $\alpha = 0$, we obtain

$$\tilde{X}_i = (A_i^T \bar{P}_i A_i)^{-1} A_i^T \bar{P}_i L_i, \tag{7.182}$$

which is a robust estimator by using only the new measurements at epoch $i$.

**Case 5**: If $\alpha = 1$, then

$$\tilde{X}_i = (A_i^T \bar{P}_i A_i + P_{\underline{X}_i})^{-1}(A_i^T \bar{P}_i L_i + P_{\underline{X}_i}\underline{X}_i), \tag{7.183}$$

which is an M–LS filter estimator (Yang 1997a, b).

***Further Development of the Theory***

The adaptive factor $\alpha$ was considered a diagonal matrix by Ou (2004) and grouped by the physical meaning of the parameters by Yang and Xu (2004). Since then, several advances have been made (cf. Yang and Cui 2006; Yang and Gao 2005a, b, 2006a, b, c, Yang et al. 2006).

## 7.7.5
## *Progress in Adaptively Robust Filter Theory and Application*

A new adaptively robust filtering technique for use in kinematic navigation and positioning has been systematically established and developed in recent years (Yang et al. 2013). The adaptively robust filter applies a robust estimation principle to resist the effects of measurement outliers, and introduces an adaptive factor to control the influence of dynamic model disturbances. It can thus balance the contribution of the dynamic model information and the measurements in accordance with the magnitude of their discrepancy (Yang et al. 2001a). In this section, we introduce the major advancements in the theory and application of the adaptively robust filter.

Following the development of adaptively robust filtering, four learning statistics and four adaptive factors were established based on experiences, and these have been proven effective in practical applications. An accompanying adaptive factor was created that features a three-segment descending function and a learning statistic constructed using the discrepancy between the predicted state from the kinematic model and the state estimated from the measurements. Three other types of adaptive factors have been developed: a two-segment descending function (Yang et al. 2001b), an exponential function (Yang and Gao 2005), and a zero/one function for state component adaptation (Ou et al. 2004; Ren et al. 2005). Three additional learning statistics have also been set up, which include a predicted residual statistic (Xu and Yang 2000; Yang and Gao 2006b), a variance component ratio statistic from both the measurements and the predicted states (Yang and Xu 2003), and a velocity discrepancy between the predicted velocity from the kinematic model and the velocity evaluated from the measurements (Cui and Yang 2006).

A key problem has been in constructing an adaptive factor suitable for balancing the contribution of the measurements and the predicted dynamic model information. Two optimal adaptive factors have been established that satisfy the conditions that the theoretical uncertainty of the predicted state outputted from the adaptive filtering is equal or nearly equal to its actual estimated uncertainty, or that the theoretical uncertainty of the predicted residual vector is equal or nearly equal to its actual estimated uncertainty (Yang and Gao 2006a). An adaptively robust filter with classified adaptive factors (Cui and Yang 2006) was also developed, which is more effective in tracking the disturbances of the vehicle movements. In addition, an adaptively robust filter with multi-adaptive factors (Yang and Cui 2008) was created, which is more general in theory and contains adaptively robust filters with single and classified adaptive factors.

To control the influence of the measurement outliers and disturbances of the dynamic model, an adaptively robust filter based on the current statistical model (Gao et al. 2006b) was developed. In addition, an adaptively robust filter based on a neural network (Gao et al. 2007a, b) was studied to solve the construction of the dynamic model. The adaptively robust filter can also be integrated with error detection, identification, and application (DIA). To control the nonlinear disturbances of the dynamic model, an adaptive unscented Kalman filter (UKF) algorithm for improving the generalization of neural networks (Gao et al. 2008) and an adaptively robust filter based on the Bancroft algorithm (Zhang et al. 2007) have been derived.

In terms of applications, the adaptively robust filter has been successfully applied to satellite orbit determination (Yang and Wen 2004) and data processing in repeated observations of geodetic networks (Sui et al. 2007). An adaptively robust filter with constraints has also been studied for navigation applications (Yang et al. 2011). In integrated navigation applications, an adaptive Kalman filtering algorithm for the IMU/GPS integrated navigation system (Gao et al. 2006a) and a two-step adaptively robust Kalman filtering algorithm for a GPS/INS integrated navigation system (Wu and Yang 2010) have been developed. A comparison of several adaptive filtering algorithms for controlling the influence of coloured noise was analysed in order to simultaneously control the influence of coloured noise and dynamic model disturbances (Cui et al. 2006). In research on the estimation and prediction of the satellite clock offset, an adaptively robust sequential adjustment with opening window classified adaptive factors (Huang et al. 2011) and an adaptively robust Kalman filter with classified adaptive factors for real-time estimation of satellite clock offset (Huang and Zhang 2012) were derived. Improvements in adaptive filtering have also been made with regard to estimation of deformation parameters in relation to geometric measurements and geophysical models (Yang and Zeng 2009).

## 7.7.6
### *A Brief Introduction to the Intelligent Kalman Filter*

Considering the filtering methods applied in kinematic navigation, the motion state models of the moving vehicles are described and set empirically without exception. However, the actual motion rules of the moving carriers are unpredictable. In Kalman filtering, the unknown motion is described by an a priori empirical model, while GNSS observations are used to obtain the unknown motion. For this case, a method called intelligent Kalman filtering is proposed in this section, for the purpose of upgrading and extending the adaptive filter theory. The original concept of the intelligent Kalman filter was introduced in 2007 by Guochang Xu, and was funded for study by the Chinese Natural Science Foundation in 2012.

The purpose of this so-called intelligent Kalman filter is to apply the Doppler observation information in constructing the system equation. The system descriptions—which until now, without exception, have used a few empiric system equations—will be upgraded using co-determined Doppler measurements, thus providing more realistic descriptions. Because of the additional velocity information (nearly as much as the positioning information), the much more objective description of the system, and the more reasonable and precise estimation of the error disturbances, intelligent Kalman filtering can provide for greater stability and can yield more accurate results. Furthermore, the additional velocity information will be considered in determining a more reasonably adaptive factor, which is a new and advanced extension in adaptive filtering. Application of the intelligent Kalman filter in kinematic GNSS navigation and positioning is ongoing, and its application for autonomous orbit determination and manoeuvring in particular is expected to yield outstanding results.

## 7.8
## A Priori Constrained Least Squares Adjustment

Thus far in the chapter, we have discussed several adjustment and filtering methods, all of which are suitable for full-rank linear equation problems. A full-rank quadratic matrix can be inverted to obtain its inversion. A rank-deficient linear equation system is sometimes referred to as an over-parameterised problem. Except for the conditional least squares adjustment method, none of the methods discussed above can be directly used for solving a rank-deficient problem. The conditional least squares adjustment method with extra conditions can make the problem solvable. The conditions, of course, should be mathematically well formulated and physically well reasoned. In other words, the conditions are considered as exactly known. In practice, the conditions are quite often known with certain a priori precision. Adjustment that uses such a priori information as constraints is called an a priori constrained adjustment, which will be discussed in this section.

## 7.8.1
## A Priori Parameter Constraints

1. A linearised observation equation system can be represented by

$$V = L - AX, \quad P_L, \tag{7.184}$$

where
   $P_L$ symmetric and definite weight matrix of dimension $m \times m$.
2. The corresponding a priori condition equation system can be written as

$$U = W - BX, \quad P_W, \tag{7.185}$$

where
   $B$ coefficient matrix of dimension $r \times n$,
   $W$ constant vector of dimension $r$,
   $U$ residual vector of dimension $r$,
   $P_W$ a priori (symmetric and definite) weight matrix of dimension $r \times r$, and
   $r$ number of condition equations; $r < n$.
3. One may interpret the constraints of Eq. 7.185 as additional pseudo-observations or as fictitious observations. This leads to the total observation equations

$$\begin{pmatrix} V \\ U \end{pmatrix} = \begin{pmatrix} L \\ W \end{pmatrix} - \begin{pmatrix} A \\ B \end{pmatrix} X, \quad P = \begin{pmatrix} P_L & 0 \\ 0 & P_W \end{pmatrix}. \tag{7.186}$$

The least squares normal equations are then well known, as (see, e.g., Sect. 7.2.1)

$$(A^T \quad B^T) \begin{pmatrix} P_L & 0 \\ 0 & P_W \end{pmatrix} \begin{pmatrix} A \\ B \end{pmatrix} X = (A^T \quad B^T) \begin{pmatrix} P_L & 0 \\ 0 & P_W \end{pmatrix} \begin{pmatrix} L \\ W \end{pmatrix}$$

or

$$(A^T P_L A + B^T P_W B)X = (A^T P_L L + B^T P_W W). \tag{7.187}$$

For convenience, a factor $k$ (here $k = 1$) is introduced in Eq. 7.187:

$$(A^T P_L A + k B^T P_W B)X = (A^T P_L L + k B^T P_W W). \tag{7.188}$$

Equation 7.188 shows that the a priori information constraints can be added to the original least squares normal equations. In other words, the a priori information can be used for solving the rank-deficient problem and makes it possible to invert the normal matrix. Of course, these a priori information constraints should be reasonable and realistic; otherwise, the solutions could be disturbed by more serious a priori constraints. In the case of $k = 0$, the normal Eq. 7.188 turns out to be the original one, and will yield the free solution (without any a priori constraints).

The solution to the a priori constrained least squares solution is then

$$X = (A^T P_L A + k B^T P_W B)^{-1} (A^T P_L L + k B^T P_W W), \tag{7.189}$$

where $k = 1$. Generally, the a priori weight matrix is given by covariance matrix $Q_W$ and

$$P_W = Q_W^{-1}. \tag{7.190}$$

The a priori constraints cause only two additional terms in both sides of the normal equations; therefore, all the adjustment and filtering methods discussed above can be directly used for solving the a priori constrained problem.

## 7.8.2
## A Priori Datum

Suppose the $B$ matrix in the a priori constraints of Eq. 7.185 is an identity matrix, and the parameter vector $W$ is just a coordinate sub-vector of the total parameter vector. This results in a special case called a priori datum. The observation equations and a priori constraints may be rewritten as

$$V = L - (A_1 \quad A_2)\begin{pmatrix} X_1 \\ X_2 \end{pmatrix}, \quad P_L \quad \text{and} \tag{7.191}$$

$$U = \bar{X}_2 - X_2, \quad P_2, \tag{7.192}$$

where $\bar{X}_2$ is the "observed" parameter sub-vector, $P_2$ is the weight matrix with respect to the parameter sub-vector $X_2$ and is generally a diagonal matrix, and $U$ is a residual vector that has the same property as $V$. Generally, $\bar{X}_2$ is "observed" independently, so $P_2$ is a diagonal matrix. If $X_2$ is a sub-vector of station coordinates, then the constraint of Eq. 7.192 is called the datum constraint (this is also the reason for the name "a priori datum").

The least squares normal equation of problems 7.191 and 7.192 can then be formed (similar to what discussed in Sect. 7.8.1) as

$$\begin{pmatrix} M_{11} & M_{12} \\ M_{21} & M_{22} \end{pmatrix} \begin{pmatrix} X_1 \\ X_2 \end{pmatrix} = \begin{pmatrix} B_1 \\ B_2 \end{pmatrix} \quad (7.193)$$

or

$$M_{11}X_1 + M_{12}X_2 = B_1 \quad \text{and} \quad (7.194)$$

$$M_{21}X_1 + M_{22}X_2 = B_2, \quad (7.195)$$

where

$$M_{11} = A_1^T P_L A_1, \quad (7.196)$$

$$M_{12} = M_{21}^T = A_1^T P_L A_2, \quad (7.197)$$

$$M_{22} = A_2^T P_L A_2 + P_2, \quad (7.198)$$

$$B_1 = A_1^T P_L L, \quad \text{and} \quad (7.199)$$

$$B_2 = A_2^T P_L L + P_2 \bar{X}_2. \quad (7.200)$$

The least squares principle used here is

$$V^T P_L V + U^T P_2 U = \min. \quad (7.201)$$

The normal Eq. 7.193 can be also derived by differentiating Eq. 7.201 with respect to $X$, and then letting it equal zero and taking Eq. 7.192 into account. In practice, the sub-vector $\bar{X}_2$ is usually a zero vector; this can be achieved through careful initialisation by forming the observation Eq. 7.191. Comparing the normal equation system of the a priori datum problem of Eqs. 7.191 and 7.192 with the normal equation of Eq. 7.191, the only difference is that the a priori weight matrix $P_2$ has been added to $M_{22}$. This indicates that the a priori datum problem can be dealt with simply by adding $P_2$ to the normal equation of the observation Eq. 7.191.

If some diagonal components of the weight matrix $P_2$ are set to zero, then the related parameters ($X_2$) are free parameters (or free datum) of the adjustment problem (without a priori constraints). Otherwise, parameters with a priori constraints are called a priori datum. Large weight values indicate strong constraint and small weight values indicate soft constraint. The strongest constraint is keeping the datum fixed.

## 7.8.3
### Zhou's Theory: Quasi-Stable Datum

The quasi-stable datum method was proposed by Zhou et al. (1997). Its basic premise is that the network is dynamic, i.e. most parameters are changing all the time. However, a few points are relatively stable, or their geometric centre is relatively stable. All assumptions and observation equations are the same as in Sect. 7.8.2:

$$V = L - \begin{pmatrix} A_1 & A_2 \end{pmatrix} \begin{pmatrix} X_1 \\ X_2 \end{pmatrix}, \quad P_L \quad \text{and} \tag{7.202}$$

$$U = \bar{X}_2 - X_2, \quad P_2. \tag{7.203}$$

The least squares principles for the quasi-stable datum are

$$V^T P_L V = \min \tag{7.204}$$

and

$$U^T P_2 U = \min. \tag{7.205}$$

Equation 7.204 is the same as the original least squares principle. From Eq. 7.204, one has the normal equation

$$\begin{pmatrix} M_{11} & M_{12} \\ M_{21} & M_{22} \end{pmatrix} \begin{pmatrix} X_1 \\ X_2 \end{pmatrix} = \begin{pmatrix} B_1 \\ B_2 \end{pmatrix}, \tag{7.206}$$

where

$$\begin{aligned} M_{11} &= A_1^T P_L A_1, \\ M_{12} &= M_{21}^T = A_1^T P_L A_2, \\ M_{22} &= A_2^T P_L A_2, \\ B_1 &= A_1^T P_L L, \quad \text{and} \\ B_2 &= A_2^T P_L L. \end{aligned} \tag{7.207}$$

Even if Eq. 7.206 is a rank-deficient equation, one may first solve Eq. 7.206 to get an explicit expression for $X_2$. Recalling the discussion in Sect. 7.5, one gets a normal equation related to $X_2$:

$$M_2 X_2 = R_2, \tag{7.208}$$

where

$$M_2 = M_{22} - M_{21}M_{11}^{-1}M_{12} \quad \text{and}$$
$$R_2 = B_2 - M_{21}M_{11}^{-1}B_1^{\mathrm{T}}. \tag{7.209}$$

The new condition can be considered by forming

$$F = U^{\mathrm{T}}P_2 U + 2K^{\mathrm{T}}(M_2 X_2 - R_2)$$

and

$$\frac{\partial F}{\partial X} = 2U^{\mathrm{T}}P_2 + 2K^{\mathrm{T}}M_2 = 0.$$

Considering the symmetry of $M_2$, we have

$$U = -P_2^{-1}M_2 K. \tag{7.210}$$

Substituting Eq. 7.210 into 7.203, one gets

$$X_2 = \bar{X}_2 + P_2^{-1}M_2 K \tag{7.211}$$

or

$$M_2 X_2 = M_2 \bar{X}_2 + M_2 P_2^{-1} M_2 K. \tag{7.212}$$

Substituting Eq. 7.208 into 7.212, one has

$$K = (M_2 P_2^{-1} M_2)^{-1}(M_2 \bar{X}_2 - R_2). \tag{7.213}$$

Thus,

$$X_2 = \bar{X}_2 + P_2^{-1}M_2 K, \tag{7.214}$$

$$X_1 = M_{11}^{-1}(A_1^{\mathrm{T}}P_L L - M_{12}X_2), \quad \text{and} \tag{7.215}$$

$$m_0 = \sqrt{\frac{V^{\mathrm{T}}P_L V}{n-r}}, \tag{7.216}$$

where $m_0$ is the standard deviation, $n$ is the number of observations, and $r$ is the summation of the both ranks of the matrices $A_1$ and $A_2$.

## 7.9
## Summary

In this chapter, we have outlined the most applicable and necessary algorithms for static and kinematic as well as dynamic GPS data processing.

Least squares adjustment is the most basic adjustment method. It starts by establishing observation equations and forming normal equations, and then solves the unknowns. The sequential application of least squares adjustment by accumulating the sequential normal equations makes applications of least squares adjustment more effective. Normal equations can be formed epoch-wise and then accumulated. This method can be used not only for ultimately solving the problem, but also for obtaining epoch-wise solutions. It is suitable for static GPS data processing. The equivalent sequential least squares adjustment, which can be found in various publications, was also derived. This is an epoch-wise solving method and thus is generally not suitable for static GPS data processing. Xu (author) and Morujao (Coimbra University, Portugal) have independently reported that results obtained by applying such an algorithm will differ from those obtained by the accumulation method. The differences increase with time and are generally non-negligible. Therefore, when this method is used, the numerical process must be carefully examined to avoid the accumulation of numerical errors.

If there are constraints that have to be taken into account, a conditional least squares adjustment is needed. The commonly used least squares ambiguity search criterion is derived from this principle (cf. Sect. 8.3.4), and the general criterion of integer ambiguity search is also based on this theory (cf. Sect. 8.3.5). This method is typically applied in GPS data processing to take into account the known distance of multiple kinematic antennas. The sequential application of conditional least squares adjustment was discussed here in terms of practical needs. The problem may be solved first without conditions, after which conditions may be applied. Constraints such as the known distances of multiple antennas fixed on an aircraft must be considered for every epoch.

We also discussed lock-wise least squares adjustment for separating the unknowns into two groups—for example, one group of time-dependent parameters such as kinematic coordinates, and the other a group of time-independent parameters such as ambiguities. The sequential application of block-wise least squares adjustment makes it possible to give up some unknowns (say, out-of-date unknowns, such as past coordinates) and to keep the information related to the common unknowns during processing. This method avoids problems that may be caused by a rapid increase in the number of unknowns. There are two ways to keep the solution equivalent to a solution that is not sequential. One is to use the time-independent unknowns at the end of data processing as known, and to then go back to process the data again. The other is to remember all sequential normal equations until the best solution of the time-independent unknowns is obtained, after which the coordinates can be recomputed. A special application of block-wise least squares adjustment was discussed for a code–phase combination model. Of course, the two observables must be suitably scaled and weighted.

We discussed the equivalently eliminated observation equation system for eliminating some nuisance parameters. This method is nearly the same as block-wise least squares adjustment if one carefully compares the normal equations of the second group of unknowns (see Sect. 7.5) and the eliminated normal equations (see Sect. 7.6). However, the most important point here is that the equivalently eliminated observation equations have been derived. Instead of solving the original problem, one may directly solve the equivalently eliminated observation equations, where the unknowns are greatly reduced, whereas the original observation vector and weight matrix remain (i.e. the problem remains uncorrelated). The precision estimation can also be made more easily by using the formulas derived in least squares adjustment. The derivation of such an equivalent observation equation was first described by Zhou (1985) and was then applied in GPS theory by Xu (2002). The unified GPS data processing method is derived using this principle (cf. Sect. 6.8). Based on the derivation of the equivalent equation, a diagonaliation algorithm of the normal equation and the observation equation was presented. The diagonalisation algorithm can be used for separating one adjustment problem into two sub-problems.

The classic Kalman filter was also discussed. It is suitable for real-time applications. A key problem of the classic Kalman filter is the divergence caused by the inexact description of system equations and its statistical properties as well as the inhomogeneity of the data. Furthermore, the solutions can be strongly dependent on the given initial values. The sequential least squares adjustment method as a special case of Kalman filtering was outlined.

Efforts have been made to modify the performance of classic Kalman filtering. In the classic Kalman filter, the weight matrix $P$ of observables is static, i.e. $P$ is assumed to be a definitive defined matrix. Taking the residuals of Kalman filtering into account, one may adjust the weight $P$ of the observables accordingly; this process is called robust Kalman filtering (Koch and Yang 1998a, b). This principle can be also used for controlling the outliers of observations (Yang 1999). This idea indeed can be also used in all of the adjustment methods. The weight of an observation is usually either one (be accepted) or zero (be rejected). In robust Kalman filtering, a continuous weight between one and zero was defined and introduced. A modified weight function was also discussed and given for use. Generally speaking, the robust weighting method may modify the convergence process of the filtering procedure.

As soon as the system is defined, the Kalman filter also obtains memory abilities. However, if the system makes a discontinuous change (for example, aircraft that is static begins to run), the Kalman filter should be able to forget a part of the updated parameters. A robust Kalman filter with the addition of this ability is called an adaptively robust Kalman filter (Yang et al. 2001a, b), and was discussed in detail.

A priori constrained least squares adjustment was discussed in Sect. 7.8 for solving the rank-deficient problems, and a general discussion on the a priori parameter constraints was provided. This method makes it possible to form the observation equations in a general way, and then a priori information can be added to keep some references fixed, such as the clock error of the reference satellite and the coordinates of the reference station. As a special case of the a priori parameter constraints, a so-called a priori datum method was discussed. The advantage of this method is that the a priori constraints just change the normal equation by adding a term (the a priori weight matrix), so that all discussed least squares adjustment and filtering methods can be directly used for solving the rank-deficient problems. Linear conditions related to the coordinate parameters can be introduced using this method. A quasi-stable datum method was also discussed. From the point of view of the dynamic earth, none station is fixed. The quasi-stable datum method takes such dynamic behaviour of the stations into account.

# References

Cui X, Yang Y (2006) Adaptively Robust Filtering with Classified Adaptive Factors. Progress in Natural Science 16(8):846–851
Cui X, Yang Y, Gao W (2006) Comparison of Adaptive Filter Arithmetics in Controlling Influence of Colored Noises. Geomatics and Information Science of Wuhan University 31(8):731-735
Cui X, Yu Z, Tao B, Liu D (1982) Adjustment in surveying. Surveying Press, Peking, (in Chinese)
Gao W, Feng X, Zhu D (2007a) GPS/INS Adaptively Integrated Navigation Algorithm Based on Neural Network. Journal of Geodesy and Geodynamics 27(2):64-67
Gao W, Yang Y, Cui X, Zhang S (2006a) Application of Adaptive Kalman Filtering Algorithm in IMU/GPS Integrated Navigation System. Geomatics and Information Science of Wuhan University 31(5):466-469
Gao W, Yang Y, Zhang S (2006b) Adaptive Robust Kalman Filter Based on the Current Statistical Model. Acta Geodaetica et Cartographica Sinica 35(1):15-18
Gao W, Yang Y, Zhang T (2007b) Neural Network Aided Adaptive Filtering for GPS/INS Integrated Navigation. Acta Geodaetica et Cartographica Sinica 36(1):26-30
Gao W, Yang Y, Zhang T (2008) An Adaptive UKF Algorithms for Improving the Generalizaiton of Neural Network. Geomatics and Information Science of Wuhan University 33(5):500-503
Gotthardt E (1978) Einführung in die Ausgleichungsrechnung. Herbert Wichmann Verlag, Karlsruhe
Huang G, Yang Y, Zhang Q (2011) Estimate and Predict Satellite Clock Error Using Adaptively Robust Sequential Adjustment with Classified Adaptive Factors Based on Opening Windows. Acta Geodaetica et Cartographica Sinica 40(1):15-21
Huang G, Zhang Q (2012) Real-time estimation of satellite clock offset using adaptively robust Kalman filter with classified adaptive factors. GPS Solutions 16(4):531-539
Huber PJ (1964) Robust estimation of a location parameter. Ann Math Stat 35:73–101
Jazwinski AH (1970) Stochastic processes and filtering theory. In: Mathematics in science and engineering, Vol. 64. Academic Press, New York and London
Koch KR (1986) Maximum likelihood estimate of variance components. Bulletin Géodésique, 60:329–338
Koch KR (1988) Parameter estimation and hypothesis testing in linear models. Springer-Verlag, Berlin
Koch KR, Yang Y (1998a) Konfidenzbereiche und Hypothesentests für robuste Parameterschätzungen. ZfV 123(1):20–26

Koch KR, Yang Y (1998b) Robust Kalman filter for rank deficient observation model. J Geodesy 72: 436–441
Masreliez CJ, Martin RD (1977) Robust Bayesian estimation for the linear model and robustifying the Kalman filter. IEEE T Automat Contr AC-22:361–371
Mohamed AH, Schwarz KP (1999) Adaptive Kalman filtering for INS/GPS. J Geodesy 73: 193–203
Ou J, Chai Y, Yuan Y (2004) Adaptive filtering for kinematic positioning by selection of the parameter weights. In: Zhu, Y. and Sun, H. (eds) Progress in Geodesy and Geodynamics. Hubei Science & Technology Press, Hubei, 816–823 (in Chinese)
Ou JK (2004) Private communication
Ren C, Ou J, Yuan Y (2005) Application of adaptive filtering by selecting the parameter weight factor in precise kinematic GPS positioning. Prog. Nat. Sci., 15(1), 41–46
Salzmann M (1995) Real-time adaptation for model errors in dynamic systems. B Geod 69:81–91
Schaffrin B (1991) Generating robustified Kalman filters for the integration of GPS and INS. Techni-cal Report, No. 15, Institute of Geodesy, University of Stuttgart
Schwarz K-P, Cannon ME, Wong RVC (1989) A Comparison of GPS kinematic models for the determination of position and velocity along a trajectory. Manuscr Geodaet 14:345–353
Sui L, Liu Y, Wang W (2007) Adaptive Sequential Adjustment and Its Application. Geomatics and Information Science of Wuhan University 32(1):51-54
Teunissen P (1990) An integrity and quality control procedure for use in multi sensor integration. In: Proceedings ION GPS90, pp. 513–522
Wang G, Chen Z, Chen W, Xu G (1988) The principle of GPS precise positioning system. Surveying Press, Peking, ISBN 7-5030-0141-0/P.58, 345 p, (in Chinese)
Wu F, Yang Y (2010) A New Two-Step Adaptive Robust Kalman Filtering in GPS/INS Integrated Navigation System. Acta Geodaetica et Cartographica Sinica 39(5):522-533
Xu G (2002) GPS data processing with equivalent observation equations, GPS Solutions, Vol. 6, No. 1-2, 6:28-33
Xu G (2003) GPS – Theory, Algorithms and Applications, Springer Heidelberg, ISBN 3-540-67812-3, 315 pages, in English
Xu G (2007) GPS – Theory, Algorithms and Applications, 2nd Ed. Springer Heidelberg, ISBN 978-3-540-72714-9, 350 pages
Xu G, Qian Z (1986) The application of block elimination adjustment method for processing of the VLBI Data. Crustal Deformation and Earthquake, Vol. 6, No. 4, (in Chinese)
Xu T, Yang Y (2000) The Improved Method of Sage Adaptive Filtering. Science of Surveying and Mapping 25(3):22-24
Yang M, Tang CH, Yu TT (2000) Development and assessment of a medium-range real-time kinematic GPS algorithm using an ionospheric information filter. Earth Planets Space 52 (10):783–788
Yang Y (1991) Robust Bayesian estimation. B Geod 65:145–150
Yang Y (1997a) Estimators of covariance matrix at robust estimation based on influence functions. ZfV 122(4):166–174
Yang Y (1997b) Robust Kalman filter for dynamic systems. Journal of Zhengzhou Institute of Surveying and Mapping 14:79–84
Yang Y (1999) Robust estimation of geodetic datum transformation. J Geodesy 73:268–274
Yang Y, Chai H, Song L (1999) Approximation for Contaminated Distribution and Its Applications. Acta Geodaetica et Cartographic Sinica 28(3):209–214
Yang Y, Cui X (2006) Adaptively Robust Filter with Multi Adaptive Factors. J. Surv. Eng.
Yang Y, Cui X (2008) Adaptively Robust Filter with Multi Adaptive Factors. Survey Review 40 (309):260-270
Yang Y, Gao W (2006a) A New Learning Statistic for Adaptive Filter Based on Predicted Residuals. Progress in Natural Science 16(8):833-837
Yang Y, Gao W (2006b) An Optimal Adaptive Kalman Filter. Journal of Geodesy 80(4):177-183
Yang Y, Gao W (2005) Comparison of Adaptive Factors on Navigation Results. The J. Navigation, 2005, 58: 471-478.

Yang Y, Gao W (2005) Influence comparison of adaptive factors on navigation results. Journal of Navigation 58, 471–478

Yang Y, Gao W (2006c) Optimal Adaptive Kalman Filter with Applications in Navigation. J Geodesy

Yang Y, He H, Xu G (2001a) Adaptively robust filtering for kinematic geodetic positioning. J Geodesy 75:109–116

Yang Y, Ren X, Xu Y (2013) Main Progress of Adaptively Robust Filter with Application in Navigation. Journal of Navigation and Positioning 1(1):9-15

Yang Y, Tang Y, Li Q and Zou Y (2006) Experiments of Adaptive Filters for Kinematic GPS Positioning Applied in Road Information Updating in GIS. J. Surv. Eng. (in press)

Yang Y, Wen Y (2004) Synthetically adaptive robust filtering for satellite orbit determination. Science in China Series D Earth Sciences 47(7):585-592

Yang Y, Xu T (2003) An Adaptive Kalman Filter Based on Sage Windowing Weights and Variance Components. Journal of Navigation 56(2):231-240

Yang Y, Xu T (2004) An Adaptively Regularization Method with Combination of Priori and Posterior Information. In: Zhu, Y. and Sun, H. (eds) Progress in Geodesy and Geodynamics. Hubei Science & Technology Press, Hubei

Yang Y, Xu T, He H (2001b) On adaptively kinematic filtering. Selected Papers for English of Acta Geodetica et Cartographica Sinica, pp. 25–32

Yang Y, Zeng A (2009) Adaptive Filtering for Deformation Parameter Estimation in Consideration of Geometrical Measurements and Geopgysical Models. Science in China Series D Earth Sciences 52(8):1216-1222

Yang Y, Zhang X, Xu J (2011) Adaptively Constrained Kalman Filtering for Navigation Applications. Survey Review 43(322):370-381

Zhang S, Yang Y, Zhang Q (2007) An Adaptively Robust Filter Based on Bancroft Algorithm in GPS Navigation. Geomatics and Information Science of Wuhan University 32(4):309-311

Zhou J (1985) On the Jie factor. Acta Geodaetica et Geophysica 5 (in Chinese)

Zhou J, Huang Y, Yang Y, Ou J (1997) Robust least squares method. Publishing House of Huazhong University of Science and Technology, Wuhan

# Chapter 8
# Cycle Slip Detection and Ambiguity Resolution

Ambiguity problems can arise during phase measurement when the receiver loses its lock on the signal, and phase measurement must be reinitiated. This phenomenon is called cycle slip, i.e. the cycle count must begin again because of a signal interruption. The consequence of a cycle slip is an observable jump by an integer number of cycles in the adjacent carrier phase, and a new ambiguity parameter is required in the related observation model. Accurate cycle slip detection thus ensures correct ambiguity parameterisation. Here, we begin with a discussion of cycle slip detection, after which we will focus on integer ambiguity resolution, including integer ambiguity search criteria. We also provide an outline and discussion of the historical ambiguity function method.

## 8.1
## Cycle Slip Detection

Recalling the discussions in Sect. 6.5, several methods of cycle slip detection can be summarised as follows.

1. ***Phase–Code Comparison***

   Using the first equation of 6.88

   $$\Delta_t R_j = \lambda_j \Delta_t \Phi_j - \lambda_j \Delta_t N_j + \varepsilon, \tag{8.1}$$

   cycle slips of the phase observable in working frequency $j$ can be detected. $\Delta_t$, $R_j$, $\Phi_j$, $N_j$, $\lambda_j$, $\varepsilon$, and $j$ are the time difference operator, code range, phase, ambiguity, wavelength, residual, and index of the frequency, respectively. In the case of no cycle slips, the time difference of the ambiguity is zero, i.e. $\Delta_t N_j = 0$. Because the noise level of the code range is much higher than that of the phase, this method can be used only for the detection of large cycle slips.

## 2. Phase–Phase Ionospheric Residual

Using Eq. 6.80

$$\lambda_1 \Delta_t \Phi_1(t_j) - \lambda_2 \Delta_t \Phi_2(t_j) = \lambda_1 \Delta_t N_1 - \lambda_2 \Delta_t N_2 - \Delta_t \Delta \delta_{\text{ion}}(t_j) + \Delta_t \Delta \varepsilon_p, \qquad (8.2)$$

cycle slips of the two phase observables in frequencies 1 and 2 can be detected. $\Delta_t \Delta \delta_{\text{ion}}(t_j)$ is the so-called ionospheric residual. Generally speaking, the computed ionospheric residual of the two adjacent epochs should be very small. Any unusual change of the ionospheric residual may indicate cycle slips in one or two phases. However, two special cycle slips, $\Delta N_1$ and $\Delta N_2$, can lead also to a very small combination of $\lambda_1 \Delta_t N_1 - \lambda_2 \Delta_t N_2$. Examples of such combinations can be found, for example, in Hofmann-Wellenhof et al. (1997). Therefore, a large ionospheric residual indicates cycle slips, but a small ionospheric residual does not guarantee the absence of cycle slips. Another shortcoming of this method is that the ionospheric residual itself provides no means of determining the specific phase in which the cycle slip occurs.

## 3. Doppler Integration

Using Eq. 6.87

$$\Delta_t N_j = \Delta_t \Phi_j - \int_{t_{j-1}}^{t_j} D_j dt + \varepsilon, \quad j = 1, 2, 5, \qquad (8.3)$$

cycle slips of the phase observable in working frequency $j$ can be detected. $D_j$ is the Doppler observable of frequency $j$. Recalling the discussions made in Chap. 4, the phase is measured by keeping track of the partial phase and accumulating the integer count. If there is any loss of lock of the signal during this time, the integer accumulation will be inaccurate, i.e. a cycle slip occurs. Therefore, external instantaneous Doppler integration is a good choice for cycle slip detection. The integration can be made first by fitting the Doppler data with a polynomial of suitable order, and then integrating that within the desired time interval. Polynomial fitting and numerical integration methods can be found in Sects. 11.5.2 and 3.4.

## 4. Differential Phases (of Time)

Using the first equation of 6.86

$$\lambda_j \Delta_t \Phi_j = \Delta_t \rho - \Delta_t (\delta t_r - \delta t_e) c + \lambda_j \Delta_t N_j + \varepsilon_p, \quad j = 1, 2, \qquad (8.4)$$

cycle slips can be detected. Except for the ambiguity term, all other terms on the right side are of low variation. Any cycle slips will lead to a sudden jump of the time difference of the phases. The differenced data may be fitted with polynomials, and the polynomials can be used for interpolating or extrapolating the data at the checking epoch; the computed and differenced data then can be compared to decide if there are any cycle slips.

## 8.2
## Method of Dealing with Cycle Slips

As soon as the cycle slips have been detected, there are two ways to deal with them. One is to repair the cycle slips; the other is to set a new ambiguity unknown parameter in the GPS observation equations. To repair the cycle slips, the cycle slips have to be known exactly. Any incorrect reparation will affect all later observations. Setting a new unknown ambiguity parameter after a cycle slip is a more secure method, as there will be more unknowns in the observation equations. However, there exists a condition between the former ambiguity parameter $N(1)$ and the new one $N(2)$, i.e.

$$N(1,i,j,k) = N(2,i,j,k) + I(i,j,k), \tag{8.5}$$

where $I$ is an integer constant, and $i$, $j$, and $k$ are indices of the receiver, satellite, and observing frequency, respectively. For any solution of $N(1)$ and $N(2)$ with good qualities, the integer constant should be able to be easily distinguished. If $I = 0$, then no cycle slips have actually occurred.

If instrumental biases have not been modelled, the biases may destroy the integer property of the original ambiguity parameters. However, in such a case, the double-differenced ambiguities are still integers.

## 8.3
## A General Criterion of Integer Ambiguity Search

In this section, we propose an integer ambiguity search method based on conditional adjustment theory. By taking the coordinate and ambiguity residuals into account, a general criterion for ambiguity searching is derived. The search can be carried out in both ambiguity and coordinate domains. The optimality and uniqueness properties of the general criterion are also discussed. A numerical explanation of the general criterion is outlined. An equivalent criterion of the general criterion is derived based on a diagonalised normal equation, which shows that the commonly used least squares ambiguity search (LSAS) criterion is just one of the terms of the equivalent general criterion. Numerical examples are given to illustrate the two components of the equivalent criterion.

### *8.3.1*
### *Introduction*

It is well known that ambiguity resolution is a major problem that must be solved in GPS precise positioning. Some well-derived ambiguity fixing and search algorithms have been published over the last ten years. Four types of methods have been categorised. The first includes Remondi's static initialisation approach (cf., e.g., Remondi 1984; Wang et al. 1988; Hofmann-Wellenhof et al. 1997), which requires a static survey time to solve the ambiguity unknowns even after a complete loss of lock. Normally, the results are good enough to take a round up ambiguity fixing. The second type includes the so-called phase–code combined methods (cf., e.g., Remondi 1984; Han and Rizos 1997; Sjoeberg 1999); the phase and code must be used in the derivation as though they have the same precision, and in the case of anti-spoofing (AS), the C/A code must be used. A search process is still needed in this case. The third type is the so-called ambiguity function method (Remondi 1984; Han and Rizos 1997); its search domain is geometric. The fourth type includes approaches in which the search domain is only in the domain of ambiguity, including some optimal algorithms for reducing the search area and accelerating the search process (cf., e.g., Euler and Landau 1992; Teunissen 1995; Cannon et al. 1997; Han and Rizos 1997). Because of the statistical nature of validation criteria, it is possible that no valid result will have been obtained at the end of the search process. Gehlich and Lelgemann (1997) separated the ambiguities from the other parameters; this is similar to the equivalent method (cf. Sect. 6.7).

The effort to develop KSGsoft (**K**inematic/**S**tatic **GPS** **Soft**ware) at the GeoForschungsZentrum (GFZ) in Potsdam was initiated at the beginning of 1994 due to requirements of kinematic GPS positioning in aerogravimetry applications (Xu et al. 1998). An optimal ambiguity resolution method was needed in order to integrate it into the software; however, selecting the published algorithms has proven to be a difficult task. This has led to the independent development of this so-called integer ambiguity search method, which appears to be a very promising algorithm. This general criterion can be used to search for and find an optimal solution vector. The search result is optimal under the least squares principle and integer ambiguity property.

In the following sections, a brief summary of the conditional adjustment is given for convenience of discussion. Ambiguity searches in the ambiguity domain and both ambiguity and coordinate domains are then discussed. Properties of the general criterion are discussed, and an equivalent criterion of the general criterion is derived. Numerical examples, conclusions, and comments are provided.

## 8.3.2
### Summary of Conditional Least Squares Adjustment

The principle of least squares adjustment with condition equations can be summarised as follows (for details cf. Sect. 7.4; Gotthardt 1978; Cui et al. 1982):

1. The linearised observation equation system can be represented by

$$V = L - AX, \quad P, \tag{8.6}$$

where $L$ is the observation vector of dimension $m$, $A$ is the coefficient matrix of dimension $m \times n$, $X$ is the unknown vector of dimension $n$, $V$ is the residual vector of dimension $m$, $n$, and $m$ are numbers of unknowns and observations, and $P$ is the symmetric and quadratic weight matrix of dimension $m \times m$.

2. The condition equation system can be written as

$$CX - W = 0, \tag{8.7}$$

where $C$ is the coefficient matrix of dimension $r \times n$, $W$ is the constant vector of dimension $r$, and $r$ is the number of conditions.

3. The least squares criterion for solving the observation equations with condition equations is well-known as

$$V^T PV = \min, \tag{8.8}$$

where $V^T$ is the transpose of the related vector $V$.

4. The solution of the conditional problem in Eqs. 8.6 and 8.7 under the least squares principle of Eq. 8.8 is then

$$\begin{aligned} X_c &= (A^T PA)^{-1}(A^T PL) - (A^T PA)^{-1} C^T K \\ &= (A^T PA)^{-1}(A^T PL - C^T K) \end{aligned} \tag{8.9}$$

and

$$K = (CQC^T)^{-1}(CQW_1 - W), \tag{8.10}$$

where $A^T$ and $C^T$ are the transpose matrices of $A$ and $C$, superscript $^{-1}$ is an inversion operator, $Q = (A^T PA)^{-1}$, $K$ is a gain vector (of dimension $r$), index c is used to denote the variables related to the conditional solution, and $W_1 = A^T PL$.

5. The precisions of the solutions are then

$$p[i] = s_d \sqrt{Q_c[i][i]}, \tag{8.11}$$

where $i$ is the element index of a vector or a matrix, $\sqrt{\phantom{x}}$ is the square root operator, $s_d$ is the standard deviation (or sigma) of unit weight, $p[i]$ is the $i$th element of the precision vector, $Q_c[i][i]$ is the $i$th diagonal element of the quadratic matrix $Q_c$, and

$$Q_c = Q - QC^T Q_2 CQ, \tag{8.12}$$

$$Q_2 = (CQC^T)^{-1}, \tag{8.13}$$

$$s_d = \sqrt{\frac{(V^T PV)_c}{m - n + r}}, \quad \text{if } (m > n - r). \tag{8.14}$$

6. For recursive convenience, $(V^T PV)_c$ can be calculated by using

$$(V^T PV)_c = L^T PL - (A^T PL)^T X_c - W^T K. \qquad (8.15)$$

Above are the complete formulas of conditional least squares adjustment. The application of such an algorithm for the purpose of integer ambiguity search will be further discussed in later sections.

## 8.3.3
## Float Solution

GPS observation equation can be represented with Eq. 8.6. Considering the case without condition (Eq. 8.7), i.e. $C = 0$ and $W = 0$, the least squares solution of Eq. 8.6 is

$$X_0 = Q(A^T PL) = QW_1, \qquad (8.16)$$

and

$$(V^T PV)_0 = L^T PL - (A^T PL)^T X_0, \qquad (8.17)$$

$$s_d = \sqrt{\frac{(V^T PV)_0}{m-n}}, \quad \text{if } (m > n) \text{ and} \qquad (8.18)$$

$$p[i] = s_d \sqrt{Q[i][i]}, \qquad (8.19)$$

where index 0 is used for convenience to denote the variables related to the least squares solution without conditions. $X_0$ is the complete unknown vector including coordinates and ambiguities and is called a float solution later on. Solution $X_0$ is the optimal one under the least squares principle. However, because of the observation and model errors as well as method limitations, float solution $X_0$ may not be exactly the right one, e.g., the ambiguity parameters are real numbers and do not fit to the integer property. Therefore, one sometimes needs to search for a solution, say $X$, which not only fulfils some special conditions, but also meanwhile keeps the deviation of the solution as small as possible (minimum). This can be represented by

$$V_x^T PV_x = \min, \qquad (8.20)$$

or equivalently by a symmetric quadratic form of (cf. also Eq. 8.35 derived later)

$$(X_0 - X)^T Q^{-1}(X_0 - X) = \min. \qquad (8.21)$$

In Eq. 8.20, $V_x$ is the residual vector in the case of solution $X$. For simplification, let

$$X = \begin{pmatrix} Y \\ N \end{pmatrix}, \quad Q = \begin{pmatrix} Q_{11} & Q_{12} \\ Q_{21} & Q_{22} \end{pmatrix}, \quad W_1 = A^T PL = \begin{pmatrix} W_{11} \\ W_{12} \end{pmatrix},$$
$$M = A^T PA = \begin{pmatrix} M_{11} & M_{12} \\ M_{21} & M_{22} \end{pmatrix}, \quad M = Q^{-1},$$
(8.22)

where $Y$ is the coordinate vector, $N$ is the ambiguity vector (generally, a real vector). The float solution is denoted by

$$X_0 = \begin{pmatrix} Y_0 \\ N_0 \end{pmatrix} = \begin{pmatrix} Q_{11} W_{11} + Q_{12} W_{12} \\ Q_{21} W_{11} + Q_{22} W_{12} \end{pmatrix},$$

where $X_0$ is the solution of Eq. 8.6 without Condition 8.7.

## 8.3.4
## Integer Ambiguity Search in Ambiguity Domain

To use the conditional adjustment algorithm for integer ambiguity searching in the ambiguity domain, the condition shall be selected as $N = W$; here $W$ of course is an integer vector. Generally, letting $C = (0, E)$, then Condition 8.7 becomes

$$N = W. \tag{8.23}$$

Using the definitions of $C$ and $Q$, one has

$$CQ = (Q_{21} \quad Q_{22}) \quad \text{and}$$
$$CQC^T = Q_{22}.$$

The gain $K_N$ can be computed by using Eq. 8.10:

$$K_N = Q_{22}^{-1}(CQW_1 - W) = Q_{22}^{-1}(N_0 - W). \tag{8.24}$$

So under Condition 8.23, the conditional least squares solution in Eq. 8.9 can be written as

$$X_c = \begin{pmatrix} Y_c \\ N_c \end{pmatrix} = \begin{pmatrix} Q_{11} & Q_{12} \\ Q_{21} & Q_{22} \end{pmatrix} \begin{pmatrix} W_{11} \\ W_{12} - K_N \end{pmatrix} = \begin{pmatrix} Y_0 \\ N_0 \end{pmatrix} - \begin{pmatrix} Q_{12} \\ Q_{22} \end{pmatrix} K_N. \tag{8.25}$$

Simplifying Eq. 8.25, one gets

$$Y_c = Y_0 - Q_{12}K_N \tag{8.26}$$

and

$$N_c = N_0 - Q_{22}K_N = N_0 - Q_{22}Q_{22}^{-1}(N_0 - W) = W. \tag{8.27}$$

The precision computing formulas under Condition 8.23 can be derived as follows:

$$Q_c = Q - QC^T Q_{22}^{-1} CQ = \begin{pmatrix} Q_{11} - Q_{12}Q_{22}^{-1}Q_{21} & 0 \\ 0 & 0 \end{pmatrix} \text{ and } \tag{8.28}$$

$$\begin{aligned}(V^T PV)_c &= L^T PL - (A^T PL)^T X_c - W^T K_N \\ &= L^T PL - (A^T PL)^T X_0 + (A^T PL)^T \begin{pmatrix} Q_{12} \\ Q_{22} \end{pmatrix} K_N - W^T K_N \\ &= (V^T PV)_0 + \begin{pmatrix} W_1^T & W_2^T \end{pmatrix} \begin{pmatrix} Q_{12} \\ Q_{22} \end{pmatrix} K_N - W^T K_N \\ &= (V^T PV)_0 + (N_0 - W)^T K_N \\ &= (V^T PV)_0 + (N_0 - W)^T Q_{22}^{-1}(N_0 - W) \end{aligned} \tag{8.29}$$

where $(V^T PV)_0$ is the value obtained without Condition 8.23. The second term on the right side of the last line in Eq. 8.29 is the often-used LSAS criterion for an integer ambiguity search in the ambiguity domain, which can be expressed as

$$\delta(dN) = (N_0 - N)^T Q_{22}^{-1}(N_0 - N). \tag{8.30}$$

It indicates that any ambiguity fixing will cause an enlargement of the standard deviation. However, one may also notice that here only the enlargement of the standard deviation caused by ambiguity parameter changing has been considered. Furthermore, Condition 8.23 does not really exist. Ambiguities are integers; however, they are unknowns. The formula to compute the accuracy vector of the ambiguity also does not exist, because the ambiguity condition is considered exactly known in conditional adjustment.

## 8.3.5
### Integer Ambiguity Search in Coordinate and Ambiguity Domains

In order to see the enlargement of the standard deviation caused by the fixed solution, the condition shall be selected as $X = W$; here, $W$ consists of two sub-vectors (coordinate and ambiguity parameter-related sub-vectors), and only the ambiguity parameter-related sub-vector is an integer type. Letting $C = E$, Condition 8.7 is then

$$X = W. \tag{8.31}$$

One has

$$CQ = CQC^T = Q.$$

Denote $X_0 = QW_1$; here $X_0$ is the solution of Eq. 8.6 without Condition 8.31. The gain $K$ can be computed using Eq. 8.10:

$$K = Q^{-1}(CQW_1 - W) = Q^{-1}(X_0 - W). \tag{8.32}$$

Thus, under Condition 8.31, the conditional least squares solution in Eq. 8.9 can be written as

$$X_c = X_0 - QK = X_0 - QQ^{-1}(X_0 - W) = W. \tag{8.33}$$

Precision computing formulas under Condition 8.31 can be derived as follows:

$$Q_c = 0,$$

$$\begin{aligned}(V^T PV)_c &= L^T PL - (A^T PL)^T X_c - W^T K \\ &= L^T PL - (A^T PL)^T X_0 + (A^T PL)^T (X_0 - X_c) - W^T K \\ &= (V^T PV)_0 + W_1^T QK - W^T K \\ &= (V^T PV)_0 + (X_0 - W)^T K \\ &= (V^T PV)_0 + (X_0 - W)^T Q^{-1}(X_0 - W)\end{aligned} \tag{8.34}$$

where $(V^T PV)_0$ is the value obtained without Condition 8.31.

Condition 8.31 will force the observation Eq. 8.6 to take the condition $W$ as the solution and will take the zero value as the precision of the conditional solution (i.e. the precision is undefined). The reason for this is that the condition is considered exactly known in conditional adjustment. The second term on the right side of Eq. 8.34 is denoted as

$$\delta = (X_0 - X)^T Q^{-1}(X_0 - X). \tag{8.35}$$

This term in Eq. 8.34 indicates that any solution vector $X$ that is different from the float solution vector $X_0$, will enlarge the weighted squares residuals. It is well known that the float solution is the optimal solution under the least squares principle. Therefore, statistically, the optimal solution $X$ shall be the $X$ that takes the minimum value of $\delta$ in Eq. 8.35. Mathematically speaking, Eq. 8.35 is the "distance" between vector $X$ and $X_0$ in the solution space (of dimension $n$). If one considers $n = 3$ and $Q^{-1}$ to be a diagonal matrix, then $\delta$ is the geometric distance of point $X$ and $X_0$ in a cubic space. So Eq. 8.35 can be used as a general criterion to express the nearness of the two vectors. By using criterion of Eq. 8.35, one may search for solution $X$ in the area being searched so that the value of $\delta$ reaches the minimum. Under such a criterion, the deviation of the result vector $X$ related to the float vector $X_0$ is homogenously considered.

Furthermore, Condition 8.31 is considered exactly known in conditional adjustment. However, in integer ambiguity, searching, we just know the ambiguities are integers, but their values are indeed not known, or they are known with uncertainty (precision) within an area around the float solution. So the best solution shall be searched for. For computing the precision of the searched $X$, the formulas of least squares adjustment shall be further used, and meanwhile the enlarged residuals shall be taken into account by

$$p[i] = s_d \sqrt{Q[i][i]},$$
$$s_d = \sqrt{\frac{(V^T P V)_c}{m-n}}, \quad \text{if } (m > n) \quad \text{and} \tag{8.36}$$
$$(V^T P V)_c = (V^T P V)_0 + \delta.$$

In other words, the original $Q$ matrix and $(V^T P V)_0$ of the least squares problem in Eq. 8.6 are further used. The $\delta$ has the function of enlarging the standard deviation. The precision computing formulas have nothing to do with the conditions. Searching for a minimum $\delta$ leads to a minimum of standard deviation $s_d$, and therefore the best precision values.

Equation 8.35 is called the general criterion (Xu's general criterion) of an integer ambiguity search, which may be used for searching for the optimal solution in the ambiguity domain, or both coordinate and ambiguity domains (cf. Xu 2002a). In most cases, the search will be started from the ambiguity domain. An integer vector $N$ can be selected in the search area, then the related coordinate vector $Y$ can be computed using the consistent relation of $Y$ and $N$ (cf. Eqs. 8.26 and 8.24). The optimal solution searched shall be that $X$ which leads Eq. 8.35 to a minimum value.

In the case of searching in the ambiguity domain, $X$ consists of the selected sub-vector of $N_c$ in Eq. 8.27 and the computed coordinate sub-vector $Y_c$ in Eq. 8.26, i.e.

$$W = \begin{pmatrix} Y_c \\ N_c \end{pmatrix}. \tag{8.37}$$

## 8.3.6
### Properties of Xu's General Criterion

1. **Equivalence of the Two Search Scenarios**

It should be emphasised that the same search criterion of Eq. 8.35 and the same formulas of precision estimation in Eq. 8.36 are used in the two integer ambiguity search scenarios. And the same normal equation of 8.6 is used to compute the $Y_c$ using the selected $N_c$ if necessary. The two search processes indeed deal with the same problem, just with different methods of searching.

Suppose that by searching in the ambiguity domain, the vector $X = (Y_c \ N_c)^T$ is found so that $\delta$ reaches the minimum, where $N_c$ is the selected integer sub-vector and $Y_c$ is computed. And in the case of searching in both coordinate and ambiguity domains, a candidate vector $X = (Y \ N)^T$ is selected so that $\delta$ reaches the minimum, where $N$ is the selected integer sub-vector and $Y$ is the selected coordinate vector. Because of the optimality and uniqueness properties of the vector $X$ in Eq. 8.35 (please refer to 2, which is discussed next), here the selected $(Y \ N)^T$ must be equal to $(Y_c \ N_c)^T$. Thus the theoretical equivalence of the two search processes is confirmed.

2. **Optimality and Uniqueness Properties**

The float solution $X_0$ is the optimal and unique solution of Eq. 8.6 under the principle of least squares. A minimum of $\delta$ in Eq. 8.35 will lead to a minimum of $(V^T P V)_c$ in Eq. 8.36. Therefore, using the criterion of Eq. 8.35 analogously, the searched vector $X$ is the optimal solution of Eq. 8.6 under the least squares principle and integer ambiguity properties. The uniqueness property is obvious. If $X_1$ and $X_2$ are such that $\delta(X_1) = \delta(X_2) = \min$ or $\delta(X_1) - \delta(X_2) = 0$, then by using Eq. 8.35, one may assume that $X_1$ must be equal to $X_2$.

3. **Geometric Explanation of Xu's General Criterion**

Geometrically, $\delta = (X_0 - X)^T Q^{-1}(X_0 - X)$ is the "distance" between the vector $X$ and float vector $X_0$. The distance contributed to enlarge the standard deviation $s_d$ (cf. Eq. 8.36). Ambiguity searching is then the search for the solution vector with the integer ambiguity property and with the minimum distance to the float solution vector.

## 8.3.7
## An Equivalent Ambiguity Search Criterion and Its Properties

Suppose the undifferenced GPS observation equation and related LS normal equation are

$$V = L - \begin{pmatrix} A_1 & A_2 \end{pmatrix} \begin{pmatrix} X_1 \\ X_2 \end{pmatrix}, \quad P \tag{8.38}$$

$$\begin{pmatrix} M_{11} & M_{12} \\ M_{21} & M_{22} \end{pmatrix} \begin{pmatrix} X_1 \\ X_2 \end{pmatrix} = \begin{pmatrix} W_1 \\ W_2 \end{pmatrix}, \tag{8.39}$$

where

$$\begin{pmatrix} A_1^T P A_1 & A_1^T P A_2 \\ A_2^T P A_1 & A_2^T P A_2 \end{pmatrix} = \begin{pmatrix} M_{11} & M_{12} \\ M_{21} & M_{22} \end{pmatrix} = M, \quad M^{-1} = Q = \begin{pmatrix} Q_{11} & Q_{12} \\ Q_{21} & Q_{22} \end{pmatrix},$$
$$W_1 = A_1^T P L \quad \text{and} \quad W_2 = A_2^T P L.$$
(8.40)

Where all symbols have the same meanings as that of Eqs. 7.117 and 7.118. Equation 8.39 can be diagonalised as (cf. Sect. 7.6.1)

$$\begin{pmatrix} M_1 & 0 \\ 0 & M_2 \end{pmatrix} \begin{pmatrix} X_1 \\ X_2 \end{pmatrix} = \begin{pmatrix} B_1 \\ B_2 \end{pmatrix}, \tag{8.41}$$

where

$$\begin{aligned} Q_{11} &= M_1^{-1}, & Q_{22} &= M_2^{-1} \\ Q_{12} &= -M_{11}^{-1}(M_{12}Q_{22}), & Q_{21} &= -M_{22}^{-1}(M_{21}Q_{11}) \end{aligned}. \tag{8.42}$$

The related equivalent observation equation of the diagonal normal Eq. 8.41 can be written (cf. Sect. 7.6.1)

$$\begin{pmatrix} U_1 \\ U_2 \end{pmatrix} = \begin{pmatrix} L \\ L \end{pmatrix} - \begin{pmatrix} D_1 & 0 \\ 0 & D_2 \end{pmatrix} \begin{pmatrix} X_1 \\ X_2 \end{pmatrix}, \quad \begin{pmatrix} P & 0 \\ 0 & P \end{pmatrix}, \tag{8.43}$$

where all symbols have the same meanings as that of Eqs. 7.140 and 7.142.

Suppose the GPS observation equation is Eq. 8.38 and the related least squares normal equation is Eq. 8.39, where $X_2 = N$ ($N$ is the ambiguity sub-vector) and $X_1 = Y$ ($Y$ is the other unknown sub-vector). The general criterion is (cf. Eq. 8.35)

$$\delta(dX) = (X_0 - X)^T Q^{-1} (X_0 - X), \tag{8.44}$$

where $X = (Y \ N)^T$, $X_0 = (Y_0 \ N_0)^T$, $dX = X_0 - X$ and index 0 denotes the float solution. The search process in the ambiguity domain is to find a solution $X$ (which includes $N$ in the search area and the computed $Y$) so that the value of $\delta(dX)$ reaches the minimum. The optimality property of this criterion is obvious.

For the equivalent observation Eq. 8.43, the related least squares normal equation is Eq. 8.41. The related equivalent general criterion is then (putting the diagonal cofactor of Eq. 8.41 into Eq. 8.44 and taking Eqs. 8.40 and 8.42 into account)

$$\begin{aligned} \delta_1(dX) &= (Y_0 - Y)^T Q_{11}^{-1} (Y_0 - Y) + (N_0 - N)^T Q_{22}^{-1} (N_0 - N) \\ &= \delta(dY) + \delta(dN), \end{aligned} \tag{8.45}$$

where index 1 is used to distinguish criterion of Eq. 8.45 from Eq. 8.44. The observation Eqs. 8.38 and 8.43 are equivalent, and the related normal Eqs. 8.39 and 8.41 are also equivalent. Therefore, Criterion 8.45 is called an equivalent criterion (or Xu's equivalent criterion) of Xu's general Criterion 8.44. Zhang et al. (2016) proved that the equivalent criterion was identical to the general criterion by expanding the general criterion. The equivalent criterion is clearly better and easier for use in practice than the general criterion.

Furthermore, $Y$ and $N$ shall be consistent with each other because they are presented in the same normal Eqs. 8.39 and 8.41. Using condition $W = N$ and notation of Eq. 8.42, one has from Eqs. 8.26 and 8.24

$$Y_0 - Y = Q_{12} Q_{22}^{-1} (N_0 - N). \tag{8.46}$$

Putting Eq. 8.46 into Eq. 8.45, one has

$$\delta_1(\mathrm{d}X) = (N_0 - N)^\mathrm{T} [Q_{22}^{-1}(E + Q_{21} Q_{11}^{-1} Q_{12} Q_{22}^{-1})](N_0 - N). \tag{8.47}$$

It is notable that the second term $\delta(\mathrm{d}N)$ of the equivalent criterion Eq. 8.45 is exactly the same as the commonly used LSAS criterion of Eq. 8.30 (cf., e.g., Teunissen 1995; Leick 1995; Hofmann-Wellenhof et al. 1997; Euler and Landau 1992; Han and Rizos 1997). Through Eq. 8.47 one may clearly see the differences between the criteria of Eqs. 8.30 and 8.45. When the results searched using Eq. 8.30 are different from that of using Eq. 8.45, the results from the search using Eq. 8.30 shall be only sub-optimal ones due to the optimality and uniqueness property of Eq. 8.45. The first term on the right side of Eq. 8.45 signifies an enlarging of the residuals due to the coordinate change caused by ambiguity fixing (cf. Sect. 8.3.3). The second term on the right side of Eq. 8.45 signifies an enlarging of the residuals due to the ambiguity change caused by ambiguity fixing (cf. Sect. 8.3.4). Equation 8.45 takes both effects into account.

1. *Optimality and Uniqueness Properties of Xu's Equivalent Criterion*

The float solution $X_0$ is the optimal and unique solution of Eq. 7.117 under the least squares principle. Criterion Eq. 8.45 is equivalent to criterion Eq. 8.44. A $X$ leads to the minimum of $\delta_1(\mathrm{d}X)$ in Eq. 8.45, which will lead to the minimum of $\delta(\mathrm{d}X)$ in Eq. 8.44 and consequentially the minimum of $(V^\mathrm{T} PV)_c$ in Eq. 8.36; therefore, using criterion of Eq. 8.45, analogously, the searched vector $X$ is the optimal solution of Eq. 8.38 under the least squares principle and integer ambiguity properties. The uniqueness property is obvious. If one has $X_1$ and $X_2$ so that $\delta_1(\mathrm{d}X_1) = \delta_2(\mathrm{d}X_2) = \min.$, or $\delta_1(\mathrm{d}X_1) - \delta_1(\mathrm{d}X_2) = 0$, then by using Eq. 8.45, one may assume that $X_1$ must be equal to $X_2$.

It is notable that Eqs. 8.44 and 8.45 are equivalent for use in searching; however, the equivalent one is preferred.

## 8.3.8
## Numerical Examples of the Equivalent Criterion

Several numerical examples are given here to illustrate the behaviour of the two terms of the criterion. The first and second terms on the right-hand side of Eq. 8.45 are denoted as $\delta(dY)$ and $\delta(dN)$, respectively. $\delta_1(dX) = \delta(dY) + \delta(dN)$ is the equivalent criterion of the general criterion and is denoted as $\delta(total)$. The term $\delta(dN)$ is the LSAS criterion. Of course, the search is made in the ambiguity domain. The search area is determined by the precision vector of the float solution. All possible candidates are tested one by one, and the related $\delta_1(dX)$ are compared with each other to find the minimum.

In the first example, precise orbits and dual-frequency GPS data of 15 April 1999 at station Brst (N 48.3805°, E 355.5034°) and Hers (N 50.8673°, E 0.3363°) are used. The session length is 4 h. The total search candidate number is 1020. Results of the two delta components are illustrated as 2-D graphics with the first axis of search number and the second axis of delta in Fig. 8.1. The red and blue lines represent $\delta(dY)$ and $\delta(dN)$, respectively. $\delta(dY)$ reaches the minimum at search no. 237, and $\delta(dN)$ at no. 769. $\delta(total)$ is plotted in Fig. 8.2, and it shows that the general criterion reaches the minimum at search no. 493. For more detail, a portion of the results are listed in Table 8.1.

$\delta(dN)$ reaches the second minimum at search no. 771. This example shows that the minimum of $\delta(dN)$ may not lead to the minimum of total delta, because the related $\delta(dY)$ is large. If the delta ratio criterion is used in this case, the LSAS

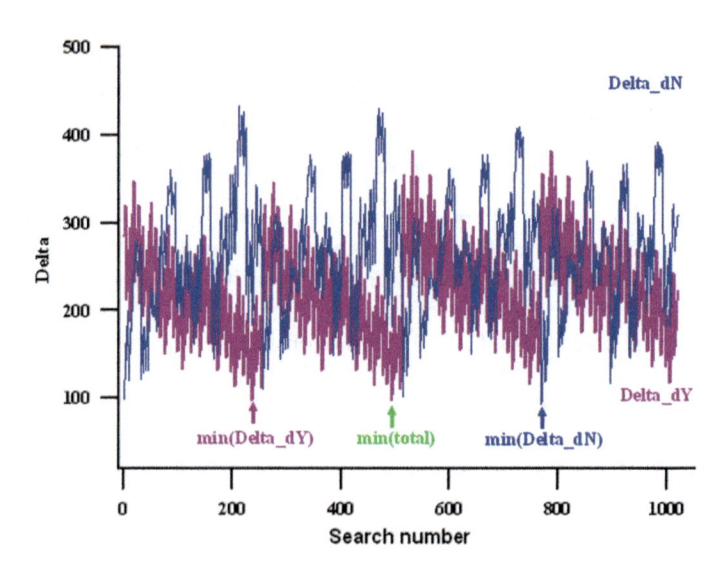

**Fig. 8.1** Two components of the equivalent ambiguity search criterion

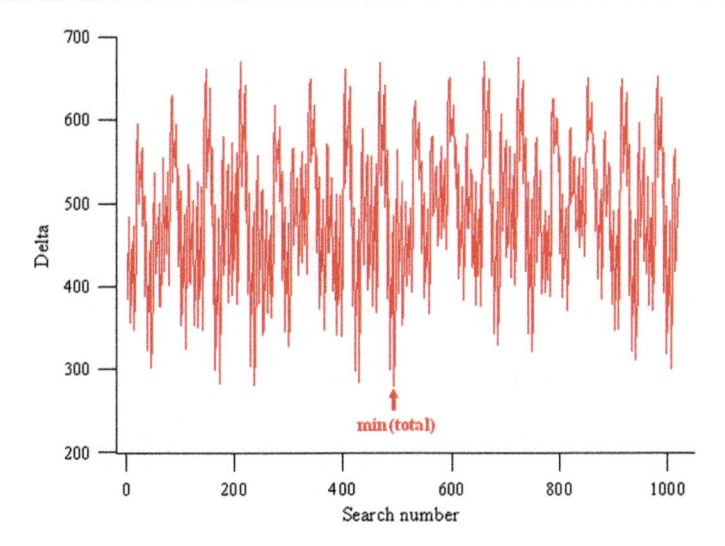

**Fig. 8.2** Equivalent ambiguity search criterion

**Table 8.1** Delta values of search process

| Search No. | $\delta(dN)$ | $\delta(dY)$ | $\delta(total)$ |
|---|---|---|---|
| 237 | 183.0937 | 97.8046 | 280.8984 |
| 493 | 181.7359 | 97.9494 | 279.6853 |
| 769 | 93.3593 | 315.2760 | 408.6353 |
| 771 | 96.0678 | 343.5736 | 439.6414 |

method will reject the found minimum and explain that no significant ambiguity fixing can be made. However, because of the uniqueness principle of the general criterion, the search reaches the total minimum uniquely.

The second example is very similar to the first. The delta values of the search process are plotted in Fig. 8.3, where $\delta(dY)$ is much smaller than $\delta(dN)$. $\delta(dN)$ reaches the minimum at search no. 5 and $\delta(dY)$ at 171. $\delta(total)$ reaches the minimum at search no. 129. The total 11 ambiguity parameters are fixed and listed in Table 8.2. Two ambiguity fixings have just one cycle difference at the 6th ambiguity parameter. The related coordinate solutions after the ambiguity fixings are listed in Table 8.3. The coordinate differences at component $x$ and $z$ are about 5 mm. Even the results are very similar; however, two criteria do give different results.

In the third example, real GPS data of 3 October 1997 at station Faim (N 38.5295°, E 331.3711°) and Flor (N 39.4493°, E 328.8715°) are used. The delta values of the search process are listed in Table 8.4. Both $\delta(dN)$ and $\delta(total)$ reach the minimum at search no. 5. This indicates that the LSAS criterion may sometimes reach the same result as that of the equivalent criterion being used.

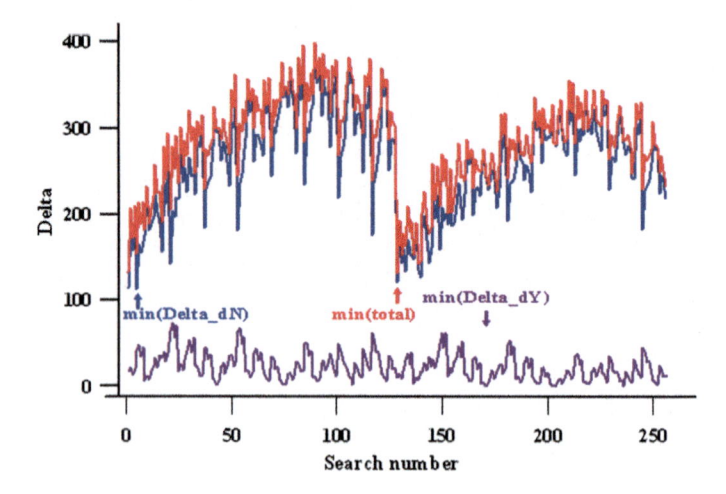

**Fig. 8.3** Example of equivalent ambiguity search criterion

**Table 8.2** Two kinds of ambiguity fixing due to two criteria

| Ambiguity No. | 1 | 2 | 3 | 4 | 5 | 6 | 7 | 8 | 9 | 10 | 11 |
|---|---|---|---|---|---|---|---|---|---|---|---|
| LSAS fixing | 0 | 0 | 1 | 0 | 0 | 0 | −1 | 0 | 0 | −1 | −1 |
| General fixing | 0 | 0 | 1 | 0 | 0 | −1 | −1 | 0 | 0 | −1 | −1 |

**Table 8.3** Ambiguity fixed coordinate solutions (in meters)

| Coordinates | x | y | z |
|---|---|---|---|
| LSAS fixing | 0.2140 | −0.0449 | 0.1078 |
| General fixing | 0.2213 | −0.0465 | 0.1127 |

**Table 8.4** Delta values of the ambiguity search process

| Search No. | $\delta(dN)$ | $\delta(dY)$ | $\delta(total)$ |
|---|---|---|---|
| 1 | 248.5681 | 129.0555 | 377.6236 |
| 2 | 702.6925 | 58.9271 | 761.6195 |
| 3 | 889.5496 | 107.9330 | 997.4825 |
| 4 | 452.1952 | 42.3226 | 494.5178 |
| 5 | 186.7937 | 112.3030 | 299.0967 |
| 6 | 739.0487 | 55.9744 | 795.0231 |
| 7 | 931.4125 | 89.9074 | 1021.3199 |
| 8 | 592.1887 | 38.0969 | 630.2856 |

## 8.3.9
## Conclusions and Comments

### 1. Conclusions

A general criterion and its equivalent criterion of integer ambiguity searching are proposed in this section. Using these two criteria, the searched result is optimal and unique under the least squares minimum principle and under the condition of integer ambiguities. The general criterion has a clear geometrical explanation. The theoretical relationship between the equivalent criterion and the commonly used LSAS criterion is obvious. It shows that the LSAS criterion is just one of the terms of the equivalent criterion of the general criterion (this does not take into account the residual enlarging effect caused by coordinate change due to ambiguity fixing). Numerical examples show that a minimum $\delta(dN)$ may have a relatively large $\delta(dY)$, and therefore, a minimum $\delta(dN)$ may not guarantee a minimum $\delta$(total). For an optimal search, the equivalent criterion or the general criterion shall be used.

2. **Comments**

The float solution is the optimal solution to the GPS problem under the least squares minimum principle. Using the equivalent general criterion, the searched solution is the optimal solution under the least squares minimum principle and under the condition of integer ambiguities. However, the ambiguity search criterion is just a statistical criterion. Statistical correctness does not guarantee correctness in all applications. Ambiguity fixing makes sense only with sufficiently good GPS observables and accurate data processing models.

## 8.4
## Ambiguity Resolution Approach Based on the General Criterion

A general criterion and its equivalent criterion of integer ambiguity searching are proposed in Sect. 8.3. In addition, the optimal criterion for ambiguity search is also discussed in Xu et al. (2010). In this section we highlight an ambiguity resolution approach called General Criterion Cascading Ambiguity Resolution (GECCAR), proposed by Morujao and Mendes (2008), which selects the integer set of ambiguities using the general ambiguity search criterion.

The GECCAR is designed to instantaneous resolve the ambiguities for GPS and galileo and it uses (1) a cascading procedure; (2) an a priori transformation to decorrelate the ambiguities; (3) a search algorithm, where each ambiguity is constrained with the values of previously selected ambiguities; and (4) the general ambiguity search criterion or the equivalent general criterion for integer ambiguity selection.

The cascading procedure for the three-frequency systems was introduced by Forsell et al. (1997), who developed the Three-Carrier Ambiguity Resolution (TCAR) method for the galileo carrier phase ambiguity resolution, and by Jung (1999), who suggested the Cascade Integer Resolution (CIR) method for the modernised GPS. Although proposed for different systems, both methods are based on the idea of wide-laning to take advantage of the stepwise-improved precision in carrier phase ranges from the longest wavelength to the shortest wavelength. Both methods are geometry-free, instantaneous integer ambiguity resolution methods, using integer rounding. Other approaches have been developed based on this principle, such as the Integrated Three-Carrier Ambiguity Resolution (ITCAR) (Vollath et al. 1998) and the geometry-based cascading ambiguity resolution methods (Zhang et al. 2003).

The linear combinations used in the implemented algorithm are based on the set of frequencies established for modernised GPS (L1, L2, L5) and for galileo (E1, E5a, E5b).

The full GECCAR procedure consists of three steps: (1) The EWL (extra-wide lane) ambiguities are estimated using the most precise pseudorange available and the EWL phase combination as observables. (2) With the ranges based on the results obtained in the first step, the WL (wide lane) or the ML (medium lane) ambiguities are estimated using the most precise pseudorange and the WL or the ML phase combination as observables. (3) The L1/E1 ambiguities are estimated— the observables used are the L1/E1, L2/E5b, and L5/E5a carrier phases, and the unknown ambiguities are just the L1/E1 ambiguities as the L2/E5b and L5/E5a ambiguities may be written in function of L1/E1, ML, WL, and EWL ambiguities and the ML, WL, and EWL ambiguities have been estimated in the second and first steps, respectively.

In each step, we follow the decorrelation process proposed by Teunissen (1993). Before estimating the ambiguities as integers, the float ambiguities $\hat{a}$ are transformed into an equivalent but less correlated set of ambiguities $\hat{z}$, and the corresponding variance-covariance matrix $Q_{\hat{a}}$ is transformed into the variance-covariance matrix $Q_{\hat{z}}$, using the Z-transformation:

$$\hat{z} = Z^T \hat{a}, \quad Q_{\hat{z}} = Z^T Q_{\hat{z}} Z, \tag{8.48}$$

This Z-matrix needs to fulfil certain requirements in order to be admissible, as the integer nature of the ambiguities should be maintained. Therefore, all entries in the Z-matrix should be integers and the inverse of the Z-matrix should exist and its entries should be integers as well (Teunissen 1994). The Z-transformation should aim for maximum possible decorrelation of the ambiguities to make the search algorithm more efficient.

After the decorrelation, the search is performed over the search space region to estimate the correct values of the ambiguity parameter vector. This is done by constraining each ambiguity candidate on the values of the previously selected ambiguities. In order to select the correct ambiguity set $\breve{a}$ using the equivalent general criterion, for each candidate $a$, the following should be performed: (1) the computation of $da$; (2) the computation of $b$ and $db$; (3) the computation of $da + db$; (4) the selection of $\breve{a}$ that minimises $da + db$, where $da$ represents enlargement of the residuals due to the ambiguity change caused by ambiguity fixing, and $db$ represents an enlargement of the residuals due to the coordinate changes caused by ambiguity fixing.

To verify the effectiveness of the algorithm, real GPS data and simulated data from a modernised GPS-only system, a galileo-only system, and both systems were used m. Details can be found in Morujao and Mendes (2008). Results showed that the general ambiguity search criterion represented a clear improvement in the selection of the correct set of ambiguities. Thus we may conclude that the GECCAR approach is a promising algorithm for instantaneous ambiguity resolution. Simulation runs have shown that single-epoch ambiguity resolution was possible for 99 % of the considered epochs when the three frequencies from both systems were used together.

## 8.5
## Ambiguity Function

It is well known that in GPS precise positioning, ambiguity resolution is one of the key problems that must be solved. Some well-derived ambiguity fixing and search algorithms have been published in the past. One of these methods is the ambiguity function (AF) method, which can be found in many standard publications (Remondi 1984; Wang et al. 1988; Han and Rizos 1995; Hofmann-Wellenhof et al. 1997).

The principle of the ambiguity function method is to use the single-differenced phase observation

$$\Phi_j(t_k) = \frac{1}{\lambda} \rho_j(t_k) + N_j - \gamma(t_k), \tag{8.49}$$

to form an exponential complex function

$$e^{i2\pi[\Phi_j(t_k) - \rho_j(t_k)/\lambda]} = e^{i2\pi[N_j - \gamma(t_k)]} \tag{8.50}$$

or

$$e^{i2\pi[\Phi_j(t_k) - \rho_j(t_k)/\lambda]} = e^{-i2\pi\gamma(t_k)}, \tag{8.51}$$

where $\Phi$ is the phase observable, $\rho$ is the geometric distance of the signal transmitting path, $\lambda$ is the wavelength, index $j$ denotes the observed satellite, $t_k$ is the $k$th observational time, $N$ is ambiguity, $\gamma$ is the model of the receiver clock errors, and $i$ is the imaginary unit. All terms in Eq. 8.49 have the units of cycles and are single-differenced terms. Property

$$e^{i2\pi N_j} = 1$$

is used in order to get Eq. 8.51.

Making a summation over all satellites and then taking the modulus operation, one has

$$\left| \sum_{j=1}^{n_j} e^{i2\pi[\Phi_j(t_k) - \rho_j(t_k)/\lambda]} \right| = n_j(k), \tag{8.52}$$

where property

$$\left|e^{-i2\pi\gamma(t_k)}\right| = 1$$

is used, $n_j$ is the satellite number and $n_j(k)$ is the observed satellite number at epoch $k$. Making a summation of Eq. 8.52 over all the observed time epochs, one has

$$\sum_{k=1}^{n_k} \left| \sum_{j=1}^{n_j} e^{i2\pi[\Phi_j(t_k)-\rho_j(t_k)/\lambda]} \right| = \sum_{k=1}^{n_k} n_j(k), \quad (8.53)$$

where $n_k$ is the total epochs number. The left side of Eq. 8.53 is called the ambiguity function, where unknowns are the coordinates of the remote station. The values of the ambiguity function have to be computed for all candidates of coordinates, and the optimum solution is found if the function reaches the maximum, i.e.

$$\sum_{k=1}^{n_k} \left| \sum_{j=1}^{n_j} e^{i2\pi[\Phi_j(t_k)-\rho_j(t_k)/\lambda]} \right| \Rightarrow \text{maximum}. \quad (8.54)$$

The search area can be determined by the standard deviations ($\sigma$) of the initial coordinates (e.g., a cube with side lengths of $3\sigma$ or a sphere with a radius of $3\sigma$). The AF method is indeed an ambiguity free method. The ambiguity can be computed using the optimal coordinate solution of Eq. 8.54.

Further discussion on the AF method is given in the next sub-section.

## 8.5.1
## Xu's Conjecture: Maximum Property of Ambiguity Function

The ambiguity function is discussed in Sect. 8.5. Here a numerical study of the maximum property of the ambiguity function (AF) is given. It seems that the maximum value of the AF trends to be reached at the boundary of any given search area. Numerical examples are given to illustrate the conclusion. However, a theoretical proof has still not yet been found; even the author tried to find one, but failed.

**Numerical Examples**

Several numerical examples are given here to illustrate the behaviours of the ambiguity function criterion. The GPS data of the EU AGMASCO project (cf., e.g., Xu et al. 1997a) are used. Data are combined with the data of IGS network and solved for precise coordinates as references. The station Faim (N 38.5295°, E 331.3711°) is used as the reference and Flor (N 39.4493°, E 328.8715°) is used as the remote station. The baseline length is about 240 km. The data length is about 4 h on 3 October 1997. KSGsoft (Xu et al. 1998) is used for computing a static solution of the coordinates of Flor. The differences of the KSGsoft solution and IGS solution are (0.26,1.93,1.37) cm in the global Cartesian coordinate system. Related standard deviations of the KSGsoft solution are (0.04,0.04,0.02) cm. The differences are caused partly by the different data lengths. This assures a good standard for the software being used.

The search step is selected as 1 mm. Tropospheric and ionospheric effects are corrected. In the first example, 3 h of data are used. The search area is a 3-D cube with side lengths of $\pm(0.7,0.7,0.4)$ cm in $(x, y, z)$. Results show that the AF maximum is reached at point $(-0.7,0.7,0.4)$ cm, which is on the boundary of the area being searched.

A search process (with a search area of $\pm 7$ mm and 1 h of data) is illustrated in 2-D graphics with the first axis containing search numbers and the second axis containing AF values in Fig. 8.4. The graphic looks like a 3-D AF projection of the cubic search area (the picture could be quite different in other examples). Figure 8.4 clearly shows the boundary maximum effect of the AF criterion. Expanding the searched area (and, of course, its boundary), the maximum is reached on the new boundary (of the new cubic surface).

Alternatively, the search may be made on a spherical surface with an expanding radius. The results of such an example are illustrated in Fig. 8.5, where only radii of 1,2,...,10 mm are given. As the radius expands, the AF maximum becomes greater and is always reached over the spherical surface with the maximum radius.

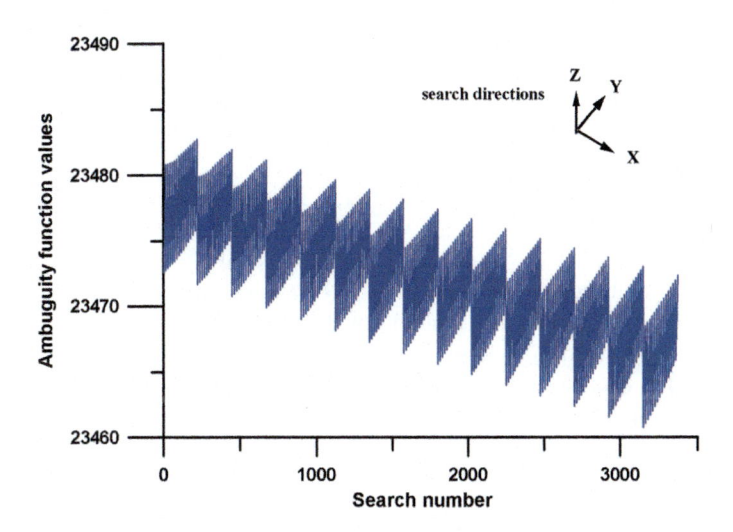

**Fig. 8.4** 3-D coordinate search using ambiguity function

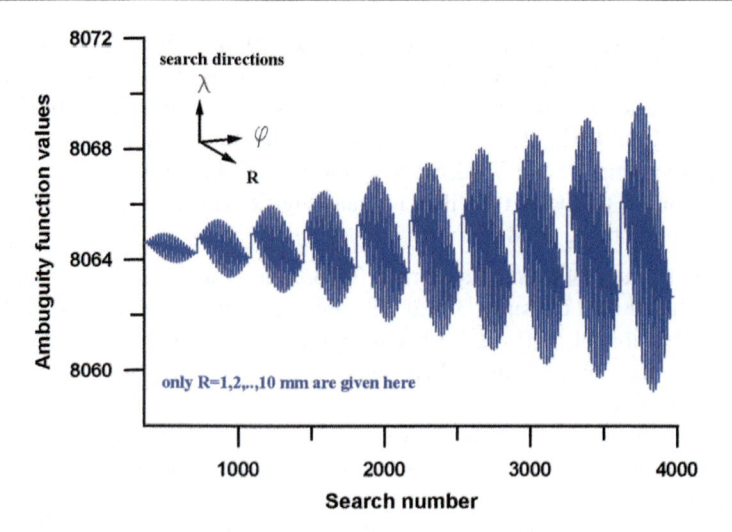

**Fig. 8.5** Spherical coordinate search using ambiguity function

**Theoretical Indications**

The AF Eq. 8.54 is rewritten as

$$\sum_{k=1}^{n_k} G(t_k) \Rightarrow \max, \qquad (8.55)$$

$$G(t_k) = |S_k|, \quad S_k = \sum_{j=1}^{n_j} e^{i2\pi \cdot v_j(t_k)}, \qquad (8.56)$$

$$v_j(t_k) = \Phi_j(t_k) - \rho_j(t_k)/\lambda, \quad Y \in \Omega \quad \text{and}, \qquad (8.57)$$

$$e^{i2\pi \cdot v_j(t_k)} = \cos(2\pi v_j(t_k)) + i\,\sin(2\pi v_j(t_k)), \qquad (8.58)$$

where $Y$ is the coordinate vector, $\Omega$ is to be the searched coordinate area and is a closed area (i.e. it includes the boundary $\Gamma$), $v_j(t_k)$ are the residuals of GPS observation equations (a continuous function of $Y$), $S_k$ is a complex function of $Y$, and $G(t_k)$ is the modulus of $S_k$.

If the GPS data sampling intervals are sufficiently close and the numerical integration error is negligible (cf. Xu 1992), then one has

$$\frac{1}{n_k}\sum_{k=1}^{n_k} G(t_k) = \frac{1}{T}\int_{t_1}^{t_e} G(t)\cdot dt, \qquad (8.59)$$

where $T = t_e - t_1$, $t_e = t(n_k)$, and $t_1$ and $t_e$ are the beginning and end time of the observations. According to the *middle value theorem* of the integration (cf., e.g., Bronstain and Semendjajew 1987; Wang et al. 1979) (such a theorem can be found in all integration related books), one has a time point $\xi(t_1 < \xi < t_e)$ so that

$$\frac{1}{T} \int_{t_1}^{t_e} G(t) \cdot dt = G(\xi), \tag{8.60}$$

i.e. the AF can be represented by a unique $G(t)$ at time $\xi$ (the constant factor is omitted here). Equation 8.54 turns out to be

$$G(\xi) \Rightarrow \max. \tag{8.61}$$

Because of the definition of AF, $G(\xi)$ is a modulus of a complex function.

In complex function analysis theory, there is a so-called maximum theorem (cf., e.g., Bronstain and Semendjajew 1987; Wang et al. 1979):

*Maximum Modulus Theorem:* if complex function $f(z)$ is analytic within a limited area Z and is continuous over the closed Z, then modulus $|f(z)|$ reaches the maximum on the boundary $\Gamma$ of Z.

However, this theorem cannot be directly used for Eq. 8.61 because it is valid only for the analytic complex function defined over a complex plane, whereas function $G(\xi)$ is a complicated three-dimensional complex function.

Perhaps the interested reader will consider this in detail and determine a theoretical proof.

## 8.6
## Ionosphere-Free Ambiguity Fixing

Ionosphere-free ambiguity fixing is called float ambiguity fixing, and will be discussed in this section. The theory proposed here was reported by Lemmens in 2004 as "NEW" in a published book review on GPS (Xu 2003).

### 8.6.1
### Introduction

It is well known that in GPS data processing, the longer the baselines or the larger the measured area, the greater the number of problems caused by ambiguity resolution that will be encountered (Wang et al. 1988; Hofmann-Wellenhof et al. 1997). This phenomenon indicates that the ambiguity resolution is influenced by certain effects, and those amounts increase with the increasing length of the baselines (Klobuchar 1996; Wanninger 1995a, b). Clock errors and multipath effects can be excluded from the candidate list because they have nothing to do with the length of the baseline. Ionospheric and tropospheric effects are different from station to station. In general, the longer the baseline, the larger the differences will be. Because the effects of ionospheric delay are usually ten times those of the troposphere, the main difficulty in ambiguity resolution is caused by the ionospheric effects.

For larger-area GPS applications, ionosphere-free combination $L_c$ must be used (Hofmann-Wellenhof et al. 1997). As a consequence of the $L_c$ combination, ambiguities obtained are no longer a set of integers. However, numerically speaking, only float numbers of the combination $N_1 - N_2(f_2/f_1)$ are possible solutions, where $N_1$ and $N_2$ are integers. Practically, research has shown that $N_c$ ambiguity fixing is necessary for precise applications (Xu et al. 1997b). The most common method involves the use of the wide lane and other linear combinations to compute and fix the ambiguity $N_1$ and $N_2$, and then to form the $N_c$; however, in this case, the residual ionospheric effects are not modelled. The solved ambiguity will include a constant part of the ionospheric effects, and this may make ambiguity fixing more complicated (Hofmann-Wellenhof et al. 1997). For fixing integer ambiguities, many methods have been developed (Han and Rizos 1995, 1997; Merbart 1995; Teunissen 1995; Euler and Landau 1992; cf. Sect. 8.3).

To gain a better understanding of the influence of the ionospheric effects on ambiguity resolution and ways to fix ionosphere-free ambiguity, an intensive study was carried out during the extensive AGMASCO GPS data processing (Xu et al. 1997b). Ionospheric effects that are not well modelled or well eliminated will be distributed to all parameters that are obtained through adjustment or filtering. However, because of the constant properties of the ambiguity parameters, only the constant parts of the residuals of ionospheric effects will affect the ambiguity results. Numerical results have shown that these constant parts of ionospheric effects on ambiguity parameters can be eliminated in any case through an ionosphere-free ambiguity combination. In other words, the ionosphere-free ambiguities can be obtained by combining the non-ionosphere-free ambiguity parameters. The influence of the higher-order ionospheric effects, which has not been eliminated by the ionosphere-free combination, can be separated into two parts. The constant part will go into the ambiguities; the time variance part will affect the coordinate or other parameters. Therefore, for precise ambiguity resolution, a constant ionospheric correction or model is necessary. After such correction or model parameterisation, the ambiguity resolution is independent from the length of the baselines.

In the following section, a concept of ionospheric ambiguity correction, determination of the ionospheric ambiguity correction using ambiguity-ionospheric equations, methods of integer and float ambiguity fixing, and a condition for an integer ambiguity under the second-order ionosphere-free combination are discussed.

## 8.6.2
## Concept of Ionospheric Ambiguity Correction

In double-difference GPS data processing, ionosphere-free ambiguities can be computed from ambiguities obtained by using any non-ionosphere-free combinations without modelling the ionospheric effects. For example, without modelling the ionospheric effects and by using $\Phi_1$ and $\Phi_2$ solutions, one gets ambiguity vectors $N_1$ and $N_2$, respectively. One can form the ionosphere-free ambiguity $N_c$ by using $N_1$ and $N_2$, and solve for the ionosphere-free ambiguity $N_{cs}$ separately. The numerical results shown that the relation $N_c = N_{cs}$ is always valid. This indicates that the influence of ionospheric effects on the ambiguity parameters can only be a set of constant values, and for these values the ionosphere-free formula is always valid, i.e. when these values are set into the ionosphere-free combination, a zero value results in any case. The time variation part of the ionospheric effects does not affect the ambiguity solution.

The mathematical model of the carrier phase observations can be written as

$$\Phi_k^{ji} = \rho^{ji}/\lambda_k + N_k^{ji} + \alpha d\rho_k(\text{ion})/\lambda_k + d\rho(\text{others})/\lambda_k \qquad (8.62)$$

or

$$\lambda_k \Phi_k^{ji} = \rho^{ji} + \lambda_k N_k^{ji} + \alpha d\rho_k(\text{ion}) + d\rho \ (\text{others}), \qquad (8.63)$$

where $\Phi$ is the measured carrier phase in the cycle, $\lambda$ is the wavelength, $\rho$ is the distance between the satellite $j$ and station $i$, $N$ is the ambiguity parameter, $d\rho_k(\text{ion})$ is the ionospheric effects, $d\rho(\text{others})$ is the summation of other effects, $k$ is the observational frequency index, frequency $f_1 = 154 f_0$, and $f_2 = 120 f_0$, where $f_0$ is the fundamental frequency (10.23 MHz), the future frequency (announced by U.S. Vice President Gore, March 1998, see ION homepage http://www.ion.org/) $f_3 = 115 f_0$. $\alpha$ is a switch; $\alpha = 1,0$ indicates whether or not the residuals of ionospheric effects are modelled. The ambiguity parameter $N$ is a set of integers, physically. However, if there are any constant effects not modelled or not well modelled, these effects will be absorbed by the ambiguities. Therefore, integer ambiguity fixing then makes sense only if all constant errors are well modelled or separated. One choices for avoiding the separation problem of the constant ionospheric effects and the ambiguities is through parameterisation of the ionospheric residuals.

Recalling the discussions in Sect. 5.1.2, the first-order ionosphere-free observation can be formed by

$$\frac{f_1^2 \lambda_1 \Phi_1^{ji} - f_2^2 \lambda_2 \Phi_2^{ji}}{f_1^2 - f_2^2} \qquad (8.64)$$

or

$$\lambda_c \left( \Phi_1^{ji} - \Phi_2^{ji} \frac{f_2}{f_1} \right), \tag{8.65}$$

where

$$\lambda_c = \frac{f_1^2 \lambda_1}{f_1^2 - f_2^2}. \tag{8.66}$$

The above combination has units of length. $\lambda_c$ is the wavelength of the combined phase. The second-order ionospheric model can be presented as (units: length)

$$d\rho_k(\text{ion}) = F \left( \frac{A_1}{f_k^2} + \frac{A_2}{f_k^3} \right), \tag{8.67}$$

where $A_1$ and $A_2$ are unknowns and have the physical meaning of the total electron content in the zenith direction and $F$ is the mapping function discussed in Sect. 5.1.4 to map the zenith delay to the path direction. Forming the first-order ionosphere-free combination (Eq. 8.64), one gets the second-order ionospheric residuals

$$\begin{aligned}
\frac{f_1^2 d\rho_1(\text{iono}) - f_2^2 d\rho_2(\text{iono})}{f_1^2 - f_2^2} &= F \left( \frac{1}{f_1} - \frac{1}{f_2} \right) \frac{A_2}{f_1^2 - f_2^2} \\
&= -\lambda_c F (f_1 - f_2) \frac{A_2}{f_2 f_1^3 \lambda_1}, \\
&= -\lambda_c F \frac{17}{60} \frac{A_2}{f_1^3 \lambda_1}
\end{aligned} \tag{8.68}$$

The amount of the second-order ionospheric effects is only about 1/1000 of the first-order effects (Syndergaard 1999). Thus the second-order effects can reach a level of a few centimetres, and therefore must be taken into account or modelled. The observation equation of the first-order ionosphere-free combination can be written as

$$\lambda_c \left( \Phi_1^{ji} - \Phi_2^{ji} \frac{f_2}{f_1} \right) = \lambda_c \left\{ \begin{array}{l} \rho^{ji} \left( \frac{1}{\lambda_1} - \frac{\frac{f_2}{f_1}}{\lambda_2} \right) + \left( N_1^{ji} - \frac{60}{77} N_2^{ji} \right) \\ + D_c + \alpha d\rho(\text{others}) \left( \frac{1}{\lambda_1} - \frac{\frac{f_2}{f_1}}{\lambda_2} \right) \end{array} \right\} \tag{8.69}$$

or

$$\lambda_c \Phi_c^{ji} = \rho^{ji} + \lambda_c N_c^{ji} + F\alpha\lambda_c D_c + d\rho \text{ (others)}, \qquad (8.70)$$

where

$$\begin{aligned} N_c^{ji} &= N_1^{ji} - \frac{60}{77} N_2^{ji}, \\ D_c &= -\frac{17}{60} \frac{A_2}{f_1^3 \lambda_1} \quad \text{and} \\ \Phi_c^{ji} &= \Phi_1^{ji} - \Phi_2^{ji} \frac{f_2}{f_1}. \end{aligned} \qquad (8.71)$$

$N_c^{ji}$ is the (first order) ionosphere-free ambiguity, $\lambda_c D_c$ is the ionospheric residuals in the zenith, and $\Phi_c^{ji}$ is the phase of the ionosphere-free combination (in cycles).

Because of the integer property of the ambiguities, the ionospheric effects can only affect the ambiguities with a constant part; therefore, a so-called ionospheric ambiguity correction can be given as

$$dN = \int_T F\lambda_c \left(-\frac{17}{60}\right) \frac{A_2}{f_1^3 \lambda_1} \frac{dt}{T}, \qquad (8.72)$$

where the integration time interval $T$ is the same as that of the individual ambiguity parameters. This term also explains why sometimes a very good float solution has ambiguity values that cannot easily be rounded up. Such a term also destroys the integer property of the ambiguity parameter. Without this correction, ambiguity is not separable from the constant part of ionospheric residuals. The ambiguity parameters will absorb all similar constant effects. If these constant effects have not been separated from the ambiguity parameters, ambiguity fixing cannot be carried out correctly. In such cases, a float solution is preferred. This also explains why float solutions are typically used for precise applications such as deformation or plate movement studies.

In the case of $\alpha = 0$ (i.e. the remaining ionospheric error is not modelled; this is the case for most GPS software), an ionospheric ambiguity correction must be applied to the ambiguity solutions before ambiguity fixing. After ambiguity fixing, such correction must again be added to the ambiguity parameters, because it presents the constant effects of the ionospheric residuals, and therefore must be taken into account.

Taking such a correction into account or modelling the ionospheric residuals, the problem of the dependence of the ambiguity resolution on the distances of the baselines can been solved.

## 8.6.3
## Determination of the Ionospheric Ambiguity Correction

To determine the ionospheric ambiguity correction, the ambiguity-ionospheric equations are suggested for use. Recalling the discussion of Sect. 6.5.2, there only the first-order ionospheric effects are modelled in the so-called ambiguity-ionospheric equations. Using Eq. 8.68, the ambiguity-ionospheric equations (cf. Sect. 6.5.2) can be written as

$$R_1 - R_2 \approx \Delta\delta_{\text{ion}} + \Delta\varepsilon_c = F\left(\frac{1}{f_1^2} - \frac{1}{f_2^2}\right)A_1 + F\left(\frac{1}{f_1^3} - \frac{1}{f_2^3}\right)A_2 + \Delta\varepsilon_c, \quad (8.73)$$

$$\lambda_1 \Phi_1 - \lambda_2 \Phi_2 = \lambda_1 N_1 - \lambda_2 N_2 - \Delta\delta_{\text{ion}} + \Delta\varepsilon_p \quad \text{and} \quad (8.74)$$

$$\lambda_j \Phi_j - R_j \approx \lambda_j N_j - 2\delta_{\text{ion}}(j) + \Delta\varepsilon_{\text{pc}}, \quad j = 1, \quad (8.75)$$

where

$$\delta_{\text{ion}}(j) = \mathrm{d}\rho_j(\text{ion}) = F\left(\frac{A_1}{f_j^2} + \frac{A_2}{f_j^3}\right) = Fa_2, \quad j = 1, 2 \quad \text{and}$$

$$\Delta\delta_{\text{ion}} = \delta_{\text{ion}}(1) - \delta_{\text{ion}}(2) = F\left(\frac{1}{f_1^2} - \frac{1}{f_2^2}\right)A_1 + F\left(\frac{1}{f_1^3} - \frac{1}{f_2^3}\right)A_2 = Fa_1. \quad (8.76)$$

The weight matrix can be found in Sect. 6.5.2. In a dual-frequency case, for every observed satellite, there are two ambiguity unknowns and a total of two ionospheric parameters ($a_1$ and $a_2$). Under usual conditions, if there are four satellites observed (12 observables), then there are eight ambiguities and two ionospheric parameters. Obviously the problem is solvable. The solution to the ambiguity-ionospheric equations was already mentioned in Sect. 8.6.1 and will be discussed again in Chap. 9 in detail.

Total electron content $A_1$ and $A_2$ can be determined by using the ionospheric parameters $a_1$ and $a_2$ afterward. No matter what kind of value $A_1$ has been obtained, it will be eliminated through the first-order ionosphere-free combination. However, the $A_2$ will be needed for computing the ionospheric ambiguity correction. After the ionospheric ambiguity correction, the ionosphere-free ambiguity can be then fixed and will be outlined in next section. Physically, $A_2$ should be very small compared with $A_1$, and the $A_1$ of all stations and all time intervals should have negative values (cf. Eqs. 5.16 and 5.18).

## 8.6.4
## Integer Ambiguity Fixing Through Ambiguity-Ionospheric Equations

It is advantageous to fix the integer ambiguities through the ambiguity-ionospheric equations. The whole equation system includes only the ambiguity and the ionospheric parameters. Even the code observables are also used; through a suitable weight matrix derived in Sect. 6.5.2, the problem can be solved very accurately. Because original (or zero-difference, un-combined) ambiguity parameters are used here, the integer ambiguity search methods discussed in Sect. 8.3 can be directly used for fixing the integer ambiguities $N_1$ and $N_2$.

## 8.6.5
## Float Ambiguity Fixing

Using the first-order ionosphere-free combination, the first-order ionosphere-free ambiguity $N_{cs}$ can be obtained. Through the ambiguity fixing, the ionosphere-free ambiguity can be computed by $N_c(=N_1-N_2 60/77)$. The ionospheric ambiguity correction $dN$ (cf. Eq. 8.72) can be also computed. Then according to the above discussions, the float ambiguity $N_{cs}$ should be fixed to the value of $N_c + dN$.

The $dN$ represents the constant part of ionospheric effects, which are not modelled by an ionosphere-free combination and have been absorbed into the ambiguity parameters.

## 8.7
## PPP Ambiguity Fixing

In traditional precise point positioning (PPP), the zero-difference ambiguity of a satellite-receiver pair is naturally not an integer value due to the existence of a fractional part of the uncalibrated phase delays (FCBs) originating in the receiver and satellite (Collins 2008; Ge et al. 2008; Mercier and Laurichesse 2008). Research in PPP ambiguity resolution has been a focus throughout the world over the past decade, since it can significantly improve positioning quality, especially for the east component (Blewitt 1989; Dong and Bock 1989; Geng et al. 2010). In this section, we will briefly describe these research efforts. The basic principle of PPP ambiguity resolution is that the a priori information, which recovers the integer property of PPP zero-difference ambiguity, is first obtained through a regional network solution, and thus the rover station can then apply the a priori information as constraints to fix the PPP ambiguity. The a priori constraints obtained from regional network solutions could be FCBs products, wide-lane and narrow-lane ambiguity, or double-differenced ambiguity.

## 1. *Ambiguity constraint based on the single-difference-between-satellites method*

The single-difference-between-satellites method for PPP integer ambiguity resolution was proposed by Ge et al. (2008). This method decomposes zero-difference ambiguity into wide-lane and narrow-lane, and applies the difference between satellites to remove the receiver-dependent FCBs. Based on a network of reference stations, the wide-lane FCBs can be determined by averaging the fractional parts of all pertinent wide-lane ambiguities derived from the Melbourne-Wübbena combination measurements (Melbourne 1985; Wübbena 1985). Similarly, narrow-lane FCBs can be determined by averaging the fractional parts of all pertinent narrow-lane ambiguities derived from the wide-lane ambiguities and ionosphere-free combination ambiguities. For a single rover station, these wide-lane and narrow-lane FCBs are used as a priori constraints to recover the integer property of the PPP ambiguities. The constraints can be represented as

$$N_{3,m}^{i,j} = \frac{f_1 f_2}{f_1^2 - f_2^2} n_{w,m}^{i,j} + \frac{f_1}{f_1 + f_2} \left( n_{1,m}^{i,j} + \Delta \phi_{1,m}^{i,j} \right), \tag{8.77}$$

where the superscript $(i, j)$ denotes the single-difference between satellites $i$ and $j$, $f_1$, and $f_2$ are the frequencies of carrier phase, $N_{3,m}^{i,j}$ is the ionosphere-free combination ambiguities at single rover station $m$, $n_{w,m}^{i,j}$ and $n_{1,m}^{i,j}$ are the wide-lane integer ambiguities and narrow-lane integer ambiguities, $\Delta \phi_{1,m}^{i,j}$ is the narrow-lane FCBs, which can be determined from regional net work solution.

## 2. *Ambiguity constraint based on integer phase clock estimates*

Integer phase clock estimation was proposed by Laurichesse et al. (2009), who applied the same decomposition as that of the single-difference-between-satellites method, but directly fixed the zero-difference ambiguities to integers. Therefore, an arbitrary value should be assigned to the FCBs of a specific receiver to obtain the satellite-dependent FCBs. Their wide-lane FCB determination is the same as that of the single-difference-between-satellites method. Nevertheless, the narrow-lane FCBs are not determined but assimilated into the clock estimates. Based on a network of reference stations, the narrow-lane ambiguities can be fixed to integers before estimating the clocks. Thus, the clocks containing the narrow-lane FCBs are called integer-recovery clocks and can be obtained. For the single rover station, the integer-recovery clocks are used to guarantee the integer property of the narrow-lane ambiguities. Therefore, PPP ambiguity resolution can be achieved through the constraints of wide-lane FCBs and integer-recovery clocks.

## 3. *Ambiguity constraint based on double-difference integer ambiguities method*

The double-difference integer ambiguities method was proposed by Bertiger et al. (2010). This method calculates the float ambiguities, including the FCBs, directly. For the network of the reference stations, the float wide-lane ambiguities can be derived from the Melbourne-Wübbena combination measurements, while the float ionosphere-free combination ambiguities can be obtained through a traditional PPP algorithm. The single rover station utilises the same method as that of the reference station to calculate its float wide-lane ambiguities and float ionosphere-free combination ambiguities. The double-difference integer wide-lane ambiguities between reference station and rover station can then be formed and fixed, since the receiver-dependent FCBs and satellite-dependent FCBs both can be eliminated via double difference. Therefore, the double-difference ionosphere-free ambiguities between reference station and rover station can be formed and fixed and set as the double-difference constraints in PPP ambiguity resolution.

With the integer ambiguities resolved, the precision of PPP can be significantly improved over that with a float solution. Moreover, the convergence time in PPP can be reduced and the practicability of PPP consequently promoted.

# References

Bertiger W, Desai S, Haines B, Harvey N, Moore A, Owen S, Weiss J (2010) Single receiver phase ambiguity resolution with GPS data. J Geod 84(5):327-337.

Blewitt G (1989) Carrier phase ambiguity resolution for the global positioning system applied to geodetic baselines up to 2000 km. J Geophys Res 94(B8):10187–10203.

Bronstein IN, Semendjajew KA (1987) Taschenbuch der Mathematik.B. G. Teubner Verlagsgesellschaft, Leipzig, ISBN 3-322-00259-4.

Cannon ME, Lachapelle G, Goddard TW (1997) Development and results of a precision farming sys-tem using GPS and GIS technologies. Geomatica 51,1:9–19.

Collins P (2008) Isolating and estimating undifferenced GPS integer ambiguities. In: Proceedings of ION national technical meeting, San Diego, US, pp 720–732.

Cui X, Yu Z, Tao B, Liu D (1982) Adjustment in surveying. Surveying Press, Peking, (in Chinese).

Dong D, Bock Y (1989) Global positioning system network analysis with phase ambiguity resolution applied to crustal deformation studies in California. J Geophys Res 94(B4):3949–3966.

Euler H-J, Landau H (1992) Fast GPS ambiguity resolution on-the-fly for real-time applications. In: Proceedings of 6th Int. Geod. Symp. on Satellite Positioning. Columbus, Ohio, pp 17–20.

Forsell B, Martín-Neira M, Harris R (1997) Carrier phase ambiguity resolution in GNSS-2. Proceedings of ION GPS-97, The 10th International Technical Meeting of the Satellite Division of the Institute of Navigation, Kansas City, Missouri, September 16-19, pp. 1727-1736.

Ge M, Gendt G, Rothacher M, Shi C, Liu J (2008) Resolution of GPS carrier-phase ambiguities in precise point positioning (PPP) with daily observations. J Geod 82(7):389–399.

Gehlich U, Lelgemann D (1997) Zur Parametrisierung von GPS-Phasenmessungen. ZfV 6:262–270.

Geng J, Meng X, Dodson A, Teferle F (2010) Integer ambiguity resolution in precise point positioning: method comparison. J Geod 84:569-581.

Gotthardt E (1978) Einführung in die Ausgleichungsrechnung. Herbert Wichmann Verlag, Karlsruhe.

Han S, Rizos C (1995) On-the-fly ambiguity resolution for long range GPS kinematic positioning. In: GPS Trends in Precise Terrestrial, Airborne, and Spaceborne Applications: 21st IUGG General Assembly, IAG Symposium No. 115, Boulder, USA, July 3–4, 1995. Springer-Verlag, Berlin, pp 290–294.

Han S, Rizos C (1997) Comparing GPS ambiguity resolution techniques. GPS World 8(10):54–61.

Hofmann-Wellenhof B, Lichtenegger H, Collins J (1997, 2001) GPS theory and practice. Springer-Press, Wien.
Jung, J (1999) High integrity carrier phase navigation for future LAAS using multiple civilian GPS signals. Proceedings of ION GPS-99, The 12$^{th}$ International Technical Meeting of the Satellite Division of the Institute of Navigation, Nashville, USA, September 14-17, pp. 727-736.
Klobuchar JA (1996) Ionospheric effects on GPS. In: Parkinson BW, Spilker JJ (eds) Global Position-ing System: Theory and applications, Vol. I, Chapter 12.
Laurichesse D, Mercier F, Berthias J, Broca P, Cerri L (2009) Integer ambiguity resolution on undifferenced GPS phase measurements and its application to PPP and satellite precise orbit determination. Navig J Inst Navig 56(2):135–149.
Leick A (1995) GPS satellite surveying. John Wiley & Sons Ltd., New York.
Melbourne W (1985) The case for ranging in GPS-based geodetic systems. In: Proceedings of first international symposium on precise positioning with the global positioning system, Rockville, US, pp 373–386.
Merbart L (1995) Ambiguity resolution techniques in geodetic and geodynamic applications of the Global Positioning System. Dissertation an der Philosophisch-naturwissenschaftlichen Fakultät der Universität Bern.
Mercier F, Laurichesse D (2008) Zero-difference ambiguity blocking properties of satellite/receiver widelane biases. In: Proceedings of European navigation conference, Toulouse, France.
Morujao D, Mendes V (2008) Investigation of Instantaneous Carrier Phase Ambiguity Resolution with the GPS/GALILEO Combination using the General Ambiguity Search Criterion. J. GPS, 7(1): 35-45.
Remondi B (1984) Using the Global Positioning System (GPS) phase observable for relative geodesy: Modelling, processing, and results. University of Texas at Austin, Center for Space Research.
Sjoeberg LE (1999) Unbiased vs biased estimation of GPS phase ambiguities from dual-frequency code and phase observables. J Geodesy 73:118–124.
Syndergaard S (1999) Retrieval analysis and methodologies in atmospheric limb sounding using the GNSS radio occultation technique. Dissertation, Niels Bohr Institute for Astronomy, Physics and Geophysics, Faculty of Science, University of Copenhagen.
Teunissen P (1993) Least squares estimation of the integer GPS ambiguities. Invited lecture, Section IV: Theory and methodology, IAG General Meeting, Beijing, China, August 1993. Also in LGR-Series n.6, Delft Geodetic Computing Centre, Delft University of Technology, Delft.
Teunissen P (1994) The invertible GPS ambiguity transformations. Manuscripta Geodaetica, 20 (6): 489-497.
Teunissen P (1995) The least-squares ambiguity decorrelation adjustment: A method for fast GPS integer ambiguity estimation. J Geodesy 70(1–2):65–82.
Vollath U, Birnbach S, Landau H (1998) An analysis of three-carrier ambiguity resolution (TCAR) technique for precise relative positioning in GNSS-2. Proceedings of ION GPS-98, The 11$^{th}$ International Technical Meeting of the Satellite Division of the Institute of Navigation, Nashville, Tennessee, USA, September 15-18, pp. 417-426.
Wang LX, Fang ZD, Zhang MY, Lin GB, Gu LK, Zhong TD, Yang XA, She DP, Luo ZH, Xiao BQ, Chai H, Lin DX (1979) Mathematic handbook. Educational Press, Peking, ISBN 13012-0165.
Wang G, Chen Z, Chen W, Xu G (1988) The principle of GPS precise positioning system. Surveying Press, Peking, ISBN 7-5030-0141-0/P.58, 345 p, (in Chinese).
Wanninger L (1995) Enhancing differential GPS using regional ionospheric models. B Geod 69:283–291.
Wanninger L (1995) Einfluß ionosphärischer Störungen auf präzise GPS-Messungen in Mitteleuropa. Schriftenreihe des Deutschen Vereins für Vermessungswesen, Bd. 18, Stuttgart, pp 218–232.

Wuebbena G (1985) Software developments for geodetic positioning with GPS using TI-4100 code and carrier measurements. In: Proceedings of first international symposium on precise positioning with the global positioning system, Rockville, US, pp 403–412.

Xu G (1992) Spectral analysis and geopotential determination (Spektralanalyse und Erdschwerefeldbestimmung). Dissertation, DGK, Reihe C, Heft Nr. 397, Press of the Bavarian Academy of Sciences, ISBN 3-7696-9442-2, 100 p, (with very detailed summary in German).

Xu G (2002a) A general criterion of integer ambiguity search, J. GPS, Vol. 1 No.2: 122-131.

Xu G (2003) GPS – Theory, Algorithms and Applications, Springer Heidelberg, ISBN 3-540-67812-3, 315 pages, in English.

Xu G, Bastos L, Timmen L (1997) GPS kinematic positioning in AGMASCO campaigns – Strategic goals and numerical results. In: Proceedings of ION GPS-97 meeting in Kansas City, September 16–19, 1997, pp 1173–1184.

Xu G, Fritsch J, Hehl K (1997) Results and conclusions of the carborne gravimetry campaign in northern Germany. Geodetic Week Berlin '97, Oct. 6–11, 1997, electronic version published in http://www.geodesy.tu-berlin.de.

Xu G, Schwintzer P, Reigber Ch (1998) KSGSoft – Kinematic/Static GPS Software – Software user manual (version of 1998). Scientific Technical Report STR98/19 of GeoForschungsZentrum (GFZ) Potsdam.

Xu G, Shen Y, Yang Y, Sun H, Zhang Q, Guo J, Yeh T (2010) Equivalence of GPS Algorithms and Its Inference. Sciences of Geodesy-I: Advances and Future Directions, Springer, Heidelberg, Berlin.

Zhang FZ, Xu Y, Jiang CH (2016) An alternative derivation of the equivalent criterion, Surveying Journal of China, in review.

Zhang W, Cannon M, Julien O, Alves P (2003) Investigation of combined GPS/GALILEO cascading ambiguity resolution schemes. Proceedings of ION GPS/GNSS 2003, The 16th International Technical Meeting of the Satellite Division of the Institute of Navigation, Portland, OR, USA, September 9-12, pp. 2599-2610.

# Chapter 9
# Parameterisation and Algorithms of GPS Data Processing

The parameterisation problems of the bias parameters in the GPS observation model are outlined in Sect. 12.1 of the first edition of this book. The problems are then mostly solved, and the theory will be addressed here in detail (cf. Xu 2004; Xu et al. 2006b). The equivalence properties of the algorithms of GPS data processing are described, and the standard algorithms are outlined.

## 9.1
## Parameterisation of the GPS Observation Model

The commonly used GPS data processing methods are the so-called uncombined and combining, and the undifferenced and differencing algorithms (e.g., Hofmann-Wellenhof et al. 2001; Leick 2004; Remondi 1984; Seeber 1993; Strang and Borre 1997; Blewitt 1998). The observation equations of the combining and differencing methods can be obtained by carrying out linear transformations of the original (uncombined and undifferenced) equations. As soon as the weight matrix is similarly transformed according to the law of variance–covariance propagation, all methods are theoretically equivalent. The equivalences of combining and differencing algorithms are discussed in Sects. 6.7 and 6.8, respectively. The equivalence of the combining methods is exact, whereas the equivalence of the differencing algorithms is slightly different (Xu 2004, cf. Sect. 9.2). The parameters are implicitly expressed in the discussions; therefore, the parameterisation problems of the equivalent methods have not been discussed in detail. At that time, this topic was considered one of the remaining GPS theoretical problems (Xu 2003, pp. 279–280; Wells et al. 1987, p. 34), and it will be discussed in the next subsection.

Three pieces of evidence of the parameterisation problem of the undifferenced GPS observation model are given first. Then the theoretical analysis and numerical derivation are made to show how to parameterise the bias effects of the undifferenced GPS observation model independently. A geometry-free illustration and a correlation analysis in the case of a phase–code combination are discussed. At the end, conclusions and comments are given.

## 9.1.1
## Evidence of the Parameterisation Problem of the Undifferenced Observation Model

### Evidence from Undifferenced and Differencing Algorithms

Suppose the undifferenced GPS observation equation and the related LS normal equation are

$$V = L - \begin{pmatrix} A_1 & A_2 \end{pmatrix} \begin{pmatrix} X_1 \\ X_2 \end{pmatrix}, \quad P \tag{9.1}$$

$$\begin{pmatrix} M_{11} & M_{12} \\ M_{21} & M_{22} \end{pmatrix} \begin{pmatrix} X_1 \\ X_2 \end{pmatrix} = \begin{pmatrix} W_1 \\ W_2 \end{pmatrix}, \tag{9.2}$$

where all symbols have the same meanings as that of Eqs. 7.117 and 7.118. Equation 9.2 can be diagonalised as (cf. Sect. 7.6.1)

$$\begin{pmatrix} M_1 & 0 \\ 0 & M_2 \end{pmatrix} \begin{pmatrix} X_1 \\ X_2 \end{pmatrix} = \begin{pmatrix} B_1 \\ B_2 \end{pmatrix}. \tag{9.3}$$

The related equivalent observation equation of the diagonal normal Eq. 9.3 can be written (cf. Sect. 7.6.1)

$$\begin{pmatrix} U_1 \\ U_2 \end{pmatrix} = \begin{pmatrix} L \\ L \end{pmatrix} - \begin{pmatrix} D_1 & 0 \\ 0 & D_2 \end{pmatrix} \begin{pmatrix} X_1 \\ X_2 \end{pmatrix}, \quad \begin{pmatrix} P & 0 \\ 0 & P \end{pmatrix}, \tag{9.4}$$

where all symbols have the same meanings as that of Eqs. 7.142 and 7.140. If $X_1$ is the vector containing all clock errors, then the second equation of Eq. 9.3 is the equivalent double-differencing GPS normal equation. It is well known that in a double-differencing algorithm, the ambiguity sub-vector contained in $X_2$ must be the double-differencing ambiguities; otherwise, the problem will be generally singular. It is notable that $X_2$ is identical with that of in the original undifferenced observation Eq. 9.1. Therefore, the ambiguity sub-vector contained in $X_2$ (in Eq. 9.1) must be a set of double-differencing ambiguities (or an equivalent set of ambiguities). This is the first piece of evidence (or indication) of the singularity of the undifferenced GPS observation model in which the undifferenced ambiguities are used.

### Evidence from Uncombined and Combining Algorithms

Suppose the original GPS observation equation of one viewed satellite is (cf. Eq. 6.134)

$$\begin{pmatrix} R_1 \\ R_2 \\ \lambda_1 \Phi_1 \\ \lambda_2 \Phi_2 \end{pmatrix} = \begin{pmatrix} 0 & 0 & f_s^2/f_1^2 & 1 \\ 0 & 0 & f_s^2/f_2^2 & 1 \\ 1 & 0 & -f_s^2/f_1^2 & 1 \\ 0 & 1 & -f_s^2/f_2^2 & 1 \end{pmatrix} \begin{pmatrix} \lambda_1 N_1 \\ \lambda_2 N_2 \\ B_1 \\ C_\rho \end{pmatrix}, \quad P; \qquad (9.5)$$

then the uncombined or combining algorithms have the same solution of (cf. Eq. 6.138)

$$\begin{pmatrix} \lambda_1 N_1 \\ \lambda_2 N_2 \\ B_1 \\ C_\rho \end{pmatrix} = \begin{pmatrix} 1-2a & -2b & 1 & 0 \\ -2a & 2a-1 & 0 & 1 \\ 1/q & -1/q & 0 & 0 \\ a & b & 0 & 0 \end{pmatrix} \begin{pmatrix} R_1 \\ R_2 \\ \lambda_1 \Phi_1 \\ \lambda_2 \Phi_2 \end{pmatrix}, \qquad (9.6)$$

where all symbols have the same meanings as that of Eqs. 6.134 and 6.138. Then one notices that the ionosphere ($B_1$) and geometry ($C_\rho$) are functions of the codes ($R_1$ and $R_2$) and are independent from phases ($\Phi_1$ and $\Phi_2$) in Eq. 9.6. In other words, the phase observables contribute nothing to the ionosphere and geometry, and this is not possible. Such an illogical conclusion is caused by the parameterisation of the ambiguities given in the observation model of Eq. 9.5. If one takes the first evidence discussed above into account, and defines that for each station one of the satellites in view must be selected as reference, and the related ambiguity must be merged into the clock parameter, then the phases do contribute to ionosphere and geometry. One can see again that parameterisation is a very important topic and must be discussed more specifically. An improper parameterisation of the observation model will lead to incorrect conclusions through the derivation from the model.

*Evidence from Practice*

Without using a priori information, a straightforward programming of the GPS data processing using an undifferenced algorithm leads to no results (i.e. the normal equation is singular, cf. Xu 2004). Therefore, an exact parameterisation description is necessary and will be discussed in the next section.

## 9.1.2
## A Method of Uncorrelated Bias Parameterisation

Here we restrict our discussion to the parameterisation problem of the bias parameters (or constant effects, i.e. the clock errors and ambiguities).

Recall the discussions of the equivalence of undifferenced and differencing algorithms in Sect. 6.8. The equivalence property is valid under three conditions: observation vector $L$ used in Eq. 9.1 is identical; parameterisation of $X_2$ is identical; and $X_1$ is able to be eliminated (cf. Sect. 6.8).

The first condition is necessary for the exactness of the equivalence because, through forming differences, the unpaired data will be cancelled out in the differencing.

The second condition states that the parameterisation of the undifferenced and differencing model should be the same. This may be interpreted as the following: the rank of the undifferenced and differencing equations should be the same if the differencing is formed by a full rank linear transformation. If only the differencing equations are taken into account, then the rank of the undifferenced model should equal the rank of the differencing model plus the number of eliminated independent parameters.

It is well known that one of the clock error parameters is linearly correlated with the others. This may be seen in the proof of the equivalence property of the double differences, where the two receiver clock errors of the baseline may not be separated from each other and have to be transformed to one parameter and then eliminated (Xu 2002, Sect. 6.8). This indicates that if in the undifferenced model all clock errors are modelled, the problem will be singular (i.e. rank defect). Indeed, Wells et al. (1987) noticed that the equivalence is valid if measures are taken to avoid rank defect in the bias parameterisation. Which clock error has to be kept fixed is arbitrary. Because of the different qualities of the satellite and receiver clocks, a good choice is to fix a satellite clock error (the clock is called a reference clock). In practice, the clock error is an unknown; therefore, there is no way to keep that fixed except to fix it to zero. In this case, the meaning of the other bias parameters will be changed and may represent the relative errors between the other biases.

The third condition is important to ensure a full-ranked parameterisation of the parameter vector $X_1$ which is going to be eliminated.

The undifferenced Eq. 9.1 is solvable if the parameters $X_1$ and $X_2$ are not over-parameterised. In the case of single differences, $X_1$ includes satellite clock errors and is able to be eliminated. Therefore, to guarantee that the undifferenced model Eq. 9.1 is not singular, $X_2$ in Eq. 9.1 must be not over-parameterised. In the case of double differences, $X_1$ includes all clock errors except the reference one. Here we notice that the second observation equation of 9.1 is equivalent to the double-differencing observation equation and the second equation of 9.2 is the related normal equation. In a traditional double-differencing observation equation, the ambiguity parameters are represented by double-differencing ambiguities. Recall that for the equivalence property, the number (or rank) of ambiguity parameters in $X_2$ that are not linearly correlated must be equal to the number of the double-differencing ambiguities. In the case of triple differences, $X_1$ includes all clock errors and ambiguities. The fact that $X_1$ should able to be eliminated leads again to the conclusion that the ambiguities should be linearly independent.

The two equivalent linear equations should have the same rank. Therefore, if all clock errors except the reference one are modelled, the number of independent undifferenced ambiguity parameters should be equal to the number of double-differencing ambiguities. According to the definition of the double-differencing ambiguity, one has for one baseline

$$N_{i1,i2}^{k1,k2} = N_{i2}^{k2} - N_{i1}^{k2} - N_{i2}^{k1} + N_{i1}^{k1}$$
$$N_{i1,i2}^{k1,k3} = N_{i2}^{k3} - N_{i1}^{k3} - N_{i2}^{k1} + N_{i1}^{k1}$$
$$N_{i1,i2}^{k1,k4} = N_{i2}^{k4} - N_{i1}^{k4} - N_{i2}^{k1} + N_{i1}^{k1} \qquad (9.7)$$
$$\ldots\ldots$$
$$N_{i1,i2}^{k1,kn} = N_{i2}^{kn} - N_{i1}^{kn} - N_{i2}^{k1} + N_{i1}^{k1},$$

where $i1$ and $i2$ are station indices, $kj$ is the $j$th satellite's identification, $n$ is the common observed satellite number and is a function of the baseline, and $N$ is ambiguity. Then there are $n - 1$ double-differencing ambiguities and $2n$ undifferenced ambiguities. Taking the connection of the baselines into account, there are $n - 1$ double-differencing ambiguities and $n$ new undifferenced ambiguities for any further baseline. If $i1$ is defined as the reference station of the whole network and $k1$ as the reference satellite of station $i2$, then undifferenced ambiguities of the reference station cannot be separated from the others (i.e. they are linearly correlated with the others). The undifferenced ambiguity of the reference satellite of station $i2$ cannot be separated from the others (i.e. it is linearly correlated with the others). That is, the ambiguities of the reference station cannot be determined, and the ambiguities of the reference satellites of non-reference stations cannot be determined. Either they should not be modelled or they should be kept fixed. A straightforward parameterisation of all undifferenced ambiguities will lead to rank defect, and the problem will be singular and not able to be solved.

Therefore, using the equivalence properties of the equivalent equation of GPS data processing, we come to the conclusion that the ambiguities of the reference station and ambiguities of the reference satellite of every station are linearly correlated with the other ambiguities and clock error parameters. However, a general method of parameterisation should be independent of the selection of the references (station and satellite). Therefore, we use a two-baseline network to further our analysis. The original observation equation can be written as follows:

$$L_{i1}^{k1} = \cdots \delta_{i1} + \delta_{k1} + N_{i1}^{k1} + \cdots$$
$$L_{i1}^{k2} = \cdots \delta_{i1} + \delta_{k2} + N_{i1}^{k2} + \cdots$$
$$L_{i1}^{k3} = \cdots \delta_{i1} + \delta_{k3} + N_{i1}^{k3} + \cdots$$
$$L_{i1}^{k4} = \cdots \delta_{i1} + \delta_{k4} + N_{i1}^{k4} + \cdots \qquad (9.8)$$
$$L_{i1}^{k5} = \cdots \delta_{i1} + \delta_{k5} + N_{i1}^{k5} + \cdots$$
$$L_{i1}^{k6} = \cdots \delta_{i1} + \delta_{k6} + N_{i1}^{k6} + \cdots$$

$$L_{i2}^{k1} = \cdots \delta_{i2} + \delta_{k1} + N_{i2}^{k1} + \cdots$$
$$L_{i2}^{k2} = \cdots \delta_{i2} + \delta_{k2} + N_{i2}^{k2} + \cdots$$
$$L_{i2}^{k3} = \cdots \delta_{i2} + \delta_{k3} + N_{i2}^{k3} + \cdots$$
$$L_{i2}^{k4} = \cdots \delta_{i2} + \delta_{k4} + N_{i2}^{k4} + \cdots \qquad (9.9)$$
$$L_{i2}^{k5} = \cdots \delta_{i2} + \delta_{k5} + N_{i2}^{k5} + \cdots$$
$$L_{i2}^{k7} = \cdots \delta_{i2} + \delta_{k7} + N_{i2}^{k7} + \cdots$$

$$L_{i3}^{k2} = \cdots \delta_{i3} + \delta_{k2} + N_{i3}^{k2} + \cdots$$
$$L_{i3}^{k3} = \cdots \delta_{i3} + \delta_{k3} + N_{i3}^{k3} + \cdots$$
$$L_{i3}^{k4} = \cdots \delta_{i3} + \delta_{k4} + N_{i3}^{k4} + \cdots \qquad (9.10)$$
$$L_{i3}^{k5} = \cdots \delta_{i3} + \delta_{k5} + N_{i3}^{k5} + \cdots$$
$$L_{i3}^{k6} = \cdots \delta_{i3} + \delta_{k6} + N_{i3}^{k6} + \cdots$$
$$L_{i3}^{k7} = \cdots \delta_{i3} + \delta_{k7} + N_{i3}^{k7} + \cdots,$$

where only the bias terms are listed and $L$ and $\delta$ represent observable and clock error, respectively. Observation equations of station $i1$, $i2$, and $i3$ are Eqs. 9.8, 9.9, and 9.10. Define that the baseline 1, 2 are formed by station $i1$ and $i2$, as well as $i2$ and $i3$, respectively. Select $i1$ as the reference station and then keep the related ambiguities fixed (set to zero for simplification). For convenience of later discussion, select $\delta_{i1}$ as the reference clock (set to zero, too) and select $k1$, $k2$ as reference satellites of the station $i2$, i3 (set the related ambiguities to zero), respectively. Then Eqs. 9.8–9.10 become

$$L_{i1}^{k1} = \cdots \delta_{k1} + \cdots$$
$$L_{i1}^{k2} = \cdots \delta_{k2} + \cdots$$
$$L_{i1}^{k3} = \cdots \delta_{k3} + \cdots$$
$$L_{i1}^{k4} = \cdots \delta_{k4} + \cdots \qquad (9.11)$$
$$L_{i1}^{k5} = \cdots \delta_{k5} + \cdots$$
$$L_{i1}^{k6} = \cdots \delta_{k6} + \cdots$$

$$L_{i2}^{k1} = \cdots \delta_{i2} + \delta_{k1} + \cdots$$
$$L_{i2}^{k2} = \cdots \delta_{i2} + \delta_{k2} + N_{i2}^{k2} + \cdots$$
$$L_{i2}^{k3} = \cdots \delta_{i2} + \delta_{k3} + N_{i2}^{k3} + \cdots$$
$$L_{i2}^{k4} = \cdots \delta_{i2} + \delta_{k4} + N_{i2}^{k4} + \cdots \qquad (9.12)$$
$$L_{i2}^{k5} = \cdots \delta_{i2} + \delta_{k5} + N_{i2}^{k5} + \cdots$$
$$L_{i2}^{k7} = \cdots \delta_{i2} + \delta_{k7} + N_{i2}^{k7} + \cdots$$

## 9.1 Parameterisation of the GPS Observation Model

$$\begin{aligned}
L_{i3}^{k2} &= \cdots \delta_{i3} + \delta_{k2} + \cdots \\
L_{i3}^{k3} &= \cdots \delta_{i3} + \delta_{k3} + N_{i3}^{k3} + \cdots \\
L_{i3}^{k4} &= \cdots \delta_{i3} + \delta_{k4} + N_{i3}^{k4} + \cdots \\
L_{i3}^{k5} &= \cdots \delta_{i3} + \delta_{k5} + N_{i3}^{k5} + \cdots \\
L_{i3}^{k6} &= \cdots \delta_{i3} + \delta_{k6} + N_{i3}^{k6} + \cdots \\
L_{i3}^{k7} &= \cdots \delta_{i3} + \delta_{k7} + N_{i3}^{k7} + \cdots
\end{aligned} \qquad (9.13)$$

Differences can be formed through linear operations. The total operation is a full rank linear transformation, which does not change the least squares solution of the original equations. Single differences can be formed by the following (Eq. 9.11 remains unchanged and, therefore, will not be listed again):

$$\begin{aligned}
L_{i2}^{k1} - L_{i1}^{k1} &= \cdots \delta_{i2} + \cdots \\
L_{i2}^{k2} - L_{i1}^{k2} &= \cdots \delta_{i2} + N_{i2}^{k2} + \cdots \\
L_{i2}^{k3} - L_{i1}^{k3} &= \cdots \delta_{i2} + N_{i2}^{k3} + \cdots \\
L_{i2}^{k4} - L_{i1}^{k4} &= \cdots \delta_{i2} + N_{i2}^{k4} + \cdots \\
L_{i2}^{k5} - L_{i1}^{k5} &= \cdots \delta_{i2} + N_{i2}^{k5} + \cdots \\
L_{i2}^{k7} &= \ldots \delta_{i2} + \delta_{k7} + N_{i2}^{k7} + \cdots
\end{aligned} \qquad (9.14)$$

$$\begin{aligned}
L_{i3}^{k2} - L_{i2}^{k2} &= \cdots \delta_{i3} - \delta_{i2} - N_{i2}^{k2} + \cdots \\
L_{i3}^{k3} - L_{i2}^{k3} &= \cdots \delta_{i3} - \delta_{i2} + N_{i3}^{k3} - N_{i2}^{k3} + \cdots \\
L_{i3}^{k4} - L_{i2}^{k4} &= \cdots \delta_{i3} - \delta_{i2} + N_{i3}^{k4} - N_{i2}^{k4} + \cdots \\
L_{i3}^{k5} - L_{i2}^{k5} &= \cdots \delta_{i3} - \delta_{i2} + N_{i3}^{k5} - N_{i2}^{k5} + \cdots \\
L_{i3}^{k6} &= \cdots \delta_{i3} + \delta_{k6} + N_{i3}^{k6} + \cdots \\
L_{i3}^{k7} - L_{i2}^{k7} &= \cdots \delta_{i3} - \delta_{i2} + N_{i3}^{k7} - N_{i2}^{k7} + \cdots
\end{aligned} \qquad (9.15)$$

where two observations are unpaired due to the baseline definitions. Double differences can be formed by

$$\begin{aligned}
L_{i2}^{k1} - L_{i1}^{k1} &= \cdots \delta_{i2} + \cdots \\
L_{i2}^{k2} - L_{i1}^{k2} - L_{i2}^{k1} + L_{i1}^{k1} &= \cdots N_{i2}^{k2} + \cdots \\
L_{i2}^{k3} - L_{i1}^{k3} - L_{i2}^{k1} + L_{i1}^{k1} &= \cdots N_{i2}^{k3} + \cdots \\
L_{i2}^{k3} - L_{i1}^{k3} - L_{i2}^{k1} + L_{i1}^{k1} &= \cdots N_{i2}^{k3} + \cdots \\
L_{i2}^{k4} - L_{i1}^{k4} - L_{i2}^{k1} + L_{i1}^{k1} &= \cdots N_{i2}^{k4} + \cdots \\
L_{i2}^{k5} - L_{i1}^{k5} - L_{i2}^{k1} + L_{i1}^{k1} &= \cdots N_{i2}^{k5} + \cdots \\
L_{i2}^{k7} - L_{i2}^{k1} + L_{i1}^{k1} &= \cdots \delta_{k7} + N_{i2}^{k7} + \cdots
\end{aligned} \qquad (9.16)$$

$$
\begin{aligned}
L_{i3}^{k2} - L_{i2}^{k2} &= \cdots \delta_{i3} - \delta_{i2} - N_{i2}^{k2} + \cdots \\
L_{i3}^{k3} - L_{i2}^{k3} - L_{i3}^{k2} + L_{i2}^{k2} &= \cdots N_{i3}^{k3} - N_{i2}^{k3} + N_{i2}^{k2} + \cdots \\
L_{i3}^{k4} - L_{i2}^{k4} - L_{i3}^{k2} + L_{i2}^{k2} &= \cdots N_{i3}^{k4} - N_{i2}^{k4} + N_{i2}^{k2} + \cdots \\
L_{i3}^{k5} - L_{i2}^{k5} - L_{i3}^{k2} + L_{i2}^{k2} &= \cdots N_{i3}^{k5} - N_{i2}^{k5} + N_{i2}^{k2} + \cdots \\
L_{i3}^{k6} &= \cdots \delta_{i3} + \delta_{k6} + N_{i3}^{k6} + \cdots \\
L_{i3}^{k7} - L_{i2}^{k7} - L_{i3}^{k2} + L_{i2}^{k2} &= \cdots N_{i3}^{k7} - N_{i2}^{k7} + N_{i2}^{k2} + \cdots
\end{aligned}
\qquad (9.17)
$$

Using Eqs. 9.16 and 9.11, Eq. 9.17 can be further modified to

$$
\begin{aligned}
L_{i3}^{k2} - L_{i2}^{k2} + (L_{i2}^{k1} - L_{i1}^{k1}) + (L_{i2}^{k2} - L_{i1}^{k2} - L_{i2}^{k1} + L_{i1}^{k1}) &= \cdots \delta_{i3} + \cdots \\
L_{i3}^{k3} - L_{i2}^{k3} - L_{i3}^{k2} + L_{i2}^{k2} + (L_{i2}^{k3} - L_{i1}^{k3} - L_{i2}^{k1} + L_{i1}^{k1}) - (L_{i2}^{k2} - L_{i1}^{k2} - L_{i2}^{k1} + L_{i1}^{k1}) &= \cdots N_{i3}^{k3} + \cdots \\
L_{i3}^{k4} - L_{i2}^{k4} - L_{i3}^{k2} + L_{i2}^{k2} + (L_{i2}^{k4} - L_{i1}^{k4} - L_{i2}^{k1} + L_{i1}^{k1}) - (L_{i2}^{k2} - L_{i1}^{k2} - L_{i2}^{k1} + L_{i1}^{k1}) &= \cdots N_{i3}^{k4} + \cdots \\
L_{i3}^{k5} - L_{i2}^{k5} - L_{i3}^{k2} + L_{i2}^{k2} + (L_{i2}^{k5} - L_{i1}^{k5} - L_{i2}^{k1} + L_{i1}^{k1}) - (L_{i2}^{k2} - L_{i1}^{k2} - L_{i2}^{k1} + L_{i1}^{k1}) &= \cdots N_{i3}^{k5} + \cdots \\
L_{i3}^{k6} - L_{i1}^{k6} &= \cdots \delta_{i3} + N_{i3}^{k6} + \cdots \\
L_{i3}^{k7} - L_{i2}^{k7} - L_{i3}^{k2} + L_{i2}^{k2} + (L_{i2}^{k7} - L_{i1}^{k1} + L_{i1}^{k1}) - (L_{i2}^{k2} - L_{i1}^{k2} - L_{i2}^{k1} + L_{i1}^{k1}) &= \cdots - \delta_{k7} + N_{i3}^{k7} + \cdots
\end{aligned}
$$
(9.18)

or

$$
\begin{aligned}
L_{i3}^{k2} - L_{i1}^{k2} &= \cdots \delta_{i3} + \cdots \\
L_{i3}^{k3} - L_{i1}^{k3} - L_{i3}^{k2} + L_{i1}^{k2} &= \cdots N_{i3}^{k3} + \cdots \\
L_{i3}^{k4} - L_{i1}^{k4} - L_{i3}^{k2} + L_{i1}^{k2} &= \cdots N_{i3}^{k4} + \cdots \\
L_{i3}^{k5} - L_{i1}^{k5} - L_{i3}^{k2} + L_{i1}^{k2} &= \cdots N_{i3}^{k5} + \cdots \\
L_{i3}^{k6} - L_{i1}^{k6} - L_{i3}^{k2} + L_{i1}^{k2} &= \cdots N_{i3}^{k6} + \cdots \\
L_{i3}^{k7} - L_{i3}^{k2} + L_{i1}^{k2} &= \cdots - \delta_{k7} + N_{i3}^{k7} + \cdots
\end{aligned}
\qquad (9.19)
$$

From the last equation of Eqs. 9.16 and 9.19, it is obvious that the clock error and the ambiguities of satellite $k7$, which is not observed by the reference station, are linearly correlated. Keeping one of the ambiguities of the satellite $k7$ at station $i2$ or $i3$ is necessary and equivalent. Therefore, for any satellite that is not observed by the reference station, one of the related ambiguities should be kept fixed (station selection is arbitrary). In other words, one of the ambiguities of all satellites has to be kept fixed. In this way, every transformed equation includes only one bias parameter and the bias parameters are linearly independent (regular). Furthermore, the differencing cannot be formed for the unpaired observations of every baseline. However, in the case of an undifferenced adjustment, the situation would be different. We notice that the equation for $k6$ in Eq. 9.18 can be transformed to a double-differencing one in Eq. 9.19. If more data is used in the undifferenced algorithm than in the differencing method, the number of undifferenced ambiguity parameters will be larger than that of double-differencing parameters. Therefore, we must drive the so-called data condition to guarantee that the data are able to be differenced, or equivalently, we must extend the method of double-difference formation such that the differencing will not be limited by special baseline design. Both will be discussed in Sect. 9.2.

The meanings of the parameters are changed by independent parameterisation, and they can be read from Eqs. 9.11–9.13. The clock errors of the satellites observed by the reference station include the errors of receiver clock and ambiguities. The receiver clock errors include the error of ambiguity of the reference satellite of the same station. Due to the inseparable property of the bias parameters, the clock error parameters no longer represent pure clock errors, and the ambiguities represent no longer pure physical ambiguity. Theoretically speaking, the synchronisation applications of GPS may not be realised using the carrier-phase observations. Furthermore, Eq. 9.19 shows that the undifferenced ambiguities of $i3$ have the meaning of double-differencing ambiguities of the station $i3$ and $i1$ in this case.

Up to now, we have discussed the correlation problem of the bias parameters and found a method of how to parameterise the GPS observations regularly to avoid the problem of rank defect. Of course, many other ways to parameterise the GPS observation model can be similarly derived. However, the parameter sets should be equivalent to each other and can be transformed from one set to another uniquely as long as the same data is used.

## 9.1.3
### Geometry-Free Illustration

The reason why the reference parameters have to be fixed lies in the nature of range measurements, which cannot provide information of the datum origin (cf., e.g., Wells et al. 1987, p. 9). Suppose $d$ is the direct measurement of clock errors of satellite $k$ and receiver $i$, i.e. $d_i^k = \delta_i + \delta_k$, no matter how many observations were made and how the indices were changed, one parameter (i.e. reference clock) is inseparable from the others and has to be fixed. Suppose $h$ is the direct measurement of ambiguity $N$ and clock errors of satellite $k$ and receiver $i$, i.e. $h_i^k = \delta_i + \delta_k + N_i^k$, the number of over-parameterised biases is exactly the number of total observed satellites and used receivers. This ensures again that our parameterisation method to fix the reference clock and one ambiguity of every satellite as well as one ambiguity of the reference satellite of every non-reference station is reasonable. The case of combination of $d$ and $h$ (as code and phase observations) will be discussed in the next section.

## 9.1.4
## Correlation Analysis in the Case of Phase–Code Combinations

A phase–code combined observation equation can be written by (cf. Sect. 7.5.2)

$$\begin{pmatrix} V_1 \\ V_2 \end{pmatrix} = \begin{pmatrix} L_1 \\ L_2 \end{pmatrix} - \begin{pmatrix} A_{11} & A_{12} \\ A_{11} & 0 \end{pmatrix} \begin{pmatrix} X_1 \\ X_2 \end{pmatrix} \quad \text{and} \quad P = \begin{pmatrix} w_p P_0 & 0 \\ 0 & w_c P_0 \end{pmatrix}, \quad (9.20)$$

where $L_1$ and $L_2$ are the observational vectors of phase (scaled in length) and code, respectively; $V_1$ and $V_2$ are related residual vectors; $X_2$ and $X_1$ are unknown vectors of ambiguity and others; $A_{12}$ and $A_{11}$ are related coefficient matrices; $P_0$ is a symmetric and definite weight matrix; and $w_p$ and $w_c$ are weight factors of the phase and code observations.

The phase, code and phase–code normal equations can be formed respectively by

$$\begin{pmatrix} N_{11} & N_{12} \\ N_{21} & N_{22} \end{pmatrix} \begin{pmatrix} X_1 \\ X_2 \end{pmatrix} = \begin{pmatrix} R_1 \\ R_2 \end{pmatrix},$$

$$N_{11} X_1 = R_c, \quad \text{and} \quad (9.21)$$

$$\begin{pmatrix} M_{11} & M_{12} \\ M_{21} & M_{22} \end{pmatrix} \begin{pmatrix} X_1 \\ X_2 \end{pmatrix} = \begin{pmatrix} B_1 \\ B_2 \end{pmatrix},$$

where

$$\begin{aligned} M_{11} &= (w_p + w_c) A_{11}^T P_0 A_{11} = (w_p + w_c) N_{11}, \\ M_{12} &= M_{21}^T = w_p A_{11}^T P_0 A_{12} = w_p N_{12}, \\ M_{22} &= w_p A_{12}^T P_0 A_{12} = w_p N_{22}, \\ B_1 &= A_{11}^T P_0 (w_p L_1 + w_c L_2) = w_p R_1 + w_c R_c, \quad \text{and} \\ B_2 &= w_p A_{12}^T P_0 L_1 = w_p R_2. \end{aligned} \quad (9.22)$$

The covariance matrix $Q$ is denoted

$$Q = \begin{pmatrix} M_{11} & M_{12} \\ M_{21} & M_{22} \end{pmatrix}^{-1} = \begin{pmatrix} Q_{11} & Q_{12} \\ Q_{21} & Q_{22} \end{pmatrix}, \quad (9.23)$$

where (Gotthardt 1978; Cui et al. 1982)

$$\begin{aligned}
Q_{11} &= (M_{11} - M_{12}M_{22}^{-1}M_{21})^{-1}, \\
Q_{22} &= (M_{22} - M_{21}M_{11}^{-1}M_{12})^{-1}, \\
Q_{12} &= M_{11}^{-1}(-M_{12}Q_{22}) \quad \text{and} \\
Q_{21} &= M_{22}^{-1}(-M_{21}Q_{11}).
\end{aligned} \quad (9.24)$$

i.e.

$$\begin{aligned}
Q_{11} &= ((w_p + w_c)N_{11} - w_p N_{12}N_{22}^{-1}N_{21})^{-1}, \\
Q_{22} &= (w_p N_{22} - w_p^2(w_p + w_c)^{-1}N_{21}N_{11}^{-1}N_{12})^{-1} \quad \text{and} \\
Q_{21} &= -N_{22}^{-1}N_{21}((w_p + w_c)N_{11} - w_p N_{12}N_{22}^{-1}N_{21})^{-1}.
\end{aligned} \quad (9.25)$$

Thus the correlation coefficient $C_{ij}$ is a function of $w_p$ and $w_c$, i.e.

$$C_{ij} = f(w_p, w_c), \quad (9.26)$$

where indices $i$ and $j$ are the indices of unknown parameters in $X_1$ and $X_2$. For $w_c = 0$ (only phase is used, $X_1$ and $X_2$ are partly linear correlated) and $w_c = w_p$ ($X_1$ and $X_2$ are uncorrelated), there exists indices $ij$, so that

$$C_{ij} = f(w_p, w_c = 0) = 1 \quad \text{and} \quad C_{ij} = f(w_p, w_c = w_p) = 0. \quad (9.27)$$

In other words, there exists indices $i$ and $j$, the related unknowns are correlated if $w_c = 0$ and uncorrelated if $w_c = w_p$. In the case of a phase–code combination, $w_c = 0.01 w_p$ can be selected, and one has

$$C_{ij} = f(w_p, w_c = 0.01 w_p) \quad (9.28)$$

whose value is very close to 1 (strongly correlated) in the discussed case. Equations 9.26, 9.27, and 9.28 indicate that for the correlated unknown pair $ij$ the correlation situation may not change much by combining the code to the phase because of the lower weight of the code related to the phase. A numerical test confirmed this conclusion (Xu 2004).

## 9.1.5
## Conclusions and Comments

In this section, the singularity problem of the undifferenced GPS data processing is pointed out and an independent parameterisation method is proposed for bias parameters of the GPS observation model. The method is implemented into software, and the results confirm the correctness of the theory and algorithm. Conclusions can be summarised by the following:

1. Bias parameterisation of undifferenced GPS phase observations with all clock errors except the reference one, and all undifferenced ambiguities are linearly correlated. The linear equation system of undifferenced GPS is then singular and cannot be solved theoretically.
2. A linear independent bias parameterisation can be reached by fixing the reference clock of the reference station, fixing one of the ambiguities of every satellite of arbitrary station (called reference station of every satellite), and fixing the ambiguities of the reference satellite of every non-reference station. The selections of the references are arbitrary; however, the selections are not allowed to be duplicated.
3. The linear independent ambiguity parameter set is equivalent to the parameter set of double-differencing ambiguities, and they can be transformed from one to another uniquely if the same data is used.
4. The physical meanings of the bias parameters are varied depending on the way of parameterisation. Because of the inseparable property of the bias parameters, the synchronisation applications of GPS may not be realised using the carrier-phase observations.
5. The phase–code combination does not change the correlation relation between the correlated biases significantly.

It is noteworthy to comment on the use of the undifferenced algorithm:

1. In the undifferenced algorithm, the observation equation is rank defect if the over-parameterisation problem has not been taken into account. The numerical inexactness introduced by eliminating the clock error parameters and the use of a priori information of some other parameters are the reason why the singular problem is solvable in practice thus far.
2. Using the undifferenced and differencing methods, solutions of the common parameters must be the same if the undifferenced GPS data modelling is really an equivalent one and not over-parameterised.
3. A singular undifferenced parameterisation may become regular by introducing conditions or by fixing some of the parameters through introducing a priori information.

## 9.2
## Equivalence of the GPS Data Processing Algorithms

The equivalence theorem, an optimal method for forming an independent baseline network, and a data condition, as well as the equivalent algorithms using secondary observables are discussed in this section (cf. Xu et al. 2006c).

## 9.2.1
## Equivalence Theorem of GPS Data Processing Algorithms

In Sect. 6.7, the equivalence properties of uncombined and combining algorithms of GPS data processing are given. Whether uncombined or combining algorithms are used, the results obtained are identical and the precisions of the solutions are identical, too. It is notable that the parameterisation is very important. The solutions depend on the parameterisation. For convenience, the original GPS observation equation and the solution are listed as (cf. Sect. 6.7)

$$\begin{pmatrix} R_1 \\ R_2 \\ \lambda_1 \Phi_1 \\ \lambda_2 \Phi_2 \end{pmatrix} = \begin{pmatrix} 0 & 0 & f_s^2/f_1^2 & 1 \\ 0 & 0 & f_s^2/f_2^2 & 1 \\ 1 & 0 & -f_s^2/f_1^2 & 1 \\ 0 & 1 & -f_s^2/f_2^2 & 1 \end{pmatrix} \begin{pmatrix} \lambda_1 N_1 \\ \lambda_2 N_2 \\ B_1 \\ C_\rho \end{pmatrix}, \quad P = \begin{pmatrix} \sigma_c^2 & 0 & 0 & 0 \\ 0 & \sigma_c^2 & 0 & 0 \\ 0 & 0 & \sigma_p^2 & 0 \\ 0 & 0 & 0 & \sigma_p^2 \end{pmatrix}^{-1}, \quad (9.29)$$

and

$$\begin{pmatrix} \lambda_1 N_1 \\ \lambda_2 N_2 \\ B_1 \\ C_\rho \end{pmatrix} = \begin{pmatrix} 1-2a & -2b & 1 & 0 \\ -2a & 2a-1 & 0 & 1 \\ 1/q & -1/q & 0 & 0 \\ a & b & 0 & 0 \end{pmatrix} \begin{pmatrix} R_1 \\ R_2 \\ \lambda_1 \Phi_1 \\ \lambda_2 \Phi_2 \end{pmatrix} \quad (9.30)$$

where the meanings of the symbols are the same as that of Eqs. 6.134 and 6.138.

In Sect. 6.8, the equivalence properties of undifferenced and differencing algorithms of GPS data processing are given. Whether undifferenced or differencing algorithms are used, the results obtained are identical and the precisions of the solutions are equivalent. It is notable that the equivalence here is slightly different from the equivalence in combining algorithms. To distinguish them, we call the equivalence in the differencing case a soft equivalence. The soft equivalence is valid under three so-called conditions. The first is a data condition, which guarantees that the data used in undifferenced or differencing algorithms are the same. The data condition will be discussed in the next section. The second is a parameterisation condition, i.e. the parameterisation must be the same. The third is the elimination condition, i.e. the parameter set to be eliminated should be able to be eliminated. (Implicitly, the parameter set of the problem should be a regular one). Because of the process of elimination, the cofactor matrices of the undifferenced and differencing equations are different. If the cofactor of an undifferenced normal equation has the form of

$$\begin{pmatrix} M_{11} & M_{12} \\ M_{21} & M_{22} \end{pmatrix}^{-1} = Q = \begin{pmatrix} Q_{11} & Q_{12} \\ Q_{21} & Q_{22} \end{pmatrix}, \quad (9.31)$$

then we call the diagonal part of the cofactor

$$Q_e = \begin{pmatrix} M_1 & 0 \\ 0 & M_2 \end{pmatrix}^{-1} = \begin{pmatrix} Q_{11} & 0 \\ 0 & Q_{22} \end{pmatrix} \quad (9.32)$$

an equivalent cofactor. The equivalent cofactor has the same diagonal element blocks as the original cofactor matrix $Q$ and guarantees that the precision relation between the unknowns remains the same. The soft equivalence is defined as follows: the solutions are identical and the covariance matrices are equivalent. This definition is implicitly used in the traditional block-wise least squares adjustment. It is notable that the parameterisation is very important and the rank of the normal equation of the undifferenced observation equation must be equal to the rank of the normal equation of the differencing observation equation plus the number of the eliminated independent parameters. For convenience, the original GPS observation equation and the equivalent differencing equation can be generally written as (cf. Eqs. 9.1 and 9.4)

$$V = L - \begin{pmatrix} A_1 & A_2 \end{pmatrix} \begin{pmatrix} X_1 \\ X_2 \end{pmatrix}, \quad P \quad (9.33)$$

$$\begin{pmatrix} U_1 \\ U_2 \end{pmatrix} = \begin{pmatrix} L \\ L \end{pmatrix} - \begin{pmatrix} D_1 & 0 \\ 0 & D_2 \end{pmatrix} \begin{pmatrix} X_1 \\ X_2 \end{pmatrix}, \quad \begin{pmatrix} P & 0 \\ 0 & P \end{pmatrix}. \quad (9.34)$$

In Sect. 9.1 the way to parameterise the GPS observables independently is proposed. A correct and reasonable parameterisation is the key to a correct conclusion by combining and differencing derivations. An example is given in Sect. 6.7 where an illogical conclusion is derived due to the inexact parameterisation.

For any GPS survey with a definitive space-time configuration, observed GPS data can be parameterised (or modelled) in a suitable way and listed together in a form of linear equations for processing. Combining and differencing are two linear transformations. Because the uncombined and combining data (or equations) are equivalent, differencing the uncombined or combining equations is (soft) equivalent. Inversely, the combining operator is an invertible transformation; making or not making the combination operation on the equivalent undifferenced or differencing equations (Eqs. 9.33 and 9.34) is equivalent. That is, the mixtures of the combining and differencing algorithms are also equivalent to the original undifferenced and uncombined algorithms. The equivalence properties can be summarised in a theorem as follows.

*Equivalence Theorem of GPS Data Processing Algorithms*

Under the three so-called equivalence conditions and the definition of the so-called soft equivalence, for any GPS survey with definitive space-time configuration, GPS data processing algorithms—uncombined and combining algorithms, undifferenced and differencing algorithms, as well as their mixtures—are at least soft equivalent. That is, the results obtained by using any algorithm or any mixture of the algorithms are identical. The diagonal elements of the covariance matrix are identical. The ratios of the precisions of the solutions are identical. None of the algorithms are preferred in view of the results and precisions. Suitable algorithms or mixtures of the algorithms will be specifically advantageous for special kinds of data dealings.

The implicit condition of this theorem is that the parameterisation must be the same and regular. The parameterisation depends on different configurations of the GPS surveys and strategies of the GPS data processing. The theorem says that if the data used are the same and the model is parameterised identically and regularly, then the results must be identical and the precision should be equivalent. This is a guiding principle for the GPS data processing practice.

## 9.2.2
### Optimal Baseline Network Forming and Data Condition

It is well known that for a network with $n$ stations there are $n - 1$ independent baselines. An independent baseline network can be stated in words: all stations are connected through these baselines, and the shortest way from one station to any other stations is unique. Generally speaking, a shorter baseline leads to a better common view of the satellites. Therefore, the baseline should be formed so that the length of the baseline falls as short as possible. For a network, an optimal choice should be that the summation of weighted lengths of all independent baselines should be minimal. This is a specific mathematic problem called a minimum spanning tree (cf., e.g., Wang et al. 1979).

Algorithms exist to solve this minimum spanning tree problem with software. Therefore, we will just show an example here. An IGS network with ca. 100 stations and the related optimal and independent baseline tree is shown in Fig. 9.1. The average length of the baselines is ca. 1300 km. The maximum distance is ca. 3700 km.

In the traditional double-differencing model, the unpaired GPS observations of every designed baseline have to be omitted because of the requirement of differencing (in the example of Sect. 1.2, two observations of $k6$ will be omitted. However, if the differencing is not limited by baseline design, no observations have to be cancelled out). Therefore, an optimal means of double differencing should be based on an optimal baseline design to form the differencing first, then, without limitation of the baseline design, to check for the unpaired observations in order to form possible differencing. This measure is useful for raising the rate of data used by the differencing method. An example of an IGS network with 47 stations and a 1-day observations has shown (Xu 2004) that 87.9 % of all data is used in difference forming based on the optimal baseline design, whereas 99.1 % of all data is used in the extended method of difference forming without limitation of the

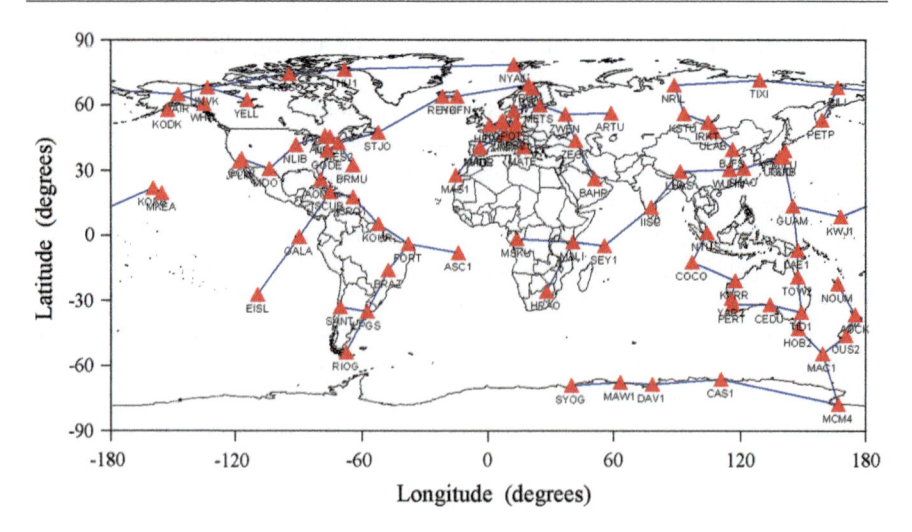

**Fig. 9.1** Independent and optimal IGS GPS baseline network (100 stations)

baseline design. That is, the original data may be nearly 100 % used for such a means of double differencing.

In the undifferenced model, in order to be able to eliminate the clock error parameters, it is sufficient that every satellite is observed at least at two stations (for eliminating the satellite clock errors) and at every station there is a satellite combined with one of the other satellites that are commonly viewed by at least one of the other stations (for eliminating the receiver clock errors). The condition ensures that extended double differencing can be formed from the data. The data has to be cancelled out if the condition is not fulfilled or the ambiguities including in the related data have to be kept fixed.

For convenience, we state the data condition as follows.

***Data Condition***: All satellites must be observed at least twice (for forming single differences) and one satellite combined with one of the other satellites should be commonly viewed by at least one of the other stations (for forming double differences).

It is notable that the data condition above is valid for single and double differencing. For triple differencing and user defined differencing the data condition may be similarly defined. The data condition is one of the conditions of the equivalence of the undifferenced and differencing algorithms. The data condition is derived from the difference forming; however, it is suggested to use it also in undifferenced methods to reduce the singular data. The optimal baseline network forming is beneficial for differencing methods to raise the rate of used data.

### 9.2.3
### Algorithms Using Secondary GPS Observables

As stated in Sects. 6.7 and 9.2, the uncombined and combining algorithms are equivalent. A method of GPS data processing using secondary data is outlined in Sect. 6.7.3. However, a concrete parameterisation of the observation model is only possible after the method of independent parameterisation is discussed in Sect. 9.1. The data processing using secondary observables leads to equivalent results of any combining algorithms. Therefore, the concrete parameterisation of the GPS observation model has to be specifically discussed again. The observation model of $m$ satellites viewed at one station is (cf. Eqs. 6.134 and 9.5)

$$\begin{pmatrix} R_1(k) \\ R_2(k) \\ \lambda_1 \Phi_1(k) \\ \lambda_2 \Phi_2(k) \end{pmatrix} = \begin{pmatrix} 0 & 0 & f_s^2/f_1^2 & 1 \\ 0 & 0 & f_s^2/f_2^2 & 1 \\ 1 & 0 & -f_s^2/f_1^2 & 1 \\ 0 & 1 & -f_s^2/f_2^2 & 1 \end{pmatrix} \begin{pmatrix} \lambda_1 N_1(k) \\ \lambda_2 N_2(k) \\ B_1(k) \\ C_\rho(k) \end{pmatrix}, \quad k=1,\ldots m, \quad (9.35)$$

where the relation

$$B_1^z = \frac{1}{m} \sum_{k=1}^{m} B_1(k)/F_k \qquad (9.36)$$

can be used to map the ionospheric parameters in the path directions to the parameter in the zenith direction. The meanings of the symbols are the same as stated in Sect. 6.7. Solutions of Eq. 9.35 are (similar to Eq. 9.6)

$$\begin{pmatrix} \lambda_1 N_1(k) \\ \lambda_2 N_2(k) \\ B_1(k) \\ C_\rho(k) \end{pmatrix} = \begin{pmatrix} 1-2a & -2b & 1 & 0 \\ -2a & 2a-1 & 0 & 1 \\ 1/q & -1/q & 0 & 0 \\ a & b & 0 & 0 \end{pmatrix} \begin{pmatrix} R_1(k) \\ R_2(k) \\ \lambda_1 \Phi_1(k) \\ \lambda_2 \Phi_2(k) \end{pmatrix}, \quad Q(k), \ k=1,\ldots,m,$$

(9.37)

where the covariance matrix $Q(k)$ can be obtained by variance-covariance propagation law. The vector on the left side of Eq. 9.37 is called the secondary observation vector. In the case where $K$ satellites are viewed, the traditional combinations of the observation model and the related secondary solutions are the same as the Eqs. 9.35 and 9.37, where the $m = K$. However, taking the parameterisation method into account, at least one satellite has to be selected as reference and the related ambiguities cannot be modelled. If one were to suppose that the satellite of index $K$ is the reference, then the first $m = K - 1$ observation equations are the same as Eq. 9.35. The satellite $K$ related observation equations can be written as

$$\begin{pmatrix} R_1(k) \\ R_2(k) \\ \lambda_1 \Phi_1(k) \\ \lambda_2 \Phi_2(k) \end{pmatrix} = \begin{pmatrix} 0 & 0 & f_s^2/f_1^2 & 1 \\ 0 & 0 & f_s^2/f_2^2 & 1 \\ 0 & 0 & -f_s^2/f_1^2 & 1 \\ 0 & 0 & -f_s^2/f_2^2 & 1 \end{pmatrix} \begin{pmatrix} \lambda_1 N_1(k) \\ \lambda_2 N_2(k) \\ B_1(k) \\ C_\rho(k) \end{pmatrix}, \quad k = K, \quad (9.38)$$

where the ambiguities are not modelled and the constant effects will be absorbed by the clock parameters. Solutions of Eq. 9.38 are

$$\begin{pmatrix} \lambda_1 N_1(k) \\ \lambda_2 N_2(k) \\ B_1(k) \\ C_\rho(k) \end{pmatrix} = \frac{1}{2} \begin{pmatrix} 0 & 0 & 0 & 0 \\ 0 & 0 & 0 & 0 \\ 1/q & -1/q & -1/q & 1/q \\ 1/2 & 1/2 & 1/2 & 1/2 \end{pmatrix} \begin{pmatrix} R_1(k) \\ R_2(k) \\ \lambda_1 \Phi_1(k) \\ \lambda_2 \Phi_2(k) \end{pmatrix}, \quad Q(K). \quad (9.39)$$

It is notable that the solutions of the traditional combinations are Eq. 9.37 with $m = K$, whereas for the combinations with independent bias parameterisation, the solutions are the combinations of the Eq. 9.37 with $m = K - 1$ and Eq. 9.39. It is obvious that the two solutions are different. Because the traditional observation model used is inexact, the solutions of the traditional combinations are also inexact. The bias effects (of ambiguities) that are not modelled are merged into the clock bias parameters. Because the bias effects cannot be absorbed into the non-bias parameters, only the clock error parameters will be different in the results and the clock errors will have the different meanings. Further, the ionosphere-free and geometry-free combinations are correct under the independent parameterisation.

It shows that through exact parameterisation, the combinations are no longer independent from satellite to satellite. For surveys with multiple stations, through correct parameterisation the combinations will no longer be independent from station to station. Therefore, traditional combinations will lead to incorrect results because of the inexact parameterisation.

The so-called secondary observables on the left-hand side of Eqs. 9.37 and 9.39 can be further processed. The original observables can be uniquely transformed to secondary observables. The secondary observables are equivalent and direct measurements of the ambiguities and ionosphere as well as geometry. Any further GPS data processing can be based on the secondary observables (cf. Sect. 6.7).

## 9.2.4
## Simplified Equivalent Representation of GPS Observation Equations

The GPS observation models, the data differentiation (single, double, and triple differences) and the equivalent property between un-differenced and differencing algorithms were discussed in Chap. 6. According to the theorem of equivalent property, Shen and Xu (2008) developed the simplified equations which equivalently represent the single and double-differenced observation equations using corresponding pseudo-observations in single or multi-baseline solutions. However, this study was based on the assumption of all stations tracking the same satellites with identical weights, thus the simplified equations were expanded in the case of each station tracking different satellites with elevation-dependent weights (Shen et al. 2009). The derived simplified equivalent algorithm was shown to be very convenient for programming and efficient in computation, which would potentially aid the development of efficient GNSS software and benefit the local, regional and even global GNSS multi-baseline solutions. The specific algorithm is emphatically introduced in this section.

### *Single-Differenced Simplified Equivalent Observation Equations*

The GNSS observation equations for one epoch can be symbolically expressed as

$$\varepsilon = Ax + By + Cz - l, \quad P, \tag{9.40}$$

where $y$ and $z$ are the vectors of station and satellite biases, $B$ and $C$ denote the respective coefficient matrices with full column rank; $x$ is a column vector with $t$ parameters, $A$ is its coefficient matrix also with full column rank; $l$ and $\varepsilon$ are the column vectors of observables and normally distributed observation errors; $P$ is the weight matrix of observations. Here elevation-dependent weights are used and different stations can track the different satellites, but the correlations among the observables (temporal, cross, and channel) are not considered. Thus, the weight matrix $P$ is diagonal with varying elements. Refer to Leick (2004) for the detailed interpretation of these parameters. If there are total of $k$ stations and each station only tracks the subset of the total $n$ satellites, then $y = (y_1 \ y_2 \ \ldots \ y_k)^T$ and $z = (z_1 \ z_2 \ \ldots \ z_n)^T$. The coefficient matrices, vector of observables and weight matrix are grouped with the following sub-blocks in the order of satellites as

$$A = \begin{pmatrix} A^1 \\ A^2 \\ \vdots \\ A^n \end{pmatrix}, B = \begin{pmatrix} B^1 \\ B^2 \\ \vdots \\ B^n \end{pmatrix}, C = \begin{pmatrix} e_{k_1} & & & \\ & e_{k_2} & & \\ & & \ddots & \vdots \\ & & & e_{k_n} \end{pmatrix},$$

$$l = \begin{pmatrix} l^1 \\ l^2 \\ \vdots \\ l^n \end{pmatrix}, P = \begin{pmatrix} P^1 & & & \\ & P^2 & & \\ & & \ddots & \\ & & & P^n \end{pmatrix} \tag{9.41}$$

where

$$A^j = \begin{pmatrix} a^j_{S^j(1)} \\ a^j_{S^j(2)} \\ \vdots \\ a^j_{S^j(k_j)} \end{pmatrix}, l^j = \begin{pmatrix} l^j_{S^j(1)} \\ l^j_{S^j(2)} \\ \vdots \\ l^j_{S^j(k_j)} \end{pmatrix}, P = \begin{pmatrix} P^j_{S^j(1)} & & & \\ & P^j_{S^j(2)} & & \\ & & \ddots & \\ & & & P^j_{S^j(k_j)} \end{pmatrix}.$$

The symbols $a^j_{S^j(i)}$ and $l^j_{S^j(i)}$ denote respectively the coefficient row vector and observable of the satellite $j$ tracked by the station $S^j(i)$, $P^j_{S^j(i)}$ is its weight. The symbol $S^j$ represents the set of all stations that simultaneously track the satellite $j$ and $S^j(i)$ is the order of the $i$th station in the total set. The letter $k_j$ denotes the number of stations that track the satellite $j$, $e_{k_j} = (1 \; 1 \; \cdots \; 1)^T$ is a $k_j$ vector. The coefficient for the $j$th satellite is a $k_j \times k$ matrix $B^j$ consisting of $k_j$ canonical row vectors, in each canonical row vector all elements are zeros expect the element associated with the tracking receiver is one. For example, if there are five stations and the 2nd station does not track the 3rd satellite, then the matrix $B^3$ is

$$B^3 = \begin{pmatrix} 1 & 0 & 0 & 0 & 0 \\ 0 & 0 & 1 & 0 & 0 \\ 0 & 0 & 0 & 1 & 0 \\ 0 & 0 & 0 & 0 & 1 \end{pmatrix}.$$

The satellite-specific parameter vector $z$ can be eliminated by single differencing in the station domain or by right-multiplying the original observation equations with the transformation matrix $R$. The transformation matrix is (Teunissen 1997a, b; Shen and Xu 2008)

$$R = I_{\sum k} - C(C^T P C)^{-1} C^T P = \begin{pmatrix} R_{k_1} & & & \\ & R_{k_2} & & \\ & & \ddots & \\ & & & R_{k_n} \end{pmatrix}, \quad (9.42)$$

where the dimension of identity matrix $I_{\sum k}$ is $\sum k = \sum_{j=1}^n k_j$, and

$$R_{k_j} = I_{k_j} - \frac{1}{\sum_{i \in S^j} p_i^j} e_{k_j} e_{k_j}^T P^j = I_{k_j} - \frac{1}{p \sum^j} P^j \otimes e_{k_j}^T, \quad (9.43)$$

where $p\sum^j = \sum_{i \in S^j} p_i^j$ being the sum of weights of observables for all stations that tack the $j$th satellites.

## 9.2 Equivalence of the GPS Data Processing Algorithms

Multiplying Eq. 9.40 with matrix $R$, we obtain the equivalently transformed observation equations

$$\tilde{\varepsilon} = \tilde{A}x + \tilde{B}y - \tilde{l}, \quad P, \tag{9.44}$$

where $\tilde{A} = RA$, $\tilde{B} = RB$, $\tilde{l} = Rl$, and $\tilde{\varepsilon} = R\varepsilon$. As shown in Eq. 9.42, the matrix $R$ is diagonal with sub-matrix $R_{k_j}$. Therefore, Eq. 9.44 can be further simplified as

$$\tilde{\varepsilon}^j = \tilde{A}^j x + \tilde{B}^j y - \tilde{l}^j, \quad P^j, \, j = 1, 2, \ldots, n, \tag{9.45}$$

with

$$\tilde{A}^j = R_{k_j} A^j, \quad \tilde{B}^j = R_{k_j} B^j, \quad \tilde{l}^j = R_{k_j} l^j, \tag{9.46}$$

It is obvious that $R_{k_j}$ has a rank defect of one. This means that one station-specific parameter can be linearly represented with the others, i.e. only $k - 1$ station-specific parameters can be independently parameterised.

In the single-differenced equivalent observation equations, the independent parameterised station-specific parameters are generally merged into $x$, and Eq. 9.45 becomes

$$\tilde{\varepsilon}^j = \tilde{A}^j x - \tilde{l}^j, \quad P^j, \, j = 1, 2, \ldots, n, \tag{9.47}$$

where the transformed coefficient matrix and observation vector can also be further simplified as

$$\tilde{A}^j = R_{k_j} A^j = A^j - \frac{1}{p\sum^j} e_{k_j} e_{k_j}^T P^j A^j = A^j - \delta A^j, \tag{9.48}$$

$$\tilde{l}^j = R_{k_j} l^j = l^j - \frac{1}{p\sum^j} e_{k_j} e_{k_j}^T P^j l^j = l^j - \delta l^j, \tag{9.49}$$

with

$$\delta A^j = \frac{1}{p\sum^j} e_{k_j} e_{k_j}^T P^j A^j = \frac{1}{p\sum^j} e_{k_j} \sum_{i \in S^j} (p_i^j a_i^j) = \frac{1}{p\sum^j} e_{k_j} [a^j], \tag{9.50}$$

$$\delta l^j = \frac{1}{p\sum^j} e_{k_j} e_{k_j}^T P^j l^j = \frac{1}{p\sum^j} e_{k_j} \sum_{i \in S^j} (p_i^j l_i^j) = \frac{1}{p\sum^j} e_{k_j} [l^j], \tag{9.51}$$

where $[a^j] = \sum_{i \in S^j} (p_i^j a_i^j)$ and $[l^j] = \sum_{i \in S^j} (p_i^j l_i^j)$. Each element of the column vector $\delta l^j$ and each column vector of $\delta A^j$ are the weighted means of their corresponding column vectors. Therefore, the transformed vector $\tilde{l}^j$ is the centrobaric vector of $l^j$, and the transformed matrix $\tilde{A}^j$ is the column centrobaric matrix of $A^j$. In other words, the equivalent observation Eq. 9.47 can also be simply obtained through the centrobaric operation to the column vectors of $A^j$ and $l^j$.

In addition, the Eq. 9.47 can alternatively be expanded in the same way as described by Shen and Xu (2008) in the form of pseudo-observations,

$$\tilde{\varepsilon}^j = \tilde{A}^j x - l^j, \quad P^j, j = 1, 2, \ldots, n, \tag{9.52}$$

$$[\varepsilon^j] = [a^j]x - [l^j], \quad -1/p\sum^j, \quad j = 1, 2, \ldots, n, \tag{9.53}$$

where $[\tilde{\varepsilon}^j]$ denotes the residual of the $j$th sum pseudo-observation. The same normal equations can be obtained by the equivalent observation Eqs. 9.47, 9.52, and 9.53. Once the unknown parameter vector $\hat{x}$ is solved, the residual vector is computed by

$$v^j = \tilde{A}^j \hat{x} - \tilde{l}^j, \quad P^j, j = 1, 2, \ldots, n, \tag{9.54}$$

### *Double-Differenced Simplified Equivalent Observation Equations*

If there are more than two stations and each station tracks a subset of the total $n$ satellites, the double-differenced equivalent observation equations for multi-baseline solutions will be much more complicated than single-differenced ones. In order to derive the simplified double-differenced equivalent observation equations, we rearrange Eq. 9.44 with the sub-blocks in the order of receivers and use the same symbols as used in Eq. 9.44 to represent the rearranged single-differenced observation equations as

$$\tilde{\varepsilon} = \tilde{A}x + \tilde{B}y - \tilde{l}, \quad P, \tag{9.55}$$

where

$$\tilde{A} = \begin{pmatrix} \tilde{A}_1 \\ \tilde{A}_2 \\ \vdots \\ \tilde{A}_k \end{pmatrix}, \quad \tilde{A}_i = \begin{pmatrix} \tilde{a}_i^{S_i(1)} \\ \tilde{a}_i^{S_i(2)} \\ \vdots \\ \tilde{a}_i^{S_i(n_i)} \end{pmatrix} = \begin{pmatrix} a_i^{S_i(1)} - [a^{S_i(1)}]/p\sum^{S_i(1)} \\ a_i^{S_i(2)} - [a^{S_i(2)}]/p\sum^{S_i(2)} \\ \vdots \\ a_i^{S_i(n_i)} - [a^{S_i(n_i)}]/p\sum^{S_i(n_i)} \end{pmatrix}.$$

$n_i$ is the number of satellites tracked by the station $i$ and $S_i$ denotes a set comprising these $n_i$ satellites. $S_i(l)$ is the order of the $l$th satellite in the total set, that

## 9.2 Equivalence of the GPS Data Processing Algorithms

$$\tilde{l} = \begin{pmatrix} \tilde{l}_1 \\ \tilde{l}_2 \\ \vdots \\ \tilde{l}_k \end{pmatrix}, \tilde{l}_i = \begin{pmatrix} \tilde{l}_i^{S_i(1)} \\ \tilde{l}_i^{S_i(2)} \\ \vdots \\ \tilde{l}_i^{S_i(n_i)} \end{pmatrix},$$

$$P = \begin{pmatrix} P_1 & & & \\ & P_2 & & \\ & & \ddots & \\ & & & P_k \end{pmatrix}, P_i = \begin{pmatrix} p_i^{S_i(1)} & & & \\ & p_i^{S_i(2)} & & \\ & & \ddots & \\ & & & p_i^{S_i(n_i)} \end{pmatrix}$$

and $\tilde{B} = (\tilde{b}_2 \ \tilde{b}_3 \ \ldots \ \tilde{b}_k)$. The first element in $y$ is fixed to zero to achieve independent parameterisation. According to Eqs. 9.45 and 9.46, we can determine the rearranged column vector $\tilde{b}_i$ as

$$\tilde{b}_i = \left( -(Q_1 G_i \alpha_i)^T \ \ldots \ -(Q_{i-1} G_i \alpha_i)^T \ (Q_1(e_n - \alpha_i))^T \ -(Q_{i+1} G_i \alpha_i)^T \ \ldots \ -(Q_k G_i \alpha_i)^T \right)^T \quad (9.56)$$

where $\alpha_i = \left( \dfrac{p_i^1}{p\sum_1} \ \dfrac{p_i^2}{p\sum_2} \ \ldots \ \dfrac{p_i^n}{p\sum_n} \right)^T$, $G_i$ is a $n \times n$ diagonal matrix and its diagonal element is equal to either one (corresponding to tracked satellite) or zero (corresponding to non-tracked satellite). The $n_i$ non-zero row vectors of $G_i$ construct the $n_i \times n$ matrix $Q_i$. If there are six satellites and the 3rd station does not track the 2nd and 5th satellites, the matrices $G_3$ and $Q_3$ are expressed as

$$G_3 = \begin{pmatrix} 1 & & & & & \\ & 0 & & & & \\ & & 1 & & & \\ & & & 1 & & \\ & & & & 0 & \\ & & & & & 1 \end{pmatrix}, \quad Q_3 = \begin{pmatrix} 1 & 0 & 0 & 0 & 0 & 0 \\ 0 & 0 & 1 & 0 & 0 & 0 \\ 0 & 0 & 0 & 1 & 0 & 0 \\ 0 & 0 & 0 & 0 & 0 & 1 \end{pmatrix}, \quad (9.57)$$

$$G_i = Q_i^T Q_i, \quad Q_i G_i = Q_i, \quad G_i = G_i^T, \quad G_i G_i = G_i, \quad (9.58)$$

In order to determine the transformation matrix $\tilde{R}$ for eliminating station-specific parameters, the following matrix is computed

$$\tilde{B}^{\mathrm{T}}P\tilde{B} = \begin{pmatrix} \tilde{b}_2^{\mathrm{T}}P\tilde{b}_2 & \tilde{b}_2^{\mathrm{T}}P\tilde{b}_3 & \cdots & \tilde{b}_2^{\mathrm{T}}P\tilde{b}_k \\ \tilde{b}_2^{\mathrm{T}}P\tilde{b}_2 & \tilde{b}_3^{\mathrm{T}}P\tilde{b}_3 & \cdots & \tilde{b}_3^{\mathrm{T}}P\tilde{b}_k \\ \vdots & \vdots & \ddots & \vdots \\ \tilde{b}_k^{\mathrm{T}}P\tilde{b}_2 & \tilde{b}_k^{\mathrm{T}}P\tilde{b}_3 & \cdots & \tilde{b}_k^{\mathrm{T}}P\tilde{b}_k \end{pmatrix}, \qquad (9.59)$$

According to Eqs. 9.56 and 9.58, the expressions for the submatrices of $\tilde{B}^{\mathrm{T}}P\tilde{B}$ are

$$\tilde{b}_i^{\mathrm{T}}P\tilde{b}_i = p\sum{}^i - \sum_{l \in S_i}\frac{p_i^l p_i^l}{p\sum{}^l}, \qquad (9.60)$$

$$\tilde{b}_i^{\mathrm{T}}P\tilde{b}_j = -\sum_{l \in S_{ij}}\frac{p_i^l p_j^l}{p\sum{}^l}, \qquad (9.61)$$

where $p\sum{}^i = \sum_{j \in S_i} p_i^j$ is the sum of weights of observables for all satellites tracked by the $i$th station. $S_{ij}$ is an intersection set of $S_i$ and $S_j$, denoted by $S_{ij} = S_i \cap S_j$ and refers to the set of satellites that are simultaneously tracked by both station $i$ and station $j$. The matrix $\tilde{B}^{\mathrm{T}}P\tilde{B}$ can be efficiently computed by Eqs. 9.60 and 9.61, but its inverse is rather complicated and not symbolically expressible. Therefore, the transformation matrix is numerically computed by

$$\tilde{R} = I_{\sum k} - \tilde{B}(\tilde{B}^{\mathrm{T}}P\tilde{B})^{-1}\tilde{B}^{\mathrm{T}}P = I_{\sum k} - \tilde{J}, \qquad (9.62)$$

Analogously, multiplying the transformation matrix $\tilde{R}$ by Eq. 9.55, the double-differenced equivalent equations are obtained as

$$\bar{\tilde{\varepsilon}} = \bar{\tilde{A}}x - \bar{\tilde{l}}, \quad P, \qquad (9.63)$$

with

$$\bar{\tilde{A}} = \tilde{R}\tilde{A} = \tilde{A} - \tilde{J}\tilde{A}, \qquad (9.64)$$

$$\bar{\tilde{l}} = \tilde{R}\tilde{l} = \tilde{l} - \tilde{J}\tilde{l}, \qquad (9.65)$$

where the arrays $\tilde{A}$ and $\tilde{l}$ consist of all sub-matrices $\tilde{A}^j$ and sub-vectors $\tilde{l}^j$, respectively, and can be very efficiently computed by centrobaric operation to their column vectors. The $(k-1) \times (k-1)$ square matrix $\tilde{B}^{\mathrm{T}}P\tilde{B}$ and its inverse are needed to determine the transformation matrix $\tilde{R}$. The matrix $\tilde{B}^{\mathrm{T}}P\tilde{B}$ can be efficiently implemented by Eqs. 9.60 and 9.61, its inverse matrix can be trivially computed, which is certainly more efficient than computing the weight matrix of double-differenced observables for multi-baseline solutions. Once the least squares solution to parameter vector $\hat{x}$ is obtained, the residuals can be exactly computed by

$$v = \bar{\bar{A}}\hat{x} - \bar{\bar{l}}, \tag{9.66}$$

The simplified equivalent algorithm and its efficiency for computation have been verified by numerical examples; the details can be referred to Shen et al. (2009).

## 9.3
## Non-equivalent Algorithms

As stated in the equivalence theorem of GPS algorithms, the equivalence properties are valid for GPS surveys with definitive space-time configuration. As long as the measures are the same and the parameterisation is identical and regular, the GPS data processing algorithms are equivalent. It is notable that if the surveys and the parameterisation are different, then the algorithms are not equivalent to each other. For example, algorithms of single point positioning and multi-points positioning, algorithms of orbit-fixed and orbit co-determined positioning, algorithms of static and kinematic, as well as dynamic applications, etc., are non-equivalent algorithms.

## 9.4
## Reference Changing in GPS Difference Algorithm

### 9.4.1
### Changing Reference Satellite

The single baseline solution with single reference station is the most simplified and commonly used processing mode in the GPS kinematic relative positioning. Compared to network solutions with multiple reference stations, a single baseline solution has advantages of fewer unknown parameters, simple weight determination method, no baseline correlation, and a small amount of data processing. However, it can hardly meet the requirement of long endurance and long range airborne GNSS kinematic positioning. The problem of reference satellite changing is inevitable in the long time airborne relative positioning. This problem was studied by Wang et al. (2010) and Wang (2013). The details of the method are introduced in this section.

The main idea of the reference satellite changing method is that when the old reference satellite is disappeared or in the case that the reference satellite should be changed since cycle slip happens in its observation data, the double-differenced ambiguity after reference satellite changing can be obtained through multiplying the double-differenced ambiguity before changing reference satellite by a transform matrix.

Assuming the relationship between un-differenced ambiguity and double-differenced ambiguity before changing reference satellite can be expressed as

$$\nabla \Delta N = A N_0, \qquad (9.67)$$

where $\nabla \Delta N$ denotes the double-differenced ambiguity before changing reference satellite, $N_0$ denotes the un-differenced ambiguity, $A$ is the transform matrix.

The double-differenced ambiguity after changing the reference satellite can be expressed as

$$\nabla \Delta \bar{N} = B N_0, \qquad (9.68)$$

where $\nabla \Delta \bar{N}$ denotes the new double-differenced ambiguity after changing reference satellite, $B$ is the new transform matrix.

Thus, the relationship between the new double-differenced ambiguity and the old double-differenced ambiguity can be assumed as:

$$\nabla \Delta \bar{N} = C \nabla \Delta N, \qquad (9.69)$$

where $C$ is the transform matrix between old and new double-differenced ambiguities.

Therefore, how to get the matrix $C$ is the key problem to deal with the reference satellite changing. Substituting Eqs. 9.67 and 9.68 into Eq. 9.69 and dividing out $N_0$, one has

$$CA = B, \qquad (9.70)$$

Multiplying the both sides of Eq. 9.70 by $A^T(AA^T)^{-1}$, one has

$$C = B A^T (AA^T)^{-1}, \qquad (9.71)$$

Therefore it can be inferred that through multiplying the double-differenced ambiguity before changing reference satellite by the transform matrix of Eq. 9.71, the double-differenced ambiguity after reference satellite changing can be obtained. The numerical example of this method can be referred to Wang (2013).

## 9.4.2
### Changing Reference Station

Although the single baseline solution has been widely applied in the kinematic relative positioning, it is generally not appropriate for the case of long range airborne positioning. Because of the long distance between the reference station and kinematic station, many types of common errors cannot be eliminated directly by

## 9.4 Reference Changing in GPS Difference Algorithm

the difference method. And the number of common satellites will be reduced with the increase of baseline length. If a closer reference station can be used in place of the original reference station, these problems may be well solved. Therefore, a method for adaptively changing reference station for long distance airborne GPS applications developed by Wang et al. (2010, 2011) is introduced in this section.

The basic idea of the adaptively changing reference station method is that the positioning model always keeps the single baseline model during the whole solution. When the distance between kinematic station and reference station is longer than the maximum distance, which is defined by the user, the new reference station is used to replace the old one. At the same time, all information of old observation equation including covariance matrix will be transferred to the new observation equation based on the equivalent eliminated parameter method. The calculation steps of adaptively changing reference station are described below.

Suppose that the observation equations before and after changing reference station can be respectively expressed as follows:

$$L - A \begin{bmatrix} X_1 \\ X_2 \\ \nabla \Delta N_{i1,i2} \end{bmatrix} = V, \quad P, \tag{9.72}$$

$$L' - B \begin{bmatrix} X_2 \\ X_3 \\ \nabla \Delta N_{i3,i2} \end{bmatrix} = V', \quad P', \tag{9.73}$$

where $L$ and $L'$ are the double-differenced observations, $A$ and $B$ are the design matrices, $X_1$ and $X_3$ are the position parameters of the old and new reference stations, respectively, $X_2$ is the position parameter of the kinematic station, $\nabla \Delta N_{i1,i2}$ is the double-differenced ambiguities between old reference station $i1$ and kinematic station $i2$, $\nabla \Delta N_{i3,i2}$ is the double-differenced ambiguities between new reference station $i3$ and kinematic station $i2$, $V, V', P, P'$ are the residual vectors and weight matrices respectively.

The Eq. 9.72 can be rewritten as

$$L - \begin{bmatrix} A_1 & A_2 \end{bmatrix} \begin{bmatrix} X_2 \\ \bar{X} \end{bmatrix} = V, \quad P, \tag{9.74}$$

where $\bar{X}$ includes $X_1$ and $\nabla \Delta N_{i1,i2}$.

The normal equation of Eq. 9.74 can be formed as

$$\begin{bmatrix} M_{11} & M_{12} \\ M_{21} & M_{22} \end{bmatrix} \begin{bmatrix} X_2 \\ \bar{X} \end{bmatrix} = \begin{bmatrix} U_1 \\ U_2 \end{bmatrix}, \tag{9.75}$$

where

$$\begin{bmatrix} M_{11} & M_{12} \\ M_{21} & M_{22} \end{bmatrix} = \begin{bmatrix} A_1^T P A_1 & A_1^T P A_2 \\ A_2^T P A_1 & A_2^T P A_2 \end{bmatrix}, \begin{bmatrix} U_1 \\ U_2 \end{bmatrix} = \begin{bmatrix} A_1^T P L \\ A_2^T P L \end{bmatrix}, \quad (9.76)$$

The equivalently eliminated equation (cf. Sect. 7.6) of Eq. 9.75 can be formed as

$$\begin{bmatrix} M_1 & 0 \\ M_{21} & M_{22} \end{bmatrix} \begin{bmatrix} X_2 \\ \bar{X} \end{bmatrix} = \begin{bmatrix} R_1 \\ U_2 \end{bmatrix}, \quad (9.77)$$

where $M_1 = A_1^T(E-J)^T P(E-J) A_1$, $R_1 = A_1^T(E-J)^T PL$, $J = A_2 M_{22}^{-1} A_2^T P$. Let $D_1 = (E-J)A_1$, the first equation of Eq. 9.77 can be expressed as

$$D_1^T P D_1 X_2 = D_1^T P L, \quad (9.78)$$

Therefore the equivalent observation equation of Eq. 9.78 is

$$L - D_1 X_2 = V, \quad P, \quad (9.79)$$

The normal equation of Eq. 9.73 can be formed as

$$B^T P' B \begin{bmatrix} X_2 \\ X_3 \\ \nabla \Delta N_{i3,i2} \end{bmatrix} = B^T P' L', \quad (9.80)$$

Since the Eqs. 9.78 and 9.80 have the same position parameter $X_2$, the corresponding element of the two normal equations can be accumulated directly, then one has

$$\bar{B}^T \bar{P}' \bar{B} \begin{bmatrix} X_2 \\ X_3 \\ \nabla \Delta N_{i3,i2} \end{bmatrix} = \bar{B}^T \bar{L}' \bar{L}', \quad (9.81)$$

where $\bar{B}$ and $\bar{L}'$ are the design matrix and observation matrix after accumulation, $\bar{P}'$ is the weight matrix.

Therefore using the sequential least squares, the position parameter and ambiguity parameter of the kinematic station can be estimated based on the Eq. 9.81.

The numerical example for validation of this method can be found in Wang et al. (2011).

## 9.5
## Standard Algorithms of GPS Data Processing

### 9.5.1
### Preparation of GPS Data Processing

Preparation of GPS data processing can be carried out either in pre-processing or in the main data processing. It depends on the strategy and the purpose of the data processing. Only in the case of data post-processing (i.e. data are available before the processing) is pre-processing possible. In the case of data quasi real time or real time processing, usually data are only available up to the instantaneous epoch. Data availability also causes different strategies of the data processing.

Data preparation may include raw GPS data decoding. ASCII code data are usually given in RINEX format (Gurthner 1994). Even in the unified format, different decoders may work a little bit differently from one another. This has to be noted only if one is going to process the data decoded by using different decoders. Usually, most GPS data processing software has its own internal input data format. Transforming the data from the RINEX format (maybe also from multiple stations) into the internal input data format should be no principle problem.

Cycle slip detection is one of the most important works in data preparation. Marks are given for further use in the data where the cycle slips are detected. There are two types of cycle slips; one is repairable, and another is not repairable. Non-repairable cycle slips have to be modelled by new ambiguity unknowns. Repairing and setting new unknowns are equivalent if the repair is made correctly and the new unknown is well-solved. By real time data processing, this process has to be done in the main data processing process.

Orbit data are also needed. Depending on the purposes of the data processing, broadcast navigation data, IGS precise orbits, and IGS predicted orbits can be used where the satellite clock error model is also included. In broadcast data, there is also an ionospheric model available. Even for the GPS precise orbit determination, initial orbits are still needed.

Further preparations depend on the organisation and purpose of the data processing. Generally speaking, standard tropospheric models are needed for use (cf. Sect. 5.2). An ionospheric model (from broadcast) can be used as an initial model (cf. Sect. 5.1) if the non-ionosphere-free combination is used. An ionospheric model can be also obtained from the ambiguity-ionospheric equations (see discussions in Sect. 6.5.2). Earth tide and ocean loading tide, as well as relativistic effects have to be computed for use (cf. Sect. 5.4).

In the case of orbit determination and/or geopotential determination, an initial geopotential model is needed. The initial models of the solar radiation and air drag have to be computed. All corrections can be computed in real time or in advance and then listed in tables for use. Coordinate transformations between the ECEF system and the ECSF system are also needed.

## 9.5.2
## Single Point Positioning

Single point positioning is a sub-process of GPS data processing, which is needed in almost all GPS data processing. Station coordinates and receiver clock error are determined with such a sub-process. Depending on the accuracy requirement, single point positioning can be done with single frequency code or phase data, dual-frequency code or phase data, and combined code-phase data. Generally speaking, single point positioning has a lower accuracy than that of relative positioning, where systematic errors are reduced (through keeping the reference fixed). However, the receiver clock bias determined by single point positioning is accurate enough to correct the second type of clock error influence (the influence scaled by the velocity of the satellite, cf. Sect. 5.5).

### Code Data Single Point Positioning

The GPS code pseudorange model is (cf. Sect. 6.1):

$$R_i^k(t_r, t_e) = \rho_i^k(t_r, t_e) - (\delta t_r - \delta t_k)c + \delta_{\text{ion}} + \delta_{\text{trop}} + \delta_{\text{tide}} + \delta_{\text{rel}} + \varepsilon, \quad (9.82)$$

where $R$ is the observed pseudorange, $t_e$ denotes the GPS signal emission time of the satellite $k$, $t_r$ denotes the GPS signal reception time of the receiver $i$, $c$ is the speed of light, subscript $i$ and superscript $k$ denote the receiver and satellite, and $\delta t_r$ and $\delta t_k$ are the clock errors of the receiver and satellite at the times $t_r$ and $t_e$, respectively. The terms $\delta_{\text{ion}}$, $\delta_{\text{trop}}$, $\delta_{\text{tide}}$, and $\delta_{\text{rel}}$ denote the ionospheric, tropospheric, tidal, and relativistic effects, respectively. The multipath effect is omitted here. The remaining error is denoted as $\varepsilon$. $\rho_i^k$ is the geometric distance. The computed value (denoted as $C$) of the pseudorange is

$$C = \rho_i^k(t_r, t_e) + \delta t_k c + \delta_{\text{ion}} + \delta_{\text{trop}} + \delta_{\text{tide}} + \delta_{\text{rel}}, \quad (9.83)$$

where the clock error of the satellites can be interpolated from the IGS orbit data or broadcast navigation message, models of other effects can be found in Chap. 5, and the initial value of receiver clock error is assumed to be zero. It should be emphasised that the earth rotation correction has to be taken into account by the geometric distance computation no matter if it is done in the Earth or space fixed coordinate systems (cf. Sect. 5.3.2).

The linearised observation Eq. 9.82 is then (cf. Sects. 6.2 and 6.3)

$$l_k = \frac{-1}{\rho_i^k(t_r, t_e)}(x_k - x_{i0} \quad y_k - y_{i0} \quad z_k - z_{i0})\begin{pmatrix}\Delta x \\ \Delta y \\ \Delta z\end{pmatrix} - \Delta t + v_k, \quad (9.84)$$

where $l_k$ is the so-called O–C (observed minus computed pseudorange), $v_k$ is the residual, vector $(\Delta x \; \Delta y \; \Delta z)^{\text{T}}$ is the difference between the coordinate vector $(x_i \; y_i \; z_i)^{\text{T}}$

## 9.5 Standard Algorithms of GPS Data Processing

and the initial coordinate vector $(x_{i0}\ y_{i0}\ z_{i0})^T$, $\Delta t$ is the receiver clock error in length (i.e. $\Delta t = \delta t_r c$), and the initial coordinate vector is used for computing the geometric distance. Equation 9.84 can be written in a more general form as

$$l_k = \begin{pmatrix} a_{k1} & a_{k2} & a_{k3} & -1 \end{pmatrix} \begin{pmatrix} \Delta x \\ \Delta y \\ \Delta z \\ \Delta t \end{pmatrix} + v_k, \qquad (9.85)$$

where $a_{kj}$ is the related coefficient given in Eq. 9.84. Putting all of the equations from all observed satellites together, we find the single point positioning equation system has a general form of

$$L = AX + V, \quad P, \qquad (9.86)$$

where $L$ is called the observation vector, $X$ is the unknown vector, $A$ is the coefficient matrix, $V$ is the residual vector, and $P$ is the weight matrix of the observation vector. The least squares solution of observational Eq. 9.86 is then (cf. Sect. 7.2)

$$X = \left(A^T P A\right)^{-1} A^T P L. \qquad (9.87)$$

The formulas for computing the precision vector of the solved $X$ can be found in Sect. 7.2. It is notable that the coefficients of the equation are computed using the initial coordinate vector, and the initial coordinate vector is usually not (exactly) known; therefore, an iterative process has to be carried out to solve the single point positioning problem. For the given initial vector, a modified one can be obtained by solving the above problem; the modified initial vector can be used in turn as the initial vector to form the equations, and the problem can be solved again until the process converges. Because there are four unknowns in the single point positioning equation, at least four observables are needed to make the problem solvable. In other words, as soon as four or more satellites are observed, single point positioning is always possible.

For static reference stations, as soon as the coordinates are known with sufficient accuracy, the unknown vector $(\Delta x\ \Delta y\ \Delta z)^T$ can be considered zero. Then the Eq. 9.85 turns out to be

$$l_k = -\Delta t + v_k, \qquad (9.88)$$

and the receiver clock error can be computed directly by

$$\Delta t = \frac{-1}{K} \sum_{k=1}^{K} l_k, \qquad (9.89)$$

where $K$ is the total number of observed satellites at this epoch. Equation 9.89 can be used to compute the receiver clock error of the static reference.

## Dual-Code Ionosphere-Free Single Point Positioning

The above-mentioned single point positioning (using single frequency code data) is accurate enough for correcting the second type of clock error influence (the influence scaled by the velocity of the satellite). For more precise single point positioning, dual-frequency code data can be used to form the ionosphere-free combinations (cf. Sect. 6.5). Assuming that for frequencies 1 and 2, the single point positioning equation of Eq. 9.86 can be formed as

$$L_1 = AX + V_1, \quad P_1 \tag{9.90}$$

$$L_2 = AX + V_2, \quad P_2,$$

then the ionosphere-free combination can be formed by (cf. Sect. 6.5.1)

$$\frac{f_1^2}{f_1^2 - f_2^2} L_1 - \frac{f_2^2}{f_1^2 - f_2^2} L_2 = AX + V, \quad P, \tag{9.91}$$

where

$$P = Q^{-1}, \quad Q = \left(\frac{f_1^2}{f_1^2 - f_2^2}\right)^2 P_1^{-1} + \left(\frac{f_2^2}{f_1^2 - f_2^2}\right)^2 P_2^{-1},$$

and $V$ is the residual vector. Because the ionospheric effects have been cancelled out of Eq. 9.91, the ionospheric model can be also omitted by computing $L_1$ and $L_2$ in Eq. 9.90. The solution of Eq. 9.91 is then the solution to the dual-code ionosphere-free single point positioning problem.

## Phase Single Point Positioning

GPS carrier phase model is (cf. Sect. 6.1)

$$\lambda \Phi_i^k(t_r, t_e) = \rho_i^k(t_r, t_e) - (\delta t_r - \delta t_k)c + \lambda N_i^k - \delta_{\text{ion}} + \delta_{\text{trop}} + \delta_{\text{tide}} + \delta_{\text{rel}} + \varepsilon, \tag{9.92}$$

where $\lambda \Phi$ is the observed phase in length, $\Phi$ is the phase in cycle, wave length is denoted as $\lambda$, and $N_i^k$ is the ambiguity related to receiver $i$ and satellite $k$, except for the ambiguity term and the sign difference of the term of ionospheric effect; other terms are the same as that of the pseudorange discussed at the beginning of this section.

The computed value (denoted as $C$) of phase is

$$C = \rho_i^k(t_r, t_e) + \delta t_k c + \lambda N_{i0}^k - \delta_{\text{ion}} + \delta_{\text{trop}} + \delta_{\text{tide}} + \delta_{\text{rel}}, \tag{9.93}$$

where $N_{i0}^k$ is the initial ambiguity parameter related to the receiver $i$ and satellite $k$. Scaling the ambiguity parameter in length and denoting

$$\Delta N_i^k = \lambda N_i^k - \lambda N_{i0}^k, \tag{9.94}$$

the phase single point positioning equation is (very similar to Eq. 9.85)

$$l_k = \begin{pmatrix} a_{k1} & a_{k2} & a_{k3} & -1 \end{pmatrix} \begin{pmatrix} \Delta x \\ \Delta y \\ \Delta z \\ \Delta t \end{pmatrix} + \Delta N_i^k + v_k. \tag{9.95}$$

Putting all equations related to all observed satellites together, the single point positioning equation system has a general form of

$$L = AX + EN + V, \quad P, \tag{9.96}$$

where $L$ is called the observation vector, $X$ is the unknown vector of coordinates and clock error, $A$ is the $X$ related coefficient matrix, $E$ is an identity matrix of order $K$, $K$ is the number of observed satellites, $N$ is the unknown vector of ambiguity parameters $\Delta N_i^k$, $V$ is the residual vector, and $P$ is the weight matrix. If $K$ satellites are observed, then there are $K$ ambiguity parameters, three coordinate parameters and one clock parameter, so that the phase single point positioning problem is not solvable at the first few epochs. Using the ambiguity parameters obtained from the ambiguity-ionospheric equations (cf. Sect. 6.5) as the initial ambiguity values, $N$ is then zero (can be cancelled), and Eq. 9.96 has the same form as that of Eq. 9.86. In this way, the equation system of single-frequency phase point positioning can be formed and solved every epoch. Even the codes are used in the ambiguity-ionospheric equations, ambiguity parameters can be obtained with high accuracy through a reasonable weight and instrumental bias model (cf. Sects. 6.7 and 9.2).

### Dual-Phase Ionosphere-Free Single Point Positioning

The single point positioning equation of the dual-phase observables for frequencies 1 and 2 can be formed as

$$\begin{aligned} L1 &= AX + EN1 + V1, \quad P1 \text{ and} \\ L2 &= AX + EN2 + V2, \quad P2. \end{aligned} \tag{9.97}$$

Then the ionosphere-free combinations can be formed by (cf. Sect. 6.5.1)

$$\frac{f_1^2}{f_1^2 - f_2^2} L_1 - \frac{f_2^2}{f_1^2 - f_2^2} L_2 = AX + EN_c + V, \quad P, \tag{9.98}$$

where

$$N_c = \frac{f_1^2}{f_1^2 - f_2^2} N_1 - \frac{f_2^2}{f_1^2 - f_2^2} N_2 \quad \text{and} \tag{9.99}$$

$$P = Q^{-1}, \quad Q = \left(\frac{f_1^2}{f_1^2 - f_2^2}\right)^2 P_1^{-1} + \left(\frac{f_2^2}{f_1^2 - f_2^2}\right)^2 P_2^{-1}. \tag{9.100}$$

$V$ is the residual vector, and index $c$ is used to denote the ionosphere-free combinations. Equation 9.98 is the dual-phase ionosphere-free single point positioning equation system. The solution of Eq. 9.98 is then the solution of the dual-phase ionosphere-free single point positioning problem.

### *Phase–Code Combined Single Point Positioning*

Phase and code ionosphere-free single point positioning Eqs. 9.98 and 9.91 can be written in more compact form as

$$\begin{aligned} L_p &= A_{11} X_1 + A_{12} N + V_p, \quad P_p \quad \text{and} \\ L_c &= A_{11} X_1 + V_c, \quad P_c, \end{aligned} \tag{9.101}$$

where index p and c denote the phase and code related variables, $X_1$ is the vector of the coordinate and receiver clock error, $N$ is the ambiguity vector, $P$ is the weight matrix, and $V$ is the residual vector. To guarantee the same coefficient matrix $A_{11}$ for both the phase and code observation equations, data of commonly observed satellites have to be used.

Usually the code single point positioning problem (second equation system of Eq. 9.101) is always solvable (as soon as more than four satellites are observed). And the ambiguity parameter number is equal to the number of phase observables. Therefore, the phase–code combined single point positioning problem in Eq. 9.101 is usually solvable at every epoch.

Block-wise least squares adjustment for solving the phase–code combined problem has been discussed in Sect. 7.5.2. The algorithm can be used directly to solve the combined Eq. 9.101.

### *Precise Point Positioning*

The availability of precise GPS satellite orbit and clock products from the International GNSS Service (IGS) has enabled) the development of a positioning technology known as precise point positioning (PPP). Based on the processing of un-differenced pseudorange and carrier phase observations from a single GPS receiver, this approach effectively eliminates the inter-station limitations introduced by differential GPS processing as no base station is necessary. As a result, it offers an alternative to differential GPS that is logistically simpler and almost as accurate (Zumberge et al. 1997; Kouba and Héroux 2001). Although PPP does not require any base station, it requires accurate knowledge of the GPS satellite coordinates and the state of their clocks. The algorithm of PPP is described in the following.

With a single GPS dual-frequency receiver, the following ionosphere-free combinations can be applied to facilitate PPP positioning using un-differenced observations.

$$P_{IF} = \frac{f_1^2 \cdot P_1 - f_2^2 \cdot P_2}{f_1^2 - f_2^2} = \rho + c \cdot dt + d_{trop} + dm_{IF} + \varepsilon_{P_{IF}}, \quad (9.102)$$

$$\Phi_{IF} = \frac{f_1^2 \cdot \Phi_1 - f_2^2 \cdot \Phi_2}{f_1^2 - f_2^2} = \rho + c \cdot dt + d_{trop} + \frac{cf_1 N_1 - cf_2 N_2}{f_1^2 - f_2^2} + \delta m_{IF} + \varepsilon_{\Phi_{IF}},$$
$$(9.103)$$

where $P_{IF}$ is the ionosphere-free code observation, $\Phi_{IF}$ is the ionosphere-free phase observation, $P_i$ and $\Phi_i$ ($i = 1, 2$) are the pseudorange observation and phase observation on $L_i$, respectively, $f_i$ is the frequency of $L_i$, $\rho$ is the geometric distance between the satellite and the receiver, $c$ is the speed of light, $dt$ denotes the receiver clock error, $d_{trop}$ denotes the tropospheric delay, $N_i$ is the integer phase ambiguity on $L_i$, $dm_{IF}$, and $\delta m_{IF}$ denote a series of error corrections including relativistic effect, earth tide, ocean tide, and hardware delay in pseudorange observation and phase observation, respectively, $\varepsilon_P$ and $\varepsilon_\Phi$ denote the remaining errors not modelled such as multipath and observation noise of code and phase, respectively. Satellite orbit and clock errors are not present in Eqs. 9.102 and 9.103 since they can be removed by the use of precise orbit and clock products. The remaining receiver clock and tropospheric delays in Eqs. 9.102 and 9.103 will be estimated in PPP.

The potential impact of PPP on the positioning community is expected to be significant. It brings great flexibility to field operations and also reduces labour and equipment cost and simplifies operational logistics by eliminating the need for base stations. The performance of PPP for positioning determination has been demonstrated in various papers—for example, Zumberge et al. (1997), Kouba and Héroux (2001), Gao and Shen (2002), Gao et al. (2003)—using post-mission precise orbit and clock from IGS. Following the availability of real-time precise GPS satellite orbit and clock products, PPP has also been applied to real-time kinematic positioning (Gao and Chen 2004; Chen et al. 2013).

## 9.5.3
### Standard Un-differential GPS Data Processing

In single point positioning, un-differenced GPS data are used. Usually, only four unknowns are solved for, as discussed in Sect. 5.2. Single point positioning has also a speciality of epoch-wise solution. Based on the algorithms of single point positioning, standard static un-differential GPS data processing should take more unknown models and more station data into account. In a kinematic case, because of the movement of the receiver, coordinates of the receiver are time variables; therefore, model parameters are usually pre-determined or determined with another algorithm in order to reduce the number of the unknowns.

The GPS code pseudorange and carrier phase are modelled as (cf. Sect. 6.1, Eqs. 6.1 and 6.2, or Eqs. 9.82 and 9.92)

$$R_i^k(t_r, t_e) = \rho_i^k(t_r, t_e) - (\delta t_r - \delta t_k)c + \delta_{\text{ion}} + \delta_{\text{trop}} + \delta_{\text{tide}} + \delta_{\text{rel}} + \varepsilon_c \quad \text{and} \quad (9.104)$$

$$\lambda \Phi_i^k(t_r, t_e) = \rho_i^k(t_r, t_e) - (\delta t_r - \delta t_k)c + \lambda N_i^k - \delta_{\text{ion}} + \delta_{\text{trop}} + \delta_{\text{tide}} + \delta_{\text{rel}} + \varepsilon_p. \tag{9.105}$$

Except for the ambiguity parameter and the sign of the ionospheric effect term, the other terms on the right sides of Eqs. 9.104 and 9.105 are the same.

For any standard data combinations (cf. Sect. 6.5 for details) as given in Eqs. 6.48 and 6.51, the above models of Eqs. 9.104 and 9.105 are still valid. Of course, on the left sides of Eqs. 9.104 and 9.105 the combined pseudorange and combined phase (scaled by wavelength) are used, and on the right side the ambiguity and ionospheric effect are combined ones respectively. Exactly, for combinations of

$$R = \frac{n_1 R_1 + n_2 R_2}{n_1 + n_2}, \tag{9.106}$$

$$\Phi = n_1 \Phi_1 + n_2 \Phi_2, \tag{9.107}$$

or

$$\lambda \Phi = \frac{1}{f}(n_1 f_1 \lambda_1 \Phi_1 + n_2 f_2 \lambda_2 \Phi_2), \tag{9.108}$$

where the combined signal has the frequency and wavelength

$$f = n_1 f_1 + n_2 f_2, \quad \text{and} \quad \lambda = c/f, \tag{9.109}$$

the combined ambiguity and ionospheric effects are

$$N_{\text{com}} = n_1 N_1 + n_2 N_2, \tag{9.110}$$

$$\delta_{\text{ion\_come}} = \frac{n_1 \delta_{\text{ion1}} + n_2 \delta_{\text{ion2}}}{n_1 + n_2} \quad \text{and}$$

$$\delta_{\text{ion\_comp}} = \frac{-1}{f}(n_1 f_1 \delta_{\text{ion1}} + n_2 f_2 \delta_{\text{ion2}}), \tag{9.111}$$

where $n_1$ and $n_2$ are the selected real constants, indices 1 and 2 are referred to frequencies 1 and 2, and indices comc and comp denote the code and phase combined terms.

The computed pseudorange and phase range are

$$C_c = \rho_i^k(t_r, t_e) - (\delta t_r - \delta t_k)c + \delta^0_{\text{ion\_comc}} + \delta^0_{\text{trop}} + \delta^0_{\text{tide}} + \delta_{\text{rel}} \quad \text{and} \quad (9.112)$$

$$C_p = \rho_i^k(t_r, t_e) - (\delta t_r - \delta t_k)c + \lambda N^k_{i0\_\text{com}} - \delta^0_{\text{ion\_comp}} + \delta^0_{\text{trop}} + \delta^0_{\text{tide}} + \delta_{\text{rel}}, \quad (9.113)$$

where superscript 0 denotes the initial values of individual models, indices c and p denote the terms related to the code and phase measurements, and index com denotes the combined terms. In the case of ionosphere-free combinations, the ionospheric effect terms will vanish. Otherwise, we should assume that the ionospheric effects are known by the given model or by the ambiguity-ionospheric equations.

The linearisation of GPS observation equations is generally discussed in Sect. 6.2, and the related partial derivatives are given in Sect. 6.3. Equations 9.106 and 9.108 can be linearised as

$$\begin{aligned} L_c &= A_{11}X_{\text{coor}} + A_{12}X_{\text{clock}} + A_{13}X_{\text{trop}} + A_{14}X_{\text{tide}} + V_c, \quad P_c \text{ and} \\ L_p &= A_{11}X_{\text{coor}} + A_{12}X_{\text{clock}} + A_{13}X_{\text{trop}} + A_{14}X_{\text{tide}} + A_{15}N + V_p, \quad P_p, \end{aligned} \quad (9.114)$$

where $X_{\text{coor}}$ is the coordinate vector, $X_{\text{clock}}$ is clock error vector, indices trop and tide are used to denote the related unknown vectors, $N$ is the ambiguity vector, $P$ is the weight matrix, $V$ is the residual vector, and $A$ is the related coefficient matrix. The data of commonly observed satellites have to be used to guarantee the common coefficient matrices $A$ for both phase and code observation equations.

To process the data of more stations, Eq. 9.114 shall be formed station by station and then combine them together. It is notable that some of the parameters are common ones for all stations, such as satellite clock errors and Love numbers of the earth tide. In the case of orbit determination (cf. Chap. 11 for details), the orbit parameters and force model parameters are also common ones. The total observation equations of the un-differential GPS can then be written symbolically as

$$\begin{aligned} L_c &= A_1X_1 + A_4X_4 + V_c, \quad P_c \text{ and} \\ L_p &= A_1X_1 + A_4X_4 + A_5X_5 + V_p, \quad P_p \end{aligned} \quad (9.115)$$

where $X_1$ is a sub-vector of the common variables of the both equations, $X_4$ is the other variable vector of the both equations, and $X_5$ is the ambiguity vector. Adding $0X_5$ to the first equation and denoting $X_2 = [X_4 \quad X_5]^T$, Eq. 9.115 can be further simplified as

$$L_c = A_1X_1 + A_2X_2 + V_c, \quad P_c \text{ and}$$
$$L_p = A_1X_1 + A_3X_2 + V_p, \quad P_p. \tag{9.116}$$

Equation 9.116 can be considered an epoch-wise formed observation equation or observation equation of all observed epochs. Most adjustment algorithms discussed in Chap. 7 can be used directly to solve the above equation system.

## 9.5.4
### Equivalent Method of GPS Data Processing

As already discussed in Sect. 6.8, the equivalently eliminated equations of Eq. 9.16 can be formed as (cf. Sects. 6.8 and 7.6 for details)

$$U_c = L_c - (E - J_c)A_2X_2, \quad P_c \text{ and}$$
$$U_p = L_p - (E - J_p)A_3X_2, \quad P_p, \tag{9.117}$$

where

$$J_c = A_1 M_{11c}^{-1} A_1^T P_c,$$
$$J_p = A_1 M_{11p}^{-1} A_1^T P_p,$$
$$M_{11c} = A_1^T P_c A_1, \quad \text{and} \tag{9.118}$$
$$M_{11p} = A_1^T P_p A_1.$$

$E$ is an identity matrix of size $J$, $L$, and $P$ are the original observation vector and weight matrix, and $U$ is the residual vector, which has the same statistic property as $V$ in Eq. 9.116. As soon as the $X_1$ in Eq. 9.116 is able to be eliminated, the equivalent Eq. 9.117 can be formed whether Eq. 9.116 is an epoch-wise equation or an all epoch equation.

Equation 9.117 is the zero-difference (un-differential) GPS observation equation system if the variable vector $X_1$ in Eq. 9.116 is considered a zero vector.

Equation 9.117 is the equivalent single-difference GPS observation equation system if the variable vector $X_1$ in Eq. 9.116 is considered an unknown vector of satellite clock errors.

Equation 9.117 is the equivalent double-difference GPS observation equation system if the variable vector $X_1$ in Eq. 9.116 is considered an unknown vector of satellite and receiver clock errors.

The second equation of 9.117 is the equivalent triple-difference GPS observation equation system if the variable vector $X_1$ in the second equation of 9.116 is considered an unknown vector of all clock errors and ambiguities.

The un-differential and differential GPS data processing can be dealt with in an equivalent and unified way. The advantages of this method are:

1. The weight remains the original one, so one does not have to deal with the correlation problem;
2. The original data are used, so one does not need to form the differences;
3. The un-differential and differential GPS data processing can be easily selected by a switch or can be used in a combined way, so that the number of unknowns (i.e. matrix size) of the whole adjustment and filtering problem can be greatly reduced.

The combinative way of using the equivalent method can be realised as follows. First, equivalent triple differences are used to determine the unknowns other than the clock error and ambiguity parameters. Taking these parameters as known, the observation equation system 9.116 can be reduced so that only the clock error and ambiguity parameters are included. Then second, equivalent double differences are used to determine the ambiguity vector. Again, taking the ambiguity vector as known, Eq. 9.116 can be further reduced so that only the clock error parameters are included. Then third, equivalent single differences are used to determine the receiver clock errors. At the end, Eq. 9.116 can be reduced so that only satellite clock errors are included in the equations, and they can be determined. The last two steps can be also done together in one step.

By the way, the ambiguity parameters are usually dealt with in an un-differential form for all methods, so that the problems caused by changing the reference satellite in a double-difference case can be avoided. This is especially important for kinematic GPS applications.

## 9.5.5
## Relative Positioning

Relative positioning is traditionally carried out with differential positioning. The key point of relative positioning is to keep the coordinates of the reference station fixed. In other words, the initial coordinate values of the reference station are considered true values so that the related unknowns are either not necessary to be adjusted or equal to zero. Therefore, the following two methods outline how relative positioning can be done. (1) Cancelling the reference coordinate unknowns out of Eq. 9.116; (2) The a priori datum method discussed in Sects. 7.8.2 and 6.8.6 is used to keep the coordinates fixed on the initial values. Both methods are equivalent. The a priori datum method (cf. Sects. 7.8.2 and 6.8.6) can be also used to keep some of the un-differential ambiguity parameters and clock parameters fixed. Keeping the reference coordinates fixed in relative positioning may lead to a better determination of the other parameters in the reference-related equations and, therefore, may lead to an indirect reduction of the residuals.

## 9.5.6
## Velocity Determination

### Single Point Velocity Determination

Analogous to the single point positioning discussed in Sect. 5.2, single point velocity determination can be carried out by using Doppler data. The GPS Doppler observation is modelled as (cf. Eq. 6.46)

$$D = \frac{d\rho_i^k(t_r, t_e)}{\lambda dt} - f\frac{d(\delta t_r - \delta t_k)}{dt} + \delta_{rel\_f} + \varepsilon, \qquad (9.119)$$

where $D$ is the observed Doppler measurement, $t_e$ denotes the GPS signal emission time of the satellite $k$, $t_r$ denotes the GPS signal reception time of the receiver $i$, subscript $i$ and superscript $k$ denote receiver and satellite, and $\delta t_r$ and $\delta t_k$ denote the clock errors of the receiver and satellite at the time $t_r$ and $t_e$, respectively. The remaining error is denoted as $\varepsilon$, $f$ is the frequency, wavelength is denoted as $\lambda$, $\delta_{rel\_f}$ is the frequency correction of the relativistic effects, $\rho_i^k$ is the geometric distance, and $d_i^k/dt$ denotes the time derivation of the radial distance between satellite and receiver at the time $t_r$.

The computed value (denoted as $C$) of Doppler is

$$C = \frac{d\rho_i^k(t_r, t_e)}{\lambda dt} + f\frac{d(\delta t_k)}{dt} + \delta_{rel\_f}, \qquad (9.120)$$

where the first term on the right-hand side can be computed by using Eqs. 6.14 and 6.15.

The time derivative of the satellite clock error and the satellite position as well as velocity can be computed from the IGS orbit data or broadcast navigation message; the relativistic effect on frequency can be found in Chap. 5. It is obvious that the initial position of the receiver is also needed for computing Eq. 9.120. Initial velocity of the receiver is assumed zero. It should be emphasised that the earth rotation correction has to be taken into account by the geometric distance computation (cf. Sect. 5.3.2).

The linearised observation Eq. 9.120 is then (cf. Sects. 6.2 and 6.3 as well as partial derivative Eq. 6.20)

$$l_k = \frac{-1}{\lambda \rho_i^k(t_r, t_e)} (x_k - x_i \quad y_k - y_i \quad z_k - z_i) \begin{pmatrix} \dot{x}_i \\ \dot{y}_i \\ \dot{z}_i \end{pmatrix} - \Delta D + v_k, \qquad (9.121)$$

where $l_k$ is the O–C (observed minus computed Doppler), $v_k$ is the residual, the receiver's velocity vector is $(\dot{x}, \dot{y}, \dot{z})^T$, $(x \quad y \quad z)^T$ is the coordinate vector with index $k$ for satellite and $i$ for receiver. $\Delta D$ is the receiver clock drift in cycle/second (i.e. $\Delta D = f(d\delta t_r/dt)$). Equation 9.121 can be written in a more general form as

$$l_k = \begin{pmatrix} a_{k1} & a_{k2} & a_{k3} & -1 \end{pmatrix} \begin{pmatrix} \dot{x} \\ \dot{y} \\ \dot{z} \\ \Delta D \end{pmatrix} + v_k, \qquad (9.122)$$

where $a_{kj}$ is the related coefficient given in Eq. 9.122. If one puts all of the equations that are related to all of the observed satellites together, the equation system of single point velocity determination has a general form of

$$L = AX + V, \quad P, \qquad (9.123)$$

where $L$ is called the observation vector, $X$ is the unknown velocity vector including clock drift, $A$ is the coefficient matrix, $V$ is the residual vector, and $P$ is the weight matrix of observation vector. The least squares solution of observation Eq. 9.123 is then (cf. Sect. 7.2)

$$X = (A^T P A)^{-1} A^T P L. \qquad (9.124)$$

The formulas for computing the precision vector of the solved $X$ can be found in Sect. 7.2. It is notable that the coefficients of the equation are computed using the initial velocity vector, and the initial velocity vector is usually not known; therefore, an iterative process has to be carried out to solve the single point velocity determining problem. For the given initial velocity vector, a modified one can be obtained by solving the problem; the modified initial velocity vector can be used in turn to form the equation and solve it again until the process converges. This iterative process is needed if the kinematic motion is very fast. Because there are four unknowns in the single velocity determining equation, at least four observables are needed to make the problem solvable; in other words, when four or more satellites are observed, it is always possible to determine the single point velocity.

For static stations, the unknown velocity vector $(\dot{x}, \dot{y}, \dot{z})^T$ can be considered the zero one. Then the Eq. 9.121 turns out to be

$$l_k = -\Delta D + v_k, \qquad (9.125)$$

and the receiver frequency error can be computed directly by

$$\Delta D = \frac{-1}{K} \sum_{k=1}^{K} l_k, \qquad (9.126)$$

where $K$ is the total number of observed satellites. Equation 9.126 can be used to compute the frequency drift of the static reference receiver. The frequency drift of kinematic receiver can be also computed by static initialisation.

## Differential Doppler Data Processing

A more general model of Doppler data processing takes the satellite clock frequency bias (clock drift) into account:

$$l_k = \begin{pmatrix} a_{k1} & a_{k2} & a_{k3} & -1 \end{pmatrix} \begin{pmatrix} \dot{x} \\ \dot{y} \\ \dot{z} \\ \Delta D_i \end{pmatrix} + \Delta D_k + v_k, \qquad (9.127)$$

where index $i$ and $k$ denote the receiver and satellite, and $\Delta D$ is the related frequency bias. For the satellite frequency bias, the initial value from the IGS data or navigation data can be used. If one puts together all of the equations related to all observed satellites of all of the stations, Eq. 9.127 has a general form of

$$L_D = A_1 X_1 + A_2 X_2 + V_D, \quad P_D. \qquad (9.128)$$

where $X_1$ is a sub-vector of the common variables, $X_2$ is the vector of the other variable, and $A$ is the related coefficient matrix. The equivalently eliminated equations of Eq. 9.128 can be formed as (cf. Sect. 6.8 for details)

$$U_D = L_D - (E - J_D) A_2 X_2, \quad P_D, \qquad (9.129)$$

where

$$\begin{aligned} J_D &= A_1 M_{11D}^{-1} A_1^T P_D \quad \text{and} \\ M_{11D} &= A_1^T P_D A_1. \end{aligned} \qquad (9.130)$$

$E$ is an identity matrix of size $J_D$, $L$, and $P$ are the original observation vector and weight matrix, and $U$ is the residual vector, which has the same property as $V$ in Eq. 9.128.

Equation 9.129 is the equivalent single-difference GPS Doppler observation equation if the variable vector $X_1$ in Eq. 9.128 is considered a vector of satellite clock frequency bias.

Equation 9.129 is the equivalent double-difference GPS Doppler observation equation if the variable vector $X_1$ in Eq. 9.128 is considered a vector of the satellite and receiver clock frequency bias.

## Relative Velocity Determination

Relative velocity determining is usually carried out with a differential method. The key point of relative velocity determination is to keep the velocity of the reference station as fixed or zero. Therefore, relative velocity determination can be done the following two ways: (1) Cancel the reference velocity unknowns out of the Eq. 9.128; (2) Use the method of a priori datum discussed in Sect. 7.8.2 to keep the reference velocity fixed on the initial values.

## 9.5.7
## Kalman Filtering Using Velocity Information

As already discussed in Sect. 6.5.5, velocity information from the differential Doppler can be used to describe the system that is needed in Kalman filtering. Whether the receiver is moving or resting, the differential Doppler includes information about the motion state of the receiver. Therefore, using velocity information as a system description should be better than any empirical model.

The principle of Kalman filtering using velocity information can be outlined as follows (cf. also Sect. 7.7).

For the initial (or predicted) vector Z, the normal equation of the phase observation equation can be formed by

$$M_z Z = B_z, \quad Z = \begin{pmatrix} X \\ N \end{pmatrix}, \tag{9.131}$$

where $M_z$ is the normal matrix, and $B_z$ is the vector on the right side of the equation. These are formed by using initial vector Z; Z includes sub-vector X (coordinates) and N (ambiguities). The estimated solution of Eq. 9.131 is then

$$\tilde{Z} = \tilde{Q}_z B_z, \quad \tilde{Q}_z = M_z^{-1}. \tag{9.132}$$

The normal equation of the differential Doppler observation equation (cf. Eq. 9.129, only the velocity vector is unknown) can be formed by

$$M_{\dot{x}} \dot{X} = B_{\dot{x}}, \tag{9.133}$$

where $\dot{X}$ is the velocity vector of the receiver; it is also used as an index to denote the related normal matrix and vector on the right side of the equation. The solution of Eq. 9.133 is then

$$\dot{X} = Q_{\dot{x}} B_{\dot{x}}, \quad Q_{\dot{x}} = M_{\dot{x}}^{-1}. \tag{9.134}$$

Thus for the next epoch, denoted as k, the predicted vector turns out to be

$$\bar{Z}(k) = \tilde{Z}(k-1) + \dot{Z}(k-1) \cdot \Delta t, \tag{9.135}$$

where $\Delta t$ is the time interval of the epoch k–1 and k, and

$$\dot{Z}(k-1) = \begin{pmatrix} \dot{X}(k-1) \\ 0 \end{pmatrix}. \tag{9.136}$$

Equation 9.135 indicates that the differential Doppler has to be used in Eq. 9.134 as observations, because the velocity is considered an average one here. The related covariance matrix of the predicted vector is then

$$\bar{Q}_z(k) = \tilde{Q}_z(k-1) + (\Delta t)^2 \begin{pmatrix} Q_{\dot{x}} & 0 \\ 0 & 0 \end{pmatrix}. \tag{9.137}$$

The weight matrix is

$$\bar{P}_z(k) = \bar{Q}_z^{-1}(k). \tag{9.138}$$

The normal Eq. 9.131 of epoch $k$ is

$$M_z(k)Z(k) = B_z(k), \tag{9.139}$$

and the Kalman filter solution of Eq. 9.139 is then

$$\tilde{Z}(k) = \tilde{Q}_z(k)B_z(k), \quad \tilde{Q}_z(k) = (M_z(k) + \bar{P}_z(k))^{-1}. \tag{9.140}$$

It is notable that the normal equation 9.139 must be computed using the predicted vector $Z(k)$ of Eq. 9.135.

Repeating the steps from Eqs. 9.133 to 9.140 for the further epoch is a process of Kalman filtering using velocity information. The algorithm outlined above is suitable both for the kinematic and static data processing. This is true especially for static data processing, because the station has not been exactly assumed as fixed (as described by Eq. 9.133); such an algorithm will modify the property of the strong dependency on the initial value of the Kalman filter. The forming of normal Eq. 9.133 is an iterative process (cf. Sect. 5.6), i.e. the velocity information has to be used for forming the equation. Equation 9.89 represents a realistic system description.

## 9.6
## Accuracy of the Observational Geometry

Recalling the discussions made in the adjustment of Chap. 7, the precision vector of the solved vector is usually represented as (cf., e.g., Eq. 7.8)

$$p[i] = m_0 \sqrt{Q[i][i]} \quad \text{and} \tag{9.141}$$

$$m_0 = \sqrt{\frac{V^T P V}{m - n}}, \quad \text{if } (m > n).$$

where $i$ is the element index, $m_0$ is the so-called standard deviation (or sigma), $p[i]$ is the $i$th element of the precision vector, $Q[i][i]$ is the $i$th diagonal element of the quadratic matrix $Q$ (the inverse of the normal matrix), $V$ is the residual vector, superscript $T$ is the transpose of the vector, $P$ is the weight matrix, $n$ is the unknown number, and $m$ is the observation number.

Equation 9.141 is used to describe the precision of the individual parameter of the unknown vector $X$. The parameters can be usually classified into several groups according to their physical properties, e.g., position unknowns and clock unknowns; in turn the position unknowns can be classified by stations, and the clock errors can be classified by satellites and receivers, etc. To describe the precision of a group of unknowns, a so-called mean-squares-root precision can be defined as

$$p_{jJ} = \sqrt{\frac{1}{n}\sum_{i=j}^{J} p[i]^2}, \qquad (9.142)$$

where $j$ is the first index and $J$ is the last index of the parameters of the discussed group, and $n$ is the total parameter number of the group. Of course, here we assume the parameters are ordered in groups. Putting Eq. 9.141 into above, one has

$$p_{jJ} = \frac{m_0}{\sqrt{n}} \text{DOP}, \quad \text{DOP} = \sqrt{\sum_{i=j}^{J} Q[i][i]}, \qquad (9.143)$$

where DOP is the shortening of the Dilution of Precision factor. So we see that the DOP factor is a very important factor to describe the precision of a group of parameters that are the same type. Supposing in the unknown vector $X[i]$, $i = 1, 2, 3$ are coordinate $x$, $y$, $z$ of a receiver, and $i = 4$ is the receiver clock error, then the Position DOP (PDOP) is defined by $j = 1$, $J = 3$ in Eq. 9.143, and the Time DOP (TDOP) is defined by $j = J = 4$ in Eq. 9.143. The Geometric DOP (GDOP) is defined by $j = 1$, $J = 4$ in Eq. 9.143 (cf. Hofmann-Wellenhof et al. 1997). For the case of multiple stations, the definition can be similarly extended.

The PDOP is a factor that indicates the factor of precision of the position. Quite often, one would prefer to express the position precision in a local coordinate system, i.e. in horizontal and vertical components. Recalling the relation between the global and local coordinates (cf. Sect. 2.3), there are

$$X_{\text{local}} = RX_{\text{global}}, \quad \text{and} \quad X_{\text{global}} = R^T X_{\text{local}}, \qquad (9.144)$$

where $X_{\text{local}}$ and $X_{\text{global}}$ are identical vectors represented in local and global coordinate systems. $R$ is the rotation matrix given in Eq. 2.11. According to the covariance propagation theorem, one has then

$$Q_{\text{local}} = RQ_{\text{global}}R^{\text{T}}, \quad \text{and} \quad Q_{\text{global}} = RTQ_{\text{local}}R, \qquad (9.145)$$

where $Q_{\text{global}}$ is the sub-matrix of $Q$, which is related to the coordinates part. Supposing in the unknown vector $X_{\text{local}}[i]$, $i = 1, 2, 3$ are coordinates of horizontal $x$, $y$, and vertical $z$ of a receiver, then the Horizontal Dilution of Precision (HDOP) and Vertical Dilution of Precision (VDOP) are defined as

$$\text{HDOP} = \sqrt{\sum_{i=1}^{2} Q_{\text{local}}[i][i]}, \quad \text{and} \quad \text{VDOP} = \sqrt{\sum_{i=3}^{3} Q_{\text{local}}[i][i]}. \qquad (9.146)$$

For many stations, the definition can be similarly given.

## 9.7 Introduction to the Real-Time Positioning System

Nowadays, real-time positioning is a hotspot in the GNSS field. Two of the main methods for precise real-time positioning, Network RTK (NRTK) and PPP-RTK, are introduced in this section.

### 9.7.1 Network RTK

In the traditional RTK a single reference station is used, and the rover station needs to work within a short range from the reference station because of the distance limit of radio communication and the spatial decorrelation of distance-dependent errors caused by the orbital ionosphere and troposphere errors. Thus, the operating area of RTK positioning is dependent on the atmospheric conditions and is usually limited to a distance of 10–20 km. Network RTK (NRTK) is a method that can overcome the restraint of the limited range of classic RTK. The range of each station in network is usually less than 100 km, and each reference station sends the observations to a processing centre, where the observations are processed with a network adjustment and both errors and corrections of observation are computed. Then the observation corrections are sent to the users through a satellite link or the Internet. Users in the coverage area of the network can mitigate their observation errors with these corrections (Mowafy 2012).

The principle of Network RTK begins with all reference stations within the RTK Network continuously streaming satellite observations to a central server running

## 9.7 Introduction to the Real-Time Positioning System

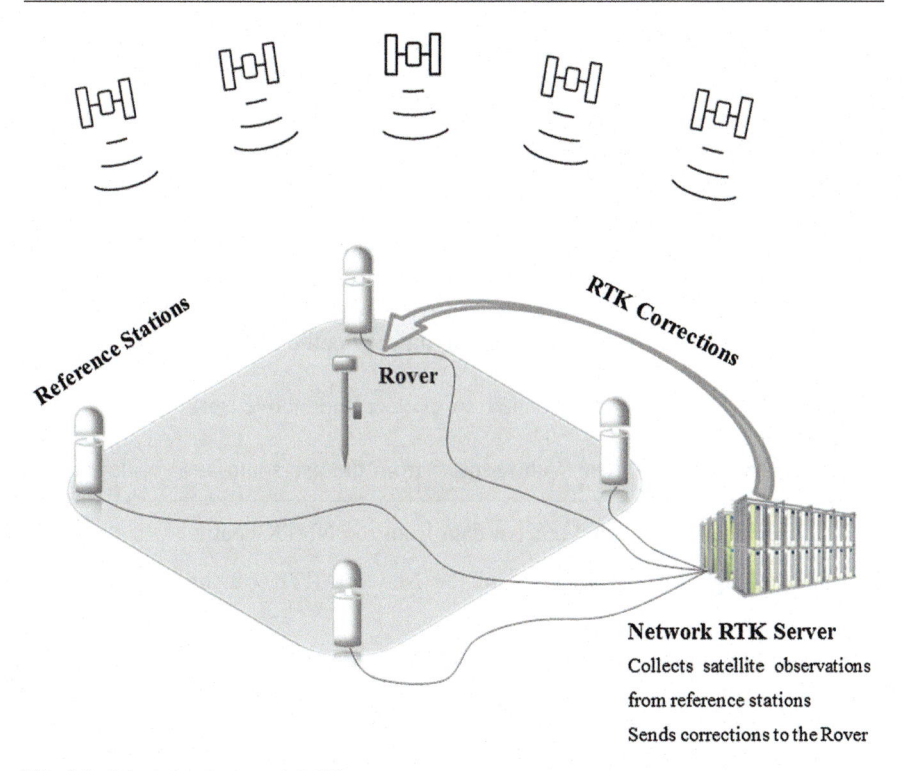

**Fig. 9.2** Principle of network RTK

Network RTK software, such as Trimble RTKNet, Leica GNSS Spider, and Geo++. The aim of Network RTK is to minimise the influence of the distance dependent errors on the rovers computed position within the bounds of the network. NRTK usually requires a minimum of three reference stations to generate corrections for the network area. In general there is no restriction concerning the network size, it can be regional, national, or even global.

In principle, the RTK network approach consists of four basic steps (cf. Fig. 9.2): data collection at the reference stations; manipulation of the data and generation of corrections at the network processing centre; broadcasting the corrections, and finally positioning at the rover using information from the NRTK. In the first step, multiple reference stations simultaneously collect GNSS satellite observations and send them to the control centre where ambiguity fixing is performed. Only observations with fixed ambiguities can be used for the precise modelling of the distance-dependent biases. The rather long distances between each reference station and the requirement to fix the ambiguities in real-time makes this step as the main challenge of Network RTK.

Normally, a NRTK server system would consist of the following components (Leica Geo. systems 2011):

- A station server managed and connected to each reference station receiver.
- A network server that acquires the data from the station servers and sends it to the processing centre.
- A cluster server that hosts the network processing software. The software performs several tasks including: quality check of data, apply antenna phase centre corrections, ambiguity fixing, modelling, and estimation of systematic errors, interpolation of errors (corrections) in some techniques (e.g., VRS, PRS) and generation of virtual observations, or model coefficients in other techniques (e.g., FKP, Mac).
- A firewall is usually established to protect the above servers from being accessed by a user.
- RTK proxy server to deal with requests from the users and send back network information.
- The user interface to send/receive data from the NRTK centre.

The most significant advantages of the Network RTK can be summarised as follows:

- Compared with single reference station RTK, cost and labour are both reduced, as there is no need to set up a base reference station for each user.
- Accuracy of the computed rover positions are more homogeneous and consistent as error mitigation refers to one processing software, which uses the same functional and stochastic modelling and assumptions and use the same data.
- Accuracy is maintained over larger distances between the reference stations and the rover.
- The same area can be covered with fewer reference stations compared to the number of permanent reference stations required using single reference RTK. The separation distances between network stations are tens of kilometres, usually kept less than 100 km.
- NRTK provides higher reliability and availability of RTK corrections with improved redundancy, such that if one station suffers from malfunctioning a solution can still be obtained from the rest of the reference stations.
- Network RTK is capable of supporting multiple users and applications.

However, Network RTK has some disadvantages, which are:

- The cost of subscription with a NRTK provider.
- The cost of wireless communication with the network (typically via a wireless mobile using for example GPRS technology).
- The dependence on an external source to provide essential information.

## 9.7.2
## PPP-RTK

PPP is limited in accuracy since the ionosphere-free linear combination is currently mandatory. Accurate ionospheric models are generally not available. Since the ionosphere-free linear combination is not based on integer coefficients and the state information currently does not preserve the integer nature of ambiguities, it is not possible to resolve ambiguities adequately to access the full GNSS carrier phase accuracy levels. Therefore, long integration or observation times are required for PPP. The limitations of PPP can be overcome with RTK (Real-Time Kinematic) networks using state space modelling. These RTK networks can consistently derive all individual GNSS errors in real time. The atmospheric GNSS effects are modelled and state information is also present for ionosphere and troposphere. The complete state information is ready for distribution to users in real time. So users are capable to resolve ambiguities and to achieve the known RTK accuracy level. This concept of precise point positioning enabling ambiguity resolution is PPP-RTK (Wübbena et al. 2005).

In a PPP mode, un-differenced observations are used and the satellite related errors are reduced by using precise satellite clock corrections and employing precise orbits to avoid the orbital errors. These precise satellite products are normally provided from a processing centre analysing global data such as the International GNSS Service (IGS). Since only one receiver is used in PPP, the ambiguities are solved as part of the unknowns with float numbers and not fixed. As a result, several minutes of data are needed when processing to achieve a reliable convergence of the solution. As the ambiguities are solved as float numbers, the PPP accuracy can only reach at the level of sub-decimetre. However, it is possible to integrate PPP and NRTK into a seamless positioning service, which can provide an accuracy of a few centimetres. The concept of PPP-RTK is to augment PPP estimation with precise un-differenced atmospheric corrections and satellite clock corrections from a reference network, so that instantaneous ambiguity fixing is available for users within the network coverage.

## References

Blewitt G (1998) GPS data processing methodology. In: Teunissen, P.J.G., and Kleusberg, A. (eds), GPS for Geodesy, Springer-Verlag, Berlin, Heidelberg, New York, 231-270.
Chen J, Li H, Wu B, Zhang Y, Wang J, Hu C (2013) Performance of Real-Time Precise Point Positioning. Marine Geodesy, 36: 98-108
Cui X, Yu Z, Tao B, Liu D (1982) Adjustment in surveying. Surveying Press, Peking, (in Chinese)
Gao Y, Shen X. (2002) A New Method for Carrier Phase Based Precise Point Positioning. Navigation, Journal of the Institute of Navigation, 49(2)
Gao Y, Chen K (2004) Performance Analysis of Precise Point Positioning Using Real-Time Orbit and Clock Products. Journal of Global Positioning Systems, 3(1-2): 95-100
Gao Y, Chen K, Shen X (2003) Real-Time Kinematic Positioning Based on Un-Differenced Carrier Phase Data Processing. Proceedings of ION National Technical Meeting, Anaheim, California, January 22-24, 2003.

Gotthardt E (1978) Einführung in die Ausgleichungsrechnung. Herbert Wichmann Verlag, Karlsruhe

Gurtner W (1994) RINEX: The receiver independent exchange format. GPS World 5(7):48–52

Hofmann-Wellenhof B, Lichtenegger H, Collins J (1997) GPS theory and practice. Springer-Press, Wien

Hofmann-Wellenhof B, Lichtenegger H, Collins J (2001) GPS theory and practice. Springer-Press, Wien

Kouba J, Heroux P (2001) Precise point positioning using IGS orbit and clock products. GPS Solutions, 5(2):12-28

Leica Geo. Systems (2011). Using Network RTK, Available from: http://smartnet.leica-geosystems.eu/spiderweb/2fNetworkRTK.html.

Leick A (2004) GPS Satellite Surveying (3rd ed.) xxiv+664 p. John Wiley, New York, NY

Mowafy A(2012). Precise real-time positioning using network RTK. J. Shuanggen (Ed.), Global Navigation Satellite Systems: Signal Theory and Applications, InTech (2012) ISBN: 978-953-307-843-4

Remondi B (1984) Using the Global Positioning System (GPS) phase observable for relative geodesy: Modelling, processing, and results. University of Texas at Austin, Center for Space Research

Seeber G (1993) Satelliten-Geodaesie. Walter de Gruyter 1989

Shen Y, Xu G (2008) Simplified equivalent representation of GPS observation equations. GPS Solut 12: 99–108

Shen Y, Li B, Xu G (2009) Simplified equivalent multiple baseline solutions with elevation-dependent weights. GPS Solut 13: 165–171

Strang G, Borre K (1997) Linear algebra, geodesy, and GPS. Wellesley-Cambridge Press

Teunissen P (1997a) GPS double difference statistics: with and without using satellite geometry. J Geod 71: 137–148

Teunissen P (1997b) Closed form expressions for the volume of the GPS ambiguity search spaces. Artificial Satellites, Journal of Planetary Geodesy 32(1):5–20

Wang LX, Fang ZD, Zhang MY, Lin GB, Gu LK, Zhong TD, Yang XA, She DP, Luo ZH, Xiao BQ, Chai H, Lin DX (1979) Mathematic handbook. Educational Press, Peking, ISBN 13012-0165

Wang Q (2013) Adaptively Changing Strategy of Reference Satellite and Reference Station for Long Endurance and Long Range Airborne GNSS Kinematic Positioning. Journal of Navigation and Positioning, 1(1): 28-33

Wang Q, Xu G, Chen Z (2010) Interpolation Method of Tropospheric Delay of High Altitude Rover Based on Regional GPS Network. Geomatics and Information Science of Wuhan University, 35(12), 1405–1408

Wang Q, Xu G, Petrovic S, Schaefer U, Meyer U, Xu T (2011) A regional tropospheric model for airborne GPS applications. Advances in Space Research 48, 362-369

Wells D, Lindlohr W, Schaffrin B, Grafarend E (1987) GPS Design: Undifferenced Carrier Beat Phase Observations and the Fundamental Differencing Theorem, University of New Brunswick

Wübbena G, Schmitz M, Bagge A (2005) PPP-RTK: Precise Point Positioning Using State-Space Representation in RTK Networks, Proceedings of the 18th International Technical Meeting of the Satellite Division of The Institute of Navigation ION GNSS 2005, Long Beach, California, September 13-16, 2005, pp. 2584-2594

Xu G (2002) GPS data processing with equivalent observation equations, GPS Solutions, Vol. 6, No. 1-2, 6:28-33

Xu G (2003) GPS – Theory, Algorithms and Applications, Springer Heidelberg, ISBN 3-540-67812-3, 315 pages, in English

Xu G (2004) MFGsoft – Multi-Functional GPS/(Galileo) Software – Software User Manual, (Version of 2004), Scientific Technical Report STR04/17 of GeoForschungsZentrum (GFZ) Potsdam, ISSN 1610-0956, 70 pages, www.gfz-potsdam.de/bib/pub/str0417/0417.pdf

Xu G, Sun H, Shen YZ (2006b) On the parameterisation of the GPS observation models

Xu G, Yang YX, Zhang Q (2006c) Equivalence of the GPS data processing algorithms

Zumberge JF, Heflin MB, Jefferson DC, Watkins MM, Webb FH (1997) Precise point positioning for the efficient and robust analysis of GPS data from large networks. J. Geophysical Res., 102 (B3): 5005-5017.

# Chapter 10
# Applications of GPS Theory and Algorithms

In this chapter, we discuss software development using GPS theory and algorithms, and present a concept of precise kinematic positioning and flight-state monitoring of an airborne remote sensing system.

## 10.1
## Software Development

GPS/Galileo software generally consists of three basic components: a functional library, a data platform, and a data processing core. The functional library provides all of the physical models, algorithms, and tools that may be needed. The data platform prepares all data that may be needed, which is performed in a time loop. The data processing core forms the observation equations, accumulates them within the time loop, and solves the problem if desired. Software can be developed using the theory and algorithms outlined in this reference and handbook.

### 10.1.1
### Functional Library

A functional library consists of physical models, algorithms, and tools. For convenience, the functions are listed below, with the references referring to the contents described in this book (cf. Figs. 10.1, 10.2 and 10.3).

**Physical Models**

1. Tropospheric models for correcting or determining the tropospheric effects (cf. Sect. 5.2)

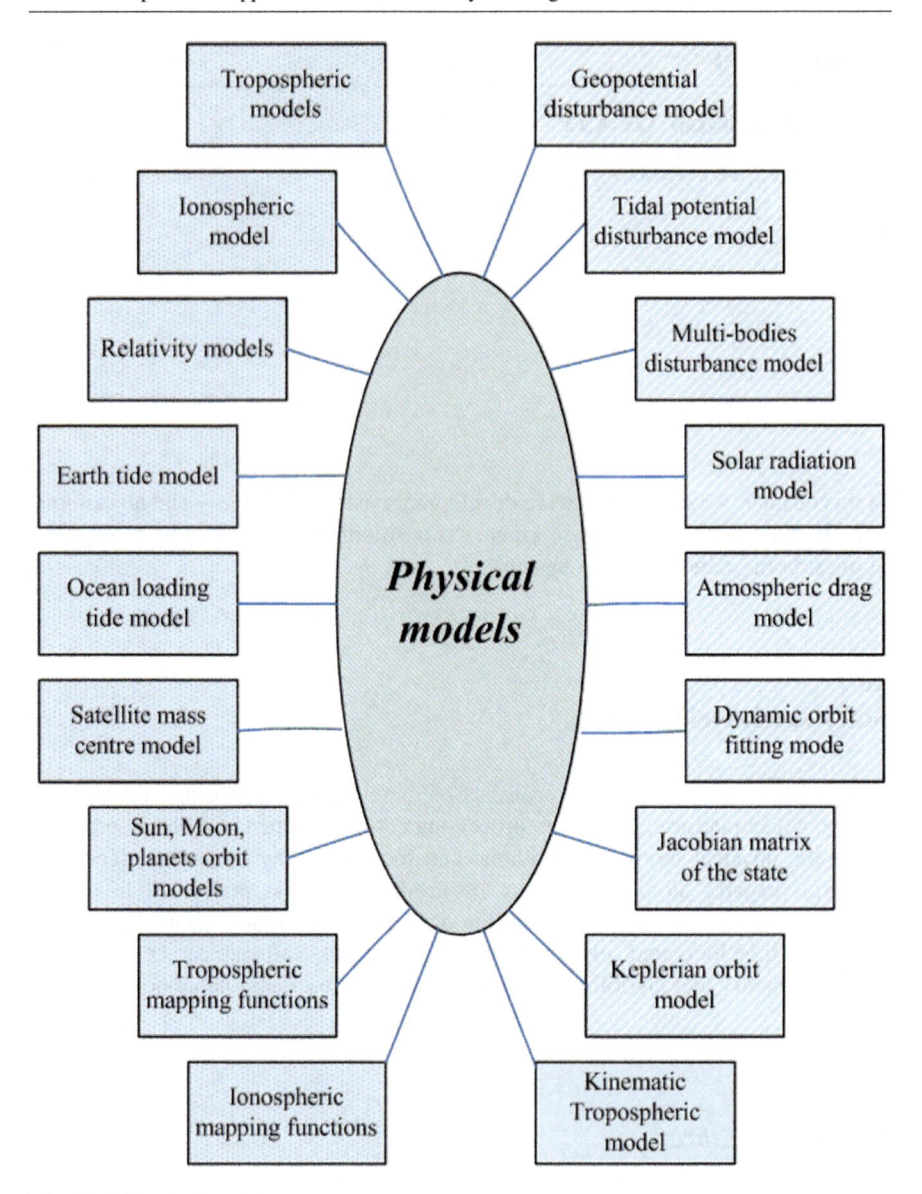

**Fig. 10.1** Physical models

2. Ionospheric model for correcting the ionospheric effects (cf. Sect. 5.1)
3. Relativity models for correcting the relativistic effects (cf. Sect. 5.3)
4. Earth tide model for the correction of the tidal displacements of the earth-fixed stations (cf. Sect. 5.4)
5. Ocean loading tide model for computing corrections of the ocean loading displacements especially for the stations near the coast (cf. Sect. 5.4)

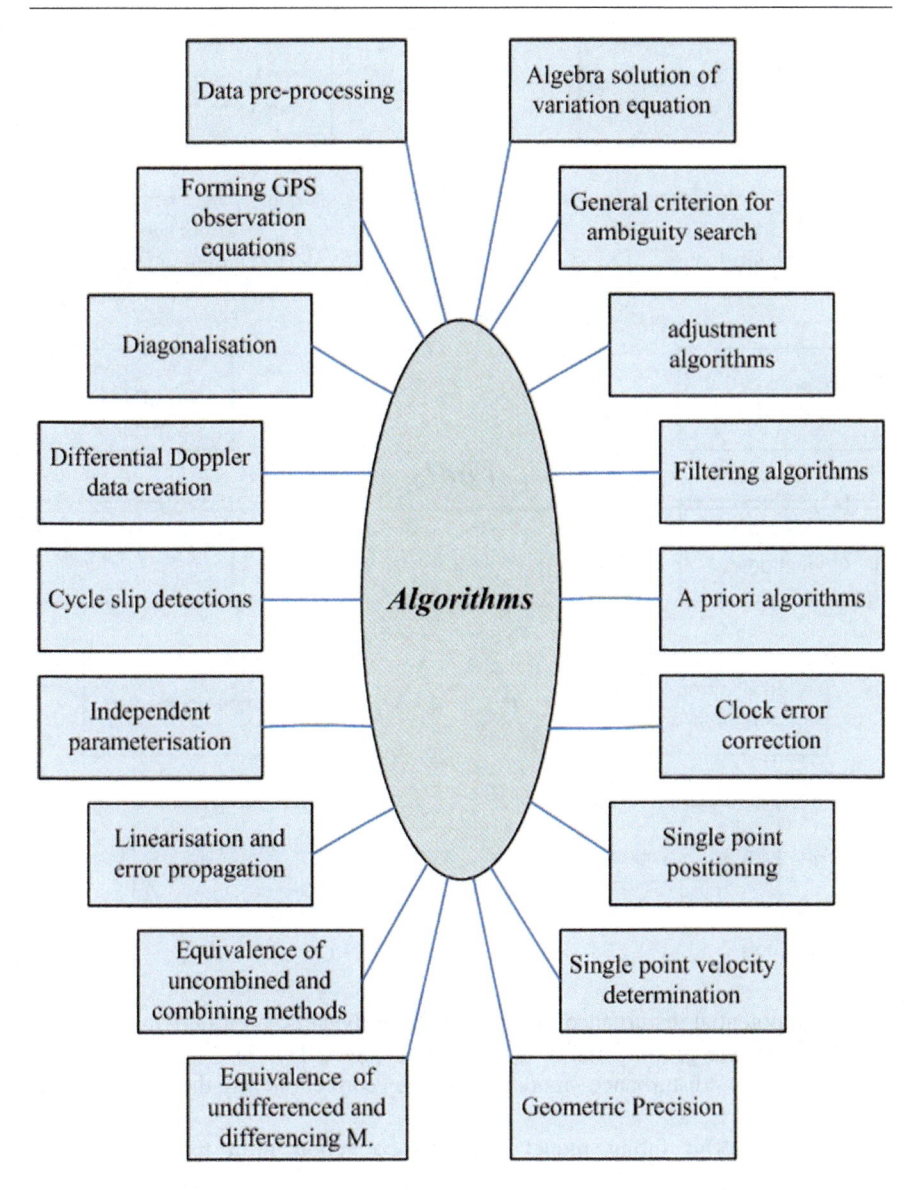

**Fig. 10.2** Algorithms

6. Satellite mass centre model for transformation between the mass centre and receiver antenna centre of the GPS satellite (cf. Sect. 5.8)
7. Solar radiation model for orbit determination (cf. Sect. 11.2.4)
8. Atmospheric drag model for LEO orbit determination (cf. Sect. 11.2.5)
9. Geopotential disturbance model for dynamic orbit determination and LEO satellite geopotential determination (cf. Sect. 11.2.1)

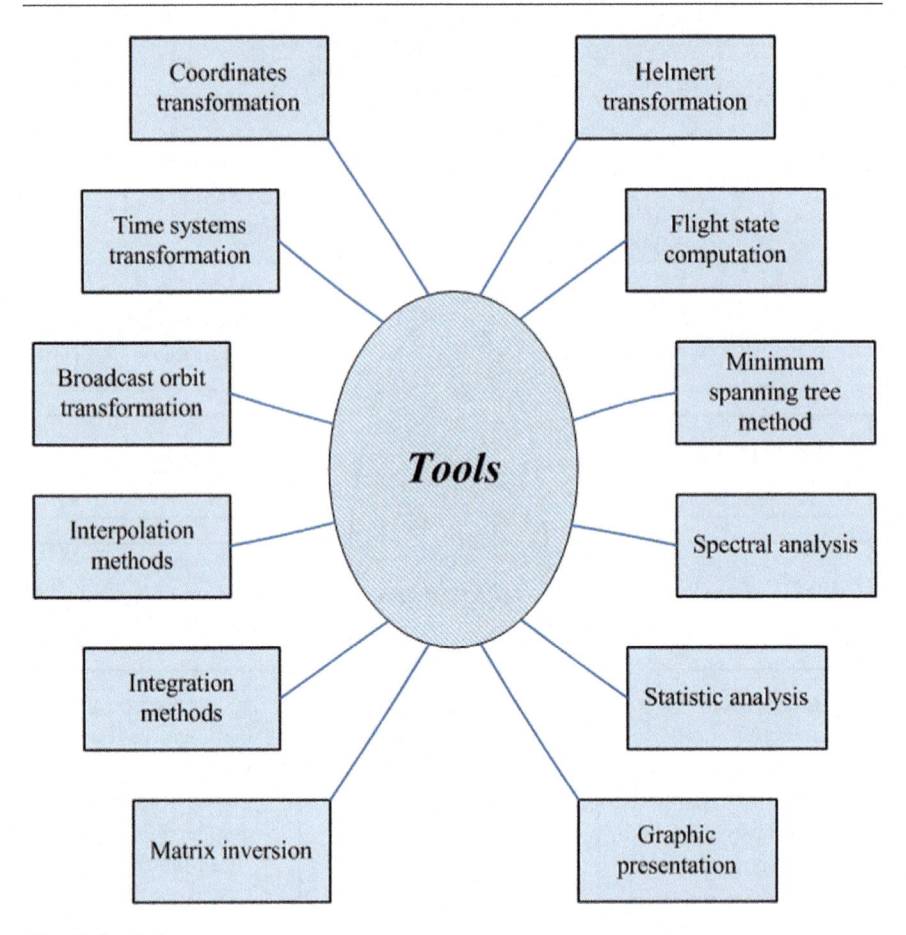

**Fig. 10.3** Tools (Coordinates transformation, Helmert transformation, Time systems transformation, Flight state computation, Broadcast orbit transformation, Minimum spanning tree method, Interpolation methods, Spectral analysis, Integration methods, Statistic analysis, Matrix inversion, Graphic presentation)

10. Tidal potential disturbance model for precise dynamic orbit determination and LEO satellite geopotential determination (cf. Sect. 11.2.3)
11. Multi-body disturbance models for the correction of the perturbations (cf. Sect. 11.2.2)
12. Dynamic orbit fitting model for orbit correction in a regional network (cf. Sect. 11.4)
13. Sun, Moon, and planet orbit models for computing the ephemerides of the Sun, Moon and planets, the multi-body disturbance and the Earth tide effects (cf. Sects. 11.2.8, 11.2.2 and 5.4)
14. Tropospheric mapping functions (cf. Sect. 5.2)
15. Ionospheric mapping functions (cf. Sect. 5.1)
16. Tropospheric model for kinematic receiver (cf. Sect. 10.3.2)
17. Keplerian orbit model (cf. Sect. 3.1.3)

18. Jacobian matrices of the Keplerian elements and the state vector of the satellite (cf. Sects. 3.1.1, 3.1.2, 3.1.3, 11.3 and 11.7)

*Algorithms*

1. Data pre-processing (cf. Sect. 9.5.1)
2. Forming GPS observation equations (cf. Sects. 4.1, 4.2, 4.3, and 6.1)
3. Differential Doppler data creation if necessary (cf. Sect. 6.5.5)
4. Cycle slip detections (cf. Sect. 8.1)
5. Independent parameterisation algorithms (cf. Sect. 9.1)
6. Linearisation and covariance propagation (cf. Sects. 6.3, 6.4, 11.5, and 11.7)
7. Equivalent algorithms of uncombined and combining methods (cf. Sects. 6.5, 6.7, and 9.2)
8. Equivalent algorithms of undifferenced and differencing algorithms (cf. Sects. 6.6, 6.8, 7.6, and 9.2)
9. Diagonalisation algorithm (cf. Sects. 7.6.1 and 9.1)
10. Algebraic solution of the variation equation (cf. Sect. 11.5.1)
11. Ambiguity search using general and equivalent criteria (cf. Sect. 8.3)
12. Classical adjustment tools (least squares adjustment [LSA], sequential LSA, conditional LSA, block-wise LSA, and equivalent algorithms, cf. Sects. 7.1–7.5)
13. Filtering algorithms (Kalman filter, robust Kalman filter, and adaptive robust Kalman filter, cf. Sects. 7.7.1, 7.7.2, and 7.7.3)
14. A priori constrained LSA (cf. Sects. 7.8.1 and 7.8.2)
15. Clock error corrections (cf. Sect. 5.5)
16. Single point positioning (cf. Sect. 9.5.2)
17. Single point velocity determination (cf. Sect. 9.5.6)
18. Accuracy of the observational geometry (cf. Sect. 9.6)

*Tools*

1. Coordinate transformation tools (cf. Sects. 2.1, 2.3, 2.4, and 2.5)
2. Time system transformation functions (cf. Sect. 2.6)
3. Broadcast orbit transformation in IGS format (cf. Sect. 3.3)
4. Interpolation tools (cf. Sects. 3.4, 5.4.2, and 11.6.5)
5. Integration methods (cf. Sects. 11.6.1, 11.6.2, 11.6.3, and 11.6.4)
6. Matrix inverse functions (Gauss-Jordan and Cholesky algorithms)
7. Helmert transformation (cf. Sect. 2.2)
8. Flight-state computation (cf. Sect. 10.3.3)
9. Minimum spanning tree method for forming optimal baseline network (cf. Sect. 9.2)
10. Spectral analysis methods (cf. Xu 1992)
11. Statistical analysis (cf. Sects. 6.4 and 7.2)
12. Graphic representation

## 10.1.2
## Data Platform

A data platform consists of three parts: the common part, the sequential time loop part, and the summary part. For convenience, the functions are listed below, with the references referring to the contents described in this book (cf. Fig. 10.4).

### Common Part

1. Program start
2. Read input parameter file for controlling the run of the software (an example of the definition of the input parameter file, cf., e.g., Xu 2004)
3. Read all possible data files necessary for the run of the software (e.g., satellite information file, station information file, geopotential data file, ocean loading coefficients, GPS orbit data file, polar motion data file)
4. Read or create the sun-moon-planet orbit data
5. Compute earth/ocean loading tide displacements
6. GPS satellite orbit data transformation if necessary
7. Data pre-processing if possible
8. Optimal baseline network construction and initialisations

### Sequential Time Loop Part

1. Sequential time loop start
2. Get the needed data for use at the related epoch (e.g., initial coordinates of the receivers)
3. Compute all possible parameters and model values for use at the related epoch (e.g., transformation matrices, interpolated orbit data, values of correcting models)
4. Read the GPS observation data and transform to a suitable form for use
5. Single point positioning (e.g., for the second type of clock error correction)
6. Single point velocity determination (velocity belongs to the state vector of the station)
7. Data processing core (cf. Sect. 10.1.3)
8. End of the sequential time loop

### Summary Part

1. Statistical analysis and spectral analysis of the results
2. Quality control and report
3. Iteration if necessary
4. Documentation and graphic representation
5. Forecast if needed
6. End of the run of the program

**Common part**

Program start
- Read input parameter file for control the run of the software
- Read all possible data files necessary for the run of the software
- Read or create the Sun-Moon-planets orbit data
- Compute Earth/ocean loading tide displacements
- GPS satellite orbit data transformation if necessary
- Data pre-processing if possible
- Optimal baseline network construction and initialisations

**Sequential time loop part**

- Sequential time loop start
- Get the needed data for use at the related epoch
- Compute all possible parameters & model values for use
- Read the GPS observation data and transformation
- Single point positioning
- Single point velocity determination
- Data processing core
- End of the sequential time loop

**Summary Part**

- Statistic and spectral analysis
- Quality control and report
- Iteration if necessary
- Documentation and graphics
- Forecast if needed
- End of the run of the program

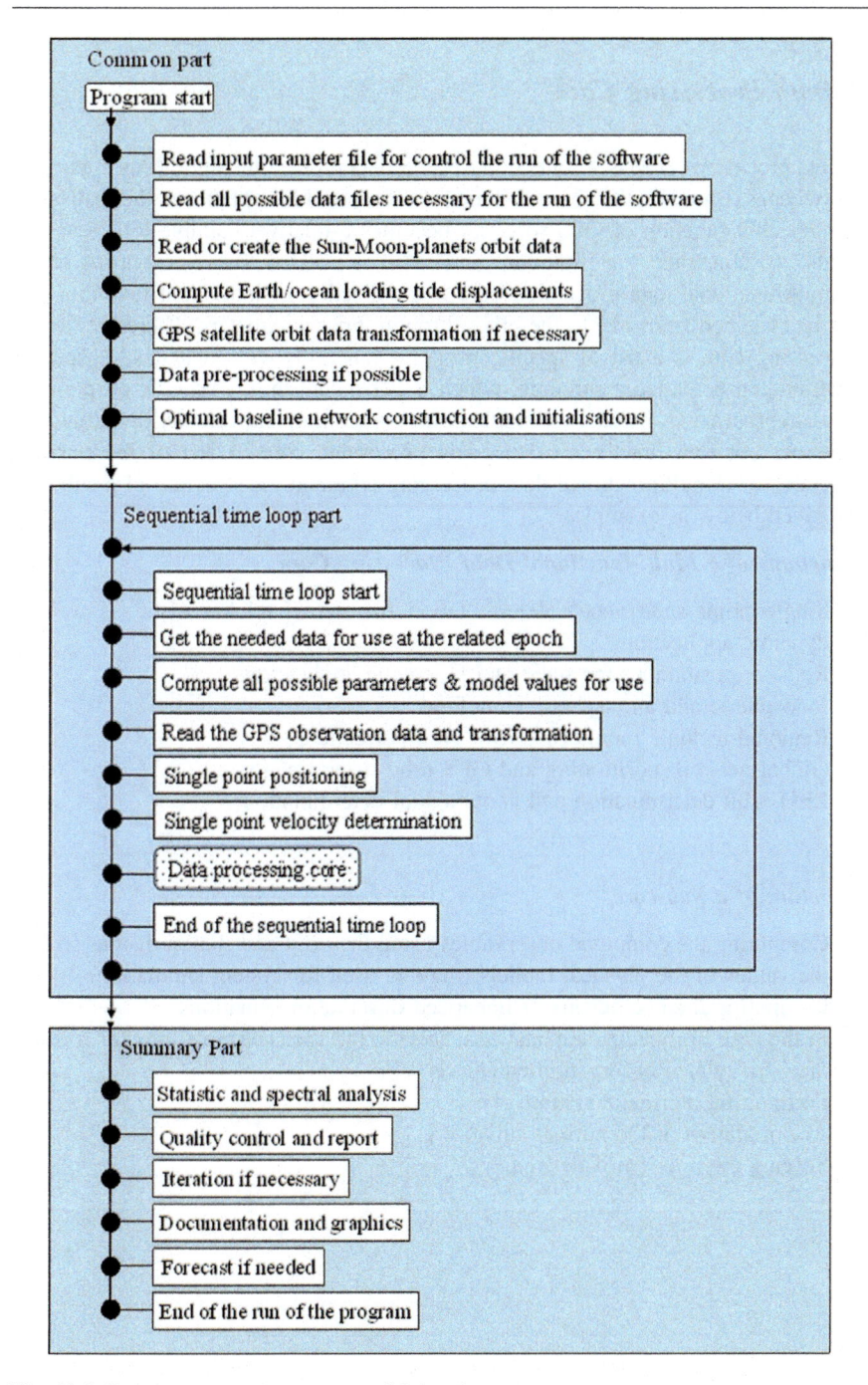

**Fig. 10.4** Program common part, sequential time loop part, summary part

## 10.1.3
## A Data Processing Core

A data processing core is a collection of GPS data processing algorithms controlled by switches. Based on the above functional library and data platform, the realisation of a specific function of GPS software becomes a relatively simple job—one just needs to construct the function and add it to the data processing core. A multifunctional data processing core is a collection of individual functions and can be switched from one to the other through input parameters. Therefore, a data processing core is a list of specific program functions with switches. A specific function can be called a sub-core, which is dependent on the specific purposes of the data processing. Indeed, the single point positioning and velocity determination functions are two functions of the data processing core. A list of the possible functions of a multifunctional data processing core and the structure of a sub-core are given below as examples.

### Functions of a Multi-functional Data Processing Core

1. Single point state vector determination for static and kinematic as well as dynamic applications
2. Relative positioning for static and kinematic applications
3. Ionospheric and atmospheric soundings
4. Regional tectonic monitoring with orbit corrections
5. Global network positioning and GPS orbit determination
6. LEO orbit determination and geopotential determination

### Structure of a Sub-core

1. Computing the computed observables using the orbit and station data as well as the values of the physical models (may be used for system simulation)
2. Computing the coefficients of linearised observation equations
3. In the case of dynamic applications, solving the variance equations for forming the orbit related observation equations
4. Forming the normal equations
5. Accumulation of the normal equations
6. Solving the problem if desired

## 10.2
## Introduction of GPS Software

In this section, we will introduce several GPS software programs commonly used around the world.

### 1. GAMIT/GLOBK/TRACK

The GAMIT, GLOBK, and TRACK (http://www-gpsg.mit.edu/~simon/gtgk/) form a comprehensive suite of programs for analysing GPS measurements primarily to study crustal deformation. The software was developed by the Massachusetts Institute of Technology (MIT), Scripps Institution of Oceanography, and Harvard University, with support from the National Science Foundation (NSF). The software is available, without written agreement or royalty fee, to individuals, universities, and governmental agencies for any non-commercial purpose. The download password and software updates can be obtained by contacting Dr. Robert W. King (rwk@chandler.mit.edu).

GAMIT is a collection of programs used for the analysis of GPS data. It uses the GPS broadcast carrier-phase and pseudorange observables to estimate three-dimensional relative positions of ground stations and satellite orbits, atmospheric zenith delays, and earth orientation parameters. The software is designed to run under any UNIX operating system. GLOBK is a Kalman filter whose primary purpose is to combine various geodetic solutions such as GPS, VLBI, and SLR experiments. It accepts as data, or "quasi-observations", the estimates and covariance matrices for station coordinates, earth orientation parameters, orbital parameters, and source positions generated from analysis of the primary observations. The input solutions are generally performed with loose a priori uncertainties assigned to all global parameters, so that constraints can be uniformly applied in the combined solution. TRACK is a GPS differential-phase kinematic positioning program. TrackRT is a real-time GPS kinematic processing program.

### 2. GIPSY–OASIS II

GIPSY-OASIS (GOA II) (https://gipsy-oasis.jpl.nasa.gov/) is a rapid automated ultra-high-precision GPS data processing software package with strict data quality control, which is supported by the National Aeronautics and Space Administration (NASA) Jet Propulsion Laboratory (JPL) in Pasadena, California.

Features of GOA II include demonstrated centimetre-level accuracy (ground and space); automated, unattended low-cost operations; innovative GPS and non-GPS analysis capabilities; real-time capability (with RTG, Real-Time GIPSY); a unique filter/smoother without equal in GPS estimation capabilities and accuracy; and adaptability to non-GPS orbiters and non-NASA programs (FAA, military, commercial). With GOA II, one can carry out orbit determination (individual spacecraft or satellite constellations); precise positioning and timing on land, sea, and air; and automated (unattended) operations, with near-real-time capability and real-time capability with RTG.

In addition, the Automatic Precise Positioning Service (APPS, replaces Auto-GIPSY) is an e-mail/ftp interface to GIPSY. It performs basic analysis of GPS data in a RINEX file. You do not need GIPSY to use APPS; all processing occurs on a computer at JPL. E-mail is used to inform APPS about the location of your data. E-mail is sent from APPS to inform you about the location of the results. Anonymous ftp is used by APPS to fetch your data, and you use anonymous ftp to fetch the results.

3. **Bernese**

The Bernese GNSS Software (http://www.bernese.unibe.ch/) is a scientific, high-precision, multi-GNSS data processing software developed at the Astronomical Institute of the University of Bern (AIUB). It is used by CODE (Centre for Orbit Determination in Europe) for its international (IGS) International GPS Service for Geodynamics (IGS) and European (EUREF/EPN) activities.

Bernese has features that include state-of-the-art modelling, detailed control over all relevant processing options, performance automation tools, adherence to up-to-date internationally adopted standards, and inherent flexibility due to a highly modular design. It is particularly well suited for rapid processing of small-size single- and dual-frequency combination surveys; automatic processing of permanent networks; processing of data from a large number of receivers; a combination of different receiver types, taking antenna phase centre variations into account; combined processing of GPS and GLONASSGLONASS observations; ambiguity resolutionAmbiguityresolution on long baselinesBaseline (2000 km and longer); ionospheric and tropospheric monitoring; clock estimation and time transfer; generation of minimum-constraint network solutions; and orbit determination and estimation of earth orientation parameters.

4. **RTKLIB**

RTKLIB (http://www.rtklib.com/) is an open source program package for GNSS positioning. RTKLIB has a number of features. It supports standard and precise positioning algorithms with GPS, GLONASS, Galileo, QZSS, BeiDou, and SBAS. It supports various positioning modes with GNSS for both real-time and post-processing (single, DGPS/DGNSS, kinematic, static, moving-baseline, fixed, PPP-kinematic, PPP-static, and PPP-fixed). It also supports many standard formats and protocols for GNSS, and several GNSS receiver proprietary messages. It supports external communication via serial, TCP/IP, NTRIP, local log file (record and playback), and FTP/HTTP (automatic download), and it provides many library functions and APIs for GNSS data processing.

## 10.3
## Concept of Precise Kinematic Positioning and Flight-State Monitoring

A concept of precise kinematic positioning and flight-state monitoring of an airborne remote sensing system is presented here, based on the practical experiences from the EU project AGMASCO. Within the project, about 2 months of kinematic GPS flight data and static reference data have been collected in Europe during four campaigns over 3 years. An independently developed GPS software package and several commercial GPS software packages have been used for data processing. In this chapter, we discuss the methods of creating the tropospheric model for the aircraft trajectory and the use of static ambiguity results as conditions in the kinematic positioning. These concepts are implemented in the kinematic/static GPS software KSGsoft, and they have demonstrated excellent performance (cf. Xu 2000).

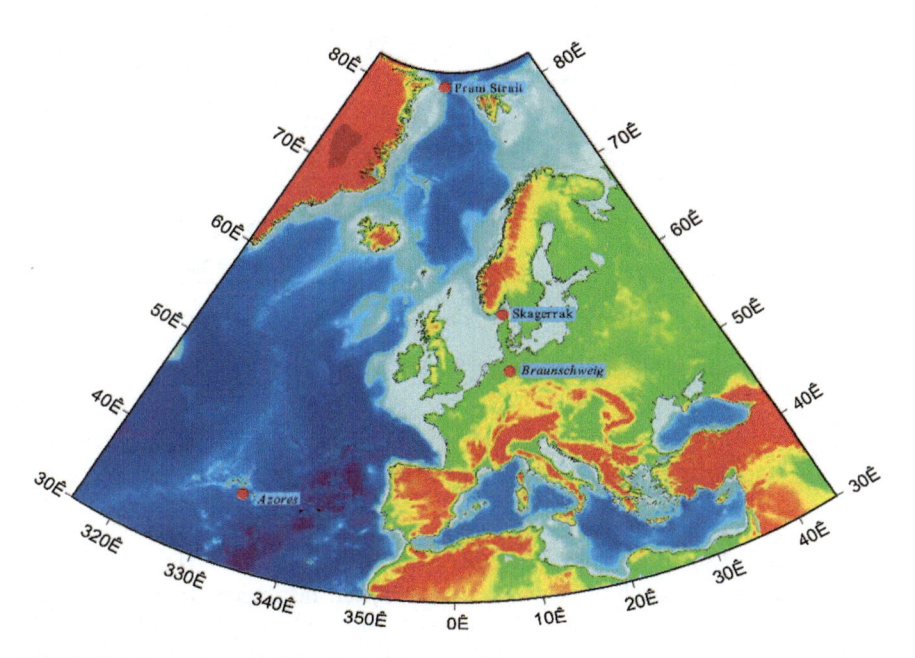

**Fig. 10.5** Measured areas of the four flight campaigns

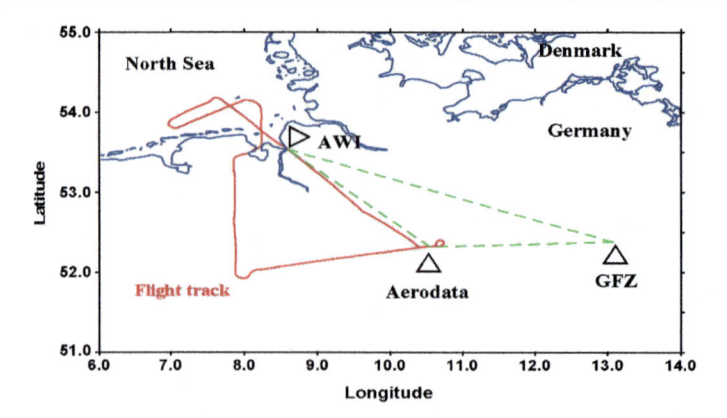

**Fig. 10.6** Flights in the Braunschweig campaign (June 1996)

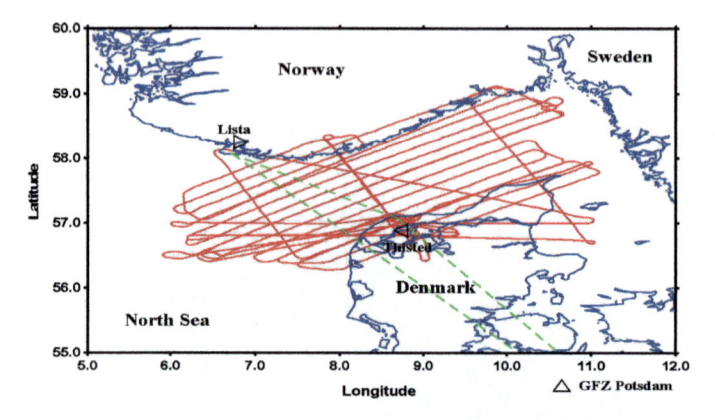

**Fig. 10.7** Flights in the Skagerrak campaign (September 1996)

## 10.3.1
## Introduction

The European Union (EU) project AGMASCO (**A**irborne **G**eoid **Ma**pping **S**ystem for **C**oastal **O**ceanography), in which five European institutions participated, has collected about 2 months of multiple static and airborne kinematic GPS data for the purpose of kinematic positioning and flight-state monitoring of an airborne remote sensing system. This system includes an aerogravimeter, accelerometer, radar and laser altimeter, INS, and datalogger. During the project, four flight campaigns were performed in Europe (Fig. 10.5): the test campaign in Braunschweig in June 1996 (Fig. 10.6), the Skagerrak campaign in September 1996 (Fig. 10.7), the Fram Strait campaign in July 1997 (Fig. 10.8), and the Azores campaign in October 1997 (Fig. 10.9). Two to three kinematic GPS antennas were mounted on the fuselage,

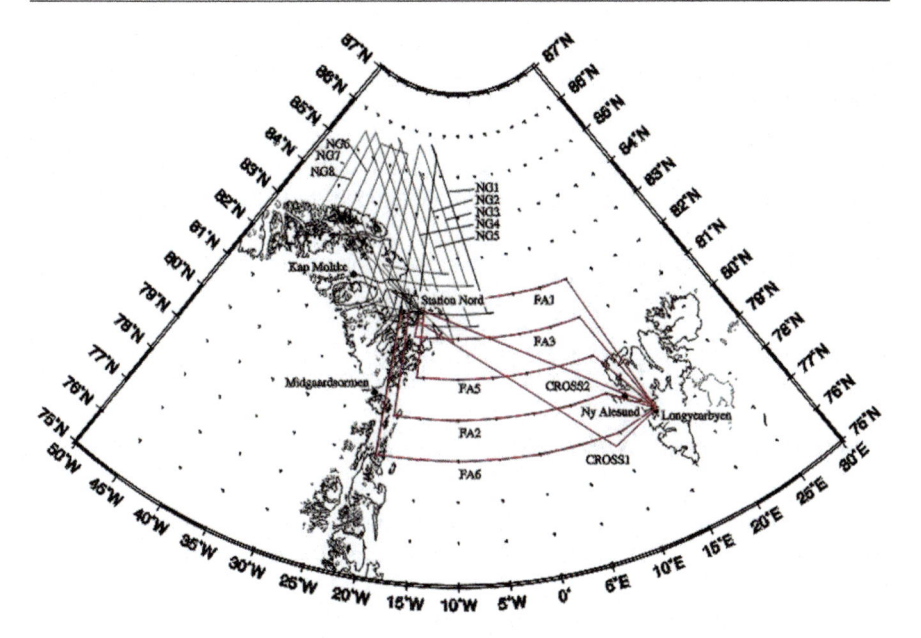

## 10.3 · Concept of Precise Kinematic Positioning and Flight-State Monitoring 325

**Fig. 10.8** Flights in the Fram Strait campaign (July 1997)

**Fig. 10.9** Flights in the Azores campaign (October 1997)

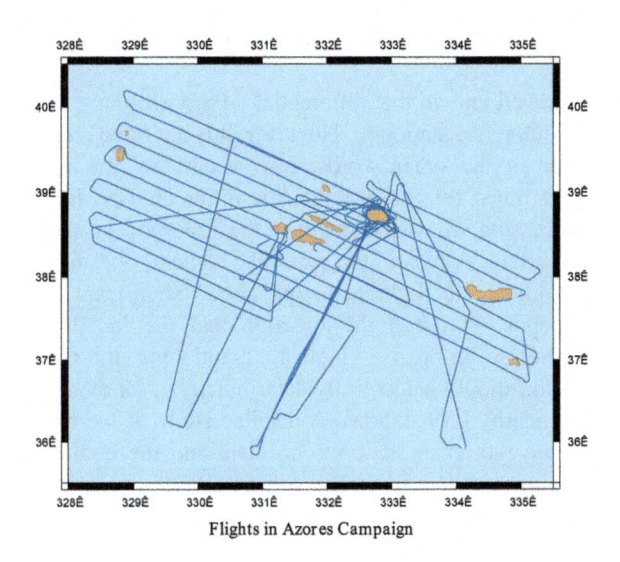

Flights in Azores Campaign

the back, and the wing of the aircraft, and at least three GPS receivers were used as static reference receivers.

The above-mentioned remote sensing system has two very important objectives: to measure the gravity acceleration of the earth and to determine the sea surface topography. Because the aerogravimeter (or accelerometer) and the altimeter are firmly attached to the aircraft, kinematic positioning and flight-state monitoring using GPS plays a key role for determining the flight acceleration, velocity and position, as well as orientation of the aircraft. The high sensitivity of the sensors requires high quality aircraft positioning and flight-state monitoring. Therefore, new strategies and methods have been studied, developed, tested, and implemented for GPS data processing.

The adopted concept of precise kinematic positioning and flight-state monitoring are discussed in Sects. 10.3.2 and 10.3.3, respectively.

## 10.3.2
## Concept of Precise Kinematic Positioning

A vast amount of literature exists on the topic of precise kinematic positioning (see, e.g., Remondi 1984; Wang et al. 1988; Schwarz et al. 1989; Cannon et al. 1997a, b; Hofmann-Wellenhof et al. 1997). Based on AGMASCO practice, a modified concept has been developed and applied to data processing.

### 10.3.2.1
### Combining the Static References with IGS Station

It is well known that differential GPS positioning results depend on the accuracy of the reference station(s). However, it is not quite clear how strong this dependency is —or in the other words, how accurately the reference coordinates should be determined for use in kinematic differential positioning. During AGMASCO data processing, it was noted that the accuracy of the reference coordinates is very important. A bias in the reference station coordinates will cause not only a bias in the kinematic flight path, but also a significant linear trend. Such a liner trend depends on the flight direction and the location of the reference receiver(s). Therefore, in precise kinematic positioning, the coordinates of the static reference station should be carefully determined by, for example, connecting these stations to the nearby IGS stations. A detailed study of the relationship between the accuracy of the reference station coordinate and the quality of kinematic and static positioning was carried out by Jensen (1999).

### 10.3.2.2
### Earth Tide and Loading Tide Corrections

A detailed study of the earth tide effects on GPS kinematic/static positioning is given in Xu and Knudsen (2000). For airborne kinematic differential GPS positioning, one needs to correct for the earth tide effects on the static reference station. Such tidal effects can reach 30 cm in Denmark and Greenland, and 60 cm at other locations in the world. Tidal effects can induce a "drift" over a few hours of measurement duration. For ground-based kinematic and static differential GPS positioning with baseline lengths less than 80 km, the impact of the earth tide effects can reach more than 5 mm. In precise application of GPS positioning, both in kinematic and static cases, the earth tide effects must therefore be taken into account even for a relatively small local GPS network.

Ocean loading tide effects could also reach up to a few cm in magnitude, in special cases (Ramatschi 1998). Generally, ocean loading tide effects should be considered at the centimetre level in coastal areas so that these effects have to be corrected for GPS data processing. However, unlike the earth tide, ocean loading tide effects can only be modelled by ocean tide models at about 60–90 % (Ramatschi 1998). Therefore, simply using a model to correct for the effects is not enough, and a detailed study of ocean loading tide effects is necessary for precise positioning. It is, however, possible to use GPS for determining the parameters of the local ocean loading tide effects (Khan 1999).

### 10.3.2.3
### Multiple Static References for Kinematic Positioning

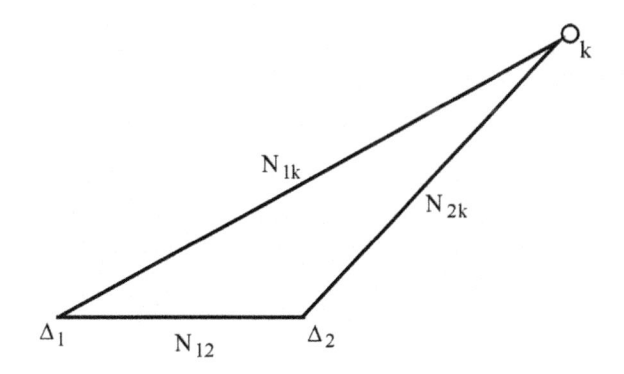

In differential GPS kinematic positioning, usually there is only one static reference used. It is obvious that if multiple static references are used, the reference station-dependent errors, such as those due to the troposphere and ionosphere as well as ocean loading tide effects, can be reduced and the geometric stability can be strengthened. For simplicity, only the case of using two static reference receivers will be discussed here. In Fig. 10.10, 1 and 2 denote the static reference receivers, and $k$ denotes the kinematic object. Suppose the two static stations are placed close by and both have the same GPS satellites in view. Using one static reference receiver for kinematic positioning, one has unknown vector $(X_k, N_{1k})$, where $X$ is the coordinate sub-vector and $N$ is the ambiguity sub-vector. Using two static

**Fig. 10.10** Multiple static references for kinematic positioning

references, one has unknown vector $(X_k\ N_{1k}\ N_{2k})$, because the unknown coordinate sub-vector $X$ is the same. The number of elements of the sub-vector $N$ compared with that of $X$ is very small in the kinematic case. Therefore, by using multiple static reference receivers for kinematic positioning, the total number of observations is increased, but the total number of unknowns remains almost the same; thus, the results will be modified.

Furthermore, according to the definition of double-differenced ambiguities, one has

$$N_{1k} = N_1^j - N_k^j - N_1^J + N_k^J, \qquad (10.1)$$

$$N_{2k} = N_2^j - N_k^j - N_2^J + N_k^J \qquad (10.2)$$

and

$$N_{12} = N_1^j - N_2^j - N_1^J + N_2^J, \qquad (10.3)$$

where $N$ on the right sides is un-differenced ambiguity, indices $j$ and $J$ denote satellites, 1 and 2 denote the static stations, and $k$ denotes the kinematic station. Then one gets

$$N_{1k} - N_{2k} = N_{12}, \qquad (10.4)$$

where $N_{12}$ is the double-difference ambiguity vector of the static baseline, which can be obtained from the static solution. Using relation Eq. 10.4, $N_{2k}$ can be represented by $(N_{1k} - N_{12})$. Thus, using two static references for kinematic positioning, one has nearly doubled the number of observables, yet the unknowns remain the same, if in addition the static results are used (typically, static measuring can be made over a longer time, and hence precise static results can be obtained).

In the case of a single difference, one has

$$N_{1k} = N_1^j - N_k^j, \qquad (10.5)$$

$$N_{2k} = N_2^j - N_k^j, \qquad (10.6)$$

and

$$N_{12} = N_1^j - N_2^j, \qquad (10.7)$$

where $N$ on the right sides is un-differenced ambiguity, index $j$ denotes the satellite, 1, 2 denote the static stations, and $k$ denotes the kinematic station. Then one gets the same relation as in the case of double difference:

$$N_{1k} - N_{2k} = N_{12}. \tag{10.8}$$

For un-differenced data processing, ambiguity vectors are $(N_1^j N_k^j)$ and $(N_2^j N_k^j)$ in kinematic data processing using a single reference. $(N_1^j N_2^j)$ is the ambiguity vector in static data processing. Regardless of how one deals with the reference-related ambiguities, the common part of the ambiguities obtained from static data processing can be used in kinematic data processing.

The accuracy of the kinematic positioning can be significantly increased by using multiple static reference receivers and by introducing the ambiguities from the static solution as conditions. An example showing the differences in the height of the front antenna determined using multiple reference receivers, with and without the use of the static ambiguity condition, is given in Fig. 10.14 (the ambiguity float solutions are used). The average and standard deviation of the differences are 27.07 and 4.34 cm, respectively. These results clearly indicate that the multiple static conditions have modified the results. A change in ambiguity resulted in not only a bias in the position solution, but also a high frequency variation. The base–base separation is about 200 km, and the length of the kinematic path is about 400 km (cf. Fig. 10.9).

For three or more static references receivers, similar arguments and improved results can be presented.

### 10.3.2.4
### Introducing Height Information as a Condition

Even after using multiple static reference receivers and static conditions, the ambiguities in kinematic positioning can still be wrong. In this case, there could be a bias and a variation in the kinematic trajectory (see Sect. 10.3.4.2 and Fig. 10.14). Therefore, introducing the height information of the aircraft at the start and/or resting point into the data processing is of great help, especially in airborne altimetry applications. The bias of the results obtained with the use of different software can then be eliminated.

### 10.3.2.5
### Creation of a Kinematic Tropospheric Model

The multiple static reference receivers can be used to determine the parameters of the tropospheric model. Using these parameters, the tropospheric model parameters for the kinematic receiver can be interpolated. Such a model, however, is generally suitable only for the footprint point of the kinematic platform. Therefore, the vertical gradient of temperature and the exponential changes in pressure and humidity (Syndergaard 1999) are introduced into the standard model to create a tropospheric model for the kinematic station in the air. Of course, this is not an ideal model, but it is a very reasonable one.

#### 10.3.2.6
#### Higher-Order Ionospheric Effect Correction

For long-distance kinematic positioning, the ionosphere-free combination must be used to eliminate the ionospheric effects. It is well known that the ionosphere-free combination is indeed only a first-order approximation (Klobuchar 1996). The second-order ionospheric effects are about 0.1 % of those of the first order (Syndergaard 1999). Therefore, the residual ionospheric effects can reach a level of a few centimetres. This must be taken into account using some form of modelling of the total ionospheric effects.

#### 10.3.2.7
#### A General Method of Integer Ambiguity Fixing

An integer ambiguity search method based on the conditional adjustment theory was proposed in Sect. 8.3. This method has been implemented in the GPS software KSGsoft (Kinematic/Static GPS Software), developed in GFZ Potsdam (Xu et al. 1998), and used extensively for real data processing in the EU project AGMASCO (Xu et al. 1997a, b). The search can be carried out in the coordinate domain, ambiguity domain, or both domains. Most other least squares ambiguity search methods (Euler and Landau 1992; Teunissen 1995; Merbart 1995; Han and Rizos 1995, 1997) are special cases of this algorithm, if only the ambiguity search domain is selected and without considering the uncertainty of the coordinates caused by ambiguity fixing. By taking the coordinate and ambiguity residuals into account, a general criterion for ambiguity searching is proposed to ensure an optimal search result. Detailed formulas are derived, and their use is described in Sect. 8.3. The theoretical relationship between the general criterion and the least squares ambiguity search criterion is also derived and illustrated by numerical examples in Sect. 8.3.

### *10.3.3*
### *Concept of Flight-State Monitoring*

For flight-state monitoring of an aircraft, several GPS antennas must be used, and the relative positions between the multiple antennas must be determined. As an example, using the method presented in Sect. 10.3.2, the position and velocity of one of the kinematic antennas can be determined. Using this point as a reference, the related position differences of other antennas can be determined. Because the distances between the multiple antennas are only a few metres, the atmospheric and ionospheric effects are nearly identical, and therefore only the single-frequency L1 observations are needed for relative positioning. In addition, due to the short ranges, this relative positioning can be performed with high accuracy.

Early stage tests of multiple kinematic GPS antennas mounted on a platform were made for checking purposes, using the known baseline length. Typically, such checks indicate that the distance has a systematic bias if the distance is computed from the two positions, and these two positions are determined separately. However, a combined solution of multiple kinematic positioning does not overcome the distance bias problem completely, because of inaccuracies in the ambiguity solution. Therefore, for precise flight-state monitoring, it is necessary to introduce the known distances between the antennas fixed on the aircraft as additional constraints in the data processing.

The distance condition can be represented as

$$\rho = \sqrt{(\Delta X)^2 + (\Delta Y)^2 + (\Delta Z)^2}, \qquad (10.9)$$

where $\Delta X$, $\Delta Y$, and $\Delta Z$ are the coordinate differences between two antennas, and $\rho$ is the distance. Because of the short distances, the linearisation of the condition cannot be done precisely in the initial step, and thus an iterative process must be used. The conditions can be used in a conditional adjustment, or they can be used for eliminating unknowns. The two methods are equivalent.

Flight state is usually represented by so-called state angles (heading, pitch, and roll). They are rotation angles between the body and the local horizontal coordinate frames of the aircraft. The axes of the local horizontal frame are selected as follows: the $x^b$ axis points out the nose, the $y^b$ axis points to the right parallel to the wing, and the $z^b$ axis points out the belly to form a right-handed coordinate system, where b denotes the body frame. The body frame can be rotated to be aligned to the local horizontal frame in a positive, right-handed sense, which is outlined in three steps. First, the body frame is rotated about the local vertical downward axis $z$ by angle $\psi$ (heading). Then the body frame is rotated about the new $y^b$ axis by angle $\theta$ (pitch). Finally, the body frame is rotated about the new $x^b$ axis by angle $\phi$ (roll). In the local horizontal coordinate system, the heading is the azimuth of axis $x^b$ of the body frame, the pitch is the elevation of axis $x^b$ of the aircraft and the roll is the elevation of axis $y^b$ of the aircraft. Note that the directions of the axis $x^b$ and the velocity vector of aircraft are usually not the same. Through kinematic positioning, the three flight-state monitoring angles can be computed (Cohen 1996).

Suppose three kinematic GPS antennas are mounted on the aircraft at the front, the back and the right wing (denoted as f, b, w), so that the $y$ components of the coordinates of front and back antennas in the body frame are zero, i.e. $y_f^b = y_b^b = 0$, and $x$, $z$ components of the coordinates of the wing antenna in the body frame are zero, i.e. $x_w^b = z_w^b = 0$. Then the coordinates of three antennas in body fixed frame are $P_f(x_f^b, 0, z_f^b)$, $P_b(x_b^b, 0, z_b^b)$, and $P_w(0, y_w^b, 0)$. Because the antennas are mounted as above supposed and because the flight state is computed by the positions of three antennas, there are pitch and roll correction angles that can be computed from the three coordinates by

$$\tan(\theta_0) = \frac{z_f^b - z_b^b}{x_f^b - x_b^b} \quad \text{and} \tag{10.10}$$

$$\tan(\phi_0) = -\frac{z_0}{y_w^b}, \tag{10.11}$$

where

$$z_0 = z_f^b - x_f^b \tan(\theta_0). \tag{10.12}$$

Through kinematic positioning and coordinate transformation, one has the coordinates of the three points in the local horizontal frame $P_f(x_f, y_f, z_f)$, $P_b(x_b, y_b, z_b)$, and $P_w(x_w, y_w, z_w)$. Then the three flight-state monitoring angles can be computed by

$$\tan(\psi) = \frac{y_f - y_b}{x_f - x_b}, \tag{10.13}$$

$$\tan(\theta - \theta_0) = \frac{z_f - z_b}{S}, \tag{10.14}$$

$$S = \sqrt{(x_f - x_b)^2 + (y_f - y_b)^2}, \tag{10.15}$$

$$\tan(\phi - \phi_0) = \frac{z_w - z_0}{s}, \quad \text{and} \tag{10.16}$$

$$s = \sqrt{(x_w - x_0)^2 + (y_w - y_0)^2}, \tag{10.17}$$

where $\sqrt{\phantom{x}}$ is the squares root operator and

$$x_0 = x_f - (x_f - x_b)K, \tag{10.18}$$

$$y_0 = y_f - (y_f - y_b)K, \tag{10.19}$$

$$z_0 = z_f - (z_f - z_b)K \quad \text{and} \tag{10.20}$$

$$K = \frac{x_f^b}{x_f^b - x_b^b}. \tag{10.21}$$

Comparisons of numerical GPS flight-state monitoring results are made with the results of INS. It is possible to use GPS to determine the heading with an accuracy

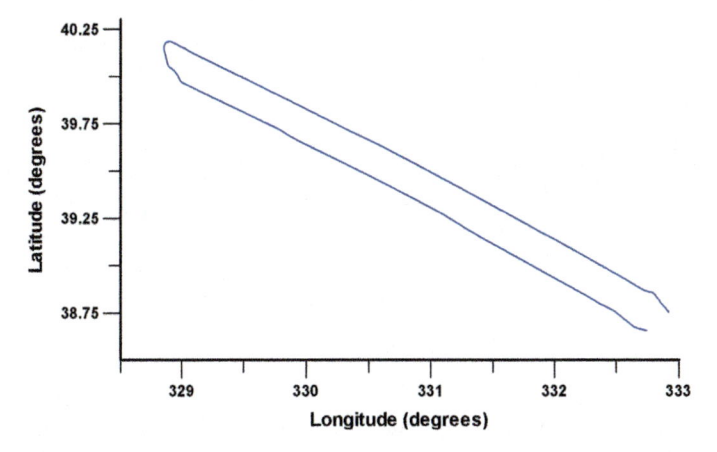

**Fig. 10.11** One flight trace determined by kinematic GPS

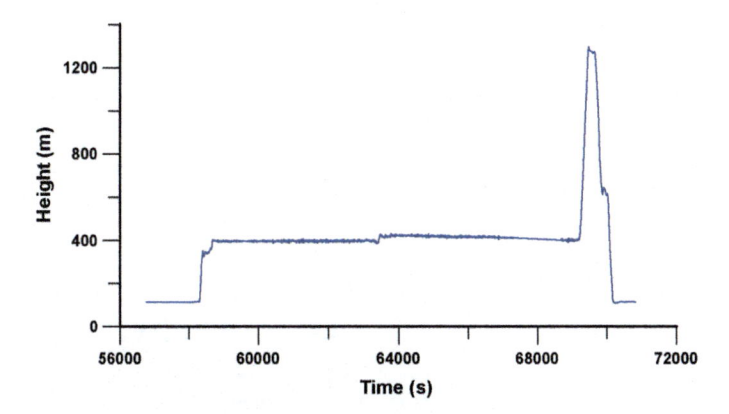

**Fig. 10.12** Height profile of one flight determined by kinematic GPS

up to 0.1°, and pitch and roll up to 0.2°. In this case, the distances between the three antennas were 5.224, 5.510, and 4.798 m.

## 10.3.4
## *Results, Precision Estimation, and Comparisons*

Examples demonstrating the above-mentioned methods are given through kinematic/static processing of a set of kinematic/static GPS data collected on 3 October 1997 at the islands of Portugal in the Atlantic Ocean within the Azores campaign. Two reference stations (Faim and Flor) served as static references, with a distance of about 239.4 km. Three antennas are fixed at the front, the back, and the

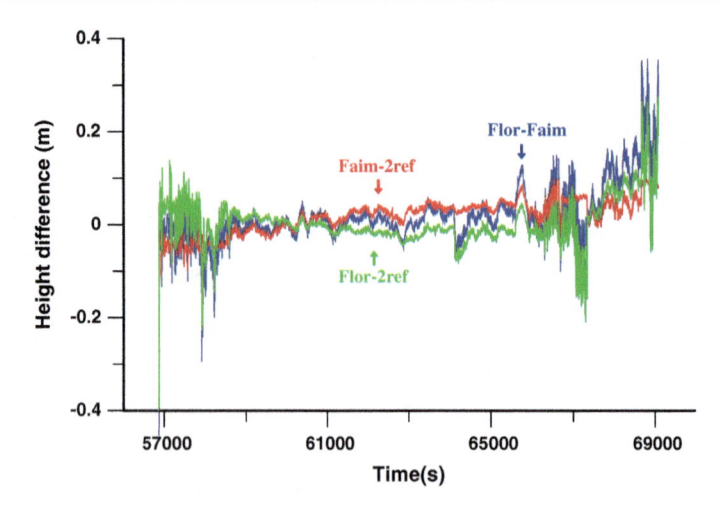

**Fig. 10.13** Height differences caused by static references

**Table 10.1** Statistics of the differences in the heights determined

| Height differences | Average | Deviation |
|---|---|---|
| Flor-2ref | 1.62 | 3.52 |
| Faim-2ref | −0.27 | 4.76 |
| Flor-Faim | 1.89 | 6.27 |

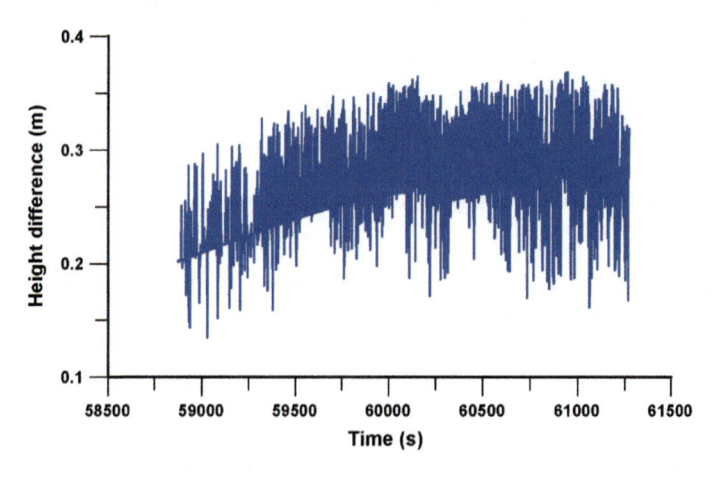

**Fig. 10.14** Height differences caused by static references

wing of the aircraft for determining the flight state. The distances between the baselines of front-back, front-wing, and back-wing are 5224, 5510, and 4798 m, respectively. The flight time is about 4 h. The length of the area is about 400 km (Fig. 10.11), and the height of the flight is about 400 m (Fig. 10.12).

## 10.3.4.1
### Multiple Static References for Kinematic Positioning

The flight trajectory determined using multiple references is a kind of weighted average path of the trajectories determined separately using a single reference. The heights of the front antenna are determined using a single reference and multiple references, respectively. The height differences of Flor-Faim, Flor-2ref, and Faim-2ref are given in Fig. 10.13, indicated by blue, red, and green lines, where "Flor-Faim" indicates the height differences between the results obtained using Flor and Faim separately as a static reference, and "2ref" indicates that two references were used. The difference statistics are given in Table 10.1 (units: cm), which shows that the multiple references helped to stabilise the results.

## 10.3.4.2
### Ambiguity of Multiple Static References as a Condition for Kinematic Positioning

In the case of multiple static references, a static solution between the static references can be made for obtaining the static ambiguity vector, which can usually be obtained with excellent quality. By introducing such results as conditions for kinematic positioning, the accuracy of the position solution can be modified. The differences of the heights of the front antenna determined by using multiple references with and without static ambiguities as conditions are given in Fig. 10.14. The average and standard deviation of the differences are 27.07 and 4.34 cm,

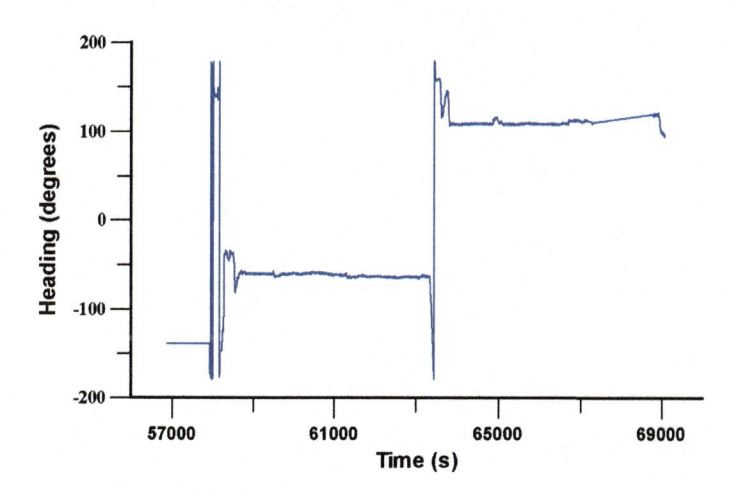

**Fig. 10.15** Heading determined by GPS

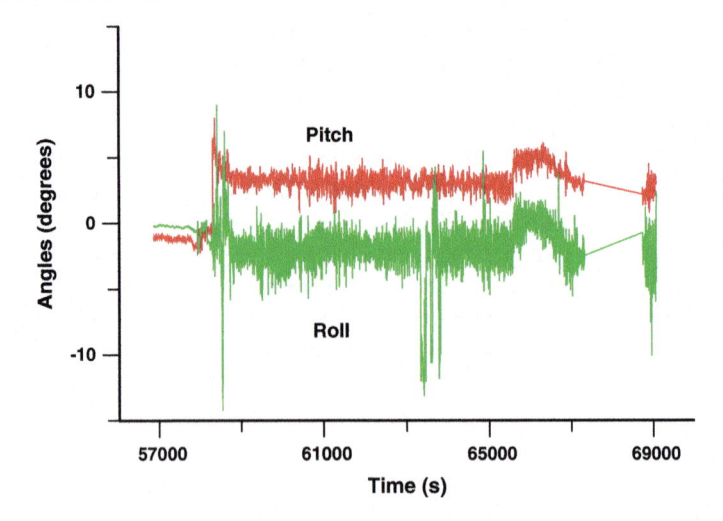

**Fig. 10.16** Pitch and roll determined by GPS

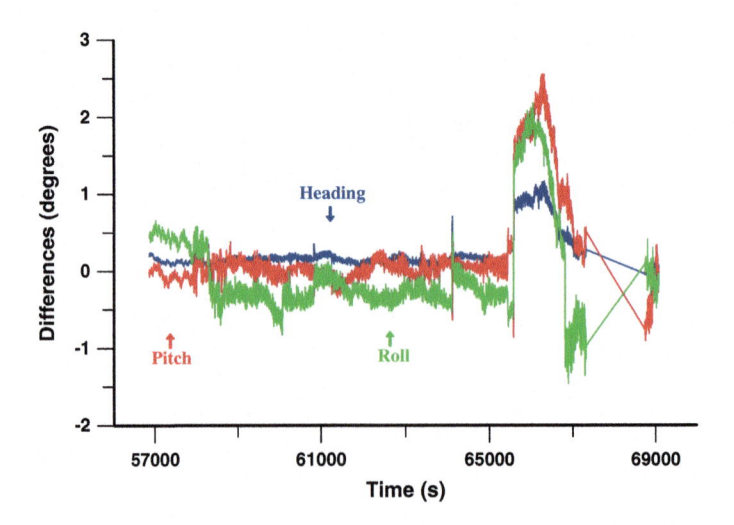

**Fig. 10.17** Differences in flight state determined by GPS and INS

**Table 10.2** Statistics of the differences in flight-state angles determined by GPS and INS

| Differences of | Average | Deviation |
| --- | --- | --- |
| Heading | 0.230 | 0.238 |
| Pitch | 0.233 | 0.596 |
| Roll | −0.249 | 0.612 |

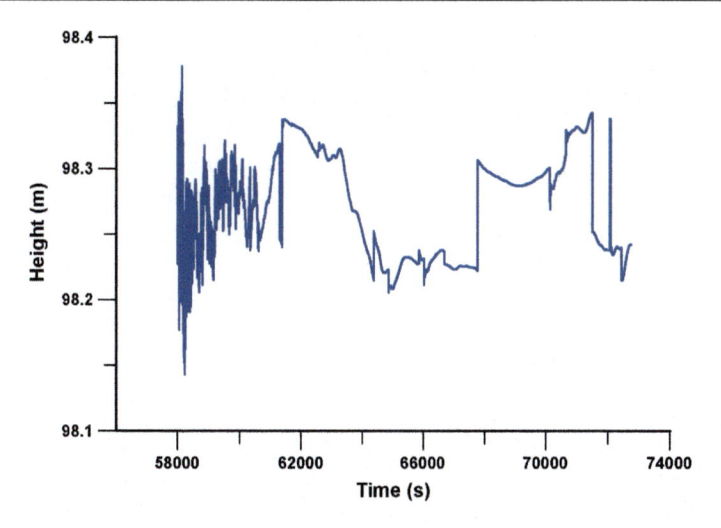

**Fig. 10.18** Height changes by static data kinematic processing

respectively. These have shown that the multiple static conditions have helped to modify the results. A change in ambiguity caused not only a bias, but also a variation in the results.

Using multiple static references and introducing the static ambiguities as conditions, accuracy of the kinematic positioning can rise significantly. However, the airborne altimetry results have shown that the GPS height solution still has a bias. Therefore, airport height information is introduced as a condition for modifying ambiguity resolution and eliminating the height bias of the GPS solution. To introduce tropospheric parameters for the aircraft, statically determined parameters and the vertical temperature gradient have been used.

### 10.3.4.3
### Multiple Kinematic GPS for Flight-State Monitoring and Its Comparison with INS

GPS-determined heading, pitch, and roll are given in Figs. 10.15 and 10.16 (with red and green lines), respectively, and the results are compared with those of INS. The differences in flight-state angles determined by GPS and INS are given in

**Table 10.3** Statistics of the differences in velocity

| Velocity difference | $dVs$ | $dVe$ | $dVh$ |
|---|---|---|---|
| Average | 0.03 | −0.01 | −0.06 |
| Deviation | 0.27 | 0.17 | 0.53 |

Fig. 10.17, with differences in the heading, pitch, and roll represented by blue, red, and green lines, respectively. The statistics of the differences in heading, pitch, and roll are given in Table 10.2 (units: degrees). The larger deviations of the pitch and roll are due to larger uncertainties of the height components determined by GPS. Considering the large deviation around epoch 66,000 and the data gap in INS around epoch 68,000, it is possible to use GPS to determine the heading with accuracy up to 0.1° and pitch and roll up to 0.2°.

#### 10.3.4.4
#### Static GPS Data Kinematic Processing

Static GPS data kinematic processing is one of the methods used to check the reliability of the GPS software working in a kinematic module. Such data processing has been used for studying earth tide effects (Xu and Knudsen 2000) and ocean loading tide effects. Faim has been used as a reference, and the position of Flor has been solved with static and kinematic modules. The static height of Flor is 98.257 m. The kinematic height average is 98.272 m, and its standard deviation is 3.8 cm. This indicates that kinematic data processing can reach an accuracy of about 4 cm with a baseline length of about 240 km. Although these results seem very good, the kinematic height graphic (see Fig. 10.18) shows a clear ambiguity problem in kinematic data processing. As soon as a satellite moves up or down, a jump will occur in the solution trajectory.

#### 10.3.4.5
#### Doppler Velocity Comparisons

Previous research (Xu et al. 1997a, b) has shown high accuracy for velocity solutions derived from Doppler measurements, and it appears that the velocity solutions are independent from the static references. A statistical analysis of the differences in velocity solved using different static references (Flor or Faim) is given in Table 10.3 (units: cm s$^{-1}$), which again confirms the previous conclusion. The nominal flight velocity is about 80 m s$^{-1}$ in the horizontal. In the vertical component, the maximum velocity is about 5 m s$^{-1}$.

## *10.3.5*
## *Conclusions*

GPS research during the AGMASCO project has revealed that GPS can be used for airborne kinematic positioning and flight-state monitoring to fulfil the needs of navigating a remote sensing system for applications in aerogravimetry and oceanography.

A methodology has been proposed for precise kinematic GPS positioning that addresses the following issues:

- Using IGS stations to obtain precise reference coordinates, and introducing earth tide and ocean loading tide corrections
- Introducing multiple static reference receivers and using static ambiguity solutions as conditions
- Introducing the initial height information as a condition, and introducing the tropospheric model to the aircraft kinematic GPS receivers
- Modelling the higher-order ionospheric effects, and using the general method of ambiguity searching in coordinate and ambiguity domains.

For flight-state monitoring, the kinematic reference and data of the single-frequency L1 are used. Known distances between the multiple kinematic antennas are used as additional constraints.

Results have shown adequate performance for this methodology.

# References

Cannon ME, Lachapelle G, Goddard TW (1997) Development and results of a precision farming sys-tem using GPS and GIS technologies. Geomatica 51,1:9–19

Cannon ME, Lachapelle G, Szarmes M, Herbert J, Keith J, Jokerst S (1997) DGPS kinematic carrier phase signal simulation analysis for precise velocity and position determination. Proceedings of ION NTM 97, Santa Monica, CA

Cohen CE (1996) Altitude determination. Parkinson BW, Spilker JJ (eds) Global Positioning System: Theory and applications, Vol. II

Euler H-J, Landau H (1992) Fast GPS ambiguity resolution on-the-fly for real-time applications. In: Proceedings of $6^{th}$ Int. Geod. Symp. on Satellite Positioning. Columbus, Ohio, pp 17–20

Han S, Rizos C (1995) On-the-fly ambiguity resolution for long range GPS kinematic positioning. In: GPS Trends in Precise Terrestrial, Airborne, and Spaceborne Applications: $21^{st}$ IUGG General Assembly, IAG Symposium No. 115, Boulder, USA, July 3–4, 1995. Springer-Verlag, Berlin, pp 290–294

Han S, Rizos C (1997) Comparing GPS ambiguity resolution techniques. GPS World 8(10):54–61

Hofmann-Wellenhof B, Lichtenegger H, Collins J (1997) GPS theory and practice. Springer-Press, Wien

Jensen A (1999) Influences of references on precise static/kinematic GPS positioning.

Khan SA (1999) Ocean loading tide effects on GPS positioning. MSc. thesis, Copenhagen University

Klobuchar JA (1996) Ionospheric effects on GPS. In: Parkinson BW, Spilker JJ (eds) Global Position-ing System: Theory and applications, Vol. I, Chapter 12

Merbart L (1995) Ambiguity resolution techniques in geodetic and geodynamic applications of the Global Positioning System. Dissertation an der Philosophisch-naturwissenschaftlichen Fakultät der Universität Bern

Ramatschi M (1998) Untersuchung von Vertikalbewegungen durch Meeresgezeitenauflasten an Referenzstationen auf Grönland. Dissertation, Technische Universität Clausthal

Remondi B (1984) Using the Global Positioning System (GPS) phase observable for relative geodesy: Modelling, processing, and results. University of Texas at Austin, Center for Space Research.

Schwarz K-P, Cannon ME, Wong RVC (1989) A Comparison of GPS kinematic models for the determination of position and velocity along a trajectory. Manuscr Geodaet 14:345–353

Syndergaard S (1999) Retrieval analysis and methodologies in atmospheric limb sounding using the GNSS radio occultation technique. Dissertation, Niels Bohr Institute for Astronomy, Physics and Geophysics, Faculty of Science, University of Copenhagen

Teunissen P (1995) The least-squares ambiguity decorrelation adjustment: A method for fast GPS integer ambiguity estimation. J Geodesy 70(1–2):65–82

Wang G, Chen Z, Chen W, Xu G (1988) The principle of GPS precise positioning system. Surveying Press, Peking, ISBN 7-5030-0141-0/P.58, 345 p, (in Chinese)

Xu G (1992) Spectral analysis and geopotential determination (Spektralanalyse und Erdschwerefeldbestimmung). Dissertation, DGK, Reihe C, Heft Nr. 397, Press of the Bavarian Academy of Sciences, ISBN 3–7696–9442–2, 100 p, (with very detailed summary in German)

Xu G (2000) A concept of precise kinematic positioning and flight-state monitoring from the AGMASCO practice. Earth Planets Space 52(10):831–836

Xu G (2004) MFGsoft – Multi-Functional GPS/(Galileo) Software – Software User Manual, (Version of 2004), Scientific Technical Report STR04/17 of GeoForschungsZentrum (GFZ) Potsdam, ISSN 1610-0956, 70 pages, www.gfz-potsdam.de/bib/pub/str0417/0417.pdf

Xu G, Knudsen P (2000) Earth tide effects on kinematic/static GPS positioning in Denmark and Greenland. Phys Chem Earth 25(A4):409–414

Xu G, Bastos L, Timmen L (1997a) GPS kinematic positioning in AGMASCO campaigns – Strategic goals and numerical results. In: Proceedings of ION GPS-97 meeting in Kansas City, September 16–19, 1997, pp 1173–1184

Xu G, Fritsch J, Hehl K (1997b) Results and conclusions of the carborne gravimetry campaign in northern Germany. Geodetic Week Berlin '97, Oct. 6–11, 1997, electronic version published in http://www.geodesy.tu-berlin.de

Xu G, Schwintzer P, Reigber Ch (1998) KSGSoft – Kinematic/Static GPS Software – Software user manual (version of 1998). Scientific Technical Report STR98/19 of GeoForschungsZentrum (GFZ) Potsdam

# Chapter 11
# Perturbed Orbit and Its Determination

Satellites are attracted not only by the central force of the earth, but also by the non-central force, the attraction forces of the sun and the moon, and the drag force of the atmosphere. They are also affected by solar radiation pressure, earth and ocean tides, general relativity effects (cf. Chap. 5), and coordinate perturbations. Equations of satellite motion must be represented by perturbed equations. In this chapter, after discussions of the perturbed equations of motion and the attraction forces, for convenience of the earth tide and ocean loading tide computations, the ephemerides of the sun and the moon are described. Orbit correction is discussed based on an analysis solution of the $\overline{C}_{20}$ perturbation. Emphasis is given to the precise orbit determination, which includes the principle of orbit determination, algebraic solution of the variation equation, numerical integration, and interpolation algorithms, as well as the related partial derivatives.

## 11.1
## Perturbed Equation of Satellite Motion

The perturbed equation of motion of the satellite is described by Newton's second law in an inertial Cartesian coordinate system as

$$m\ddot{\vec{r}} = \vec{f}, \qquad (11.1)$$

where $\vec{f}$ is the summated force vector acting on the satellite, and $\vec{r}$ is the radius vector of the satellite with mass $m$. $\ddot{\vec{r}}$ is the acceleration. Equation 11.1 is a second-order differential equation. For convenience, it can be written as two first-order differential equations as

$$\frac{d\vec{r}}{dt} = \dot{\vec{r}}$$
$$\frac{d\dot{\vec{r}}}{dt} = \frac{1}{m}\vec{f}. \tag{11.2}$$

Denoting the state vector of the satellite as

$$\vec{X} = \begin{pmatrix} \vec{r} \\ \dot{\vec{r}} \end{pmatrix}, \tag{11.3}$$

then Eq. 11.2 can be written as

$$\dot{\vec{X}} = \vec{F}, \tag{11.4}$$

where

$$\vec{F} = \begin{pmatrix} \dot{\vec{r}} \\ \vec{f}/m \end{pmatrix}. \tag{11.5}$$

Equation 11.4 is called the state equation of the satellite motion. Integrating Eq. 11.4 from $t_0$ to $t$, one has

$$\vec{X}(t) = \vec{X}(t_0) + \int_{t_0}^{t} \vec{F} dt, \tag{11.6}$$

where $\vec{X}(t)$ is the instantaneous state vector of the satellite, $\vec{X}(t_0)$ is the initial state vector at time $t_0$, and $\vec{F}$ is a function of the state vector $\vec{X}(t)$ and time $t$. Denoting the initial state vector as $\vec{X}_0$, then the perturbed satellite orbit problem turns out to be a problem of solving the differential state equation under the initial condition as

$$\begin{cases} \dot{\vec{X}}(t) = \vec{F} \\ \vec{X}(t_0) = \vec{X}_0 \end{cases}. \tag{11.7}$$

## 11.1.1
### Lagrangian Perturbed Equation of Satellite Motion

If the force $\vec{f}$ includes only the conservative forces, then there is a potential function $V$ so that

$$\frac{\vec{f}}{m} = \text{grad} V = \left( \frac{\partial V}{\partial x} \quad \frac{\partial V}{\partial y} \quad \frac{\partial V}{\partial z} \right) = \left( \frac{\partial V}{\partial r} \quad \frac{\partial V}{\partial \varphi} \quad \frac{\partial V}{\partial \lambda} \right), \tag{11.8}$$

where $(x, y, z)$ and $(r, \varphi, \lambda)$ are Cartesian coordinates and spherical coordinates, respectively. Denoting $R$ as the disturbance potential, $V_0$ as the potential of the centred force $\vec{f}_0$, then

$$R = V - V_0, \quad \frac{\vec{f} - \vec{f}_0}{m} = \text{grad} R. \tag{11.9}$$

The perturbed Eq. 11.2 of satellite motion in Cartesian coordinates is then

$$\begin{aligned}
\frac{dx}{dt} &= \dot{x} \\
\frac{dy}{dt} &= \dot{y} \\
\frac{dz}{dt} &= \dot{z} \\
\frac{d\dot{x}}{dt} &= -\frac{\mu}{r^3} x + \frac{\partial R}{\partial x} \\
\frac{d\dot{y}}{dt} &= -\frac{\mu}{r^3} y + \frac{\partial R}{\partial y} \\
\frac{d\dot{z}}{dt} &= -\frac{\mu}{r^3} z + \frac{\partial R}{\partial z},
\end{aligned} \tag{11.10}$$

where $\mu$ is the gravitational constant of the earth. The state vector $(\vec{r}, \dot{\vec{r}})$ of the satellite corresponds to an instantaneous Keplerian ellipse $(a, e, \omega, i, \Omega, M)$. Using the relationships between the two sets of parameters (cf. Chap. 3), the perturbed equation of motion 11.10 can be transformed into a so-called Lagrangian perturbed equation system (cf., e.g., Kaula 1966):

$$\begin{aligned}
\frac{da}{dt} &= \frac{2}{na} \frac{\partial R}{\partial M} \\
\frac{de}{dt} &= \frac{1-e^2}{na^2 e} \frac{\partial R}{\partial M} - \frac{\sqrt{1-e^2}}{na^2 e} \frac{\partial R}{\partial \omega} \\
\frac{d\omega}{dt} &= \frac{\sqrt{1-e^2}}{na^2 e} \frac{\partial R}{\partial e} - \frac{\cos i}{na^2 \sqrt{1-e^2} \sin i} \frac{\partial R}{\partial i} \\
\frac{di}{dt} &= \frac{1}{na^2 \sqrt{1-e^2} \sin i} \left( \cos i \frac{\partial R}{\partial \omega} - \frac{\partial R}{\partial \Omega} \right) \\
\frac{d\Omega}{dt} &= \frac{1}{na^2 \sqrt{1-e^2} \sin i} \frac{\partial R}{\partial i} \\
\frac{dM}{dt} &= n - \frac{2}{na} \frac{\partial R}{\partial a} - \frac{1-e^2}{na^2 e} \frac{\partial R}{\partial e}.
\end{aligned} \tag{11.11}$$

Based on the above equation system, Kaula derived the first-order perturbed analysis solution (cf. Kaula 1966). In the case of a small $e (e \ll 1)$, the orbit is nearly circular, so that the perigee and the related Keplerian elements $f$ and $\omega$ are not defined (this is not to be confused with the force vector $\vec{f}$ and true anomaly $f$). To overcome this problem, let $u = f + \omega$, and a parameter set of $(a, i, \Omega, \xi, \eta, \lambda)$ is used to describe the motion of the satellite, where

$$\xi = e \cos \omega$$
$$\eta = -e \sin \omega \quad (11.12)$$
$$\lambda = M + \omega.$$

Thus, one has

$$\frac{d\xi}{dt} = \frac{\xi}{e}\frac{de}{dt} + \eta\frac{d\omega}{dt}$$
$$\frac{d\eta}{dt} = \frac{\eta}{e}\frac{de}{dt} - \xi\frac{d\omega}{dt} \quad (11.13)$$
$$\frac{d\lambda}{dt} = \frac{dM}{dt} + \frac{d\omega}{dt}$$

and

$$\frac{\partial R}{\partial \omega} = \frac{\partial R}{\partial(\xi,\eta,\lambda)}\frac{\partial(\xi,\eta,\lambda)}{\partial \omega} = \frac{\partial R}{\partial(\xi,\eta,\lambda)}(\eta,-\xi,1)^{\mathrm{T}} = \eta\frac{\partial R}{\partial \xi} - \xi\frac{\partial R}{\partial \eta} + \frac{\partial R}{\partial \lambda}$$

$$\frac{\partial R}{\partial e} = \frac{\partial R}{\partial(\xi,\eta,\lambda)}\frac{\partial(\xi,\eta,\lambda)}{\partial e} = \frac{\partial R}{\partial(\xi,\eta,\lambda)}\left(\frac{\xi}{e},\frac{\eta}{e},0\right)^{\mathrm{T}} = \frac{\xi}{e}\frac{\partial R}{\partial \xi} + \frac{\eta}{e}\frac{\partial R}{\partial \eta} \quad (11.14)$$

$$\frac{\partial R}{\partial M} = \frac{\partial R}{\partial(\xi,\eta,\lambda)}\frac{\partial(\xi,\eta,\lambda)}{\partial M} = \frac{\partial R}{\partial(\xi,\eta,\lambda)}(0,0,1)^{\mathrm{T}} = \frac{\partial R}{\partial \lambda}.$$

Substituting Eq. 11.14 into Eq. 11.11 and then substituting the second, third, and sixth equations into Eq. 11.13, one has

$$\frac{da}{dt} = \frac{2}{na}\frac{\partial R}{\partial \lambda}$$

$$\frac{di}{dt} = \frac{1}{na^2\sqrt{1-e^2}\sin i}\left[\cos i\left(\eta\frac{\partial R}{\partial \xi} - \xi\frac{\partial R}{\partial \eta} + \frac{\partial R}{\partial \lambda}\right) - \frac{\partial R}{\partial \Omega}\right]$$

$$\frac{d\Omega}{dt} = \frac{1}{na^2\sqrt{1-e^2}\sin i}\frac{\partial R}{\partial i}$$

$$\frac{d\xi}{dt} = \frac{\sqrt{1-e^2}}{na^2}\frac{\partial R}{\partial \eta} - \eta\frac{\cos i}{na^2\sqrt{1-e^2}\sin i}\frac{\partial R}{\partial i} + \xi\frac{1-e^2-\sqrt{1-e^2}}{na^2e^2}\frac{\partial R}{\partial \lambda}$$

$$\frac{d\eta}{dt} = -\frac{\sqrt{1-e^2}}{na^2}\frac{\partial R}{\partial \xi} + \xi\frac{\cos i}{na^2\sqrt{1-e^2}\sin i}\frac{\partial R}{\partial i} + \eta\frac{1-e^2-\sqrt{1-e^2}}{na^2e^2}\frac{\partial R}{\partial \lambda}$$

$$\frac{d\lambda}{dt} = n - \frac{2}{na}\frac{\partial R}{\partial a} - \frac{\cos i}{na^2\sqrt{1-e^2}\sin i}\frac{\partial R}{\partial i} - \frac{1-e^2-\sqrt{1-e^2}}{na^2e^2}\left(\xi\frac{\partial R}{\partial \xi} + \eta\frac{\partial R}{\partial \eta}\right).$$

$$(11.15)$$

The new variables of Eq. 11.12 do not have clear geometric meanings. An alternative is to use the Hill variables (cf., e.g., Cui 1990).

## 11.1.2
### Gaussian Perturbed Equation of Satellite Motion

Considering the non-conservative disturbance forces such as solar radiation and air drag, no potential functions exist for use; therefore, the Lagrangian perturbed equation of motion cannot be directly used in such a case. The equation of motion perturbed by non-conservative disturbance force must be derived.

Considering any force vector $\vec{f} = (f_x f_y f_z)^T$ in ECSF coordinate system, one has

$$\begin{pmatrix} f_x \\ f_y \\ f_z \end{pmatrix} = R_3(-\Omega) R_1(-i) R_3(-u) \begin{pmatrix} f_r \\ f_\alpha \\ f_h \end{pmatrix}, \tag{11.16}$$

where $(f_r \; f_\alpha \; f_h)^T$ is a force vector with three orthogonal components in an orbital plane coordinate system, the first two components are in the orbital plane, $f_r$ is the radial force component, $f_\alpha$ is the force component perpendicular to $f_r$ and pointed in the direction of satellite motion, and $f_h$ completes a right-handed system. For convenience, the force vector may also be represented by tangential, central components in the orbital plane $(f_t, f_c)$ as well as $f_h$ (cf. Fig. 11.1). It is obvious that

$$\begin{pmatrix} f_r \\ f_\alpha \\ f_h \end{pmatrix} = R_3(-\beta) \begin{pmatrix} f_t \\ f_c \\ f_h \end{pmatrix}, \tag{11.17}$$

where

$$\tan \beta = r \frac{df}{dr} = \frac{a(1-e^2)}{1+e\cos f} \frac{df}{\frac{a(1-e^2)}{(1+e\cos f)^2} e \sin f df} = \frac{1+e\cos f}{e \sin f} \tag{11.18}$$

or

$$\sin \beta = \frac{1+e\cos f}{\sqrt{1+2e\cos f + e^2}}$$
$$\cos \beta = \frac{e \sin f}{\sqrt{1+2e\cos f + e^2}}. \tag{11.19}$$

In order to replace the partial derivatives $\partial R/\partial \sigma$ by force components, the relationships between them must be derived, where $\sigma$ is a symbol for all Keplerian elements. Using the regulation of partial derivatives, one has

**Fig. 11.1** Relation of radial and tangential forces

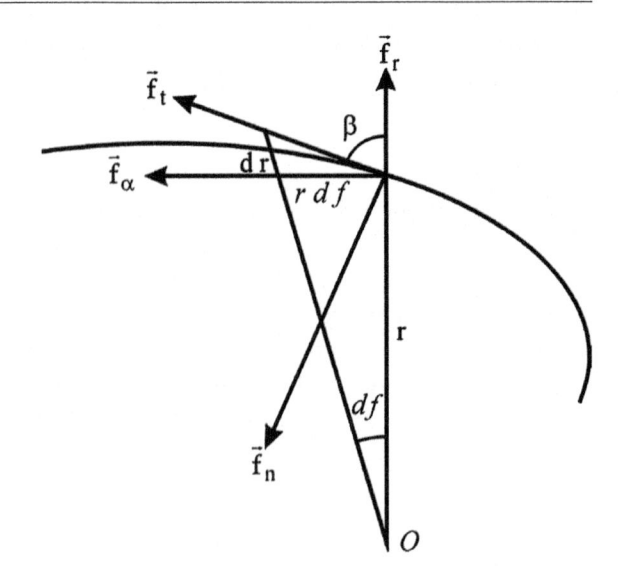

$$\frac{\partial R}{\partial \sigma} = \frac{\partial R}{\partial \vec{r}} \cdot \frac{\partial \vec{r}}{\partial \sigma} = \vec{f} \cdot \left( \frac{\partial r}{\partial \sigma} \vec{e}_r + r \frac{\partial \vec{e}_r}{\partial \sigma} \right)$$

$$= R_3(-\Omega) R_1(-i) R_3(-u) \begin{pmatrix} f_r \\ f_\alpha \\ f_h \end{pmatrix} \cdot \left( \frac{\partial r}{\partial \sigma} \vec{e}_r + r \frac{\partial \vec{e}_r}{\partial \sigma} \right), \quad (11.20)$$

where $\vec{e}_r$ is the radial identity vector of the satellite, the dot is the vector dot product, and

$$\vec{e}_r = \begin{pmatrix} \varepsilon_1 \\ \varepsilon_2 \\ \varepsilon_3 \end{pmatrix} = R_3(-\Omega) R_1(-i) R_3(-u) \begin{pmatrix} 1 \\ 0 \\ 0 \end{pmatrix} = \begin{pmatrix} \cos \Omega \cos u - \sin \Omega \cos i \sin u \\ \sin \Omega \cos u + \cos \Omega \cos i \sin u \\ \sin i \sin u \end{pmatrix}$$

$$\frac{\partial \vec{e}_r}{\partial \sigma} = \begin{pmatrix} \sin \Omega \sin i \sin u \frac{\partial i}{\partial \sigma} - \varepsilon_2 \frac{\partial \Omega}{\partial \sigma} - (\cos \Omega \sin u + \sin \Omega \cos i \cos u) \frac{\partial u}{\partial \sigma} \\ -\cos \Omega \sin i \sin u \frac{\partial i}{\partial \sigma} + \varepsilon_1 \frac{\partial \Omega}{\partial \sigma} - (\sin \Omega \sin u - \cos \Omega \cos i \cos u) \frac{\partial u}{\partial \sigma} \\ \cos i \sin u \frac{\partial i}{\partial \sigma} + \sin i \cos u \frac{\partial u}{\partial \sigma} \end{pmatrix}.$$

$$(11.21)$$

Substituting Eq. 11.21 into Eq. 11.20 and simplifying it, one has

$$\frac{\partial R}{\partial \sigma} = \frac{\partial r}{\partial \sigma} f_r + r \left( \cos i \frac{\partial \Omega}{\partial \sigma} + \frac{\partial u}{\partial \sigma} \right) f_\alpha + r \left( \sin u \frac{\partial i}{\partial \sigma} - \sin i \cos u \frac{\partial \Omega}{\partial \sigma} \right) f_h. \quad (11.22)$$

For deriving the partial derivatives of $r$ and $u(= f + \omega)$ with respect to the six Keplerian elements, the following basic relations (cf. Chap. 3) are used

$$r = \frac{a(1-e^2)}{1+e\cos f} = a(1-e\cos E)$$
$$r\cos f = a(\cos E - e)$$
$$r\sin f = a\sqrt{1-e^2}\sin E \qquad (11.23)$$
$$\tan\frac{f}{2} = \sqrt{\frac{1+e}{1-e}}\tan\frac{E}{2}$$
$$E - e\sin E = M,$$

where $E$ is a function of $(e, M)$, $f$ is a function of $(e, E)$, i.e. $(e, M)$, $r$ is a function of $(a, e, M)$, and $u$ is a function of $(\omega, f)$, i.e. $(\omega, e, M)$. Thus,

$$\frac{\partial E}{\partial(e,M)} = \left(\frac{a}{r}\sin E, \frac{a}{r}\right)$$

$$\frac{\partial f}{\partial(e,M)} = \left(\frac{2+e\cos f}{1-e^2}\sin f, \left(\frac{a}{r}\right)^2\sqrt{1-e^2}\right)$$

$$\frac{\partial r}{\partial M} = ae\sin E \frac{\partial E}{\partial M} = \frac{a^2 e}{r}\sin E = \frac{ae}{\sqrt{1-e^2}}\sin f$$

$$\frac{\partial r}{\partial(a,e,i,\Omega,\omega)} = \left(\frac{r}{a}, -a\cos f, 0, 0, 0\right) \qquad (11.24)$$

$$\frac{\partial u}{\partial e} = \frac{\partial u}{\partial f}\frac{\partial f}{\partial e} = \frac{2+e\cos f}{1-e^2}\sin f$$

$$\frac{\partial u}{\partial M} = \frac{\partial u}{\partial f}\frac{\partial f}{\partial M} = \left(\frac{a}{r}\right)^2\sqrt{1-e^2}$$

$$\frac{\partial u}{\partial(a,i,\Omega,\omega)} = (0,0,0,1).$$

Substituting Eq. 11.24 into Eq. 11.22, one has

$$\frac{\partial R}{\partial a} = \frac{r}{a}f_\mathrm{r}$$
$$\frac{\partial R}{\partial e} = -a\cos f \cdot f_\mathrm{r} + \frac{r\sin f}{1-e^2}(2+e\cos f)\cdot f_\alpha$$
$$\frac{\partial R}{\partial i} = r\sin u \cdot f_\mathrm{h}$$
$$\frac{\partial R}{\partial \Omega} = \cos i \cdot f_\alpha - r\sin i\cos u \cdot f_\mathrm{h} \qquad (11.25)$$
$$\frac{\partial R}{\partial \omega} = r \cdot f_\alpha$$
$$\frac{\partial R}{\partial M} = \frac{ae}{\sqrt{1-e^2}}\sin f \cdot f_\mathrm{r} + \frac{a(1+e\cos f)}{\sqrt{1-e^2}}\cdot f_\alpha.$$

Putting Eq. 11.25 into Lagrangian perturbed equations of motion 11.11, the so-called Gaussian perturbed equations of motion are then

$$\frac{da}{dt} = \frac{2}{n\sqrt{1-e^2}} [e \sin f \cdot f_r + (1+e \cos f) \cdot f_\alpha]$$

$$\frac{de}{dt} = \frac{\sqrt{1-e^2}}{na} [\sin f \cdot f_r + (\cos E + \cos f) \cdot f_\alpha]$$

$$\frac{di}{dt} = \frac{(1-e \cos E) \cos u}{na\sqrt{1-e^2}} \cdot f_h$$

$$\frac{d\Omega}{dt} = \frac{(1-e \cos E) \sin u}{na\sqrt{1-e^2} \sin i} \cdot f_h \qquad (11.26)$$

$$\frac{d\omega}{dt} = \frac{\sqrt{1-e^2}}{nae} \left[ -\cos f \cdot f_r + \frac{2+e \cos f}{1+e \cos f} \sin f \cdot f_\alpha \right] - \cos i \frac{d\Omega}{dt}$$

$$\frac{dM}{dt} = n - \frac{1-e^2}{nae} \left[ -\left( \cos f - \frac{2e}{1+e \cos f} \right) \cdot f_r + \frac{2+e \cos f}{1+e \cos f} \sin f \cdot f_\alpha \right].$$

The force components of $(f_r, f_\alpha, f_h)$ are used. Using Eq. 11.17, the Gaussian perturbed equations of motion can be represented by a disturbed force vector of $(f_t, f_c, f_h)$.

## 11.2
## Perturbation Forces of Satellite Motion

Perturbation forces of satellite motion will be discussed in this section. These are the gravitational forces of the earth, the attraction forces of the sun, the moon, and the planets, the drag force of the atmosphere, solar radiation pressure, and earth and ocean tides, as well as coordinate perturbations.

### 11.2.1
### *Perturbation of the Earth's Gravitational Field*

After a brief review of the earth's gravitational field, the perturbation force of the earth will be outlined here.

#### 11.2.1.1
#### The Earth's Gravitational Field

The complete real solution of the Laplace equation is called potential function $V$ of the earth. In spherical coordinates, $V$ can be expressed by Moritz (1980), Sigl (1989):

$$V = \sum_{lmi} \frac{1}{r^{l+1}} V_{lmi} = \sum_{l=0}^{\infty} \sum_{m=0}^{l} \frac{1}{r^{l+1}} P_{lm}(\sin\varphi)[C_{lm}\cos m\lambda + S_{lm}\sin m\lambda], \quad (11.27)$$

where $r$ is the radius, $\varphi$ is the latitude, and $\lambda$ is the longitude measured eastward (counterclockwise looking toward the origin from the positive end of the $z$-axis). One can, of course, use the co-latitude $\vartheta$ (or polar distance) instead of the latitude $\varphi$ ($\sin\varphi = \cos\vartheta$). The subscript $i$ in the first term denotes the $\cos m\lambda$ or $\sin m\lambda$ term. $P_{lm}(\sin\varphi)$ is the so-called associated Legendre function, $V_{lmi}$ denotes surface spherical harmonics, $C_{lm}$, $S_{lm}$ are coefficients of the spherical functions, and

$$P_{lm}(\sin\varphi) = \cos^m\varphi \sum_{t=0}^{k} T_{lmt}\sin^{l-m-2t}\varphi, \quad (11.28)$$

where $k$ is the integer part of $(l-m)/2$, and

$$T_{lmt} = \frac{(-1)^t(2l-2t)!}{2^l t!(l-t)!(l-m-2t)!}. \quad (11.29)$$

An important property of surface spherical harmonics $V_{lmi}$ is that they are orthogonal. Namely, for the integration over the surface of a sphere, there is (Heiskanen and Moritz 1967; Kaula 1966)

$$\int_{\text{sphere}} V_{LMI} V_{lmi} d\sigma = 0, \quad \text{if } L \neq l \quad \text{or } M \neq m \quad \text{or } I \neq i. \quad (11.30)$$

The integral of the square of $V_{lmi}$ for $C_{lm} = 1$ or $S_{lm} = 1$ is

$$\int_{\text{sphere}} V_{lmi}^2 d\sigma = \left[\frac{(l+m)!}{(l-m)!(2l+1)(2-\delta_{0m})}\right] 4\pi, \quad (11.31)$$

where the Kronecker delta $\delta_{0m}$ is equal to 1 for $m = 0$ and 0 for $m \neq 0$.

The normalised Legendre functions can be defined and denoted by

$$\overline{P}_{lm}(x) = P_{lm}(x) \left[\frac{(l-m)!(2l+1)(2-\delta_{0m})}{(l+m)!}\right]^{1/2}, \quad (11.32)$$

where $x = \sin\varphi = \cos\vartheta$. Recurrence formulae can be easily derived (Wenzel 1985):

$$\overline{P}_{(l+1)(l+1)}(x) = \overline{P}_{ll}(x)\left[\frac{(2l+3)}{(l+1)(2-\delta_{0l})}\right]^{1/2}(1-x^2)^{1/2},$$

$$\overline{P}_{(l+1)l}(x) = \overline{P}_{ll}(x)[2l+3]^{1/2}x \quad l \geq 1,$$

$$\overline{P}_{(l+1)m}(x) = \overline{P}_{lm}(x)\left[\frac{(2l+1)(2l+3)}{(l+m+1)(l-m+1)}\right]^{1/2}x \quad (11.33)$$

$$- \overline{P}_{(l-1)m}(x)\left[\frac{(l+m)(l-m)(2l+3)}{(l+m+1)(l-m+1)(2l-1)}\right]^{1/2}, \text{ and}$$

$$\overline{P}_{00}(x) = 1, \quad \overline{P}_{10}(x) = \sqrt{3}x, \quad \overline{P}_{11}(x) = \sqrt{3(1-x^2)}.$$

Since the first term of $V$ (i.e. $l = 0$) is represented by $GM/r$, the fully normalised geopotential function is taken as follows (Torge 1989; Rapp 1986):

$$V(r, \varphi, \lambda) = \frac{GM}{r}\left[1 + \sum_{l=2}^{\infty}\sum_{m=0}^{l}\left(\frac{a}{r}\right)^{l}\overline{P}_{lm}(\sin\varphi)[\overline{C}_{lm}\cos m\lambda + \overline{S}_{lm}\sin m\lambda]\right], \quad (11.34)$$

where $GM$ is the geocentric gravitational constant, $\overline{C}_{lm}, \overline{S}_{lm}$ are normalised coefficients and $a$ is the mean equatorial radius of the earth. The first term of $V$ is the potential of the central force of the earth. The perturbation potential of the earth is then (denoting $GM = \mu$)

$$R_{\text{geo}}(r, \varphi, \lambda) = \frac{\mu}{r}\sum_{l=2}^{\infty}\sum_{m=0}^{l}\left(\frac{a}{r}\right)^{l}\overline{P}_{lm}(\sin\varphi)[\overline{C}_{lm}\cos m\lambda + \overline{S}_{lm}\sin m\lambda]. \quad (11.35)$$

For any initial external potential of the earth

$$U(r, \varphi, \lambda) = \frac{\mu}{r}\left[1 + \sum_{l=2}^{L}\sum_{m=0}^{l}\left(\frac{a}{r}\right)^{l}\overline{P}_{lm}(\sin\varphi)\left[\overline{C}_{lm}^{N}\cos m\lambda + \overline{S}_{lm}^{N}\sin m\lambda\right]\right], \quad (11.36)$$

the disturbing potential $T$ is then

$$T = V - U = \frac{\mu}{r}\left[\sum_{l=2}^{\infty}\sum_{m=0}^{l}\left(\frac{a}{r}\right)^{l}\overline{P}_{lm}(\sin\varphi)[\Delta\overline{C}_{lm}\cos m\lambda + \Delta\overline{S}_{lm}\sin m\lambda]\right], \quad (11.37)$$

where $\overline{C}_{lm}^{N}, \overline{S}_{lm}^{N}$ are known normalised coefficients of the disturbing potential and

$$\overline{C}_{lm} = \Delta\overline{C}_{lm} - \overline{C}_{lm}^{N}, \quad \overline{S}_{lm} = \Delta\overline{S}_{lm} - \overline{S}_{lm}^{N}, \quad (l \leq L). \quad (11.38)$$

## 11.2.1.2
### Perturbation Force of the Earth's Gravitational Field

Denoting $(x', y', z')$ as three orthogonal Cartesian coordinates in ECEF system, then the force vector is

$$\vec{f}_{\text{ECEF}} = \begin{pmatrix} \frac{\partial V}{\partial x'} \\ \frac{\partial V}{\partial y'} \\ \frac{\partial V}{\partial z'} \end{pmatrix} = \begin{pmatrix} \frac{\partial V}{\partial (r,\varphi,\lambda)} \frac{\partial (r,\varphi,\lambda)}{\partial x'} \\ \frac{\partial V}{\partial (r,\varphi,\lambda)} \frac{\partial (r,\varphi,\lambda)}{\partial y'} \\ \frac{\partial V}{\partial (r,\varphi,\lambda)} \frac{\partial (r,\varphi,\lambda)}{\partial z'} \end{pmatrix} = \left( \frac{\partial V}{\partial (r,\varphi,\lambda)} \frac{\partial (r,\varphi,\lambda)}{\partial (x',y',z')} \right)^{\text{T}}. \quad (11.39)$$

From the relation between the Cartesian and spherical coordinates

$$\begin{pmatrix} x' \\ y' \\ z' \end{pmatrix} = \begin{pmatrix} r \cos\varphi \cos\lambda \\ r \cos\varphi \sin\lambda \\ r \sin\varphi \end{pmatrix}, \quad \begin{pmatrix} r = \sqrt{x'^2 + y'^2 + z'^2} \\ \varphi = \tan^{-1} \frac{z'}{\sqrt{x'^2 + y'^2}} \\ \lambda = \tan^{-1} \frac{y'}{x'} \end{pmatrix}, \quad (11.40)$$

one has

$$\frac{\partial (r,\varphi,\lambda)}{\partial (x',y',z')} = \begin{pmatrix} \cos\varphi \cos\lambda & \cos\varphi \sin\lambda & \sin\varphi \\ -\frac{1}{r}\sin\varphi \cos\lambda & -\frac{1}{r}\sin\varphi \sin\lambda & \frac{1}{r}\cos\varphi \\ -\frac{1}{r\cos\varphi}\sin\lambda & \frac{1}{r\cos\varphi}\cos\lambda & 0 \end{pmatrix}. \quad (11.41)$$

For differentiations of the associated Legendre function, from Eq. 11.33, one has similar recurrence formulae:

$$\begin{gathered}
\frac{d\overline{P}_{00}(\sin\varphi)}{d\varphi} = 0, \\
\frac{d\overline{P}_{10}(\sin\varphi)}{d\varphi} = \sqrt{3}\cos\varphi, \\
\frac{d\overline{P}_{11}(\sin\varphi)}{d\varphi} = -\sqrt{3}\sin\varphi, \\
\frac{d\overline{P}_{(l+1)(l+1)}(\sin\varphi)}{d\varphi} = -q\sin\varphi \overline{P}_{ll}(\sin\varphi) + q\cos\varphi \frac{d\overline{P}_{ll}(\sin\varphi)}{d\varphi}, \\
q = \sqrt{\frac{2l+3}{2l+2}}, \quad l \geq 1, \\
\frac{d\overline{P}_{(l+1)l}(\sin\varphi)}{d\varphi} = g\cos\varphi \overline{P}_{ll}(\sin\varphi) + g\sin\varphi \frac{d\overline{P}_{ll}(\sin\varphi)}{d\varphi}, \quad l \geq 1, \\
g = \sqrt{2l+3}, \\
\frac{d\overline{P}_{(l+1)m}(\sin\varphi)}{d\varphi} = h\cos\varphi \overline{P}_{lm}(\sin\varphi) + h\sin\varphi \frac{d\overline{P}_{lm}(\sin\varphi)}{d\varphi} - k\frac{d\overline{P}_{(l-1)m}(\sin\varphi)}{d\varphi}, \\
h = \sqrt{\frac{(2l+1)(2l+3)}{(l+m+1)(l-m+1)}}, \\
k = \sqrt{\frac{(l+m)(l-m)(2l+3)}{(l+m+1)(l-m+1)(2l-1)}}.
\end{gathered} \quad (11.42)$$

and

$$\frac{d^2\overline{P}_{00}(\sin\varphi)}{d\varphi^2} = 0,$$

$$\frac{d^2\overline{P}_{10}(\sin\varphi)}{d\varphi^2} = -\sqrt{3}\sin\varphi,$$

$$\frac{d^2\overline{P}_{11}(\sin\varphi)}{d\varphi^2} = -\sqrt{3}\cos\varphi,$$

$$\frac{d^2\overline{P}_{(l+1)(l+1)}(\sin\varphi)}{d\varphi^2} = -q\cos\varphi\overline{P}_{ll}(\sin\varphi) - 2q\sin\varphi\frac{d\overline{P}_{ll}(\sin\varphi)}{d\varphi} + q\cos\varphi\frac{d^2\overline{P}_{ll}(\sin\varphi)}{d\varphi^2},$$

$$\frac{d^2\overline{P}_{(l+1)l}(\sin\varphi)}{d\varphi^2} = -g\sin\varphi\overline{P}_{ll}(\sin\varphi) + 2g\cos\varphi\frac{d\overline{P}_{ll}(\sin\varphi)}{d\varphi} + g\sin\varphi\frac{d^2\overline{P}_{ll}(\sin\varphi)}{d\varphi^2},$$

$l \geq 1,$

$$\frac{d^2\overline{P}_{(l+1)m}(\sin\varphi)}{d\varphi^2} = -h\sin\varphi\overline{P}_{lm}(\sin\varphi) + 2h\cos\varphi\frac{d\overline{P}_{lm}(\sin\varphi)}{d\varphi}$$
$$+ h\sin\varphi\frac{d^2\overline{P}_{lm}(\sin\varphi)}{d\varphi^2} - k\frac{d^2\overline{P}_{(l-1)m}(\sin\varphi)}{d\varphi^2}.$$

(11.43)

The partial derivatives of the potential function with respect to the spherical coordinates are

$$\frac{\partial V}{\partial r} = -\frac{\mu}{r^2}\left[1 + \sum_{l=2}^{\infty}\sum_{m=0}^{l}(l+1)\left(\frac{a}{r}\right)^l\overline{P}_{lm}(\sin\varphi)\left[\overline{C}_{lm}\cos m\lambda + \overline{S}_{lm}\sin m\lambda\right]\right]$$

$$\frac{\partial V}{\partial \varphi} = \frac{\mu}{r}\sum_{l=2}^{\infty}\sum_{m=0}^{l}\left(\frac{a}{r}\right)^l\frac{d\overline{P}_{lm}(\sin\varphi)}{d\varphi}\left[\overline{C}_{lm}\cos m\lambda + \overline{S}_{lm}\sin m\lambda\right]$$

$$\frac{\partial V}{\partial \lambda} = \frac{\mu}{r}\sum_{l=2}^{\infty}\sum_{m=0}^{l}m\left(\frac{a}{r}\right)^l\overline{P}_{lm}(\sin\varphi)\left[-\overline{C}_{lm}\sin m\lambda + \overline{S}_{lm}\cos m\lambda\right].$$

(11.44)

Using the transformation formula of Eq. 2.14, the perturbation force of the earth's gravitational field in the ECSF system is then

$$\vec{f}_{\text{ECSF}} = R_P^{-1}R_N^{-1}R_S^{-1}R_M^{-1}\vec{f}_{\text{ECEF}}.$$

(11.45)

The computation process of disturbance force of the earth's gravitational field in the ECSF coordinate system may be carried out as follows:

1. Using Eq. 2.14 to transform the satellite coordinates in the ECSF system to the ECEF system;
2. Using Eq. 11.40 to compute the spherical coordinates of the satellite in the ECEF system;

3. Using Eq. 11.39 to compute the force vector in the ECEF system;
4. Using Eq. 11.45 to transform the force vector to the ECSF system.

## 11.2.2
### Perturbations of the Sun, the Moon, and the Planets

The equations of motion of two point-masses $M$ and $m$ under their mutual action can be given by

$$M\ddot{\vec{r}}_M = GMm\frac{\vec{r}_{Mm}}{r^3_{Mm}} \quad \text{and} \quad m\ddot{\vec{r}}_m = GMm\frac{\vec{r}_{mM}}{r^3_{mM}}, \tag{11.46}$$

where $r$ is the length of the vector $\vec{r}$, index $Mm$ means the vector is pointing from point-mass $M$ to $m$, and single index $M$ or $m$ means the vector is pointing to point-mass $M$ or $m$. Introducing additional point-masses $m(j)$, $j = 1, 2, \ldots$, the attractions of $m(j)$ on $M$ and $m$ can be given as equations similar to Eq. 11.46, and the total attractions may be obtained by summations

$$\begin{aligned} M\ddot{\vec{r}}_M &= GMm\frac{\vec{r}_{Mm}}{r^3_{Mm}} + \sum_j GMm(j)\frac{\vec{r}_{Mm(j)}}{r^3_{Mm(j)}} \\ m\ddot{\vec{r}}_m &= GMm\frac{\vec{r}_{mM}}{r^3_{mM}} + \sum_j Gmm(j)\frac{\vec{r}_{mm(j)}}{r^3_{mm(j)}}. \end{aligned} \tag{11.47}$$

By dividing the above two equations with $-M$ and $m$, respectively, then adding them together, one has

$$\ddot{\vec{r}}_m - \ddot{\vec{r}}_M = -G(M+m)\frac{\vec{r}_{Mm}}{r^3_{mM}} + \sum_j Gm(j)\left[\frac{\vec{r}_{mm(j)}}{r^3_{mm(j)}} - \frac{\vec{r}_{Mm(j)}}{r^3_{Mm(j)}}\right]. \tag{11.48}$$

Letting $\vec{r} = \vec{r}_m - \vec{r}_M$, i.e. using the point-mass $M$ as the origin, substituting $\vec{r}_{mm(j)} = -(\vec{r}_m - \vec{r}_{m(j)})$ in the right side of Eq. 11.48 and omitting the mass $m$ (mass of satellite), one has

$$\ddot{\vec{r}} = -GM\frac{\vec{r}}{r^3} - \sum_j Gm(j)\left[\frac{\vec{r} - \vec{r}_{m(j)}}{|\vec{r} - \vec{r}_{m(j)}|^3} + \frac{\vec{r}_{m(j)}}{r^3_{m(j)}}\right]. \tag{11.49}$$

It is obvious that the first term on the right side is the central force of the earth; therefore, the disturbance forces of multiple point-masses acting on the satellite are then

$$\vec{f}_m = -m \sum_j Gm(j) \left[ \frac{\vec{r} - \vec{r}_{m(j)}}{|\vec{r} - \vec{r}_{m(j)}|^3} + \frac{\vec{r}_{m(j)}}{r_{m(j)}^3} \right], \quad (11.50)$$

where $Gm(j)$ are the gravitational constants of the sun and the moon as well as the planets.

## 11.2.3
## Earth Tide and Ocean Tide Perturbations

As discussed in Sect. 5.4, the tidal potential generated by the moon and the sun can be written as

$$W_P = \sum_{j=1}^{2} \mu_j \sum_{n=2}^{\infty} \frac{\rho^n}{r_j^{n+1}} P_n(\cos z_j)$$

or

$$W_P = \sum_{j=1}^{2} \mu_j \sum_{n=2}^{\infty} \frac{\rho^n}{r_j^{n+1}} \left[ \begin{array}{l} P_n(\sin \varphi) P_n(\sin \delta_j) \\ + 2 \sum_{k=1}^{n} \frac{(n-k)!}{(n+k)!} P_{nk}(\sin \varphi) P_{nk}(\sin \delta_j) \cos k h_j \end{array} \right], \quad (11.51)$$

where $j$ is the index of the moon ($j = 1$) and the sun ($j = 2$), $\mu_j$ is the gravitational constant of body $j$, $\rho$ is the geocentric distance of the earth's surface (set as $a_e$), $r_j$ is the geocentric distance of the body $j$, $P_n(x)$ and $P_{nk}(x)$ are the Legendre function and associated Legendre function, $z_j$ is the zenith distance of the body $j$, $\delta_j$ and $h_j$ are the declination and local hour angle of body $j$, $h_j = H_j - \lambda$, and $H_j$ is the hour angle of $j$. The tidal deformation of the earth caused by the tidal potential can be considered a tidal deformation potential acting on the satellite by Dirichlet's theorem (Melchior 1978; Dow 1988):

$$\delta V = \sum_{j=1}^{2} \mu_j \sum_{n=2}^{\infty} k_n \left( \frac{\rho}{r} \right)^{n+1} \frac{\rho^n}{r_j^{n+1}} P_n(\cos z_j)$$

or

$$\delta V = \sum_{j=1}^{2} \mu_j \sum_{n=2}^{N} k_n \frac{a_e^{2n+1}}{r^{n+1} r_j^{n+1}} \left[ \begin{array}{l} P_n(\sin \varphi) P_n(\sin \delta_j) \\ + 2 \sum_{k=1}^{n} \frac{(n-k)!}{(n+k)!} P_{nk}(\sin \varphi) P_{nk}(\sin \delta_j) \cos k h_j \end{array} \right],$$

$$(11.52)$$

where $k_n$ is the Love number, $(r, \varphi, \lambda)$ is the spherical coordinate of the satellite in the ECEF system, and $N$ is the truncating number. The recurrence formulas of the Legendre function are (cf., e.g., Xu 1992)

$$(n+1)P_{n+1}(x) = (2n+1)xP_n(x) - nP_{n-1}(x)$$
$$(1-x^2)\frac{\mathrm{d}P_n(x)}{\mathrm{d}x} = nP_{n-1}(x) - nxP_n(x) \quad (11.53)$$
$$P_0(x) = 1 \quad P_1(x) = x.$$

The disturbing force vector of the tidal potential in the ECEF coordinate system is then

$$\vec{f}_{\text{ECEF}} = \begin{pmatrix} \frac{\partial \delta V}{\partial x'} \\ \frac{\partial \delta V}{\partial y'} \\ \frac{\partial \delta V}{\partial z'} \end{pmatrix} = \begin{pmatrix} \frac{\partial \delta V}{\partial (r,\varphi,\lambda)} \frac{\partial (r,\varphi,\lambda)}{\partial x'} \\ \frac{\partial \delta V}{\partial (r,\varphi,\lambda)} \frac{\partial (r,\varphi,\lambda)}{\partial y'} \\ \frac{\partial \delta V}{\partial (r,\varphi,\lambda)} \frac{\partial (r,\varphi,\lambda)}{\partial z'} \end{pmatrix} = \left( \frac{\partial \delta V}{\partial (r,\varphi,\lambda)} \frac{\partial (r,\varphi,\lambda)}{\partial (x',y',z')} \right)^{\mathrm{T}}, \quad (11.54)$$

where

$$\frac{\partial \delta V}{\partial r} = \sum_{j=1}^{2} \mu_j \sum_{n=2}^{N} -k_n \frac{(n+1)a_e^{2n+1}}{r^{n+2}r_j^{n+1}} \left[ \begin{array}{l} P_n(\sin \varphi)P_n(\sin \delta_j) \\ + 2 \sum_{k=1}^{n} \frac{(n-k)!}{(n+k)!} P_{nk}(\sin \varphi)P_{nk}(\sin \delta_j) \cos kh_j \end{array} \right],$$

$$\frac{\partial \delta V}{\partial \varphi} = \sum_{j=1}^{2} \mu_j \sum_{n=2}^{N} k_n \frac{a_e^{2n+1}}{r^{n+1}r_j^{n+1}} \left[ \begin{array}{l} \frac{n}{\cos \varphi}(P_{n-1}(\sin \varphi) - \sin \varphi P_n(\sin \varphi))P_n(\sin \delta_j) \\ + 2 \sum_{k=1}^{n} \frac{(n-k)!}{(n+k)!} (P_{n(k+1)}(\sin \varphi) - k \tan \varphi P_{nk}(\sin \varphi)) \\ \times P_{nk}(\sin \delta_j) \cos kh_j \end{array} \right]$$

and

$$\frac{\partial \delta V}{\partial \lambda} = \sum_{j=1}^{2} \mu_j \sum_{n=2}^{N} k_n \frac{a_e^{2n+1}}{r^{n+1}r_j^{n+1}} \left[ 2 \sum_{k=1}^{n} \frac{(n-k)!}{(n+k)!} k P_{nk}(\sin \varphi) P_{nk}(\sin \delta_j) \sin kh_j \right].$$
(11.55)

Other partial derivatives in Eq. 11.54 were given in Sect. 11.2.1. The transformation of the force vector from the ECEF to the ECSF coordinate system can be done by Eq. 11.45.

As discussed in Sect. 5.4, the ocean tidal potential generated by tide element $\sigma H \mathrm{d}s$ can be written as

$$\frac{G\sigma H \mathrm{d}s}{r'} \quad \text{or} \quad G\sigma H \mathrm{d}s \sum_{n=0}^{\infty} \frac{a_e^n}{r^{n+1}} P_n(\cos z), \quad (11.56)$$

where $H$ is the ocean tide height of the area d$s$, $G$ is the gravitational constant, $\sigma$ is the water density, $r'$ is the distance between the satellite and the water element d$s$, $r$ is the geocentric distance of the satellite, $z$ is the zenith distance of the d$s$, and $a_e$ is the radius of the earth. Using the spherical triangle

$$\cos z = \sin \varphi \sin \varphi_s + \cos \varphi \cos \varphi_s \cos(\lambda_s - \lambda),$$

where $(\varphi_s, \lambda_s)$ is the spherical coordinate of d$s$ and $(r, \varphi, \lambda)$ is the spherical coordinate of satellite in the ECEF system, Eq. 11.56 turns out to be (denoted by $Q$)

$$Q = G\sigma H \mathrm{d}s \sum_{n=0}^{\infty} \frac{a_e^n}{r^{n+1}} \left[ \begin{array}{l} P_n(\sin\varphi)P_n(\sin\varphi_s) + (2 - \delta_{0n}) \\ \times \sum_{k=0}^{n} \frac{(n-k)!}{(n+k)!} P_{nk}(\sin\varphi)P_{nk}(\sin\varphi_s) \cos k(\lambda_s - \lambda) \end{array} \right].$$

(11.57)

The direct ocean tide potential is then the integration of $Q/\mathrm{d}s$ over the ocean (denotes by $O$), including the potential of the deformation of the ocean loading. The ocean tide potential is then

$$\delta V_1 = \oiint_O G\sigma H \sum_{n=0}^{\infty} (1 + k'_n) \frac{a_e^n}{r^{n+1}} \left[ \begin{array}{l} P_n(\sin\varphi)P_n(\sin\varphi_s) + (2 - \delta_{0n}) \\ \times \sum_{k=0}^{n} \frac{(n-k)!}{(n+k)!} P_{nk}(\sin\varphi)P_{nk}(\sin\varphi_s) \cos k(\lambda_s - \lambda) \end{array} \right] \mathrm{d}s,$$

(11.58)

where $k'_n$ is the ocean loading Love number. Equation 11.58 does not include the potential changing due to the loading deformation over the continents, which may give a non-negligible contribution to the orbit motion of the satellite (cf. Knudsen et al. 2000). Recalling the discussion in Sect. 5.4, the loading deformation generated by the ocean tide can be represented as

$$u_r(\varphi, \lambda) = \oiint_{ocean} \sigma H u(z) \mathrm{d}s \quad \text{and}$$

$$u(z) = \frac{a_e h'_\infty}{2M \sin(z/2)} + \frac{a_e}{M} \sum_{n=0}^{N} (h'_n - h'_\infty) P_n(\cos z),$$

(11.59)

where $a_e$ is the radius of the earth, $M$ is the mass of the earth, $z$ is the geocentric zenith distance of the loading point (related to the computing point, see Fig. 5.11), $P_n(\cos z)$ is the Legendre function, $u(z)$ is the radial loading displacement Green's function, $h'_n$ is the loading Love number of order $n$, and $u_r$ is the radial loading deformation. Substituting $u_r$ for $H$ in Eq. 11.57 and integrating $Q/\mathrm{d}s$ over the continents (denoted by $C$), the potential of the loading deformation is then

$$\delta V_2 = \oiint_C G\sigma_e u_r \sum_{n=0}^{\infty} \frac{a_e^n}{r^{n+1}} \left[ \begin{array}{l} P_n(\sin\varphi)P_n(\sin\varphi_s) + (2-\delta_{0n}) \\ \times \sum_{k=0}^{n} \frac{(n-k)!}{(n+k)!} P_{nk}(\sin\varphi)P_{nk}(\sin\varphi_s)\cos k(\lambda_s - \lambda) \end{array} \right] ds,$$
(11.60)

where $\sigma_e$ is the density of the mass $u_r\, ds$ on the earth's surface. The total ocean tide potential disturbance is the summation of Eqs. 11.58 and 11.60. Similar to above, the disturbing force can be derived and transformed to the ECSF system. There are

$$\vec{f}_{\text{ECEF}} = \begin{pmatrix} \dfrac{\partial(\delta V_1 + \delta V_2)}{\partial x'} \\ \dfrac{\partial(\delta V_1 + \delta V_2)}{\partial y'} \\ \dfrac{\partial(\delta V_1 + \delta V_2)}{\partial z'} \end{pmatrix} = \left( \dfrac{\partial(\delta V_1 + \delta V_2)}{\partial(r,\varphi,\lambda)} \dfrac{\partial(r,\varphi,\lambda)}{\partial(x',y',z')} \right)^{\text{T}},$$
(11.61)

where

$$\frac{\partial \delta V_1}{\partial r} = \oiint_O G\sigma H \sum_{n=0}^{\infty} (1+k'_n) \frac{-(n+1)a_e^n}{r^{n+2}} \left[ \begin{array}{l} P_n(\sin\varphi)P_n(\sin\varphi_s) + (2-\delta_{0n}) \\ \times \sum_{k=0}^{n} \frac{(n-k)!}{(n+k)!} P_{nk}(\sin\varphi)P_{nk}(\sin\varphi_s)\cos k(\lambda_s - \lambda) \end{array} \right] ds,$$

$$\frac{\partial \delta V_1}{\partial \varphi} = \oiint_O G\sigma H \sum_{n=0}^{\infty} (1+k'_n) \frac{a_e^n}{r^{n+1}} \left[ \begin{array}{l} \frac{dP_n(\sin\varphi)}{d\varphi} P_n(\sin\varphi_s) + (2-\delta_{0n}) \\ \times \sum_{k=0}^{n} \frac{(n-k)!}{(n+k)!} \frac{dP_{nk}(\sin\varphi)}{d\varphi} P_{nk}(\sin\varphi_s)\cos k(\lambda_s - \lambda) \end{array} \right] ds,$$

$$\frac{\partial \delta V_1}{\partial \lambda} = \oiint_O G\sigma H \sum_{n=0}^{\infty} (1+k'_n) \frac{a_e^n}{r^{n+1}} \left[ \begin{array}{l} (2-\delta_{0n}) \\ \times \sum_{k=0}^{n} \frac{(n-k)!}{(n+k)!} P_{nk}(\sin\varphi)P_{nk}(\sin\varphi_s) k \sin k(\lambda_s - \lambda) \end{array} \right] ds,$$

$$\frac{\partial \delta V_2}{\partial r} = \oiint_C G\sigma_e u_s \sum_{n=0}^{\infty} \frac{-(n+1)a_e^n}{r^{n+2}} \left[ \begin{array}{l} P_n(\sin\varphi)P_n(\sin\varphi_s) + (2-\delta_{0n}) \\ \times \sum_{k=0}^{n} \frac{(n-k)!}{(n+k)!} P_{nk}(\sin\varphi)P_{nk}(\sin\varphi_s)\cos k(\lambda_s - \lambda) \end{array} \right] ds,$$

$$\frac{\partial \delta V_2}{\partial \varphi} = \oiint_C G\sigma_e u_s \sum_{n=0}^{\infty} \frac{a_e^n}{r^{n+1}} \left[ \begin{array}{l} \frac{dP_n(\sin\varphi)}{d\varphi} P_n(\sin\varphi_s) + (2-\delta_{0n}) \\ \times \sum_{k=0}^{n} \frac{(n-k)!}{(n+k)!} \frac{dP_{nk}(\sin\varphi)}{d\varphi} P_{nk}(\sin\varphi_s)\cos k(\lambda_s - \lambda) \end{array} \right] ds$$

and

$$\frac{\partial \delta V_2}{\partial \lambda} = \oiint_C G\sigma_e u_s \sum_{n=0}^{\infty} \frac{a_e^n}{r^{n+1}} \left[ \begin{array}{l} (2-\delta_{0n}) \\ \times \sum_{k=0}^{n} \frac{(n-k)!}{(n+k)!} P_{nk}(\sin\varphi)P_{nk}(\sin\varphi_s) k \sin k(\lambda_s - \lambda) \end{array} \right] ds.$$
(11.62)

## 11.2.4
## Solar Radiation Pressure

Solar radiation pressure is force acting on the satellite's surface caused by the sunlight. The radiation force can be represented as (cf., e.g., Seeber 1993)

$$\vec{f}_{solar} = m\gamma P_s C_r r_{sun}^2 \frac{S}{m} \frac{\vec{r} - \vec{r}_{sun}}{|\vec{r} - \vec{r}_{sun}|^3}, \tag{11.63}$$

where $\gamma$ is the shadow factor, $P_s$ is the luminosity of the Sun, $C_r$ is the surface reflectivity, $r_{sun}$ is the geocentric distance of the sun, $(S/m)$ is the surface to mass ratio of the satellite, and $\vec{r}$ and $\vec{r}_{sum}$ are the geocentric vector of the satellite and the sun. Usually, $P_s$ has the value of $4.5605 \times 10^{-6}$ N/m, $C_r$ has values from 1 to 2, 1 is for the complete absorption of the sunlight, and for aluminium, $C_r = 1.95$.

The shadow factor is defined as

$$\gamma = 1 - \frac{A_{ss}}{A_s}, \tag{11.64}$$

where $A_s$ is the sight surface of the sun viewed from the satellite, and $A_{ss}$ is the shadowed sight surface of the sun. The sunlight may be shadowed by the earth and the moon. For convenience, we will discuss both parameters that are only in the satellite-earth-sun system (cf. Fig. 11.2). It is obvious that the half sight angles of the earth and the moon as well as the sun viewed from the satellite are

$$\begin{aligned}\alpha_e &= \sin^{-1}\left(\frac{a_e}{|\vec{r}|}\right) \\ \alpha_m &= \sin^{-1}\left(\frac{a_m}{|\vec{r}_m - \vec{r}|}\right) \\ \alpha_s &= \sin^{-1}\left(\frac{a_s}{|\vec{r}_s - \vec{r}|}\right),\end{aligned} \tag{11.65}$$

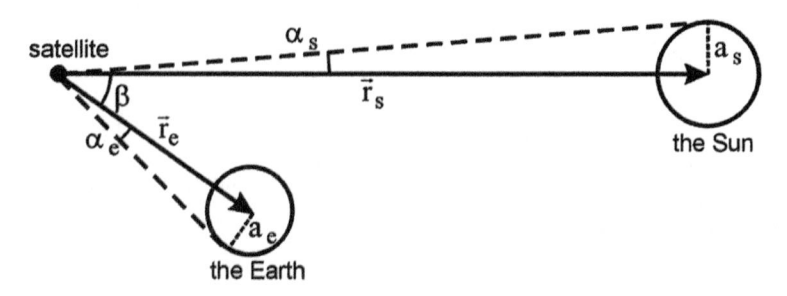

**Fig. 11.2** Satellite-Earth-Sun system

where $a_e$, $a_s$ and $a_m$ are semi-major radii of the earth, sun, and moon, respectively; $a_m = 0.272493\ a_e$, and $a_s = 959.63\ \pi/(3600 \times 180)$ (AU). For the GPS satellite, $\alpha_s < 0.3°$, $\alpha_e \approx 16.5°$ and $\alpha_m \approx \alpha_s \pm 0.03°$. Furthermore, $A_s = \alpha_s^2 \pi$ and $A_m = \alpha_m^2 \pi$. The angles between the centre of the earth and the sun, as well as the centre of the moon and the sun are

$$\beta_{es} = \cos^{-1}\left(\frac{-\vec{r}\cdot(\vec{r}_s-\vec{r})}{r|\vec{r}_s-\vec{r}|}\right)$$
$$\beta_{ms} = \cos^{-1}\left(\frac{(\vec{r}_m-\vec{r})\cdot(\vec{r}_s-\vec{r})}{|\vec{r}_m-\vec{r}|\cdot|\vec{r}_s-\vec{r}|}\right), \qquad (11.66)$$

where the vectors with indices s and m are the geocentric vectors of the sun and moon, respectively. The vector without an index is the geocentric vector of the satellite, and $r = |\vec{r}|$. If $\beta_{es} \geq \alpha_e + \alpha_s$, then the satellite is not in the shadow of the earth (i.e. $A_{ss} = 0$). If $\beta_{es} \geq \alpha_e - \alpha_s$, then the sun is not in view of the satellite (i.e. $A_{ss} = A_s$). If $\alpha_e - \alpha_s < \beta_{es} < \alpha_e + \alpha_s$, then the sunlight is partly shadowed by the earth. The formula of the shadowed surface can be derived as follows (cf. Fig. 11.3). The two circles with radius $\alpha_e$ and $\alpha_s$ cut each other at point $p$ and $q$, line $\bar{p}\,\bar{q}$ is called a chord (denoted by 2a), the chord-related central angle at origin $o_s$ is denoted by $\phi_1$, the surface area between the chord and the arc of the circle $\alpha_s$ on the right side of the chord is denoted by $A_1$. Line $\bar{p}\,\bar{q}$ cuts $\bar{O}_s\,\bar{O}_e$ at point $g$, while $\bar{O}_s\,\bar{g}$ and $\bar{g}\,\bar{O}_e$ are denoted by $b$ and $b_1$. Then one has

$$a^2 = \alpha_s^2 - b^2, \quad b_1 = \frac{\alpha_e^2 + \beta_{es}^2 - \alpha_s^2}{2\beta_{es}}$$

$$b = \begin{cases} \beta_{es} - b_1 & \text{if } b_1 \leq \alpha_e \\ b_1 - \beta_{es} & \text{if } b_1 > \alpha_e \end{cases}$$

$$\phi_1 = \begin{cases} 2\cos^{-1}\left(\frac{b}{\alpha_s}\right) & \text{if } b_1 \leq \alpha_e \\ 2\pi - 2\cos^{-1}\left(\frac{b}{\alpha_s}\right) & \text{if } b_1 > \alpha_e \end{cases} \qquad (11.67)$$

$$A_1 = \begin{cases} \frac{1}{2}\phi_1\alpha_s^2 - ab & \text{if } b_1 \leq \alpha_e \\ \frac{1}{2}\phi_1\alpha_s^2 + ab & \text{if } b_1 > \alpha_e \end{cases}.$$

Similarly, the chord-related central angle at origin $o_e$ is denoted by $\phi_2$, while the surface area between the chord and the arc of the circle $\alpha_e$ on the left side of chord is denoted by $A_2$. Then one has

$$\phi_2 = 2\cos^{-1}\left(\frac{b_1}{\alpha_e}\right), \quad A_2 = \frac{1}{2}\phi_2\alpha_e^2 - ab_1 \qquad (11.68)$$

**Fig. 11.3** Shadowed surface area

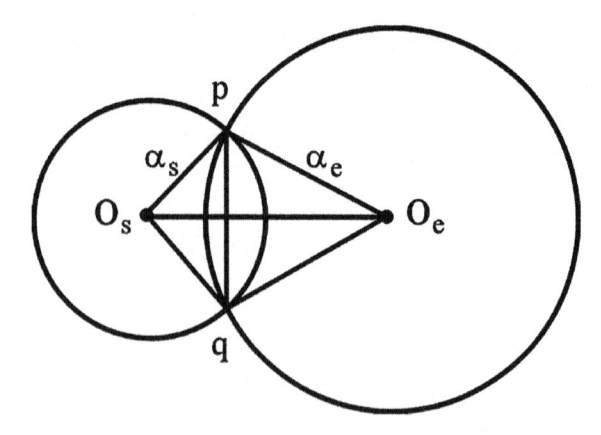

and

$$\gamma = 1 - \frac{A_1 + A_2}{\alpha_s^2 \pi}. \tag{11.69}$$

A similar discussion can be made for the moon. If $\beta_{ms} \geq \alpha_m + \alpha_s$, then the satellite is not in the shadow of the moon, i.e. $A_{ss} = 0$. If $\beta_{ms} \geq \alpha_m - \alpha_s$, then the full shadow has occurred, i.e. $A_{ss} = \min(A_s, A_m)$. If $|\alpha_m - \alpha_s| < \beta_{ms} < \alpha_m + \alpha_s$, then the sunlight is partially shadowed by the moon. The formula of the shadowed surface can be similarly derived by changing the index $e$ to $m$ in Eqs. 11.67 and 11.68. Because of the small sight angle of the moon viewed from the satellite, the shadowed time will be very short if it happens. By GPS satellite dynamic orbit determination (e.g., in IGS orbit determination), only the data that have the $\gamma$ value of 0 or 1 are used.

Because of the complex shape of the satellite, the use of constant reflectivity and homogenous luminosity of the sun, as well as the existence of indirect solar radiation (reflected from the earth's surface), the model of Eq. 11.63 discussed above is not accurate enough for precise purposes and will be used as a first-order approximation. A further model for the adjustment to fit solar radiation effects is needed.

The force vector is pointed from the sun to the satellite. The satellite fixed coordinate system is introduced in Sect. 5.8 (cf. Sect. 5.8 for details). The solar radiation force vector in the ECSF system is then

$$\begin{aligned}\vec{f}_{\text{solar}} &= m\gamma P_s C_r \frac{S}{m} \frac{r_{\text{sun}}^2}{|\vec{r} - \vec{r}_{\text{sun}}|^2} \vec{n}_{\text{sun}} \\ &= m\gamma P_s C_r \frac{S}{m} \frac{r_{\text{sun}}^2}{|\vec{r} - \vec{r}_{\text{sun}}|^2} (\sin\beta \cdot \vec{e}_x + \cos\beta \cdot \vec{e}_z),\end{aligned} \tag{11.70}$$

where

$$\vec{e}_z = -\frac{\vec{r}}{|\vec{r}|}, \quad \vec{e}_y = \frac{\vec{e}_z \times \vec{n}_{\text{sun}}}{|\vec{e}_z \times \vec{n}_{\text{sun}}|}, \quad \vec{e}_x = \vec{e}_y \times \vec{e}_z \quad \text{and} \quad \vec{n}_{\text{sun}} = \frac{\vec{r} - \vec{r}_{\text{sun}}}{|\vec{r} - \vec{r}_{\text{sun}}|}. \quad (11.71)$$

Further formulas of Eq. 11.71 can be found in Sect. 5.8. Taking the remaining error of the radiation pressure into account, the solar radiation force model can be represented as (cf. Fliegel et al. 1992; Beutler et al. 1994)

$$\vec{f}_{\text{solar-force}} = \vec{f}_{\text{solar}} + \begin{pmatrix} a_{11} & a_{12} & a_{13} \\ a_{21} & a_{22} & a_{23} \\ a_{31} & a_{32} & a_{33} \end{pmatrix} \begin{pmatrix} 1 \\ \cos u \\ \sin u \end{pmatrix}. \quad (11.72)$$

That is, nine parameters are used to model the solar radiation force error for every satellite.

An alternative adjustment model of solar radiation is given by introducing a so-called disturbance coordinate system and will be outlined below (cf. Xu 2004).

*Disturbance Coordinate System and Radiation Error Model*

The solar radiation force vector is pointed from the sun to the satellite. If the shadow factor is computed exactly, the luminosity of the sun is a constant, and the surface reflectivity of the satellite is a constant, then the length of the solar force vector can be considered as a constant, because

$$\frac{r_{\text{sun}}^2}{(r_{\text{sun}} + r)^2} \leq \frac{r_{\text{sun}}^2}{|\vec{r} - \vec{r}_{\text{sun}}|^2} \leq \frac{r_{\text{sun}}^2}{(r_{\text{sun}} - r)^2}, \quad (11.73)$$

and

$$\frac{r_{\text{sun}}^2}{(r_{\text{sun}} \pm r)^2} = \left(\frac{r_{\text{sun}}}{r_{\text{sun}} \pm r}\right)^2 \approx \left(1 \mp \frac{r}{r_{\text{sun}}} \pm \ldots\right)^2 \approx 1 \mp \frac{2r}{r_{\text{sun}}} \approx 1 \mp 3 \times 10^{-5}.$$

Any bias error in $P_s$, $C_r$ and $(S/m)$ may cause a model error of $\alpha \vec{f}_{\text{solar}}$, where $\alpha$ is a parameter. So the $\alpha \vec{f}_{\text{solar}}$ can be considered a main error model of the solar radiation. Because the ratio of the geocentric distances of the satellite and the sun is so small, the direction and distance changes of the sun-satellite vector are negligible. With the motion of the sun, the solar radiation force vector changes its direction with the time in the ECSF (Earth-Centred-Space-Fixed) coordinate system ca. 1° per day. Such an effect can only be considered a small drift, not a periodical change for the orbit determination. To model such an effect in the ECSF system, one needs three bias parameters in three coordinate axes and three drift terms instead of a few periodical parameters. It is obvious that to model this effect in the direction of $\vec{n}$, just one parameter $\alpha$ is needed. Therefore, it is very advantageous to define a so-called disturbance coordinate system as follows: the origin is the geo-centre, and the three axes are defined by $\vec{r}$ (radial vector of the satellite), $\vec{n}$ (the sun-satellite identity vector), and $\vec{p}$ (the atmospheric drag identity vector). These three axes are always in the main disturbance directions of the indirect solar radiation (reflected from the earth's surface), direct solar radiation and atmospheric drag, respectively. This coordinate system is not a Cartesian one and the axes are not orthogonal to each other. The parameters in individual axes are mainly used to model the related disturbance effects, and meanwhile to absorb the remained error of other un-modelled effects.

In the so-called disturbance coordinate system, the solar radiation pressure error model can be represented alternatively by (cf. Xu 2004)

$$\alpha \vec{f}_{\text{solar}} = \begin{pmatrix} a_1 & b_1 \\ a_2 & b_2 \\ a_3 & b_3 \end{pmatrix} \begin{pmatrix} 1 \\ t \end{pmatrix}, \tag{11.74}$$

where $b$-terms are very small. Equation 11.74 can be called Xu's adjustment model of solar radiation.

## 11.2.5
## Atmospheric Drag

Atmospheric drag is the disturbance force acting on the satellite's surface caused by the air. Air drag force can be represented as (cf., e.g., Seeber 1993; Liu and Zhao 1979)

$$\vec{f}_{\text{drag}} = -m \frac{1}{2} \left( \frac{C_d S}{m} \right) \sigma \left| \dot{\vec{r}} - \dot{\vec{r}}_{\text{air}} \right| \left( \dot{\vec{r}} - \dot{\vec{r}}_{\text{air}} \right), \tag{11.75}$$

where $S$ is the cross section (or effective area) of the satellite, $C_d$ is the drag factor, $m$ is the mass of the satellite, $\dot{\vec{r}}$ and $\dot{\vec{r}}_{\text{air}}$ are the geocentric velocity vectors of the satellite and the atmosphere, and $\sigma$ is the density of the atmosphere. Usually, $S$ has a value of one-fourth of the outer surface area of the satellite, and $C_d$ has labour values of $2.2 \pm 0.2$. The velocity vector of the atmosphere can be modelled by

$$\dot{\vec{r}}_{\text{air}} = k\vec{\omega} \times \vec{r} = k\omega \begin{pmatrix} -y \\ x \\ 0 \end{pmatrix}, \tag{11.76}$$

where $\vec{\omega}$ is the angular velocity vector of the earth's rotation, and $\omega = |\vec{\omega}|$, $k$ is the atmospheric rotation factor. For the lower layer of the atmosphere, $k = 1$, i.e. the lower layer of the atmosphere is considered to rotate with the earth. For the higher layer, $k = 1.2$, because the higher ionosphere is accelerated by the earth's magnetic field.

Gravity balanced atmospheric density model has the exponential form of (cf. Liu and Zhao 1979)

$$\sigma = \sigma_0(1+q)\exp\left(-\frac{r-\rho}{H}\right), \qquad (11.77)$$

where $\sigma_0$ is the atmospheric density at the reference point $\rho$, $q$ is the daily change factor of the density, $r$ is the geocentric distance of the satellite, and $H$ is the density-height scale factor. For the spherical and rotating ellipsoidal layer atmospheric models, one has

$$\rho = a_e + h_i \qquad (11.78)$$

and

$$\rho = (a_e + h_i)\sqrt{1-e^2}\sqrt{\frac{1+\tan^2\varphi}{1+\tan^2\varphi - e^2}}, \qquad (11.79)$$

respectively. Where $a_e$ is the semi-major radius of the earth, $h_i (i = 1, 2, \ldots)$ is a set of numbers, $\varphi$ is the geocentric latitude of the satellite, and $e$ is the eccentricity of the ellipsoid. Equations 11.78 and 11.79 are sphere with radius $a_e + h_i$ and rotating ellipsoid with semi-major axis $a_e + h_i$. Equation 11.79 can be derived from the relation of $\tan \varphi$ and the ellipsoid equation

$$z^2 = (x^2 + y^2)\tan^2\varphi$$

$$x^2 + y^2 + z^2\frac{1}{1-e^2} = (a_e + h_i)^2.$$

A reference of atmospheric densities can be read from Table 11.1, which is given by Cappellari (1976) (cf. Seeber 1993).

The density-height scale $H$ between every two layers can then be computed from the above values. It is notable that the air density may change its value up to a factor of 10 due to the radiation of the sun. The density of the atmosphere at a defined

**Table 11.1** Reference of atmospheric densities

| $h_i$ (in km) | $\sigma_0(i)$ (in g/km$^3$) | $h_i$ (in km) | $\sigma_0(i)$ (in g/km$^3$) |
|---|---|---|---|
| 100 | 497400 | 600 | 0.08–0.64 |
| 200 | 255–316 | 700 | 0.02–0.22 |
| 300 | 17–35 | 800 | 0.07–0.01 |
| 400 | 2.2–7.5 | 900 | 0.003–0.04 |
| 500 | 0.4–2.0 | 1000 | 0.001–0.02 |

point reaches its maximum value at 14 h local time and its minimum at 3.5 h. The most significant period of change is the daily change, and is represented by the daily change factor as

$$q = \frac{f-1}{f+1} \cos \psi, \tag{11.80}$$

where $f$ is the ratio of the maximum density and the minimum density, and $\psi$ is the angle between the satellite vector $\vec{r}$ and the daily maximum density direction $\vec{r}_{(m)}$. The $f$ may have the value of 3 and

$$\cos \psi = \frac{\vec{r} \cdot \vec{r}_m}{|\vec{r}| \cdot |\vec{r}_m|}, \tag{11.81}$$

where

$$\begin{pmatrix} x \\ y \\ z \end{pmatrix}_{sun} = \begin{pmatrix} r \cos \delta \cos \alpha \\ r \cos \delta \sin \alpha \\ r \sin \delta \end{pmatrix}, \quad \begin{pmatrix} r = \sqrt{x^2 + y^2 + z^2} \\ \delta = \tan^{-1} \frac{z}{\sqrt{x^2 + y^2}} \\ \alpha = \tan^{-1} \frac{y}{x} \end{pmatrix}$$

$$\vec{r}_m = \begin{pmatrix} x \\ y \\ z \end{pmatrix}_m = \begin{pmatrix} r \cos \delta \cos(\alpha + \pi/6) \\ r \cos \delta \sin(\alpha + \pi/6) \\ r \sin \delta \end{pmatrix}, \tag{11.82}$$

where $(\alpha, \delta)$ are the coordinates (right ascension, latitude) of the sun in the ECSF coordinate system.

Taking the remaining error of the atmospheric drag into account, the air drag force model can be represented as

$$\vec{f}_{air-drag} = \vec{f}_{drag} + (1+q) \Delta \vec{f}_{drag}. \tag{11.83}$$

Where the force error vector is denoted by $\Delta \vec{f}_{drag}$ and the time variation part of atmospheric density is considered in parameter $q$.

*Error Model in Disturbance Coordinate System*

In the atmospheric drag model Eq. 11.75, the velocity vector of the atmosphere is always perpendicular to the z-axis of the ECSF coordinates and the satellite velocity vector is always in the tangential direction of the orbit. The variation of the term $|\dot{\vec{r}} - \vec{r}_{air}|$ (denoted by $g$) is dominated by the direction changes of the velocity vectors of the satellite and the atmosphere. Any bias error in $S$ (effective area of the satellite), $C_d$ (drag factor), and $\sigma$ (density of the atmosphere) may cause a model error of $\mu \vec{f}_{drag}$, where $\mu$ is a parameter. So the $\mu \vec{f}_{drag}$ can be considered a main error model of the un-modelled atmospheric drag. To simplify our discussion, we consider that the velocities of the satellite and atmosphere are constants, and call the satellite positions with max(z) and -max(z) the highest and lowest points, respectively. With the satellite at the lowest point, the two velocity vectors are in the same direction, and therefore the g reaches the minimum. At the ascending node, the two vectors have the maximum angle of inclination $i$ and the $g$ reaches the maximum. Then $g$ reaches the minimum again at the highest point and reaches the maximum again at the descending node, and at the end reaches the minimum at the lowest point. It is obvious that, in addition to the constant part, $g$ has a dominant periodical component of $\cos 2f$ and $\sin 2f$, where $f$ is the true anomaly of the satellite.

In the so-called disturbance coordinate system, the atmospheric drag error model can be represented alternatively by (cf. Xu 2004)

$$\mu \vec{f}_{\text{drag}} = [a + b\varphi(2\omega)\cos(2f) + c\varphi(3\omega)\cos(3f) + d\varphi(\omega)\cos f]\vec{p}, \quad (11.84)$$

where

$$\varphi(k\omega) = \begin{cases} \sin k\omega & \text{if } \cos k\omega = 0 \\ \frac{1}{\cos k\omega} & \text{if } \cos k\omega \neq 0 \end{cases}, \quad k = 1, 2, 3 \quad (11.85)$$

where $\omega$ is the angle of perigee and $f$ is the true anomaly of the satellite, and $a$, $b$, $c$, and $d$ are model parameters to be determined. According to the simulation, the $a$-term and $b$-term are the most significant terms. The amount of $d$ is just about 1 % of the amount of $c$, and the amount of $c$ is about 1 % of that of $b$. Equation 11.84 can be called Xu's adjustment model for atmospheric drag error.

## 11.2.6
## Additional Perturbations

As mentioned above, the disturbed equation of motion of the satellite is valid only in an inertial coordinate system, or ECSF system. Therefore, the state vector and force vectors as well as the disturbing potential function have also to be represented in the ECSF system. As seen above, for some reason, the state vector and the force vectors as well as the disturbing potential function $R$ are sometimes given in the ECEF system and then transformed to the ECSF system by (cf. Sect. 11.2.4)

$$\begin{aligned} \vec{X}_{\text{ECSF}} &= R_t \cdot \vec{X}_{\text{ECEF}} \\ \vec{f}_{\text{ECSF}} &= R_t \cdot \vec{f}_{\text{ECEF}} \\ R_{\text{ECSF}} &= R(R_t^{-1} X_{\text{ECSF}}) \quad \text{for} \quad R(X_{\text{ECEF}}), \end{aligned} \quad (11.86)$$

where $R_t$ is the transformation matrix in general. Variable transformation is further denoted by $X_{\text{ECSF}} = R_t X_{\text{ECEF}}$. We have also seen that sometimes the state vectors (of the satellite, the sun, the moon) in the ECSF system must be transformed to the ECEF system for use, and then the result vectors will be transformed back to the ECSF system again. However, due to the complication of transformation $R_t^1$, quite often a simplified $R_s^1$ is used. (cf. in later discussions, for example to represent the disturbing potential function using Keplerian elements, only the earth rotation is considered). Thus,

$$R_{\text{ECSF}} = \{R(R_t^{-1}X_{\text{ECSF}}) - R(R_s^{-1}X_{\text{ECSF}})\} + R(R_s^{-1}X_{\text{ECSF}}), \tag{11.87}$$

where the first term on the right side is the correction because of the approximation using the second term. The transformations of Eqs. 11.86 and 11.87 are exact operations, and their differentiation with respect to time $t$ and the partial derivatives with respect to variable $X_{\text{ECSF}}$ are then

$$\frac{d\vec{X}_{\text{ECSF}}}{dt} = \frac{dR_t}{dt}\vec{X}_{\text{ECEF}} + R_t \frac{d\vec{X}_{\text{ECEF}}}{dt}$$

$$\frac{d\vec{f}_{\text{ECSF}}}{dt} = \frac{dR_t}{dt}\vec{f}_{\text{ECEF}} + R_t \frac{d\vec{f}_{\text{ECEF}}}{dt} \tag{11.88}$$

$$\frac{\partial R_{\text{ECSF}}}{\partial X_{\text{ECSF}}} = \frac{\partial \left[R(R_t^{-1}X_{\text{ECSF}}) - R(R_s^{-1}X_{\text{ECSF}})\right]}{\partial X_{\text{ECSF}}} + \frac{\partial R(R_s^{-1}X_{\text{ECSF}})}{\partial X_{\text{ECSF}}}.$$

That is, the time differentiations of the state vector and force vectors cannot be transformed directly like in Eq. 11.86. In other words, if the state vector and force vectors are not directly given in the ECSF system, they are not allowed to be differentiated as usual afterward. An approximated and transformed perturbing potential function will introduce an error. The first term on the right-hand side of Eq. 11.88 signifies additional perturbations, or coordinate perturbations. The order of such perturbations can be estimated by the first term on the right-hand side. If the relationship between two coordinate systems changes with time or the transformation has not been made exactly, such perturbations will occur. Recalling

$$R = R_P^{-1} R_N^{-1} R_S^{-1} R_M^{-1}$$

and their definitions (cf. Chap. 2), one has

$$\begin{aligned}\frac{dR}{dt} &= R_P^{-1} R_N^{-1} R_S^{-1} \frac{dR_M^{-1}}{dt} + R_P^{-1} R_N^{-1} \frac{dR_S^{-1}}{dt} R_M^{-1} \\ &+ R_P^{-1} \frac{dR_N^{-1}}{dt} R_S^{-1} R_M^{-1} + \frac{dR_P^{-1}}{dt} R_N^{-1} R_S^{-1} R_M^{-1},\end{aligned} \tag{11.89}$$

where

$$\frac{dR_M^{-1}}{dt} = \begin{pmatrix} 0 & 0 & -\dot{x}_p \\ 0 & 0 & \dot{y}_p \\ \dot{x}_p & -\dot{y}_p & 0 \end{pmatrix}, \quad \frac{dR_S^{-1}}{dt} = \frac{dR_3(\text{GAST})}{dt}$$

$$\frac{dR_N^{-1}}{dt} = \frac{dR_1(-\varepsilon)}{dt} R_3(\Delta\psi) R_1(\varepsilon + \Delta\varepsilon) + R_1(-\varepsilon) \frac{dR_3(\Delta\psi)}{dt} R_1(\varepsilon + \Delta\varepsilon)$$
$$+ R_1(-\varepsilon) R_3(\Delta\psi) \frac{dR_1(\varepsilon + \Delta\varepsilon)}{dt}$$

$$\frac{dR_P^{-1}}{dt} = \frac{dR_3(\zeta)}{dt} R_2(-\theta) R_3(z) + R_3(\zeta) \frac{dR_2(-\theta)}{dt} R_3(z) + R_3(\zeta) R_2(-\theta) \frac{dR_3(z)}{dt},$$

(11.90)

where all elements are defined and given in Chap. 2, $(\dot{x}_p, \dot{y}_p)$ is the polar motion rate of time, and

$$\frac{dR_1(\alpha)}{dt} = \begin{pmatrix} 0 & 0 & 0 \\ 0 & -\sin\alpha & \cos\alpha \\ 0 & -\cos\alpha & -\sin\alpha \end{pmatrix} \frac{d\alpha}{dt}$$

$$\frac{dR_2(\alpha)}{dt} = \begin{pmatrix} -\sin\alpha & 0 & -\cos\alpha \\ 0 & 0 & 0 \\ \cos\alpha & 0 & -\sin\alpha \end{pmatrix} \frac{d\alpha}{dt} \quad (11.91)$$

$$\frac{dR_3(\alpha)}{dt} = \begin{pmatrix} -\sin\alpha & \cos\alpha & 0 \\ -\cos\alpha & -\sin\alpha & 0 \\ 0 & 0 & 0 \end{pmatrix} \frac{d\alpha}{dt}.$$

Further formulas may be easily derived.

## 11.2.7
## Order Estimations of Perturbations

Perturbation forces that are scaled by the mass of the satellite are the accelerations. The accelerations caused by the forces discussed have been estimated for the GPS satellite by several authors and are summarised in Table 11.2.

If the coordinate system is used without taking precession and nutation into account, additional perturbation acceleration can reach $3 \times 10^{-10}$. Additional acceleration of gravitational potential can reach $1 \times 10^{-9}$ (cf. Liu and Zhao 1979).

**Table 11.2** Accelerations (m s$^{-2}$) caused by forces (cf. Seeber 1993; Kang 1998)

| | |
|---|---|
| Central force acceleration | 0.56 |
| Gravitational $C_2$ acceleration | $5 \times 10^{-5}$ |
| Other gravitational acceleration | $3 \times 10^{-7}$ |
| The moon's central force acceleration | $5 \times 10^{-6}$ |
| The sun's central force acceleration | $2 \times 10^{-6}$ |
| Planets' central force acceleration | $3 \times 10^{-10}$ |
| The earth's tidal acceleration | $2 \times 10^{-9}$ |
| Ocean's tidal acceleration | $5 \times 10^{-10}$ |
| Solar pressure acceleration | $1 \times 10^{-7}$ |
| Atmospheric drag acceleration (Topex) | $4 \times 10^{-10}$ |
| General relativity acceleration | $3 \times 10^{-10}$ |

## 11.2.8
### Ephemerides of the Moon, the Sun, and Planets

The ephemerides of the sun and the moon are used above for the computation of shadow functions of the sun and moon (solar radiation pressure), the tidal disturbance forces, and tidal and loading deformations (cf. Sect. 5.8). The computation of the ephemerides of the sun and the moon can be simplified by considering the orbit of the sun (indeed it is the earth!) and the moon as Keplerian motion. Consider the orbital right-handed coordinate system, the origin in the geocentre, the $xy$-plane as the orbital plane, the $x$-axis pointing to the perigee, and the $z$-axis pointing in the direction of $\vec{q} \times \dot{\vec{q}}$ where $\vec{q}$ and $\dot{\vec{q}}$ are the position and velocity vectors of the sun or the moon. The two vectors are (cf. Eqs. 3.41 and 3.42)

$$\vec{q} = \begin{pmatrix} a(\cos E - e) \\ a\sqrt{1-e^2}\sin E \\ 0 \end{pmatrix} = \begin{pmatrix} q\cos f \\ q\sin f \\ 0 \end{pmatrix}, \quad \dot{\vec{q}} = \begin{pmatrix} -\sin f \\ e + \cos f \\ 0 \end{pmatrix} \frac{na}{\sqrt{1-e^2}}, \quad (11.92)$$

where

$$q = \frac{a(1-e^2)}{1+e\cos f}. \quad (11.93)$$

The position and velocity vectors of the sun or the moon in the ECEI and ECSF coordinate systems are then (cf. Sect. 2.5 and Eq. 3.43)

$$\begin{pmatrix} \vec{p} \\ \dot{\vec{p}} \end{pmatrix} = R_3(-\Omega)R_1(-i)R_3(-\omega)\begin{pmatrix} \vec{q} \\ \dot{\vec{q}} \end{pmatrix}$$
$$\begin{pmatrix} \vec{r} \\ \dot{\vec{r}} \end{pmatrix} = R_1(-\varepsilon)\begin{pmatrix} \vec{p} \\ \dot{\vec{p}} \end{pmatrix}, \quad (11.94)$$

where $a$ and $i$ are the semi-major axis of the orbit and the inclination angle of the orbital plane of the moon or the sun in the ecliptic coordinate system (ECEI). $\Omega$ is the ecliptic right ascension of the ascending node, $e$ is the eccentricity of the ellipse, $\omega$ is the argument of perigee, $f$ is the true anomaly of the moon or the sun, and $\varepsilon$ is the mean obliquity (the formula is given in Sect. 2.4). Because the sun moves along the ecliptic and the ascending node is defined as the equinox, parameters $i$ and $\Omega$ are zero. True anomaly $f$, eccentric anomaly $E$ and mean anomaly $M$ are given by the Keplerian equation and the following formulas

$$E - e \sin E = M$$
$$q \cos f = a \cos E - ae \qquad (11.95)$$
$$q \sin f = b \sin E = a\sqrt{1 - e^2} \sin E.$$

For the moon, eccentricity $e_m = 0.05490$, inclination $i_m = 5.°145396$ and semi-major axis $a_m = 384401$ km. For the sun, eccentricity $e_s = 0.016709114 - 0.000042052T - 0.000000126T^2$ and semi-major axis $a_s = 1.0000002$ AU. AU signifies the astronomical units (AU = $1.49597870691 \times 10^8$ km). The fundamental arguments are given in the IERS Conventions (cf. McCarthy 1996) as follows:

$$l = 134.°96340251 + 1717915923.''2178T + 31.''8792T^2 + 0.''051635T^3 - 0.''00024470T^4$$
$$l' = 357.°52910918 + 129596581.''0481T - 0.''5532T^2 + 0.''000136T^3 - 0.''00001149T^4$$
$$F = 93.°27209062 + 1739527262.''8478T - 12.''7512T^2 - 0.''001037T^3 + 0.''00000417T^4$$
$$D = 297.°85019547 + 1602961601.''2090T - 6.''3706T^2 + 0.''006593T^3 - 0.''00003169T^4$$
$$\Omega = 125.°04455501 - 6962890.''2665T + 7.''4722T^2 + 0.''007702T^3 - 0.''00005939T^4,$$
$$(11.96)$$

where $l$ and $l'$ are the mean anomalies of the moon and the sun, respectively. $D$ is the mean elongation of the moon from the sun. $\Omega$ is the mean longitude of the ascending node of the moon. $F = L - \Omega$, $L$ is the mean longitude of the moon (or $L_{moon}$), and $T$ is the Julian centuries measured from epoch J2000.0. Formulas of Eq. 11.96 are the arguments used to compute the nutation. Mean angular velocities $n$ of the sun and moon are the coefficients of the linear terms of $l$ and $l'$ (units: s/century), respectively.

For computation of the ephemerides of the Sun, $l'$ is set as $M$ in Eq. 11.95, so that $E$ and $f$ of the sun can be computed. Using $D = L_{moon} - L_{sun} = F + \Omega - L_{sun}$, the mean longitude $L_{sun}$ can be computed. $\omega$ can be computed by relation $L_{sun} = \omega + f$.

For computation of the ephemerides of the moon, $l$ is set as $M$ in Eq. 11.95, so that $E$ and $f$ of the moon can be computed. $\omega$ can be computed using the spherical triangle formula:

$$\tan(\omega + f) = \tan F / \cos i_m, \qquad (11.97)$$

where angles $u$ ($= \omega + f$) and $F$ are in the same compartment.

Substituting the above values of the moon and the sun into Eqs. 11.92–11.94, respectively, ephemerides of the moon and the sun are obtained in the ECSF coordinate system. For more precise computation of the ephemerides of the moon, several corrections must be considered (cf. Meeus 1992; Montenbruck 1989). Equivalently, a correction d$F$ can be added to $F$, and the change of d$u$ in Eq. 11.97 can be considered d$f$ and added to $f$, where d$F$ has the form of (units: seconds)

$$\begin{aligned}\mathrm{d}F = {}& 22640\sin l + 769\sin(2l) + 36\sin(3l) - 125\sin D + 2370\sin(2D) - 668\sin l' \\ & - 412\sin(2F) + 212\sin(2D - 2l) + 4586\sin(2D - l) + 192\sin(2D + l) \\ & + 165\sin(2D - l') + 206\sin(2D - l - l') - 110\sin(l + l') + 148\sin(l - l').\end{aligned}$$

The orbits of the planets are given in the sun-centred ecliptic coordinate system by six Keplerian elements. They are the mean longitude ($L$) of the planet, the semi-major axis ($a$, units: AU) of the orbit of the planet, the eccentricity ($e$) of the orbit, the inclination ($i$) of the orbit to the ecliptic plane, the argument ($\omega$) of the perihelion, and the longitude ($\Omega$) of the ascending node. The orbital elements are expressed as a polynomial function of the instant of time $T$ (Julian centuries) for Mercury, Venus, Mars, Jupiter, Saturn, Uranus, and Neptune, as follows (see Meeus 1992, we use $\omega$ instead of $\pi$ in this book. The argument of perihelion, $\omega$ can be obtained through the relation: $\omega = \pi - \Omega$):

$$\begin{pmatrix}L\\a\\e\\i\\\omega\\\Omega\end{pmatrix}_{\text{Mercury}} = \begin{pmatrix}252.250906 & 149474.0722491 & 0.00030397 & -0.00000002\\0.38709831 & 0 & 0 & 0\\0.20563175 & 0.000020406 & -0.0000000284 & -0.0000000002\\7.0049860 & 0.0018215 & -0.00001809 & 0.000000053\\29.1252260 & 0.3702885 & 0.00012002 & -0.000000155\\48.3308930 & 1.1861890 & 0.00017587 & 0.000000211\end{pmatrix}\begin{pmatrix}1\\T\\T^2\\T^3\end{pmatrix},$$

$$\begin{pmatrix}L\\a\\e\\i\\\omega\\\Omega\end{pmatrix}_{\text{Venus}} = \begin{pmatrix}181.979801 & 58519.2130302 & 0.00031060 & 0.000000015\\0.72332982 & 0 & 0 & 0\\0.00677118 & -0.000047766 & 0.0000000975 & 0.00000000044\\3.3946620 & 0.00100370 & -0.00000088 & -0.000000007\\54.883787 & 0.50109980 & -0.00148002 & -0.000005235\\76.6799200 & 0.90111900 & 0.00040665 & -0.00000008\end{pmatrix}\begin{pmatrix}1\\T\\T^2\\T^3\end{pmatrix},$$

$$\begin{pmatrix}L\\a\\e\\i\\\omega\\\Omega\end{pmatrix}_{\text{Mars}} = \begin{pmatrix}355.4332750 & 19141.6964746 & 0.00031097 & 0.000000015\\1.523679342 & 0 & 0 & 0\\0.09340062 & 0.000090483 & -0.0000000806 & -0.00000000035\\1.8497260 & -0.0006010 & 0.00012760 & -0.000000006\\286.502141 & 1.0689408 & 0.00011910 & -0.000002007\\49.558093 & 0.7720923 & 0.00001605 & 0.000002325\end{pmatrix}\begin{pmatrix}1\\T\\T^2\\T^3\end{pmatrix},$$

$$\begin{pmatrix}L\\a\\e\\i\\\omega\\\Omega\end{pmatrix}_{\text{Jupiter}} = \begin{pmatrix}34.351484 & 3036.3027889 & 0.00022374 & 0.000000025\\5.202603191 & 0.0000001913 & 0 & 0\\0.04849485 & 0.000163244 & -0.0000004719 & -0.00000000197\\1.303270 & -0.00549660 & 0.00000465 & -0.000000004\\273.866868 & 0.5917118 & 0.00063010 & -0.000005138\\100.464441 & 1.0209550 & 0.00040117 & 0.000000569\end{pmatrix}\begin{pmatrix}1\\T\\T^2\\T^3\end{pmatrix}$$

and

$$\begin{pmatrix} L \\ a \\ e \\ i \\ \omega \\ \Omega \end{pmatrix}_{\text{Saturn}} = \begin{pmatrix} 50.0774710 & 1223.5110141 & 0.00051952 & -0.000000003 \\ 9.554909596 & -0.0000021389 & 0 & 0 \\ 0.05550862 & -0.000346818 & -0.0000006456 & 0.00000000338 \\ 2.488878 & -0.0037363 & -0.00001516 & 0.000000089 \\ 339.391263 & 1.0866715 & 0.00095824 & 0.000007279 \\ 113.665524 & 0.8770979 & -0.00012067 & -0.00000238 \end{pmatrix} \begin{pmatrix} 1 \\ T \\ T^2 \\ T^3 \end{pmatrix},$$

$$\begin{pmatrix} L \\ a \\ e \\ i \\ \omega \\ \Omega \end{pmatrix}_{\text{Uranus}} = \begin{pmatrix} 314.055005 & 429.8640561 & 0.00030434 & 0.000000026 \\ 19.218446062 & -0.0000000372 & 0.00000000098 & 0 \\ 0.04629590 & -0.000027337 & 0.0000000790 & 0.00000000025 \\ 0.773196 & 0.0007744 & 0.00003749 & -0.000000092 \\ 98.999212 & 0.9652526 & -0.00112532 & -0.000018083 \\ 74.005159 & 0.5211258 & 0.00133982 & 0.000018516 \end{pmatrix} \begin{pmatrix} 1 \\ T \\ T^2 \\ T^3 \end{pmatrix},$$

and

$$\begin{pmatrix} L \\ a \\ e \\ i \\ \omega \\ \Omega \end{pmatrix}_{\text{Neptune}} = \begin{pmatrix} 304.348655 & 219.8833092 & 0.00030926 & 0.000000018 \\ 30.110386869 & -0.0000001663 & 0.00000000069 & 0 \\ 0.00898809 & 0.000006408 & -0.0000000008 & -0.00000000005 \\ 1.769952 & -0.0093082 & -0.00000708 & 0.000000028 \\ 276.337634 & 0.3240620 & 0.00011912 & 0.000000633 \\ 131.784057 & 1.1022057 & 0.00026006 & -0.000000636 \end{pmatrix} \begin{pmatrix} 1 \\ T \\ T^2 \\ T^3 \end{pmatrix},$$

where except for the semi-major axis $a$ and eccentricity $e$, all elements have units of degrees. $F = L - \Omega$, and $f$ and $E$ can be computed using Eqs. 11.97 and 11.95. Mean angular velocities $n$ of the planets are the coefficients of the linear term of $L$ (units: degrees/century). The coordinate vector of the planet can then be computed using Eqs. 11.92–11.94. The results are in the sun-centred equatorial coordinate system. The results must be transformed to the ECSF coordinate system by a translation

$$\begin{pmatrix} \vec{r} \\ \dot{\vec{r}} \end{pmatrix}_{\text{ECSF}} = \begin{pmatrix} \vec{r} \\ \dot{\vec{r}} \end{pmatrix}_{\text{sun}} + \begin{pmatrix} \vec{r} \\ \dot{\vec{r}} \end{pmatrix}_{\text{SCES}}, \qquad (11.98)$$

where vectors with an index of sun and SCEF are geocentric position and velocity vectors of the sun and the planet in the sun-centred equatorial system.

Gravitational constants of the sun, the moon, and planets are given in Table 11.3.

**Table 11.3** Gravitational constants of the sun, the moon, and planets

| Gravitational constant of | Gravitational constant (m³s⁻²) |
|---|---|
| Sun | 1.3271240000000E + 20 |
| Moon | 4.9027993000000E + 12 |
| Earth | 3.9860044180000E + 14 |
| Mercury | 2.2032070000000E + 13 |
| Venus | 3.2485850000000E + 14 |
| Mars | 4.2828300000000E + 13 |
| Jupiter | 1.2671270000000E + 17 |
| Saturn | 3.7940610000000E + 16 |
| Uranus | 5.8894334680000E + 15 |
| Neptune | 6.8364650040000E + 15 |

## 11.3
## Analysis Solution of the $\overline{C}_{20}$ Perturbed Orbit

The geopotential term of $\overline{C}_{20}$ is a zonal term. Compared with other geopotential terms, $\overline{C}_{20}$ has a value that is at least 100 times larger. According to the order estimation discussed in Sect. 11.2.7, $\overline{C}_{20}$ term perturbation is one of the most significant disturbing factors. $\overline{C}_{20}$ disturbance is a perturbation of first order. The analysis solution of the $\overline{C}_{20}$ perturbation will give a clear insight of the orbit disturbance. The related perturbing potential is (cf. Sect. 11.2.1.1)

$$R_2 = \frac{\mu a_e^2}{r^3} \overline{C}_{20} \overline{P}_{20}(\sin \varphi)$$

or

$$R_2 = \frac{b}{r^3}(3\sin^2 \varphi - 1), \quad (11.99)$$

where

$$b = \frac{\sqrt{5} \mu a_e^2}{2} \overline{C}_{20}.$$

The variables $(r, \varphi, \lambda)$ of the geopotential disturbance function in the ECEF system are transformed into orbital elements in the ECSF system using the following relations (cf. Fig. 11.4, cf. Kaula 1966):

## 11.3 · Analysis Solution of the $\bar{C}_{20}$ Perturbed Orbit

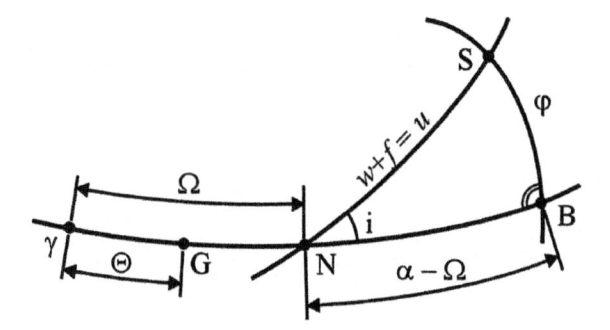

**Fig. 11.4** Orbit-equator-meridian triangle

$$\sin \varphi = \sin i \sin u$$
$$\lambda = \alpha - \Theta = \Omega - \Theta + (\alpha - \Omega)$$
$$\cos(\alpha - \Omega) = \frac{\cos u}{\cos \varphi} \tag{11.100}$$
$$\sin(\alpha - \Omega) = \frac{\sin u \cos i}{\cos \varphi}.$$

where $\alpha$ is the right ascension of the satellite, $u = \omega + f$, $\Theta$ is the Greenwich Sidereal Time, and other parameters are Keplerian elements. It is obvious that such a coordinate transformation takes only the earth's rotation into account; this will cause a coordinate perturbation (cf. Sect. 11.2.6). But such an effect can be neglected by the first-order solution. Substituting the first formula of Eq. 11.100 into Eq. 11.99 and taking the triangle formula (for reducing the order) into account, one has

$$R_2 = \frac{b}{r^3}\left[\frac{3}{2}\sin^2 i(1 - \cos 2u) - 1\right], \tag{11.101}$$

where

$$r = \frac{a(1 - e^2)}{1 + e \cos f}, \tag{11.102}$$

where $\Omega$ has not appeared in the zonal disturbance. Taking the partial derivatives of $f$ with respect to $(M, e)$ and $r$ with respect to $(a, M, e)$ into account (cf. Sect. 11.1), the derivatives of $R_2$ with respect to Keplerian elements are then

$$\frac{\partial R_2}{\partial a} = \frac{\partial R_2}{\partial r}\frac{\partial r}{\partial a} = \frac{-3}{a}R_2, \quad \frac{\partial R_2}{\partial \Omega} = 0,$$

$$\frac{\partial R_2}{\partial i} = \frac{b}{r^3}\left[\frac{3}{2}\sin 2i(1-\cos 2u)\right],$$

$$\frac{\partial R_2}{\partial \omega} = \frac{b}{r^3}\left[3\sin^2 i \sin 2u \frac{\partial u}{\partial \omega}\right] = \frac{3b}{r^3}\sin^2 i \sin 2u,$$

$$\frac{\partial R_2}{\partial e} = \frac{-3R_2}{r}\frac{\partial r}{\partial e} + \frac{b}{r^3}\left[3\sin^2 i \sin 2u \frac{\partial u}{\partial e}\right] \quad (11.103)$$

$$= \frac{3a\cos f}{r}R_2 + \frac{b}{r^3}\left[3\sin^2 i \sin 2u \frac{2+e\cos f}{1-e^2}\sin f\right] \quad \text{and}$$

$$\frac{\partial R_2}{\partial M} = \frac{-3R_2}{r}\frac{\partial r}{\partial M} + \frac{b}{r^3}\left[3\sin^2 i \sin 2u \frac{\partial u}{\partial M}\right]$$

$$= \frac{-3ae\sin f}{r\sqrt{1-e^2}}R_2 + \frac{b}{r^3}\left[3\sin^2 i \sin 2u \left(\frac{a}{r}\right)^2\sqrt{1-e^2}\right].$$

Substituting the above derivatives and $R_2$ into the equation of motion 11.103, one has

$$\frac{da}{dt} = \frac{6b\sqrt{1-e^2}}{na^4}\left\{\frac{-e}{(1-e^2)}\frac{a^4}{r^4}\sin f\left[\frac{3}{2}\sin^2 i(1-\cos 2u) - 1\right] + \frac{a^5}{r^5}\left[\sin^2 i \sin 2u\right]\right\},$$

$$\frac{de}{dt} = \frac{3b(1-e^2)^{3/2}}{na^5 e}\left\{\frac{-e}{(1-e^2)}\frac{a^4}{r^4}\sin f\left[\frac{3}{2}\sin^2 i(1-\cos 2u) - 1\right] + \frac{a^5}{r^5}\left[\sin^2 i \sin 2u\right]\right\}$$

$$- \frac{3b\sqrt{1-e^2}}{na^5 e}\frac{a^3}{r^3}\sin^2 i \sin 2u,$$

$$\frac{d\omega}{dt} = \frac{3b\sqrt{1-e^2}}{na^5 e}\left\{\frac{a^4}{r^4}\cos f\left[\frac{3}{2}\sin^2 i(1-\cos 2u) - 1\right] + \frac{a^3}{r^3}\left[\sin^2 i \sin 2u\frac{2+e\cos f}{1-e^2}\sin f\right]\right\}$$

$$- \frac{3b}{na^5\sqrt{1-e^2}}\frac{a^3}{r^3}\left[\cos^2 i(1-\cos 2u)\right],$$

$$\frac{di}{dt} = \frac{3b}{2na^5\sqrt{1-e^2}}\frac{a^3}{r^3}\sin 2i \sin 2u,$$

$$\frac{d\Omega}{dt} = \frac{3b}{na^5\sqrt{1-e^2}}\frac{a^3}{r^3}\left[\cos i(1-\cos 2u)\right] \quad \text{and}$$

$$\frac{dM}{dt} = n + \frac{6b}{na^5}\frac{a^3}{r^3}\left[\frac{3}{2}\sin^2 i(1-\cos 2u) - 1\right]$$

$$- \frac{3b(1-e^2)}{na^5 e}\left\{\begin{array}{l}\frac{a^4}{r^4}\cos f\left[\frac{3}{2}\sin^2 i(1-\cos 2u) - 1\right]\\ + \frac{a^3}{r^3}\left[\sin^2 i \sin 2u\frac{2+e\cos f}{1-e^2}\sin f\right]\end{array}\right\}.$$

(11.104)

## 11.3 · Analysis Solution of the $\overline{C}_{20}$ Perturbed Orbit

For convenience the right-hand side of the above equations will be separated into three parts:

$$\frac{d\sigma_i}{dt} = \left(\frac{d\sigma_i}{dt}\right)_0 + \left(\frac{d\sigma_i}{dt}\right)_\omega + \left(\frac{d\sigma_i}{dt}\right)_f \quad (11.105)$$

or

$$\frac{d\sigma_i}{dt} = \dot{\sigma}_{i0} + \left(\frac{d\sigma_i}{dt}\right)_\omega + \left(\frac{d\sigma_i}{dt} - \dot{\sigma}_{i0} - \dot{\sigma}_{i\omega}\right), \quad (11.106)$$

where the first term (denoted by $\dot{\sigma}_{i0}$) on the right-hand side includes all terms that are only functions of $(a, i, e)$, the second term includes all terms of $\omega$ (without $f$) (denoted by $\dot{\sigma}_{i\omega}$), and the third term includes all terms of $f$. They are denoted by the sub-index of 0, $\omega$, and $f$, respectively. Equation 11.106 is needed for later integral variable transformation. The second terms on the right-hand side of the above two equations are the same. It is notable that the $r$ is a function of $f$. The solution of the $R_2$ perturbed orbit is the integration of the above equations between initial epoch $t_0$ and any instantaneous epoch $t$. The three terms on the right side can be integrated with the integral variable of $t$, $\omega$, and $f$ respectively. The integral variable $dt$ can be changed to $df$ by

$$dt = \frac{\partial t}{\partial f} df = \frac{1}{\frac{\partial f}{\partial M}\frac{\partial M}{\partial t}} df = \left(\frac{r}{a}\right)^2 \frac{1}{\sqrt{1-e^2}} \frac{1}{n} df. \quad (11.107)$$

All terms of $\omega$ are presented in the terms of $\sin 2u$ and $\cos 2u$. Omitting the terms of $\sin 2u$ and $\cos 2u$ in Eq. 11.104, the remaining terms of $f$ are included in the following functions:

$$\left(\frac{a}{r}\right)^3, \quad \left(\frac{a}{r}\right)^4 \sin f \text{ and } \left(\frac{a}{r}\right)^4 \cos f, \quad (11.108)$$

where

$$\frac{a}{r} = \frac{1+e\cos f}{1-e^2}, \quad \left(\frac{a}{r}\right)^2 = \frac{1+0.5e^2 + 2e\cos f + 0.5e^2 \cos 2f}{(1-e^2)^2},$$

$$\left(\frac{a}{r}\right)^3 = \frac{1+1.5e^2 + (3e+0.75e^3)\cos f + 1.5e^2 \cos 2f + 0.25e^3 \cos 3f}{(1-e^2)^3},$$

$$\left(\frac{a}{r}\right)^4 = \frac{1}{(1-e^2)^4}\left[\begin{array}{l}\left(1+3e^2+\frac{3}{8}e^4\right)+(4e+3e^3)\cos f \\ +(3e^2+0.5e^4)\cos 2f + e^3 \cos 3f + \frac{1}{8}e^4 \cos 4f\end{array}\right],$$

$$\left(\frac{a}{r}\right)^4 \sin f = \frac{1}{(1-e^2)^4} \left[ \begin{array}{l} \left(1 + 1.5e^2 + \frac{1}{8}e^4\right) \sin f + (2e + e^3) \sin 2f \\ + \left(1.5e^2 + \frac{3}{16}e^4\right) \sin 3f + 0.5e^3 \sin 4f + \frac{1}{16}e^4 \sin 5f \end{array} \right],$$

$$\left(\frac{a}{r}\right)^4 \cos f = \frac{1}{(1-e^2)^4} \left[ \begin{array}{l} (2e + 1.5e^3) + \left(1 + 4.5e^2 + \frac{5}{8}e^4\right) \cos f \\ + (2e + 2e^3) \cos 2f + \left(1.5e^2 + \frac{5}{16}e^4\right) \cos 3f \\ + 0.5e^3 \cos 4f + \frac{1}{16}e^4 \cos 5f \end{array} \right],$$

(11.109)

and

$$\begin{aligned} \sin jf \sin mf &= -0.5[\cos(j+m)f - \cos(j-m)f] \\ \cos jf \cos mf &= 0.5[\cos(j+m)f + \cos(j-m)f] \\ \sin jf \cos mf &= 0.5[\sin(j+m)f + \sin(j-m)f]. \end{aligned}$$

(11.110)

Then the first term (long term perturbation) in Eq. 11.106 is

$$\begin{aligned} \left(\frac{da}{dt}\right)_0 &= \left(\frac{de}{dt}\right)_0 = \left(\frac{di}{dt}\right)_0 = 0 \\ \left(\frac{d\omega}{dt}\right)_0 &= \frac{3b}{na^5(1-e^2)^{3.5}} \left(4 \sin^2 i - 3 + \frac{15}{4} e^2 \sin^2 i - 3e^2\right) \\ \left(\frac{d\Omega}{dt}\right)_0 &= \frac{3b}{2na^5} \cos i \frac{(2+3e^2)}{(1-e^2)^{3.5}} \\ \left(\frac{dM}{dt}\right)_0 &= n + \frac{9b}{2na^5} \left(\frac{3}{2} \sin^2 i - 1\right) \frac{e^2}{(1-e^2)^3}. \end{aligned}$$

(11.111)

Due to the slow changing property of the variable $\omega$, the integral variable changing between $t$ and $\omega$ can be approximated by

$$dt = \left(\frac{d\omega}{dt}\right)_0^{-1} d\omega.$$

(11.112)

11.3 · Analysis Solution of the $\overline{C}_{20}$ Perturbed Orbit 377

The second term (long period perturbation) in Eq. 11.106 exists only in $\sin 2u$ and $\cos 2u$ related terms. All $\sin 2u$ and $\cos 2u$ terms are factorised by the following functions:

$$\left(\frac{a}{r}\right)^3, \ \left(\frac{a}{r}\right)^5, \ \left(\frac{a}{r}\right)^4 \sin f, \ \left(\frac{a}{r}\right)^4 \cos f \text{ and } \left(\frac{a}{r}\right)^3 \frac{2+e\cos f}{1-e^2} \sin f, \qquad (11.113)$$

where

$$\left(\frac{a}{r}\right)^5 = \frac{1}{(1-e^2)^5} \begin{bmatrix} \left(1+5e^2+1\frac{7}{8}e^4\right) + \left(5e+7.5e^3+\frac{5}{8}e^5\right)\cos f \\ + (5e^2+2.5e^4)\cos 2f + \left(2.5e^3+\frac{5}{16}e^5\right)\cos 3f \\ + \frac{5}{8}e^4 \cos 4f + \frac{1}{16}e^5 \cos 5f \end{bmatrix} \text{ and }$$

$$\left(\frac{a}{r}\right)^3 \frac{2+e\cos f}{1-e^2} \sin f = \frac{1}{(1-e^2)^4} \begin{bmatrix} \left(2+2.25e^2+\frac{1}{8}e^4\right)\sin f \\ + (3.5e+0.25e^3)\sin 2f \\ + \left(2.5e^2+\frac{3}{16}e^4\right)\sin 3f \\ + \frac{5}{8}e^3 \sin 4f + \frac{1}{16}e^4 \sin 5f \end{bmatrix}.$$

(11.114)

From properties of Eq. 11.110 and

$$\begin{aligned} \sin 2u &= \sin 2\Omega \cos 2f + \cos 2\Omega \sin 2f \\ \cos 2u &= \cos 2\Omega \cos 2f - \sin 2\Omega \sin 2f, \end{aligned} \qquad (11.115)$$

it is obvious that all $\omega$ terms (without $f$) may be created only by multiplying $\sin 2u$ and $\cos 2u$ by $\sin 2f$ and $\cos 2f$ in Eq. 11.113. In other words, only $\sin^2 2f$ and $\cos^2 2f$ will lead to a constant of 0.5. Therefore when seeking the $\omega$ terms (without $f$), just $\sin 2f$ and $\cos 2f$ related terms in Eq. 11.113 must be taken into account. Thus,

$$\left(\frac{da}{dt}\right)_\omega = \frac{3be^2(2+e^2)}{na^4(1-e^2)^{4.5}} \sin^2 i \sin 2\omega,$$

$$\left(\frac{de}{dt}\right)_\omega = \frac{3be(1+5e^2)}{4na^5(1-e^2)^{3.5}} \sin^2 i \sin 2\omega,$$

$$\left(\frac{d\omega}{dt}\right)_\omega = \frac{3b}{4na^5(1-e^2)^{3.5}} \left((1-5.5e^2)\sin^2 i + 3e^2 \cos^2 i\right) \cos 2\omega, \quad (11.116)$$

$$\left(\frac{di}{dt}\right)_\omega = \frac{9be^2}{8na^5(1-e^2)^{3.5}} \sin 2i \sin 2\omega,$$

$$\left(\frac{d\Omega}{dt}\right)_\omega = \frac{-9be^2}{4na^5(1-e^2)^{3.5}} \cos i \cos 2\omega \quad \text{and}$$

$$\left(\frac{dM}{dt}\right)_\omega = -\frac{3b(2+7e^2)}{8na^5(1-e^2)^3} \sin^2 i \cos 2\omega.$$

The third term of Eq. 11.106 includes all terms of $f$ and can be denoted and represented by

$$\left(\frac{d\sigma_i}{dt}\right)_f = \left(\frac{d\sigma_i}{dt} - \dot{\sigma}_{i0} - \dot{\sigma}_{i\omega}\right) = \sum_{m=1}^{m(i)} (A'''_{im} \cos mf + B'''_{im} \sin mf), \quad (11.117)$$

where $m(i)$ is the upper limit of the summation, $m(i) = (7, 7, 7, 5, 5, 7)$ for the related Keplerian elements, $A'''_{im}, B'''_{im}$ are coefficients as well as functions of $(a, e, i, \omega)$ and can be derived from Eqs. 11.104, 11.111, and 11.116. Through integral variable transformation (cf. Eq. 11.107), one has

$$\left(\frac{d\sigma_i}{dt}\right)_f \left(\frac{r}{a}\right)^2 \frac{1}{\sqrt{1-e^2}} \frac{1}{n} = \sum_{m=1}^{m(i)-2} (A''_{im} \cos mf + B''_{im} \sin mf), \quad (11.118)$$

where the upper limit of the summation is reduced by 2. $A''_{im}, B''_{im}$ are transformed coefficients. It is notable that in Eq. 11.118 the constant term ($m = 0$) doesn't exist because of the property of the short periodic term perturbations.

For the integral area of $(t_0, t)$, related areas for $\omega$ and $f$ are $(\omega_0, \omega)$ and $(f_0, f)$ respectively. For any $f$ there is an integer $k$, so that $k2\pi + f_0 \leq f \leq (k+1)2\pi + f_0$. Using the periodic property, the integration of the terms of Eq. 11.118 over the area of $(f_0, f_0 + 2k\pi)$ is zero; therefore, Eq. 11.118 just needs to be integrated over the area of $(f_0 + 2k\pi, f)$. Denoting the coefficients of $\sin 2\omega$ and $\cos 2\omega$ in Eq. 11.116 as

$$\left(\frac{d\sigma_i}{dt}\right)_{\omega s} \quad \text{and} \quad \left(\frac{d\sigma_i}{dt}\right)_{\omega c},$$

the total integration of Eq. 11.106 is then

$$\int_{t_0}^{t} d\sigma_i = \int_{t_0}^{t} \dot{\sigma}_{i0} dt + \int_{\omega_0}^{\omega} \left(\frac{d\sigma_i}{dt}\right)_\omega \left(\frac{d\omega}{dt}\right)_0^{-1} d\omega + \int_{k2\pi+f_0}^{f} \left(\frac{d\sigma_i}{dt}\right)_f \left(\frac{r}{a}\right)^2 \frac{1}{\sqrt{1-e^2}} \frac{1}{n} df$$

(11.119)

or

$$\sigma_i(t) = \sigma_i(t_0) + \dot{\sigma}_{i0}(t - t_0)$$
$$+ \frac{1}{2}\left(\frac{d\omega}{dt}\right)_0^{-1} \left[\left(\frac{d\sigma_i}{dt}\right)_{\omega c}(\sin 2\omega - \sin 2\omega_0) - \left(\frac{d\sigma_i}{dt}\right)_{\omega s}(\cos 2\omega - \cos 2\omega_0)\right]$$
$$+ \sum_{m=1}^{m(i)-2} \frac{1}{m}\left[A''_{im}(\sin mf - \sin mf_0) - B''_{im}(\cos mf - \cos mf_0)\right].$$

(11.120)

That is, the $\overline{C}_{20}$ term perturbation of the orbit has a linear term (long-term perturbation), a long periodic term (with argument of $\omega$), and a short period term (with argument of $f$). The instantaneous Keplerian elements are equal to the initial elements plus the perturbations.

Such a $\overline{C}_{20}$ disturbed orbit solution provides an indication of a general model of the perturbed orbit, which will be used as a basis for orbit correction purposes and will be discussed in the next section.

## 11.4
## Orbit Correction

When the orbit errors of GPS satellites become not negligible for special GPS applications, a process of orbit correction is the first option. Generally, orbit correction is applied to the regional or very long baseline of GPS precise positioning. Even IGS precise GPS orbits are not homogenously precise, because they are dependent on the distribution of the IGS reference stations and the length of the data used. The orbit correction is an adjustment or filtering process in which, besides the station position, the orbit errors are also modelled, determined, and corrected based on a known orbit.

Keplerian elements also describe the orbit geometry for instantaneous time. Orbit errors can be considered geometric element errors of the orbit in general. Recalling above discussions of the $\overline{C}_{20}$ perturbed orbit solution, a general orbit model can be written as

$$\sigma_j(t) = \sigma_{jc}(t) + \dot{\sigma}_{j0}(t - t_0) + A_{j\omega} \cos 2\omega + B_{j\omega} \sin 2\omega$$
$$+ \sum_{m=1}^{m(j)} \left[ A'_{jm} \cos mf + B'_{jm} \sin mf \right], \tag{11.121}$$

where $\sigma_j(t), \sigma_{jc}(t), \dot{\sigma}_{j0}$ are true orbit element at time $t$, computed element at $t$, element rate with respect to the initial epoch $t_0$, $A_{j\omega}, B_{j\omega}, A'_{jm}, B'_{jm}$ are the coefficients of the long and short periodic perturbations respectively, and $m(j)$ is the truncating integer of index m related to the $j$th Keplerian element. $\omega$ and $f$ are Keplerian elements. Generally speaking, the coefficients of $A'_{jm}, B'_{jm}$ are also functions of $\omega$, and $\omega$ can be considered in the short periodic term as a constant. Therefore, Eq. 11.121 is equivalent to

$$\sigma_j(t) = \sigma_{jc}(t) + \dot{\sigma}_{j0}(t - t_0) + A_{j\omega} \cos 2\omega + B_{j\omega} \sin 2\omega$$
$$+ \sum_{m=1}^{m(j)} \left[ A_{jm} \cos mu + B_{jm} \sin mu \right], \tag{11.122}$$

where $u = \omega + f$. The order of the polynomial term can be raised to the second order, further terms of $\omega$ may also be added, and $m(j)$ is selectable. The selection of the number of the order depends on the need and the situation of orbit errors.

In the GPS observation equations (cf. Chap. 6), the orbit state vector is presented in the range or range rate functions. It depends on the use of the GPS observables. We denote the range and range rate function generally as $\rho$; their partial derivatives with respect to the orbit state vector are given in Sect. 6.3 and have the forms of

$$\frac{\partial \rho}{\partial \vec{r}} \quad \text{and} \quad \frac{\partial \rho}{\partial \dot{\vec{r}}},$$

where the satellite state vector is $(\vec{r}, \dot{\vec{r}})$. The relations between $(\vec{r}, \dot{\vec{r}})$ and Keplerian elements $\sigma_j$ are discussed in Sect. 3.1. Also, the relations between $\sigma_j$ and the parameters of the orbit correction model are given in Eq. 11.122. Therefore, the orbit correction parts in the GPS observation equations are then

$$\frac{\partial \rho}{\partial \vec{r}} \frac{\partial \vec{r}}{\partial \vec{\sigma}} \frac{\partial \vec{\sigma}}{\partial \vec{y}} \Delta \vec{y}^{\mathrm{T}} + \frac{\partial \rho}{\partial \dot{\vec{r}}} \frac{\partial \dot{\vec{r}}}{\partial \vec{\sigma}} \frac{\partial \vec{\sigma}}{\partial \vec{y}} \Delta \vec{y}^{\mathrm{T}}, \tag{11.123}$$

where $\vec{y}, \Delta \vec{y}$ are the parameter vector in model 11.122 and the parameter correction vector of the model, and $\vec{\sigma}$ is the vector of Keplerian elements. If the initial parameter vector is selected as zero, then $\vec{y} = \Delta \vec{y}$. It is obvious that

$$\vec{y} = \left( \dot{\sigma}_{j0}, A_{j\omega}, B_{j\omega}, A_{jm}, B_{jm} \right) \tag{11.124}$$

and

$$\frac{\partial \sigma_j}{\partial(\dot{\sigma}_{j0}, A_{j\omega}, B_{j\omega}, A_{jm}, B_{jm})} = ((t-t_0), \cos 2\omega, \sin 2\omega, \cos mu, \sin mu). \quad (11.125)$$

Here, parameters $A_{jm}$, $B_{jm}$ represent symbolically the unknowns of all $m$. For the convenience of presenting the partial derivatives of the state vector with respect to the Keplerian elements, the Keplerian element vector is reordered as

$$\vec{\sigma} = (\Omega, i, \omega, a, e, M). \quad (11.126)$$

This does not affect Eq. 11.125, because the right-hand side of the equation has nothing to do with index $j$. According to the formulas in Sect. 3.1.3 (Eqs. 3.41–3.43)

$$\begin{pmatrix} \vec{r} \\ \dot{\vec{r}} \end{pmatrix} = R_3(-\Omega)R_1(-i)R_3(-\omega) \begin{pmatrix} \vec{q} \\ \dot{\vec{q}} \end{pmatrix}, \quad (11.127)$$

where

$$\vec{q} = \begin{pmatrix} a(\cos E - e) \\ a\sqrt{1-e^2} \sin E \\ 0 \end{pmatrix} = \begin{pmatrix} r \cos f \\ r \sin f \\ 0 \end{pmatrix} \quad \text{and} \quad (11.128)$$

$$\dot{\vec{q}} = \begin{pmatrix} -\sin E \\ \sqrt{1-e^2} \cos E \\ 0 \end{pmatrix} \frac{na}{1-e\cos E} = \begin{pmatrix} -\sin f \\ e+\cos f \\ 0 \end{pmatrix} \frac{na}{\sqrt{1-e^2}}, \quad (11.129)$$

one has

$$\frac{\partial \vec{r}}{\partial(\Omega, i, \omega)} = \frac{\partial R}{\partial(\Omega, i, \omega)} \vec{q} \quad \text{and} \quad \frac{\partial \dot{\vec{r}}}{\partial(\Omega, i, \omega)} = \frac{\partial R}{\partial(\Omega, i, \omega)} \dot{\vec{q}}, \quad (11.130)$$

where $(\vec{q}, \dot{\vec{q}})$ are position and velocity vectors of the satellite in the orbital plane coordinate system, and

$$R = R_3(-\Omega)R_1(-i)R_3(-\omega) \quad \text{and} \quad (11.131)$$

$$\frac{\partial R}{\partial(\Omega, i, \omega)} = \left( \frac{\partial R_3(-\Omega)}{\partial \Omega} R_1(-i)R_3(-\omega), R_3(-\Omega) \frac{\partial R_1(-i)}{\partial i} R_3(-\omega), R_3(-\Omega)R_1(-i) \frac{\partial R_3(-\omega)}{\partial \omega} \right),$$

where

$$\frac{\partial R_1(-i)}{\partial i} = \begin{pmatrix} 0 & 0 & 0 \\ 0 & -\sin i & -\cos i \\ 0 & \cos i & -\sin i \end{pmatrix},$$

$$\frac{\partial R_3(-\Omega)}{\partial \Omega} = \begin{pmatrix} -\sin \Omega & -\cos \Omega & 0 \\ \cos \Omega & -\sin \Omega & 0 \\ 0 & 0 & 0 \end{pmatrix} \quad \text{and}$$

$$\frac{\partial R_3(-\omega)}{\partial \omega} = \begin{pmatrix} -\sin \omega & -\cos \omega & 0 \\ \cos \omega & -\sin \omega & 0 \\ 0 & 0 & 0 \end{pmatrix}.$$

For the Keplerian elements in the orbital plane $(a, e, M)$, one has

$$\frac{\partial \vec{r}}{\partial(a,e,M)} = R \frac{\partial \vec{q}}{\partial(a,e,M)} \quad \text{and} \quad \frac{\partial \dot{\vec{r}}}{\partial(a,e,M)} = R \frac{\partial \dot{\vec{q}}}{\partial(a,e,M)}, \qquad (11.132)$$

where

$$\frac{\partial \vec{q}}{\partial(a,e,M)} = \begin{pmatrix} \cos E - e & \frac{-a \sin^2 E}{1-e \cos E} - a & \frac{-a \sin E}{1-e \cos E} \\ \sqrt{1-e^2} \sin E & a\sqrt{1-e^2} \left( \frac{\sin 2E}{2(1-e \cos E)} - \frac{e \sin E}{1-e^2} \right) & \frac{a\sqrt{1-e^2} \cos E}{1-e \cos E} \\ 0 & 0 & 0 \end{pmatrix}$$

and

$$\frac{\partial \dot{\vec{q}}}{\partial(a,e,M)} = \begin{pmatrix} \frac{n \sin E}{2(1-e \cos E)} & \frac{na \sin E(e - 2\cos E + e \cos^2 E)}{(1-e \cos E)^3} & \frac{na(e - \cos E)}{(1-e \cos E)^3} \\ \frac{-n\sqrt{1-e^2} \cos E}{2(1-e \cos E)} & \frac{na\left[1 + e^2 - 2e \cos E + \sin^2 E(e \cos E - 2)\right]}{\sqrt{1-e^2}(1-e \cos E)^3} & \frac{-na\sqrt{1-e^2} \sin E}{(1-e \cos E)^3} \\ 0 & 0 & 0 \end{pmatrix}.$$

The partial derivatives formulas given in Sect. 11.1 and the relation in Eq. 3.32 between $n$ and $a$ (mean angular velocity and semi-major axis of the satellite) given in Chap. 3 are used, i.e.

$$\frac{\partial E}{\partial(e,M)} = \left( \frac{a}{r} \sin E, \frac{a}{r} \right) \quad \text{and}$$

$$n^2 = \mu/a^3.$$

## 11.5
## Principle of GPS Precise Orbit Determination

Recalling the discussions in Sect. 11.1, the perturbed orbit of the satellite is the solution (or integration)

$$\vec{X}(t) = \vec{X}(t_0) + \int_{t_0}^{t} \vec{F} \, dt, \qquad (11.133)$$

which can be obtained by integrating the differential state equation under the initial condition

$$\begin{cases} \dot{\vec{X}}(t) = \vec{F} \\ \vec{X}(t_0) = \vec{X}_0 \end{cases}, \qquad (11.134)$$

where $\vec{X}(t)$ is the instantaneous state vector of the satellite, $\vec{X}(t_0)$ is the initial state vector at time $t_0$ (denoted by $\vec{X}_0$), $\vec{F}$ is a function of the state vector $\vec{X}(t)$ and time $t$, and

$$\vec{X} = \begin{pmatrix} \vec{r} \\ \dot{\vec{r}} \end{pmatrix} \quad \text{and} \quad \vec{F} = \begin{pmatrix} \dot{\vec{r}} \\ \vec{f}/m \end{pmatrix},$$

Where $\vec{f}$ is the summated force vector of all possible force vectors acting on the satellite, $m$ is the mass of satellite, and $\vec{r}, \dot{\vec{r}}$ are the position and velocity vectors of the satellite.

If the initial state vector and the force vectors are precisely known, then the precise orbits can be computed through the integration in Eq. 11.133. Expanding the integration time $t$ into the future, the so-called forecasted orbits can be obtained. Therefore, suitable numerical integration algorithms are needed (see next section).

In practice, we must determine the precise initial state vector and force models, which are related to the approximate initial state vector and force models. These can be realised through suitable parameterisation of the models in the GPS observation equations, and the parameters can then be solved by adjustment or filtering.

We denote the range and range rate function generally by $\rho$; and their partial derivatives with respect to the orbit state vector are given in Sect. 6.3 and have the forms of

$$\frac{\partial \rho}{\partial \vec{r}}, \frac{\partial \rho}{\partial \dot{\vec{r}}}, \quad \text{or} \quad \frac{\partial \rho}{\partial \vec{X}}.$$

## Chapter 11 · Perturbed Orbit and Its Determination

Therefore, the orbit parameter-related parts in the linearised GPS observation equation are

$$\frac{\partial \rho}{\partial(\vec{r},\dot{\vec{r}})} \frac{\partial(\vec{r},\dot{\vec{r}})}{\partial \vec{y}} \Delta \vec{y}^T, \quad \text{or} \quad \frac{\partial \rho}{\partial \vec{X}} \frac{\partial \vec{X}}{\partial \vec{y}} \Delta \vec{y}^T, \qquad (11.135)$$

where

$$\vec{y} = \left(\vec{X}_0, \vec{Y}\right), \quad \Delta \vec{y}^T = \left(\Delta \vec{X}_0, \Delta \vec{Y}\right)^T, \quad \frac{\partial \vec{X}}{\partial \vec{y}} = \frac{\partial \vec{X}}{\partial(\vec{X}_0, \vec{Y})}.$$

$\vec{X}, \vec{Y}$ are the state vector of satellite and the parameter vector of the force models, and index 0 denotes the related initial vectors of time $t_0$. $\vec{y}$ is the total unknown vector of the orbit determination problem, the related correction vector is $\Delta \vec{y} = \vec{y} - \vec{y}_0$, and $\Delta \vec{X}_0$ is the correction vector of the initial state vector. The partial derivatives of $\vec{X}$ with respect to $\vec{y}$ are called transition matrix, which has the dimension of $6 \times (6+n)$, where $n$ is the dimension of vector $\vec{Y}$. The partial derivatives of the equation of motion of the satellite (cf., Eq. 11.134) with respect to the vector $\vec{y}$ are

$$\frac{\partial \dot{\vec{X}}(t)}{\partial \vec{y}} = \frac{\partial \vec{F}}{\partial \vec{y}} = \frac{\partial \vec{F}}{\partial \vec{X}} \frac{\partial \vec{X}}{\partial \vec{y}} + \left(\frac{\partial \vec{F}}{\partial \vec{y}}\right)^*, \qquad (11.136)$$

where the asterisk (*) denotes the partial derivatives of $\vec{F}$ with respect to the explicit parameter vector $\vec{y}$ in $\vec{F}$, and

$$D(t) = \left(\frac{\partial \vec{F}}{\partial \vec{X}}\right) = \begin{pmatrix} 0_{3\times 3} & E_{3\times 3} \\ \dfrac{1}{m}\dfrac{\partial \vec{f}}{\partial \vec{r}} & \dfrac{1}{m}\dfrac{\partial \vec{f}}{\partial \dot{\vec{r}}} \end{pmatrix} = \begin{pmatrix} 0_{3\times 3} & E_{3\times 3} \\ A(t) & B(t) \end{pmatrix},$$

$$C(t) = \left(\frac{\partial \vec{F}}{\partial \vec{y}}\right)^* = \begin{pmatrix} 0_{3\times 6} & 0_{3\times n} \\ 0_{3\times 6} & \dfrac{1}{m}\dfrac{\partial \vec{f}}{\partial \vec{Y}} \end{pmatrix} = \begin{pmatrix} 0_{3\times(6+n)} \\ G(t) \end{pmatrix}, \qquad (11.137)$$

where $E$ is an identity matrix; the partial derivatives will be discussed and derived in detail in a later section. It is notable that the force parameters are not functions of $t$. Therefore, the order of the differentiations can be exchanged. Denoting transition matrix by $\Phi(t, t_0)$, then Eq. 11.136 turns out to be

$$\frac{d\Phi(t, t_0)}{dt} = D(t)\Phi(t, t_0) + C(t). \qquad (11.138)$$

Equation 11.138 is called a differential equation of the transition matrix or variational equation (cf., e.g., Montenbruck and Gill 2000). Denoting

$$\Phi(t, t_0) = \begin{pmatrix} \Psi(t, t_0) \\ \dot{\Psi}(t, t_0) \end{pmatrix}, \quad (11.139)$$

an alternate expression of Eq. 11.138 can be obtained by substituting Eqs. 11.139 and 11.137 into Eq. 11.138

$$\frac{d^2 \Psi(t, t_0)}{dt^2} = A(t) \Psi(t, t_0) + B(t) \frac{d\Psi(t, t_0)}{dt} + G(t). \quad (11.140)$$

The initial value matrix is (initial state vector does not depend on force parameters)

$$\Phi(t_0, t_0) = ( E_{6\times 6} \quad 0_{6\times n} ). \quad (11.141)$$

That is, in the GPS observation equation, the transition matrix must be obtained by solving initial value problem of the variation Eq. 11.138 or 11.140. The problem is solved by integration traditionally.

## 11.5.1
## Xu's Algebraic Solution to the Variation Equation

The variation equation can also be solved by numerical differentiation, first derived by Xu around 2003 (cf. preface of Xu GPS 2007).

Equation 11.140 is a matrix differential equation system of size $3 \times (6 + n)$. Because $A(t)$ and $B(t)$ are $3 \times 3$ matrices, the differential equations are independent from column to column. That is, we need only discuss the solution of the equation of a column. For column $j$, the Eqs. 11.140 and 11.141 are

$$\frac{d^2 \Psi_{ij}(t)}{dt^2} = \sum_{k=1}^{3} \left( A_{ik}(t) \Psi_{kj}(t) + B_{ik}(t) \frac{d\Psi_{kj}(t)}{dt} \right) + G_{ij}(t), \quad i = 1, 2, 3, \quad (11.142)$$

$$\begin{pmatrix} \Psi_{ij}(t_0) \\ \dot{\Psi}_{ij}(t_0) \end{pmatrix} = \begin{pmatrix} \delta_{ij} \\ \delta_{(i+3)j} \end{pmatrix}, \quad i = 1, 2, 3, \delta_{kj} = \begin{cases} 1 & \text{if } k = j \\ 0 & \text{if } k \neq j \end{cases},$$

where index $ij$ denotes the related element of the matrix. For time interval $[t_0, t]$ and differentiation step $h = (t - t_0)/m$, one has $t_n = t_0 + nh$, $n = 1, \ldots, m$ and

$$\left. \frac{d^2 \Psi_{ij}(t)}{dt^2} \right|_{t=t_n} = \frac{\Psi_{ij}(t_{n+1}) - 2\Psi_{ij}(t_n) + \Psi_{ij}(t_{n-1})}{h^2}, \quad i = 1, 2, 3,$$

$$\left. \frac{d\Psi_{ij}(t)}{dt} \right|_{t=t_n} = \frac{\Psi_{ij}(t_{n+1}) - \Psi_{ij}(t_{n-1})}{2h}, \quad \left. \Psi_{ij}(t) \right|_{t=t_n} = \Psi_{ij}(t_n), \quad i = 1, 2, 3. \quad (11.143)$$

Then Eq. 11.142 turns out to be

$$\Psi_{ij}(t_0) = \Psi_{ij}(t_0), \quad \Psi_{ij}(t_1) = \Psi_{ij}(t_0) + h\dot\Psi_{ij}(t_0), \quad i = 1,2,3.$$

$$\frac{\Psi_{ij}(t_{n+1}) - 2\Psi_{ij}(t_n) + \Psi_{ij}(t_{n-1})}{h^2} =$$

$$\sum_{k=1}^{3} \left( A_{ik}(t_n)\Psi_{kj}(t_n) + B_{ik}(t_n)\frac{\Psi_{kj}(t_{n+1}) - \Psi_{kj}(t_{n-1})}{2h} \right) + G_{ij}(t_n), \quad i = 1,2,3,$$

(11.144)

where $n = 1, 2, \ldots, m - 1$. For $i = 1, 2, 3$ and the sequential number $n$, there are three equations and three unknowns of time $t_{n+1}$; so that the initial value problem has a set of unique solutions sequentially. Equation 11.144 can be rewritten as

$$\left( \frac{E}{h^2} - \frac{B(t_n)}{2h} \right) \begin{pmatrix} \Psi_{1j}(t_{n+1}) \\ \Psi_{2j}(t_{n+1}) \\ \Psi_{3j}(t_{n+1}) \end{pmatrix} = \begin{pmatrix} R_1 \\ R_2 \\ R_3 \end{pmatrix}, \quad (11.145)$$

where

$$\begin{pmatrix} R_1 \\ R_2 \\ R_3 \end{pmatrix} = \left( \frac{2E}{h^2} + A(t_n) \right) \begin{pmatrix} \Psi_{1j}(t_n) \\ \Psi_{2j}(t_n) \\ \Psi_{3j}(t_n) \end{pmatrix}$$

$$- \left( \frac{E}{h^2} + \frac{B(t_n)}{2h} \right) \begin{pmatrix} \Psi_{1j}(t_{n-1}) \\ \Psi_{2j}(t_{n-1}) \\ \Psi_{3j}(t_{n-1}) \end{pmatrix} + \begin{pmatrix} G_{1j}(t_n) \\ G_{2j}(t_n) \\ G_{3j}(t_n) \end{pmatrix}.$$

For $n = 1, \ldots, m - 1$, above equation is solvable. It is notable that the three matrices

$$\left( \frac{E}{h^2} - \frac{B(t_n)}{2h} \right), \quad \left( \frac{2E}{h^2} + A(t_n) \right), \quad \left( \frac{E}{h^2} + \frac{B(t_n)}{2h} \right)$$

are independent from the column number $j$. The solutions of Eq. 11.145 are vectors

$$\begin{pmatrix} \Psi_{1j}(t_{n+1}) \\ \Psi_{2j}(t_{n+1}) \\ \Psi_{3j}(t_{n+1}) \end{pmatrix} \quad \text{and} \quad \begin{pmatrix} \dot\Psi_{1j}(t_{n+1}) \\ \dot\Psi_{2j}(t_{n+1}) \\ \dot\Psi_{3j}(t_{n+1}) \end{pmatrix}, \quad n = 1, \ldots, m - 1, \quad (11.146)$$

where the velocity vector can be computed using definition of Eq. 11.143. Solving the equations of all column $j$, the solutions of the initial value problem of Eqs. 11.140 and 11.141 can be obtained. It is notable that the needed values are the values of $t_n$ and can be computed by averaging the values of $t_{n+1}$ and $t_{n-1}$.

A recent study carried out by the team at Shandong University led by the first author demonstrated the successful use of the algebraic solution derived here, and will be published very soon by Nie et al. (2016).

## 11.6
## Numerical Integration and Interpolation Algorithms

The Runge–Kutta algorithm, Adams algorithm, Cowell algorithm, and mixed algorithm, as well as interpolation algorithms, are discussed in this section (cf., e.g., Brouwer and Clemence 1961; Bate et al. 1971; Herrick 1972; Xu 1994; Liu et al. 1996; Press et al. 1992).

### 11.6.1
### Runge–Kutta Algorithm

The Runge–Kutta algorithm is a method that can be used to solve the initial value problem of

$$\frac{dX}{dt} = F(t, X)$$
$$X(t_0) = X_0,$$
(11.147)

where $X_0$ is the initial value of variable $X$ at time $t_0$, and $F$ is the function of $t$ and $X$. For step size $h$, the Runge–Kutta algorithm can be used to compute $X(t_0 + h)$. By repeating such process, a series of solutions can be obtained as $X(t_0 + h)$, $X(t_0 + 2h)$, ..., $X(t_0 + nh)$, where $n$ is an integer. Denoting $t_n = t_0 + nh$, $X(t_n + h)$ can be represented by the Taylor expansion at $t_n$ by

$$X(t_n + h) = X(t_n) + h\frac{dX}{dt}\bigg|_{t=t_n} + \frac{h^2}{2}\frac{d^2X}{dt^2}\bigg|_{t=t_n} + \cdots + \frac{h^n}{n!}\frac{d^nX}{dt^n}\bigg|_{t=t_n} + \cdots, \quad (11.148)$$

where

$$\frac{dX}{dt} = F,$$
$$\frac{d^2X}{dt^2} = \frac{dF(t,X)}{dt} = \frac{\partial F}{\partial t} + \frac{\partial F}{\partial X}\frac{\partial X}{\partial t} = \frac{\partial F}{\partial t} + \frac{\partial F}{\partial X}F,$$
$$\frac{d^3X}{dt^3} = \frac{\partial^2 F}{\partial t^2} + 2\frac{\partial^2 F}{\partial t \partial X}F + \frac{\partial^2 F}{\partial t \partial X} + \frac{\partial^2 F}{\partial X^2}F^2 + \left(\frac{\partial F}{\partial X}\right)^2 F \quad \text{and}$$
$$\frac{d^4X}{dt^4} = \frac{\partial^3 F}{\partial t^3} + \frac{\partial^3 F}{\partial t^2 \partial X}(3F+1) + \frac{\partial^3 F}{\partial t \partial X^2}(5F^2 + 2F) + 2\frac{\partial^2 F}{\partial t \partial X}\frac{\partial F}{\partial t} + 4\frac{\partial^3 F}{\partial X^3}F^3$$
$$+ 2\frac{\partial^2 F}{\partial X^2}\frac{\partial F}{\partial t}F + 4\frac{\partial F}{\partial X}\frac{\partial^2 F}{\partial t \partial X}F + 6\frac{\partial F}{\partial X}\frac{\partial^2 F}{\partial X^2}F^2 + \left(\frac{\partial F}{\partial X}\right)^2\frac{\partial F}{\partial t} + \left(\frac{\partial F}{\partial X}\right)^2\frac{\partial F}{\partial X}2F.$$

...

(11.149)

The principle of the Runge–Kutta algorithm is to use a set of combinations of the first-order partial derivatives around the $(t_n, X(t_n))$ to replace the higher-order derivatives in Eq. 11.148; that is,

$$X(t_{n+1}) = X(t_n) + \sum_{i=1}^{L} w_i K_i, \tag{11.150}$$

where

$$K_1 = hF(t_n, X(t_n)) \quad \text{and}$$
$$K_i = hF\left(t_n + \alpha_i h, X(t_n) + \sum_{j=1}^{i-1} \beta_{ij} K_j\right), (i = 2, 3, \ldots), \tag{11.151}$$

where $w_i$, $\alpha_i$, and $\beta_{ij}$ are constants to be determined, and $L$ is an integer. The Taylor expansions of $K_i$ ($i = 2, 3, \ldots$) at $(t_n, X(t_n))$ to the first order are

$$K_i = hF(t_n, X(t_n)) + h^2 \alpha_i \frac{\partial F}{\partial t} + h \frac{\partial F}{\partial X} \sum_{j=1}^{i-1} \beta_{ij} K_j \text{ or} \tag{11.152}$$

$$K_2 = hF(t_n, X(t_n)) + h^2 \left(\alpha_2 \frac{\partial F}{\partial t} + \beta_{21} \frac{\partial F}{\partial X} F\right),$$

$$K_3 = hF + h^2 \left(\alpha_3 \frac{\partial F}{\partial t} + (\beta_{31} + \beta_{32}) \frac{\partial F}{\partial X} F\right) + h^3 \beta_{32} \frac{\partial F}{\partial X} \left(\alpha_2 \frac{\partial F}{\partial t} + \beta_{21} \frac{\partial F}{\partial X} F\right),$$

$$K_4 = hF + h^2 \left(\alpha_4 \frac{\partial F}{\partial t} + (\beta_{41} + \beta_{42} + \beta_{43}) \frac{\partial F}{\partial X} F\right)$$
$$+ h^3 \left[(\beta_{42}\alpha_2 + \beta_{43}\alpha_3) \frac{\partial F}{\partial X}\frac{\partial F}{\partial t} + (\beta_{42}\beta_{21} + \beta_{43}(\beta_{31} + \beta_{32})) \frac{\partial F}{\partial X}\frac{\partial F}{\partial X} F\right]$$
$$+ h^4 \beta_{43}\beta_{32} \frac{\partial F}{\partial X}\frac{\partial F}{\partial X} \left(\alpha_2 \frac{\partial F}{\partial t} + \beta_{21} \frac{\partial F}{\partial X} F\right),$$

$$K_5 = hF + h^2 \left(\alpha_5 \frac{\partial F}{\partial t} + (\beta_{51} + \beta_{52} + \beta_{53} + \beta_{54}) \frac{\partial F}{\partial X} F\right) \tag{11.153}$$
$$+ h^3 \frac{\partial F}{\partial X} \begin{pmatrix} \alpha_2 \dfrac{\partial F}{\partial t} + \beta_{21} \dfrac{\partial F}{\partial X} F + \beta_{53}\left(\alpha_3 \dfrac{\partial F}{\partial t} + (\beta_{31} + \beta_{32}) \dfrac{\partial F}{\partial X} F\right) \\ + \beta_{54}\left(\alpha_4 \dfrac{\partial F}{\partial t} + (\beta_{41} + \beta_{42} + \beta_{43}) \dfrac{\partial F}{\partial X} F\right) \end{pmatrix}$$
$$+ h^4 \frac{\partial F}{\partial X} \begin{pmatrix} (\beta_{53}\beta_{32}\alpha_2 + \beta_{54}(\beta_{42}\alpha_2 + \beta_{43}\alpha_3)) \dfrac{\partial F}{\partial X}\dfrac{\partial F}{\partial t} \\ + (\beta_{54}(\beta_{42}\beta_{21} + \beta_{43}(\beta_{31} + \beta_{32})) + \beta_{32}\beta_{21}) \dfrac{\partial F}{\partial X}\dfrac{\partial F}{\partial X} F \end{pmatrix}$$
$$+ h^5 \frac{\partial F}{\partial X} \beta_{54} \left(\beta_{43}\beta_{32} \frac{\partial F}{\partial X}\frac{\partial F}{\partial X} \left(\alpha_2 \frac{\partial F}{\partial t} + \beta_{21} \frac{\partial F}{\partial X} F\right)\right)$$

$\ldots$

## 11.6 · Numerical Integration and Interpolation Algorithms

where $F$ and the related partial derivatives have values at $(t_n, X(t_n))$. Substituting the above formulas into Eq. 11.150 and comparing the coefficients of $h^n(=1/n!)$ with Eq. 11.148, a group of equations of constants $w_i$, $\alpha_i$, and $\beta_{ij}$ can be obtained by separating them through the partial derivative combinations. For example, for $L = 4$, one has

$$w_1 + w_2 + w_3 + w_4 = 1,$$
$$w_2\alpha_2 + w_3\alpha_3 + w_4\alpha_4 = \frac{1}{2},$$
$$w_2\beta_{21} + w_3(\beta_{31} + \beta_{32}) + w_4(\beta_{41} + \beta_{42} + \beta_{43}) = \frac{1}{2},$$
$$w_3\alpha_2\beta_{32} + w_4(\alpha_2\beta_{42} + \alpha_3\beta_{43}) = \frac{1}{6}, \qquad (11.154)$$
$$w_3\beta_{21}\beta_{32} + w_4(\beta_{21}\beta_{42} + \beta_{31}\beta_{43} + \beta_{32}\beta_{43}) = \frac{1}{6},$$
$$w_4\alpha_2\beta_{43}\beta_{32} = \frac{1}{24}, \quad \text{and}$$
$$w_4\beta_{21}\beta_{43}\beta_{32} = \frac{1}{24}.$$

There are 13 coefficients in the seven equations above, so the solution set of Eq. 11.154 is not unique. Considering $w$ has the meaning of weight, and $\alpha$ is the step factor, one may set, e.g., $w_1 = w_2 = w_3 = w_4 = 1/4$, $\alpha_2 = 1/3$, $\alpha_3 = 2/3$, $\alpha_4 = 1$ into the above equations and have

$$\beta_{21} + \beta_{31} + \beta_{32} + \beta_{41} + \beta_{42} + \beta_{43} = 2,$$
$$\beta_{32} + \beta_{42} + 2\beta_{43} = 2,$$
$$\beta_{21}\beta_{32} + \beta_{21}\beta_{42} + \beta_{31}\beta_{43} + \beta_{32}\beta_{43} = \frac{2}{3},$$
$$\beta_{43}\beta_{32} = \frac{1}{2}, \quad \text{and}$$
$$\beta_{21}\beta_{43}\beta_{32} = \frac{1}{6}.$$

Letting $\beta_{32} = 1$, one has $\beta_{42} = 0$, $\beta_{31} = -1/3$, and $\beta_{41} = 1/2$. Thus, a fourth order Runge–Kutta formula is

$$X(t_{n+1}) = X(t_n) + \frac{1}{4}\sum_{i=1}^{4} K_i, \qquad (11.155)$$

where

$$K_1 = hF(t_n, X(t_n)),$$
$$K_2 = hF\left(t_n + \frac{1}{3}h, X(t_n) + \frac{1}{3}K_1\right),$$
$$K_3 = hF\left(t_n + \frac{2}{3}h, X(t_n) - \frac{1}{3}K_1 + K_2\right), \quad \text{and} \qquad (11.156)$$
$$K_4 = hF\left(t_n + h, X(t_n) + \frac{1}{2}K_1 + \frac{1}{2}K_3\right).$$

Similarly, a commonly used eighth-order Runge–Kutta formula can be derived. It is quoted as follows (cf. Xu 1994; Liu et al. 1996):

$$X(t_{n+1}) = X_n + \frac{1}{840}(41K_1 + 27K_4 + 272K_5 + 27K_6 + 216K_7 + 216K_9 + 41K_{10}),$$
(11.157)

where

$$K_1 = hF(t_n, X_n), X_n = X(t_n),$$
$$K_2 = hF\left(t_n + \frac{4}{27}h, X_n + \frac{4}{27}K_1\right),$$
$$K_3 = hF\left(t_n + \frac{2}{9}h, X_n + \frac{1}{18}K_1 + \frac{1}{6}K_2\right),$$
$$K_4 = hF\left(t_n + \frac{1}{3}h, X_n + \frac{1}{12}K_1 + \frac{1}{4}K_3\right),$$
$$K_5 = hF\left(t_n + \frac{1}{2}h, X_n + \frac{1}{8}K_1 + \frac{3}{8}K_4\right),$$
$$K_6 = hF\left(t_n + \frac{2}{3}h, X_n + \frac{1}{54}(13K_1 - 27K_3 + 42K_4 + 8K_5)\right),$$
$$K_7 = hF\left(t_n + \frac{1}{6}h, X_n + \frac{1}{4320}(389K_1 - 54K_3 + 966K_4 - 824K_5 + 243K_6)\right),$$
$$K_8 = hF\left(t_n + h, X_n + \frac{1}{20}(-231K_1 + 81K_3 - 1164K_4 + 656K_5 - 122K_6 + 800K_7)\right),$$
$$K_9 = hF\left(\begin{array}{c}t_n + \frac{5}{6}h, X_n + \frac{1}{288}(-127K_1 + 18K_3 - 678K_4 + 456K_5 \\ -9K_6 + 576K_7 + 4K_8)\end{array}\right), \quad \text{and}$$
$$K_{10} = hF\left(\begin{array}{c}t_n + h, X_n + \frac{1}{820}(1481K_1 - 81K_3 + 7104K_4 - 3376K_5 \\ + 72K_6 - 5040K_7 - 60K_8 + 720K_9)\end{array}\right).$$
(11.158)

From the derivation process, it is obvious that the Runge–Kutta algorithm is an approximation of the same-order Taylor expansions. For every step of the solution, the function values of $F$ must be computed several times. The Runge–Kutta algorithm is also called the single-step method and is commonly used for computing the start values for other multiple step methods.

Errors of the integration are dependent on the step size and the properties of function $F$. To ensure the needed accuracy of the orbit integration, a step size adaptive control is also meaningful in computing efficiency (cf. Press et al. 1992). Because of the periodical motion of the orbit, the step control just needs to be made in a few special cycles of the motion. A step doubling method is suggested by Press et al. (1992). Integration is taken twice for each step, first with a full step, then independently with two half steps. Through comparing the results, the step size can be adjusted to fit the accuracy requirement.

To apply the above formulas for solving the initial value problem of the equation of motion 11.134, Eq. 11.147 shall be rewritten as

$$\frac{dX_k}{dt} = \dot{X}_k(t, X) \quad X_k(t_0) = X_{k0}$$
$$\frac{d\dot{X}_k}{dt} = f_k(t, X)/m, \quad \dot{X}_k(t_0) = \dot{X}_{k0}, \quad k = 1, 2, 3,$$

where $X = (X_1, X_2, X_3, \dot{X}_1, \dot{X}_2, \dot{X}_3)$. Using the Runge–Kutta algorithm to solve the above problem, an additional index $k$ shall be added to all $X$ and $K$ in Eq. 11.157:

$$X_k(t_{n+1}) = X_{kn} + \frac{1}{840}(41K_{k1} + 27K_{k4} + 272K_{k5} + 27K_{k6} + 216K_{k7} + 216K_{k9} + 41K_{k10}),$$

and the same index $k$ shall be added to $K$ on the left side and $F$ on the right side of Eq. 11.158. For the last three equations, $F_k = f_k/m$, so $\ddot{X}_k$ can be computed. For the first three equations, $F_k = \dot{X}_k$, so $F_k$ can be computed through computing $\dot{X}_k$ at the needed coordinates $t$ and $X$.

## 11.6.2
## Adams Algorithms

For the initial value problem of

$$\frac{dX}{dt} = F(t, X)$$
$$X(t_0) = X_0, \qquad (11.159)$$

there exists

$$X(t_{n+1}) = X(t_n) + \int_{t_n}^{t_{n+1}} F(t, X) dt. \tag{11.160}$$

The Adams algorithm uses the Newtonian backward differential interpolation formula to present the function $F$ by

$$\begin{aligned}F(t, X) = F_n &+ \frac{t - t_n}{h} \nabla F_n + \frac{(t - t_n)(t - t_{n-1})}{2! h^2} \nabla^2 F_n + \cdots \\ &+ \frac{(t - t_n)(t - t_{n-1}) \cdots (t - t_{n-k+1})}{k! h^k} \nabla^k F_n,\end{aligned} \tag{11.161}$$

where $F_n$ is the value of $F$ at the time $t_n$, $h$ is the step size, $\nabla^k F$ is the $k$th-order backward numerical difference of $F$, and

$$\begin{aligned}\nabla F_n &= F_n - F_{n-1} \\ \nabla^2 F_n &= \nabla F_n - \nabla F_{n-1} = F_n - 2F_{n-1} + F_{n-2}, \\ &\cdots \\ \nabla^m F_n &= \sum_{j=0}^{m} (-1)^j C_m^j F_{n-j}, \quad C_m^j = \frac{m!}{j!(m-j)!},\end{aligned} \tag{11.162}$$

where $C_m^i$ is the binomial coefficient. Letting $s = (t - t_n)/h$, then $dt = h ds$, $s = 0$ if $t = t_n$, $s = 1$ if $t = t_{n+1}$, so that Eqs. 11.161 and 11.160 turn out to be

$$\begin{aligned}F(t, X) &= \sum_{m=0}^{k} C_{s+m-1}^m \nabla^m F_n \quad \text{and} \\ X(t_{n+1}) &= X(t_n) + \int_{t_n}^{t_{n+1}} \sum_{m=0}^{k} C_{s+m-1}^m \sum_{j=0}^{m} (-1)^j C_m^j F_{n-j} h ds.\end{aligned} \tag{11.163}$$

By denoting

$$\begin{aligned}\gamma_m &= \int_0^1 C_{s+m-1}^m ds \\ \beta_j &= \sum_{m=j}^{k} (-1)^j C_m^j \gamma_m,\end{aligned} \tag{11.164}$$

## 11.6 · Numerical Integration and Interpolation Algorithms

one has

$$X(t_{n+1}) = X(t_n) + h \sum_{j=0}^{k} \beta_j F_{n-j}, \tag{11.165}$$

where the sequences of the two sequential summations have been changed. For the first equation of 11.164, there is (cf. Xu 1994)

$$\gamma_0 = 1, \quad \gamma_m = 1 - \sum_{j=1}^{m} \frac{1}{j+1} \gamma_{m-j}, \quad (m \geq 1). \tag{11.166}$$

Equation 11.165 is also called the Adams–Bashforth formula. It uses the function values of $\{F_{n-j}, j = 0, \ldots, k\}$ to compute the $X_{n+1}$. When the order of the algorithm is selected, the coefficients of $\beta_j$ are constants. This makes the computation using Eq. 11.165 very simple. For every integration step, just one function value of $F_n$ must be computed. However, the Adams algorithm needs $\{F_{n-j}, j = 0, \ldots, k\}$ as initial values, whereas to compute those values, the states $\{X_{n-j}, j = 0, \ldots, k\}$ are needed. In other words, the Adams algorithm is not able to start the integration itself. The Runge–Kutta algorithm is usually used for computing the start values.

The Adams–Bashforth formula does not take the function value $F_{n+1}$ into account. Using $F_{n+1}$, the Adams algorithm is expressed by the Adams–Moulton formula. Similar to the above discussions, function $F$ can be represented by

$$\begin{aligned} F(t,X) = F_{n+1} &+ \frac{t-t_{n+1}}{h} \nabla F_{n+1} + \frac{(t-t_{n+1})(t-t_n)}{2!h^2} \nabla^2 F_{n+1} + \cdots \\ &+ \frac{(t-t_{n+1})(t-t_n) \cdots (t-t_{n-k+2})}{k!h^k} \nabla^k F_{n+1}, \end{aligned} \tag{11.167}$$

where

$$\nabla^m F_{n+1} = \sum_{j=0}^{m} (-1)^j C_m^j F_{n+1-j}. \tag{11.168}$$

If one lets $s = (t - t_{n+1})/h$, then $dt = h\,ds$, $s = -1$ if $t = t_n$, and $s = 0$ if $t = t_{n+1}$; similar formulas of Eqs. 11.165 and 11.164 can be obtained:

$$X(t_{n+1}) = X(t_n) + h \sum_{j=0}^{k} \beta_j^* F_{n+1-j}, \tag{11.169}$$

$$\beta_j^* = \sum_{m=j}^{k} (-1)^j C_m^j \gamma_m^*$$

$$\gamma_m^* = \int_{-1}^{0} C_{s+m-1}^{m} ds \qquad (11.170)$$

and (cf. Xu 1994)

$$\gamma_0^* = 1, \quad \gamma_m^* = -\sum_{j=1}^{m} \frac{1}{j+1} \gamma_{m-j}^*, \quad (m \geq 1). \qquad (11.171)$$

Because of the use of $F_{n+1}$ to approximate $F$, the Adams–Moulton formula may reach a higher accuracy than that of the Adams–Bashforth formula. However, before $X_{n+1}$ has been computed, $F_{n+1}$ might not have been computed exactly, and so an iterative process is needed to use the Adams–Moulton formula. A simplified way to use the Adams–Moulton formula is to use the Adams–Bashforth formula to compute $X_{n+1}$ and $F_{n+1}$, and then to use the Adams–Moulton formula to compute the modified $X_{n+1}$ using $F_{n+1}$. Experience shows that such a process will be accurate enough for many applications.

## 11.6.3
## Cowell Algorithms

For the initial value problem of

$$\frac{d^2 X}{dt^2} = F(t, X)$$
$$\dot{X}(t_0) = \dot{X}_0 \qquad (11.172)$$
$$X(t_0) = X_0,$$

there is

$$\dot{X}(t) = \dot{X}(t_n) + \int_{t_n}^{t} F(t, X) dt \qquad (11.173)$$

It is notable that here $X$ is the position coordinate of the satellite. In other words, the disturbing force $F$ is not the function of the velocity of the satellite.

By integrating Eq. 11.173 in areas of $[t_n, t_{n+1}]$ and $[t_n, t_{n-1}]$ respectively, one has

$$X(t_{n+1}) - X(t_n) - \dot{X}(t_n)(t_{n+1} - t_n) = \int_{t_n}^{t_{n+1}} \int_{t_n}^{t} F(t,X) dt dt \quad \text{and} \quad (11.174)$$

$$X(t_{n-1}) - X(t_n) - \dot{X}(t_n)(t_{n-1} - t_n) = \int_{t_n}^{t_{n-1}} \int_{t_n}^{t} F(t,X) dt dt, \quad (11.175)$$

where $(t_{n+1} - t_n) = h = (t_n - t_{n-1})$. Adding the equations together, one has

$$X(t_{n+1}) - 2X(t_n) + X(t_{n-1}) = \int_{t_n}^{t_{n+1}} \int_{t_n}^{t} + \int_{t_n}^{t_{n-1}} \int_{t_n}^{t} F(t,X) dt dt. \quad (11.176)$$

Similar to the Adams–Bashforth formula, function $F$ can be represented by

$$F(t,X) = F_n + \frac{t - t_n}{h} \nabla F_n + \frac{(t - t_n)(t - t_{n-1})}{2! h^2} \nabla^2 F_n + \cdots$$
$$+ \frac{(t - t_n)(t - t_{n-1}) \cdots (t - t_{n-k+1})}{k! h^k} \nabla^k F_n. \quad (11.177)$$

Substituting Eq. 11.177 into Eq. 11.176, one has (similar to the derivation of the Adams algorithms) (cf. Xu 1994)

$$X(t_{n+1}) = 2X(t_n) - X(t_{n-1}) + h^2 \sum_{j=0}^{k} \beta_j F_{n-j}, \quad (11.178)$$

where

$$\beta_j = \sum_{m=j}^{k} (-1)^j C_m^j \sigma_m,$$
$$\sigma_0 = 1, \quad \sigma_m = 1 - \sum_{j=1}^{m} \frac{2}{j+2} b_{j+1} \sigma_{m-j}, (m \geq 1), \quad (11.179)$$
$$b_j = \sum_{i=1}^{j} \frac{1}{i}.$$

Equation 11.178 is called the Stormer formula. Similar to the discussions in Adams algorithms, taking $F_{n+1}$ into account, one has

$$F(t, X) = F_{n+1} + \frac{t - t_{n+1}}{h} \nabla F_{n+1} + \frac{(t - t_{n+1})(t - t_n)}{2! h^2} \nabla^2 F_{n+1} + \cdots$$
$$+ \frac{(t - t_{n+1})(t - t_n) \cdots (t - t_{n-k+2})}{k! h^k} \nabla^k F_{n+1} \qquad (11.180)$$

and (cf. Xu 1994)

$$X(t_{n+1}) = 2X(t_n) - X(t_{n-1}) + h^2 \sum_{j=0}^{k} \beta_j^* F_{n+1-j}, \qquad (11.181)$$

where

$$\beta_j^* = \sum_{m=j}^{k} (-1)^j C_m^j \sigma_m^*,$$

$$\sigma_0^* = 1, \ \sigma_m^* = -\sum_{j=1}^{m} \frac{2}{j+2} b_{j+1} \sigma_{m-j}^*, \quad (m \geq 1), \text{ and} \qquad (11.182)$$

$$b_j = \sum_{i=1}^{j} \frac{1}{i}.$$

Equation 11.181 is called the Cowell formula. Because of the use of $F_{n+1}$ to approximate $F$, the Cowell formula may reach a higher accuracy than the Stormer formula. However, before $X_{n+1}$ has been computed, $F_{n+1}$ may not be computed exactly, and so an iterative process is needed to use the Cowell formula. A simplified way to use the Cowell formula is to use the Stormer formula to compute $X_{n+1}$ and $F_{n+1}$, and then to use the Cowell formula to compute the modified $X_{n+1}$ using $F_{n+1}$. Experience shows that such a process will be accurate enough for many applications.

## 11.6.4
### Mixed Algorithms and Discussions

Above we discussed three algorithms for solving the initial value problem of the orbit differential equation. The Runge–Kutta algorithm is a single-step method. The formulas of different order Runge–Kutta algorithms do not have simple relationships, and even for a definite order the formulas are not unique. For every step of integration, several function values of $F$ must be computed for use. The most important property of the Runge–Kutta algorithm is that the method is a self-starting one. Generally, the Runge–Kutta algorithm is often used for providing the starting values for multiple-step algorithms.

Adams algorithms are multiple-step methods. The order of the formulas can be easily raised because of their sequential relationships. However, the Adams algorithms cannot start themselves. For every step of integration, only one function value must be computed. The disturbing function is considered a function of time and the state of the satellite. Thus the Adams methods can be used in orbit determination problems without any problem with the disturbing function. In the case of a higher accuracy requirement, a mixed Adams–Bashforth method and Adams–Moulton methods can be used in an iterative process.

Cowell algorithms are also multiple-step methods, and the order can easily be changed. Cowell methods also need starting help from other methods. Analysis shows that Cowell algorithms have a higher accuracy than that of Adams algorithms when the same orders of formulas are used. However, Cowell formulas are only suitable for that kind of disturbing function $F$, which is the function of the time and the position of the satellite. It is well known that the atmospheric drag is a disturbing force, which is a function of the velocity of the satellite. Therefore, Cowell algorithms can only be used for integrating a part of the disturbing forces. A mixed Cowell method still keeps this property.

Obviously, the forces of the equation of motion must be separated into two parts, one includes the forces that are functions of the velocity of the satellite, and the other includes all remaining forces. The first part can be integrated using Adams methods, and the other can be integrated using Cowell methods. The Runge–Kutta algorithm will be used for providing the needed starting values.

The selections of the order number and step size are dependent on the accuracy requirements and the orbit conditions. Usually, the order and the step size are selectable input variables of the software, and can be properly selected after several test runs. Scheinert suggested using eighth-order Runge–Kutta algorithms, as well as 12th-order Adams and Cowell algorithms (cf. Scheinert 1996). It is notable that by order selection, it is not the higher the order is, the higher the accuracy will be. For the step size selection, it is not the smaller the step size is, the better the results will be.

## 11.6.5
## Interpolation Algorithms

Orbits are given through integration at the step points $t_0 + nh$ ($n = 0, 1, \ldots$). For GPS satellites, $h$ is usually selected as 300 s. However, GPS observations are made, usually in IGS, every 15 s. For linearisation and formation of the GPS observation equations, the orbit data must sometimes be interpolated to the needed epochs, and therefore we must discuss the method of interpolation. The often-used Lagrange interpolation algorithm was discussed in Sect. 3.4. A fifth-order polynomial interpolation method was given in Sect. 5.4.2. By deriving the Adams and Cowell algorithms, the Newtonian backward differentiation formula was used to represent the disturbing function $F$. By simply considering $F$ as a function of $t$ ($t$ is any variable), then one has

$$F(t) = F(t_n) + \frac{t - t_n}{h} \nabla F_n + \frac{(t - t_n)(t - t_{n-1})}{2!h^2} \nabla^2 F_n + \cdots \\ + \frac{(t - t_n)(t - t_{n-1}) \cdots (t - t_{n-k+1})}{k!h^k} \nabla^k F_n. \tag{11.183}$$

This is an interpolating formula of $F(t)$ using a set of function values of $\{F_{n-j}, j = 0, \ldots, k\}$.

## 11.7
## Orbit-Related Partial Derivatives

As mentioned in Sect. 11.5.1, the partial derivatives of

$$\frac{\partial \vec{f}}{\partial \vec{r}}, \quad \frac{\partial \vec{f}}{\partial \dot{\vec{r}}} \quad \text{and} \quad \frac{\partial \vec{f}}{\partial \vec{Y}} \tag{11.184}$$

will be derived in this section in detail, where the force vector is a summated vector of all disturbing forces in the ECSF coordinate system. If the force vector is given in the ECEF coordinate system, there is

$$\left( \frac{\partial \vec{f}}{\partial \vec{r}}, \frac{\partial \vec{f}}{\partial \dot{\vec{r}}} \right) = R_P^{-1} R_N^{-1} R_S^{-1} R_M^{-1} \left( \frac{\partial \vec{f}_{\text{ECEF}}}{\partial \vec{r}}, \frac{\partial \vec{f}_{\text{ECEF}}}{\partial \dot{\vec{r}}} \right). \tag{11.185}$$

Because of

$$\vec{r} = R \cdot \vec{r}_{\text{ECEF}} \\ \vec{f} = R \cdot \vec{f}_{\text{ECEF}},$$

one may have the velocity transformation formula

$$\frac{d\vec{r}}{dt} = \frac{dR}{dt} \cdot \vec{r}_{\text{ECEF}} + R \cdot \frac{d\vec{r}_{\text{ECEF}}}{dt},$$

where

$$R = R_P^{-1} R_N^{-1} R_S^{-1} R_M^{-1}.$$

Therefore, one has

$$\frac{\partial \vec{r}_{\text{ECEF}}}{\partial \vec{r}} = R^{-1},$$

$$\frac{\partial \dot{\vec{r}}_{\text{ECEF}}}{\partial \dot{\vec{r}}} = R^{-1},$$

and

$$\frac{\partial \vec{f}_{\text{ECEF}}}{\partial \vec{r}} = \frac{\partial \vec{f}_{\text{ECEF}}}{\partial \vec{r}_{\text{ECEF}}} \frac{\partial \vec{r}_{\text{ECEF}}}{\partial \vec{r}} = \frac{\partial \vec{f}_{\text{ECEF}}}{\partial \vec{r}_{\text{ECEF}}} R^{-1},$$

$$\frac{\partial \vec{f}_{\text{ECEF}}}{\partial \dot{\vec{r}}} = \frac{\partial \vec{f}_{\text{ECEF}}}{\partial \dot{\vec{r}}_{\text{ECEF}}} \frac{\partial \dot{\vec{r}}_{\text{ECEF}}}{\partial \dot{\vec{r}}} = \frac{\partial \vec{f}_{\text{ECEF}}}{\partial \dot{\vec{r}}_{\text{ECEF}}} R^{-1}.$$

1. **Geopotential Disturbing Force**

The geopotential disturbing force vector (cf. Sect. 11.2) has the form of

$$\vec{f}_{\text{ECEF}} = \begin{pmatrix} f_{x'} \\ f_{y'} \\ f_{z'} \end{pmatrix} = \begin{pmatrix} b_{11} \frac{\partial V}{\partial r} + b_{21} \frac{\partial V}{\partial \varphi} + b_{31} \frac{\partial V}{\partial \lambda} \\ b_{12} \frac{\partial V}{\partial r} + b_{22} \frac{\partial V}{\partial \varphi} + b_{32} \frac{\partial V}{\partial \lambda} \\ b_{13} \frac{\partial V}{\partial r} + b_{23} \frac{\partial V}{\partial \varphi} \end{pmatrix}, \quad (11.186)$$

where

$$\frac{\partial (r, \varphi, \lambda)}{\partial (x', y', z')} = \begin{pmatrix} b_{11} & b_{12} & b_{13} \\ b_{21} & b_{22} & b_{23} \\ b_{31} & b_{32} & b_{33} \end{pmatrix} = \begin{pmatrix} \cos \varphi \cos \lambda & \cos \varphi \sin \lambda & \sin \varphi \\ -\frac{1}{r} \sin \varphi \cos \lambda & -\frac{1}{r} \sin \varphi \sin \lambda & \frac{1}{r} \cos \varphi \\ -\frac{1}{r \cos \varphi} \sin \lambda & \frac{1}{r \cos \varphi} \cos \lambda & 0 \end{pmatrix},$$

and $(x', y', z')$ are the three orthogonal Cartesian coordinates in the ECEF system. Thus,

$$\frac{\partial \vec{f}_{\text{ECEF}}}{\partial \vec{r}} = \begin{pmatrix} \frac{\partial f_{x'}}{\partial (x', y', z')} \\ \frac{\partial f_{y'}}{\partial (x', y', z')} \\ \frac{\partial f_{z'}}{\partial (x', y', z')} \end{pmatrix} = \left( \frac{\partial (f_{x'}, f_{y'}, f_{z'})}{\partial (r, \varphi, \lambda)} \frac{\partial (r, \varphi, \lambda)}{\partial (x', y', z')} \right)^{\text{T}}. \quad (11.187)$$

Using index $j$ (=1, 2, 3) to denote index $(x', y', z')$, one has

$$\frac{\partial f_j}{\partial (r, \varphi, \lambda)} = \begin{pmatrix} \frac{\partial b_{1j}}{\partial r}\frac{\partial V}{\partial r} + \frac{\partial b_{2j}}{\partial r}\frac{\partial V}{\partial \varphi} + \frac{\partial b_{3j}}{\partial r}\frac{\partial V}{\partial \lambda} + b_{1j}\frac{\partial^2 V}{\partial r^2} + b_{2j}\frac{\partial^2 V}{\partial r \partial \varphi} + b_{3j}\frac{\partial^2 V}{\partial r \partial \lambda} \\ \frac{\partial b_{1j}}{\partial \varphi}\frac{\partial V}{\partial r} + \frac{\partial b_{2j}}{\partial \varphi}\frac{\partial V}{\partial \varphi} + \frac{\partial b_{3j}}{\partial \varphi}\frac{\partial V}{\partial \lambda} + b_{1j}\frac{\partial^2 V}{\partial r \partial \varphi} + b_{2j}\frac{\partial^2 V}{\partial \varphi^2} + b_{3j}\frac{\partial^2 V}{\partial \varphi \partial \lambda} \\ \frac{\partial b_{1j}}{\partial \lambda}\frac{\partial V}{\partial r} + \frac{\partial b_{2j}}{\partial \lambda}\frac{\partial V}{\partial \varphi} + \frac{\partial b_{3j}}{\partial \lambda}\frac{\partial V}{\partial \lambda} + b_{1j}\frac{\partial^2 V}{\partial r \partial \lambda} + b_{2j}\frac{\partial^2 V}{\partial \varphi \partial \lambda} + b_{3j}\frac{\partial^2 V}{\partial \lambda^2} \end{pmatrix}^T,$$

(11.188)

where

$$\frac{\partial}{\partial r}\begin{pmatrix} b_{11} & b_{12} & b_{13} \\ b_{21} & b_{22} & b_{23} \\ b_{31} & b_{32} & b_{33} \end{pmatrix} = \begin{pmatrix} 0 & 0 & 0 \\ \frac{1}{r^2}\sin\varphi\cos\lambda & \frac{1}{r^2}\sin\varphi\sin\lambda & \frac{-1}{r^2}\cos\varphi \\ \frac{1}{r^2\cos\varphi}\sin\lambda & \frac{-1}{r^2\cos\varphi}\cos\lambda & 0 \end{pmatrix},$$

$$\frac{\partial}{\partial \varphi}\begin{pmatrix} b_{11} & b_{12} & b_{13} \\ b_{21} & b_{22} & b_{23} \\ b_{31} & b_{32} & b_{33} \end{pmatrix} = \begin{pmatrix} -\sin\varphi\cos\lambda & -\sin\varphi\sin\lambda & \cos\varphi \\ -\frac{1}{r}\cos\varphi\cos\lambda & -\frac{1}{r}\cos\varphi\sin\lambda & -\frac{1}{r}\sin\varphi \\ -\frac{\sin\varphi}{r\cos^2\varphi}\sin\lambda & \frac{\sin\varphi}{r\cos^2\varphi}\cos\lambda & 0 \end{pmatrix} \quad \text{and}$$

$$\frac{\partial}{\partial \lambda}\begin{pmatrix} b_{11} & b_{12} & b_{13} \\ b_{21} & b_{22} & b_{23} \\ b_{31} & b_{32} & b_{33} \end{pmatrix} = \begin{pmatrix} -\cos\varphi\sin\lambda & \cos\varphi\cos\lambda & 0 \\ \frac{1}{r}\sin\varphi\sin\lambda & -\frac{1}{r}\sin\varphi\cos\lambda & 0 \\ -\frac{1}{r\cos\varphi}\cos\lambda & -\frac{1}{r\cos\varphi}\sin\lambda & 0 \end{pmatrix}$$

(11.189)

and

$$\frac{\partial^2 V}{\partial r^2} = \frac{\mu}{r^3}\left[2 + \sum_{l=2}^{\infty}\sum_{m=0}^{l}(l+1)(l+2)\left(\frac{a}{r}\right)^l \overline{P}_{lm}(\sin\varphi)[\overline{C}_{lm}\cos m\lambda + \overline{S}_{lm}\sin m\lambda]\right],$$

$$\frac{\partial^2 V}{\partial r \partial \varphi} = -\frac{\mu}{r^2}\sum_{l=2}^{\infty}\sum_{m=0}^{l}(l+1)\left(\frac{a}{r}\right)^l \frac{d\overline{P}_{lm}(\sin\varphi)}{d\varphi}[\overline{C}_{lm}\cos m\lambda + \overline{S}_{lm}\sin m\lambda],$$

$$\frac{\partial^2 V}{\partial r \partial \lambda} = -\frac{\mu}{r^2}\left[\sum_{l=2}^{\infty}\sum_{m=0}^{l}(l+1)\left(\frac{a}{r}\right)^l \overline{P}_{lm}(\sin\varphi)m[-\overline{C}_{lm}\sin m\lambda + \overline{S}_{lm}\cos m\lambda]\right],$$

$$\frac{\partial^2 V}{\partial \varphi^2} = \frac{\mu}{r}\sum_{l=2}^{\infty}\sum_{m=0}^{l}\left(\frac{a}{r}\right)^l \frac{d^2\overline{P}_{lm}(\sin\varphi)}{d\varphi^2}[\overline{C}_{lm}\cos m\lambda + \overline{S}_{lm}\sin m\lambda],$$

$$\frac{\partial^2 V}{\partial \varphi \partial \lambda} = \frac{\mu}{r}\sum_{l=2}^{\infty}\sum_{m=0}^{l}\left(\frac{a}{r}\right)^l \frac{d\overline{P}_{lm}(\sin\varphi)}{d\varphi}m[-\overline{C}_{lm}\sin m\lambda + \overline{S}_{lm}\cos m\lambda], \quad \text{and}$$

$$\frac{\partial^2 V}{\partial \lambda^2} = -\frac{\mu}{r}\sum_{l=2}^{\infty}\sum_{m=0}^{l}m^2\left(\frac{a}{r}\right)^l \overline{P}_{lm}(\sin\varphi)[\overline{C}_{lm}\cos m\lambda + \overline{S}_{lm}\sin m\lambda],$$

(11.190)

where

$$\frac{d\overline{P}_{lm}(\sin\varphi)}{d\varphi} = \beta(m)\overline{P}_{l(m+1)}(\sin\varphi) - m\tan\varphi \overline{P}_{lm}(\sin\varphi),$$

$$\frac{d^2\overline{P}_{lm}(\sin\varphi)}{d\varphi^2} = \beta(m)\frac{d\overline{P}_{l(m+1)}(\sin\varphi)}{d\varphi} - m\frac{1}{\cos^2\varphi}\overline{P}_{lm}(\sin\varphi) - m\tan\varphi\frac{d\overline{P}_{lm}(\sin\varphi)}{d\varphi}$$

$$= \beta(m)\beta(m+1)\overline{P}_{l(m+2)}(\sin\varphi) - \beta(m)\tan\varphi(2m+1)\overline{P}_{l(m+1)}(\sin\varphi)$$

$$+ \left(m^2\tan^2\varphi - m\frac{1}{\cos^2\varphi}\right)\overline{P}_{lm}(\sin\varphi),$$

$$\beta(m) = \left[\frac{1}{2}(2-\delta_{0m})(l-m)(l+m+1)\right]^{1/2} \quad \text{and}$$

$$\beta(m+1) = \left[\frac{1}{2}(l-m-1)(l+m+2)\right]^{1/2}.$$

(11.191)

Other needed functions are already given in Sect. 11.1. Because the force is not a function of velocity, it is obvious that

$$\frac{\partial \vec{f}_{\text{ECEF}}}{\partial \dot{\vec{r}}} = [0]_{3\times 3}. \tag{11.192}$$

Only non-zero partial derivatives will be given in later text.

Supposing the geopotential parameters $\overline{C}_{lm}^N, \overline{S}_{lm}^N$ are known (as initial values), $\overline{C}_{lm}, \overline{S}_{lm}$ are true values, and $\Delta \vec{\overline{C}}_{lm}, \Delta \vec{\overline{S}}_{lm}$ are searched corrections (unknowns), then the geopotential force is

$$\vec{f}_{\text{ECEF}}(\overline{C}_{lm}, \overline{S}_{lm}) = \vec{f}_{\text{ECEF}}(\overline{C}_{lm}^N, \overline{S}_{lm}^N) + \vec{f}_{\text{ECEF}}(\overline{C}_{lm}, \overline{S}_{lm}) - \vec{f}_{\text{ECEF}}(\overline{C}_{lm}^N, \overline{S}_{lm}^N)$$
$$= \vec{f}_{\text{ECEF}}(\overline{C}_{lm}^N, \overline{S}_{lm}^N) + \vec{f}_{\text{ECEF}}(\Delta \overline{C}_{lm}, \Delta \overline{S}_{lm}),$$

(11.193)

and

$$\frac{\partial \vec{f}_{\text{ECEF}}}{\partial(\Delta \overline{C}_{lm}, \Delta \overline{S}_{lm})} = \begin{pmatrix} b_{11} & b_{12} & b_{13} \\ b_{21} & b_{22} & b_{23} \\ b_{31} & b_{32} & b_{33} \end{pmatrix}^T \frac{\partial}{\partial(\Delta \overline{C}_{lm}, \Delta \overline{S}_{lm})} \begin{pmatrix} \frac{\partial V}{\partial r} \\ \frac{\partial V}{\partial \varphi} \\ \frac{\partial V}{\partial \lambda} \end{pmatrix},$$

$$\frac{\partial}{\partial(\Delta \overline{C}_{lm}, \Delta \overline{S}_{lm})}\left(\frac{\partial V}{\partial r}\right) = -\frac{\mu}{r^2}(l+1)\left(\frac{a}{r}\right)^l \overline{P}_{lm}(\sin\varphi)(\cos m\lambda \quad \sin m\lambda),$$ (11.194)

$$\frac{\partial}{\partial(\Delta \overline{C}_{lm}, \Delta \overline{S}_{lm})}\left(\frac{\partial V}{\partial \varphi}\right) = \frac{\mu}{r}\left(\frac{a}{r}\right)^l \frac{d\overline{P}_{lm}(\sin\varphi)}{d\varphi}(\cos m\lambda \quad \sin m\lambda) \quad \text{and}$$

$$\frac{\partial}{\partial(\Delta \overline{C}_{lm}, \Delta \overline{S}_{lm})}\left(\frac{\partial V}{\partial \lambda}\right) = \frac{\mu}{r}m\left(\frac{a}{r}\right)^l \overline{P}_{lm}(\sin\varphi)(-\sin m\lambda \quad \cos m\lambda).$$

## 2. Perturbation Forces of the Sun, the Moon, and the Planets

The perturbation forces of the sun, the moon, and the planets are given in Sect. 11.2.2 as (cf. Eq. 11.50)

$$\vec{f}_m = -m \sum_j \mathrm{Gm}(j) \left[ \frac{\vec{r} - \vec{r}_{m(j)}}{|\vec{r} - \vec{r}_{m(j)}|^3} + \frac{\vec{r}_{m(j)}}{r_{m(j)}^3} \right], \tag{11.195}$$

where $\mathrm{Gm}(j)$ are the gravitational constants of the sun and the moon, as well as the planets, and the vector with index $m(j)$ are the geocentric vector of the sun, the moon, and the planets. The partial derivatives of the perturbation force with respect to the satellite vector are then

$$\frac{\partial \vec{f}_m}{\partial \vec{r}} = -m \sum_j \frac{\mathrm{Gm}(j)}{|\vec{r} - \vec{r}_{m(j)}|^3} \left( E + \frac{3}{|\vec{r} - \vec{r}_{m(j)}|^2} \begin{pmatrix} x - x_{m(j)} \\ y - y_{m(j)} \\ z - z_{m(j)} \end{pmatrix} \begin{pmatrix} x - x_{m(j)} \\ y - y_{m(j)} \\ z - z_{m(j)} \end{pmatrix}^T \right), \tag{11.196}$$

where $E$ is an identity matrix of size $3 \times 3$. The partial derivatives of the force vector with respect to the velocity vector of the satellite are zero. The disturbances of the sun, moon, and planets are considered well modelled; therefore, no parameters will be adjusted. In other words, the partial derivatives of the force vector with respect to the model parameters do not exist.

## 3. Tidal Disturbing Forces

Similar to the geopotential attraction force, the tidal force (cf. Sect. 11.2.3) has the form

$$\vec{f}_{\mathrm{ECEF}} = \begin{pmatrix} f_{x'} \\ f_{y'} \\ f_{z'} \end{pmatrix} = \begin{pmatrix} b_{11} \frac{\partial V}{\partial r} + b_{21} \frac{\partial V}{\partial \varphi} + b_{31} \frac{\partial V}{\partial \lambda} \\ b_{12} \frac{\partial V}{\partial r} + b_{22} \frac{\partial V}{\partial \varphi} + b_{32} \frac{\partial V}{\partial \lambda} \\ b_{13} \frac{\partial V}{\partial r} + b_{23} \frac{\partial V}{\partial \varphi} \end{pmatrix}, \tag{11.197}$$

where $V = \delta V + \delta V_1 + \delta V_2$, it is a summation of the earth tide potential and the two parts of ocean loading tide potentials. The Eq. 11.188 is still valid for this case. Other higher-order partial derivatives can be derived as follows:

$$\frac{\partial^2 \delta V}{\partial r^2} = \sum_{j=1}^{2} \mu_j \sum_{n=2}^{N} k_n \frac{(n+1)(n+2) a_e^{2n+1}}{r^{n+3} r_j^{n+1}} \left[ P_n(\sin \varphi) P_n(\sin \delta_j) + 2 \sum_{k=1}^{n} \frac{(n-k)!}{(n+k)!} P_{nk}(\sin \varphi) P_{nk}(\sin \delta_j) \cos k h_j \right],$$

## 11.7 · Orbit-Related Partial Derivatives

$$\frac{\partial^2 \delta V}{\partial r \partial \varphi} = \sum_{j=1}^{2} \mu_j \sum_{n=2}^{N} -k_n \frac{(n+1)a_e^{2n+1}}{r^{n+2}r_j^{n+1}} \left[ \begin{array}{l} \dfrac{dP_n(\sin\varphi)}{d\varphi} P_n(\sin\delta_j) \\ + 2 \displaystyle\sum_{k=1}^{n} \dfrac{(n-k)!}{(n+k)!} \dfrac{dP_{nk}(\sin\varphi)}{d\varphi} P_{nk}(\sin\delta_j)\cos kh_j \end{array} \right],$$

$$\frac{\partial^2 \delta V}{\partial r \partial \lambda} = \sum_{j=1}^{2} \mu_j \sum_{n=2}^{N} -k_n \frac{(n+1)a_e^{2n+1}}{r^{n+2}r_j^{n+1}} \left[ 2\sum_{k=1}^{n} \frac{(n-k)!}{(n+k)!} P_{nk}(\sin\varphi) P_{nk}(\sin\delta_j) k \sin kh_j \right],$$

$$\frac{\partial^2 \delta V}{\partial \varphi^2} = \sum_{j=1}^{2} \mu_j \sum_{n=2}^{N} k_n \frac{a_e^{2n+1}}{r^{n+1}r_j^{n+1}} \left[ \begin{array}{l} \dfrac{d^2 P_n(\sin\varphi)}{d\varphi^2} P_n(\sin\delta_j) \\ + 2 \displaystyle\sum_{k=1}^{n} \dfrac{(n-k)!}{(n+k)!} \dfrac{d^2 P_{nk}(\sin\varphi)}{d\varphi^2} P_{nk}(\sin\delta_j)\cos kh_j \end{array} \right],$$

$$\frac{\partial^2 \delta V}{\partial \varphi \partial \lambda} = \sum_{j=1}^{2} \mu_j \sum_{n=2}^{N} k_n \frac{a_e^{2n+1}}{r^{n+1}r_j^{n+1}} \left[ 2\sum_{k=1}^{n} \frac{(n-k)!}{(n+k)!} \frac{dP_{nk}(\sin\varphi)}{d\varphi} P_{nk}(\sin\delta_j) k \sin kh_j \right],$$

$$\frac{\partial^2 \delta V}{\partial \lambda^2} = \sum_{j=1}^{2} \mu_j \sum_{n=2}^{N} -k_n \frac{a_e^{2n+1}}{r^{n+1}r_j^{n+1}} \left[ 2\sum_{k=1}^{n} \frac{(n-k)!}{(n+k)!} k^2 P_{nk}(\sin\varphi) P_{nk}(\sin\delta_j) \cos kh_j \right],$$

$$\frac{\partial^2 \delta V_1}{\partial r^2} = \oiint_O G\sigma H \sum_{n=0}^{\infty} (1+k'_n) \frac{(n+1)(n+2)a_e^n}{r^{n+3}} \left[ \begin{array}{l} P_n(\sin\varphi) P_n(\sin\varphi_s) + (2-\delta_{0n})\cdot \\ \displaystyle\sum_{k=1}^{n} \dfrac{(n-k)!}{(n+k)!} P_{nk}(\sin\varphi) P_{nk}(\sin\varphi_s) \cos k(\lambda_s - \lambda) \end{array} \right] ds,$$

$$\frac{\partial^2 \delta V_1}{\partial r \partial \varphi} = \oiint_O G\sigma H \sum_{n=0}^{\infty} (1+k'_n) \frac{-(n+1)a_e^n}{r^{n+2}} \left[ \begin{array}{l} \dfrac{dP_n(\sin\varphi)}{d\varphi} P_n(\sin\varphi_s) + (2-\delta_{0n})\cdot \\ \displaystyle\sum_{k=0}^{n} \dfrac{(n-k)!}{(n+k)!} \dfrac{dP_{nk}(\sin\varphi)}{d\varphi} P_{nk}(\sin\varphi_s) \cos k(\lambda_s - \lambda) \end{array} \right] ds,$$

$$\frac{\partial^2 \delta V_1}{\partial r \partial \lambda} = \oiint_O G\sigma H \sum_{n=0}^{\infty} (1+k'_n) \frac{-(n+1)a_e^n}{r^{n+2}} \left[ \begin{array}{l} (2-\delta_{0n})\cdot \\ \displaystyle\sum_{k=0}^{n} \dfrac{(n-k)!}{(n+k)!} P_{nk}(\sin\varphi) P_{nk}(\sin\varphi_s) k \sin k(\lambda_s - \lambda) \end{array} \right] ds,$$

$$\frac{\partial^2 \delta V_1}{\partial \varphi^2} = \oiint_O G\sigma H \sum_{n=0}^{\infty} (1+k'_n) \frac{a_e^n}{r^{n+1}} \left[ \begin{array}{l} \dfrac{d^2 P_n(\sin\varphi)}{d\varphi^2} P_n(\sin\varphi_s) + (2-\delta_{0n})\cdot \\ \displaystyle\sum_{k=0}^{n} \dfrac{(n-k)!}{(n+k)!} \dfrac{d^2 P_{nk}(\sin\varphi)}{d\varphi^2} P_{nk}(\sin\varphi_s) \cos k(\lambda_s - \lambda) \end{array} \right] ds,$$

$$\frac{\partial^2 \delta V_1}{\partial \varphi \partial \lambda} = \oiint_O G\sigma H \sum_{n=0}^{\infty} (1+k'_n) \frac{a_e^n}{r^{n+1}} \left[ \begin{array}{l} (2-\delta_{0n})\cdot \\ \displaystyle\sum_{k=0}^{n} \dfrac{(n-k)!}{(n+k)!} \dfrac{dP_{nk}(\sin\varphi)}{d\varphi} P_{nk}(\sin\varphi_s) k \sin k(\lambda_s - \lambda) \end{array} \right] ds,$$

$$\frac{\partial^2 \delta V_1}{\partial \lambda^2} = \oiint_O G\sigma H \sum_{n=0}^{\infty} (1+k'_n) \frac{a_e^n}{r^{n+1}} \left[ \begin{array}{l} -(2-\delta_{0n})\cdot \\ \displaystyle\sum_{k=0}^{n} \dfrac{(n-k)!}{(n+k)!} P_{nk}(\sin\varphi) P_{nk}(\sin\varphi_s) k^2 \cos k(\lambda_s - \lambda) \end{array} \right] ds,$$

$$\frac{\partial^2 \delta V_2}{\partial r^2} = \oiint_C G\sigma_e u_r \sum_{n=0}^{\infty} \frac{(n+1)(n+2)a_e^n}{r^{n+3}} \left[ \begin{array}{l} P_n(\sin\varphi)P_n(\sin\varphi_s) + (2-\delta_{0n}) \cdot \\ \sum_{k=0}^{n} \frac{(n-k)!}{(n+k)!} P_{nk}(\sin\varphi)P_{nk}(\sin\varphi_s)\cos k(\lambda_s - \lambda) \end{array} \right] ds,$$

$$\frac{\partial^2 \delta V_2}{\partial r \partial \varphi} = \oiint_C G\sigma_e u_r \sum_{n=0}^{\infty} \frac{-(n+1)a_e^n}{r^{n+2}} \left[ \begin{array}{l} \frac{dP_n(\sin\varphi)}{d\varphi} P_n(\sin\varphi_s) + (2-\delta_{0n}) \cdot \\ \sum_{k=0}^{n} \frac{(n-k)!}{(n+k)!} \frac{dP_{nk}(\sin\varphi)}{d\varphi} P_{nk}(\sin\varphi_s)\cos k(\lambda_s - \lambda) \end{array} \right] ds,$$

$$\frac{\partial^2 \delta V_2}{\partial r \partial \lambda} = \oiint_C G\sigma_e u_r \sum_{n=0}^{\infty} \frac{-(n+1)a_e^n}{r^{n+2}} \left[ \begin{array}{l} (2-\delta_{0n}) \cdot \\ \sum_{k=0}^{n} \frac{(n-k)!}{(n+k)!} P_{nk}(\sin\varphi)P_{nk}(\sin\varphi_s) k \sin k(\lambda_s - \lambda) \end{array} \right] ds,$$

$$\frac{\partial^2 \delta V_2}{\partial \varphi^2} = \oiint_C G\sigma_e u_r \sum_{n=0}^{\infty} \frac{a_e^n}{r^{n+1}} \left[ \begin{array}{l} \frac{d^2 P_n(\sin\varphi)}{d\varphi^2} P_n(\sin\varphi_s) + (2-\delta_{0n}) \cdot \\ \sum_{k=0}^{n} \frac{(n-k)!}{(n+k)!} \frac{d^2 P_{nk}(\sin\varphi)}{d\varphi^2} P_{nk}(\sin\varphi_s)\cos k(\lambda_s - \lambda) \end{array} \right] ds,$$

$$\frac{\partial^2 \delta V_2}{\partial \varphi \partial \lambda} = \oiint_C G\sigma_e u_r \sum_{n=0}^{\infty} \frac{a_e^n}{r^{n+1}} \left[ \begin{array}{l} (2-\delta_{0n}) \cdot \\ \sum_{k=0}^{n} \frac{(n-k)!}{(n+k)!} \frac{dP_{nk}(\sin\varphi)}{d\varphi} P_{nk}(\sin\varphi_s) k \sin k(\lambda_s - \lambda) \end{array} \right] ds \text{ and}$$

$$\frac{\partial^2 \delta V_2}{\partial \lambda^2} = \oiint_C G\sigma_e u_r \sum_{n=0}^{\infty} \frac{a_e^n}{r^{n+1}} \left[ \begin{array}{l} -(2-\delta_{0n}) \cdot \\ \sum_{k=0}^{n} \frac{(n-k)!}{(n+k)!} P_{nk}(\sin\varphi)P_{nk}(\sin\varphi_s) k^2 \cos k(\lambda_s - \lambda) \end{array} \right] ds,$$

(11.198)

where

$$\frac{dP_n(\sin\varphi)}{d\varphi} = \frac{n}{\cos\varphi}(P_{n-1}(\sin\varphi) - \sin\varphi P_n(\sin\varphi)) \text{ and}$$
$$\frac{dP_{nk}(\sin\varphi)}{d\varphi} = P_{n(k+1)}(\sin\varphi) - k\tan\varphi P_{nk}(\sin\varphi).$$
(11.199)

### 4. Solar Radiation Pressure

Solar radiation force acting on the satellite's surface is (cf. Sect. 11.2.4)

$$\vec{f}_{\text{solar}} = m\gamma P_s C_r r_{\text{sun}}^2 \frac{S}{m} \frac{\vec{r} - \vec{r}_{\text{sun}}}{|\vec{r} - \vec{r}_{\text{sun}}|^3};$$
(11.200)

the partial derivatives of the perturbation force with respect to the satellite vector are then

$$\frac{\partial \vec{f}_{\text{solar}}}{\partial \vec{r}} = m\gamma P_s C_r r_{\text{sun}}^2 \frac{S}{m} \frac{1}{|\vec{r}-\vec{r}_{\text{sun}}|^3}\left(E - \frac{3}{|\vec{r}-\vec{r}_{\text{sun}}|^2}\begin{pmatrix} x-x_{\text{sun}} \\ y-y_{\text{sun}} \\ z-z_{\text{sun}} \end{pmatrix}\begin{pmatrix} x-x_{\text{sun}} \\ y-y_{\text{sun}} \\ z-z_{\text{sun}} \end{pmatrix}^T\right),$$
(11.201)

where $E$ is an identity matrix of size $3 \times 3$. The partial derivatives of the force vector with respect to the velocity vector of the satellite are zero. The disturbance of the solar radiation is considered not well modelled; therefore, unknown parameters will also be adjusted. The total model is (cf. Sect. 11.2)

$$\vec{f}_{\text{solar-force}} = \vec{f}_{\text{solar}} + \begin{pmatrix} a_{11} & a_{12} & a_{13} \\ a_{21} & a_{22} & a_{23} \\ a_{31} & a_{32} & a_{33} \end{pmatrix}\begin{pmatrix} 1 \\ \cos u \\ \sin u \end{pmatrix}.$$
(11.202)

Thus,

$$\frac{\partial \vec{f}_{\text{solar-force}}}{\partial \vec{r}} = \frac{\partial \vec{f}_{\text{solar}}}{\partial \vec{r}} + \begin{pmatrix} a_{11} & a_{12} & a_{13} \\ a_{21} & a_{22} & a_{23} \\ a_{31} & a_{32} & a_{33} \end{pmatrix}\begin{pmatrix} 0 \\ -\sin u \\ \cos u \end{pmatrix}\frac{\partial u}{\partial \vec{r}},$$
(11.203)

where

$$\frac{\partial u}{\partial \vec{r}} = \frac{\partial u}{\partial(\Omega, i, \omega, a, e, M)}\frac{\partial(\Omega, i, \omega, a, e, M)}{\partial(\vec{r}, \dot{\vec{r}})}\frac{\partial(\vec{r}, \dot{\vec{r}})}{\partial \vec{r}}.$$
(11.204)

On the right-hand side of above equation there are three matrices, the first one is a $1 \times 6$ matrix (vector) and is given in Sect. 11.1.2 (cf. Eq. 11.24), the second one is given as its inverse in Sect. 11.4 (cf. Eqs. 11.130 and 11.132), and the third one is a $6 \times 3$ matrix, or

$$\frac{\partial u}{\partial(\Omega, i, \omega, a, e, M)} = \left(0, 0, 1, 0, \frac{2+e\cos f}{1-e^2}\sin f, \left(\frac{a}{r}\right)^2\sqrt{1-e^2}\right),$$

$$\frac{\partial(\Omega, i, \omega, a, e, M)}{\partial(\vec{r}, \dot{\vec{r}})} = \left(\frac{\partial(\vec{r}, \dot{\vec{r}})}{\partial(\Omega, i, \omega, a, e, M)}\right)^{-1} = \begin{pmatrix} \frac{\partial R}{\partial(\Omega, i, \omega)}\vec{q} & R\frac{\partial \vec{q}}{\partial(a, e, M)} \\ \frac{\partial R}{\partial(\Omega, i, \omega)}\dot{\vec{q}} & R\frac{\partial \dot{\vec{q}}}{\partial(a, e, M)} \end{pmatrix}^{-1} \quad \text{and}$$

$$\frac{\partial(\vec{r}, \dot{\vec{r}})}{\partial \vec{r}} = \begin{pmatrix} E_{3\times 3} \\ 0_{3\times 3} \end{pmatrix}.$$
(11.205)

$$\frac{\partial u}{\partial \vec{r}} = \frac{\partial u}{\partial (\Omega, i, \omega, a, e, M)} \frac{\partial (\Omega, i, \omega, a, e, M)}{\partial (\vec{r}, \dot{\vec{r}})} \frac{\partial (\vec{r}, \dot{\vec{r}})}{\partial \vec{r}} \quad \text{and}$$

$$\frac{\partial (\vec{r}, \dot{\vec{r}})}{\partial \vec{r}} = \begin{pmatrix} 0_{3 \times 3} \\ E_{3 \times 3} \end{pmatrix}.$$

(11.206)

The partial derivatives of the force vector with respect to the model parameters are (for $i = 1, 2, 3$)

$$\frac{\partial \vec{f}_{\text{solar-force}}}{\partial a_{ij}} = \begin{cases} 1 & \text{if } j = 1 \\ \cos u & \text{if } j = 2 \\ \sin u & \text{if } j = 3 \end{cases}.$$

(11.207)

If Xu's Model 11.74

$$\alpha \vec{f}_{\text{solar}} = \begin{pmatrix} a_1 & b_1 \\ a_2 & b_2 \\ a_3 & b_3 \end{pmatrix} \begin{pmatrix} 1 \\ t \end{pmatrix}$$

(11.208)

is used, then one has

$$\frac{\partial \vec{f}_{\text{solar-force}}}{\partial (a_i, b_i)} = (1, t), \quad i = 1, 2, 3.$$

(11.209)

### 5. Atmospheric Drag

Atmospheric drag force has a form of (cf. Sect. 11.2.5)

$$\vec{f}_{\text{drag}} = -m \frac{1}{2} \left( \frac{C_d S}{m} \right) \sigma \left| \dot{\vec{r}} - \dot{\vec{r}}_{\text{air}} \right| \left( \dot{\vec{r}} - \dot{\vec{r}}_{\text{air}} \right),$$

(11.210)

and the air drag force model is

$$\vec{f}_{\text{air-drag}} = \vec{f}_{\text{drag}} + (1 + q) \Delta \vec{f}_{\text{drag}},$$

(11.211)

where (cf. Eqs. 11.84 and 11.85)

$$\Delta \vec{f}_{\text{drag}} = [a + b \varphi(2\omega) \cos(2f) + c \varphi(3\omega) \cos(3f) + d \varphi(\omega) \cos f] \vec{p},$$

(11.212)

$$\varphi(k\omega) = \begin{cases} \sin k\omega & \text{if } \cos k\omega = 0 \\ \frac{1}{\cos k\omega} & \text{if } \cos k\omega \neq 0 \end{cases}, \quad k = 1, 2, 3$$

(11.213)

It is obvious that the partial derivatives of the air drag force with respect to the satellite position vector are zero, and

$$\frac{\partial \vec{f}_{\text{drag}}}{\partial \dot{\vec{r}}} = -m \frac{1}{2}\left(\frac{C_d S}{m}\right)\sigma \left(|\dot{\vec{r}} - \dot{\vec{r}}_{\text{air}}|E + \frac{1}{|\dot{\vec{r}} - \dot{\vec{r}}_{\text{air}}|}\begin{pmatrix}\dot{x} - \dot{x}_{\text{air}} \\ \dot{y} - \dot{y}_{\text{air}} \\ \dot{z} - \dot{z}_{\text{air}}\end{pmatrix}\begin{pmatrix}\dot{x} - \dot{x}_{\text{air}} \\ \dot{y} - \dot{y}_{\text{air}} \\ \dot{z} - \dot{z}_{\text{air}}\end{pmatrix}^T\right),$$

(11.214)

$$\frac{\partial \Delta \vec{f}_{\text{drag}}}{\partial f} = [-2b\varphi(2\omega)\sin(2f) - c\varphi(3\omega)\sin(3f) - d\varphi(\omega)\sin f]\vec{p}, \quad (11.215)$$

$$\frac{\partial \Delta \vec{f}_{\text{drag}}}{\partial \omega} = \left[b\cos(2f)\frac{\partial \varphi(2\omega)}{\partial \omega} + c\cos(3f)\frac{\partial \varphi(3\omega)}{\partial \omega} + d\cos f\frac{\partial \varphi(\omega)}{\partial \omega}\right]\vec{p}, \quad (11.216)$$

$$\frac{\partial \varphi(k\omega)}{\partial \omega} = \begin{cases} k\cos k\omega & \text{if } \cos k\omega = 0 \\ \frac{k\tan k\omega}{\cos k\omega} & \text{if } \cos k\omega \neq 0 \end{cases}, \quad k = 1, 2, 3 \quad (11.217)$$

$$\frac{\partial \Delta \vec{f}_{\text{drag}}}{\partial (\vec{r}, \dot{\vec{r}})} = \frac{\partial \Delta \vec{f}_{\text{drag}}}{\partial (\omega, f)}\frac{\partial (\omega, f)}{\partial (\Omega, i, \omega, a, e, M)}\frac{\partial (\Omega, i, \omega, a, e, M)}{\partial (\vec{r}, \dot{\vec{r}})}\frac{\partial (\vec{r}, \dot{\vec{r}})}{\partial (\vec{r}, \dot{\vec{r}})}, \quad (11.218)$$

where

$$\frac{\partial \omega}{\partial (\Omega, i, \omega, a, e, M)} = (0, 0, 1, 0, 0, 0),$$

$$\frac{\partial f}{\partial (\Omega, i, \omega, a, e, M)} = \left(0, 0, 0, 0, \frac{2 + e\cos f}{1 - e^2}\sin f, \left(\frac{a}{r}\right)^2 \sqrt{1 - e^2}\right)$$

Some of the formulas have been derived previously in this subsection. The partial derivatives of the force vector with respect to the model parameters can be obtained from Eq. 11.215.

# References

Bate RR, Mueller DD, White JE (1971) Fundamentals of astrodynamics. Dover, New York.
Beutler G, Brockmann E, Gurtner W, Hugentobler U, Mervart L, Rothacher M, Verdun A (1994) Extended orbit modelling techniques at the CODE Processing Center of the IGS: Theory and initial results. Manuscr Geodaet 19:367–386.
Brouwer D, Clemence GM (1961) Methods of celestial mechanics. Academic Press, New York.
Cappellari JO (1976) Mathematical theory of the Goddard trajectory determination system.
Cui C (1990) Die Bewegung künstlicher Satelliten im anisotropen Gravitationsfeld einer gleichmässig rotierenden starren Modellerde. Deutsche Geodätische Kommission, Reihe C: Dissertationen, Heft Nr. 357.

Dow JM (1988) Ocean tides and tectonic plate motions from Lageos. Deutsche Geodätische Kommission, Rheihe C, Dissertation, Heft Nr. 344.
Fliegel HF, Gallini TE, Swift ER (1992) Global Positioning System radiation force model for geodetic applications. J Geophys Res 97(B1):559–568.
Heiskanen WA, Moritz H (1967) Physical geodesy. W. H. Freeman, San Francisco/ London.
Herrick S (1972) Astrodynamics, Vol. II. Van Nostrand Reinhold, London.
Kang Z (1998) Präzise Bahnbestimmung niedrigfliegender Satelliten mittels GPS und die Nutzung für die globale Schwerefeldmodellierung. Scientific Technical Report STR 98/25, GeoForschungsZentrum (GFZ) Potsdam.
Kaula WM (1966, 2001) Theory of satellite geodesy. Blaisdell Publishing Company, Dover Publications, New York.
Knudsen P, Andersen O, Khan SA, Hoeyer JL (2000) Ocean tide effects on GRACE gravimetry. IAG Symposia.
Liu L, Zhao D (1979) Orbit theory of the Earth satellite. Nanjing University Press, (in Chinese).
Liu DJ, Shi YM, Guo JJ (1996) Principle of GPS and its data processing. TongJi University Press, Shanghai, (in Chinese).
McCarthy DD (1996) International Earth Rotation Service. IERS conventions, Paris, 95 pp. IERS Technical Note No. 21.
Meeus J (1992) Astronomische Algorithmen. Johann Ambrosius Barth.
Melchior P (1978) The tides of the planet Earth. Pergamon Press.
Montenbruck O (1989) Practical Ephemeris calculations. Springer-Verlag, Heidelberg.
Montenbruck O, Gill E (2000) Satellite Orbits: Models, Methods and Applications. Springer.
Moritz H (1980) Advanced physical geodesy. Herbert Wichmann Verlag, Karlsruhe.
Nie WF, Du YJ, Gao F, Ji CN, Wang TH, Xu G (2016) Validation of the numerical algebra solution for the state transition matrix in precise orbit determination. In review.
Press WH, Teukolsky SA, Vetterling WT, Flannery BP (1992) Numerical recipes in C, $2^{nd}$ Ed. Cam-bridge University Press, New York.
Rapp RH (1986) Global geopotential solutions. In: Sunkel H (ed) Mathematical and numerical tech-niques in physical geodesy. Lecture Notes inEarth Sciences, Vol. 7, Springer-Verlag, Heidelberg.
Scheinert M (1996) Zur Bahndynamik niedrigfliegender Satelliten. DGK, Reihe C, Heft 435, Verlag der Bayerischen Akademie der Wissenschaften,, DGK, Reihe C, Heft 435.
Seeber G (1993) Satelliten-Geodaesie. Walter de Gruyter 1989.
Sigl R (1989) Einführung in die Potentialtheorie. Wichmann Verlag, Karlsruhe.
Torge W (1989) Gravimetrie. Walter de Gruyter, Berlin.
Wenzel H-G (1985) Hochauflösende Kugelfunktionsmodelle für das Gravitationspotential der Erde. Wissenschaftliche Arbeiten der TU Hannover, Nr. 137.
Xu G (1992) Spectral analysis and geopotential determination (Spektralanalyse und Erdschwerefeldbestimmung). Dissertation, DGK, Reihe C, Heft Nr. 397, Press of the Bavarian Academy of Sciences, ISBN 3-7696-9442-2, 100 p, (with very detailed summary in German).
Xu QF (1994) GPS navigation and precise positioning. Army Press, Peking, ISBN 7-5065-0855-9/P.4, (in Chinese).
Xu G (2004) MFGsoft – Multi-Functional GPS/(Galileo) Software – Software User Manual, (Version of 2004), Scientific Technical Report STR04/17 of GeoForschungsZentrum (GFZ) Potsdam, ISSN 1610-0956, 70 pages, www.gfz-potsdam.de/bib/pub/str0417/0417.pdf.

# Chapter 12
# Singularity-Free Orbit Theory

The previous chapter (Chap. 11) of this book covered the most important content regarding numerical satellite orbit determination theory and algorithms. In this chapter, the emphasis will be on singularity-free orbit theory. We begin with a brief historical review of the singularity problem. We further discuss the problem of singularity of equations of motion, singularity-free theory, singularity criteria, analytical solutions, and the application of analytical orbit solutions.

## 12.1
## A Brief Historical Review of the Singularity Problem

The universal gravitational force and equation of motion and the $N$-body problem were first introduced by Isaac Newton (1687) in his *Principles*. Although it is believed that Newton found the solution to the two-body problem, this solution (i.e. Keplerian orbit) was published by the Swiss mathematician Bernoulli (1710) (Goldstein 1980; Landau and Lifshitz 1976). It was quickly recognized that finding general solutions to the three-body problem was very difficult. The Swiss mathematician and physicist Euler (1767) and the French mathematician and astronomer Lagrange (1772) solved the three-body problem under planar and spatial restrictions. The French astronomer and mathematician Delaunay (1860, 1867) introduced solutions for the sun-earth-moon three-body case (Tisserand 1894; Hagihara 1970). King Oscar II of Sweden and Norway established an award in 1885 to seek the solution to the three-body problem (Mittag-Leffler 1885). The French mathematician and theoretical physicist Poincare received the award in 1887 received the award—even though he failed to solve the problem. He provided the proof that there is no general closed-formed solution to the $N$-body problem when $N$ is greater than 2 (Poincare 1992; Barrow-Green 1996; Dvorak and Lhotka 2013). Based on the contribution of Poincare, the theoretical study of the $N$-body problem was concentrated in two directions: studying the singularity problem (including the problem of collisions; Xia 1992) and deriving a series solution (Aarseth et al. 2008). The Finnish mathematician Sundman (1906, 1909) obtained an analytical solution to the three-body problem (Sundman 1912); however, it included singularity problems in the solution. Also, the series converged very slowly and thus cannot be considered practical (Diacu 1996). The American mathematician Wang (1991) extended Sundman's solution to $N$ greater than 3 by ignoring singular equations (cf. also Diacu 1996). A historical review of the $N$-body problem was provided by Diacu (1992, 1996) and Diacu and Holmes (1996). There are also numerical studies on the $N$-body problem (e.g., Havel 2008; Aarseth 2003).

Lagrangian and Gaussian equations of satellite (and/or planet) motion describe the two-body problem disturbed by the potential function and the non-conservative force, respectively. The singularity problems exist in cases of circular, equatorial, and circular and equatorial orbits. The critical inclination problem exists in cases of long-periodic disturbances which are functions of $\omega$. Celestial mechanics textbooks typically recommend the approach of Keplerian variable transformation (including canonical transformations) in the Lagrange equations (Boccaletti and Pucacco 1998; Kaula 1966, 2001; Liu and Zhao 1979; Xu and Xu 2013b; Brouwer and Clemence 1961; Chobotov 1991; Herrick 1972; Vallado 2007). After transformation, the singularity disappears, and solutions can then be derived. Philosophically speaking, after the transformed equations are solved, the solutions must be transformed inversely back to Keplerian elements, and either the singularity will still exist or the inverse transformation will not exist (Xu and Xu 2012, 2013b). In other words, the variable transformation does not really solve the singularity problem, and it persists. Furthermore, the transformed variables no longer have clear geometric properties. And ignoring all singular equations will degrade a three-dimensional perturbation problem to a two-dimensional problem, and this may not be correct (Xu 2010a, b). The singularity problem in equations of satellite motion is one that has existed since the science of celestial mechanics was established and the satellite era began.

The satellite orbit of the earth is studied using the so-called two-body disturbing theory based on Lagrangian and Gaussian equations of motion, because the perturbations are relatively small compared with the central force of the earth (Battin 1999). Studies are mostly concentrated on the solution of geopotential disturbance which is of the first order (Kaula 1966, 2001; Cui 1990; Shapiro 1962; Wnuk 1990).

Xu (2008, 2010a, b) suggested an alternative method to simplify the equations in the case of singularity (cf. Xu and Xu 2013b, Sect. 10.1). However, the method is similar to that used for extending Sundman's three-body solutions to the $N$-body by omitting the singular equations (Sundman 1912; Wang 1991). On one hand, this is reasonable, because the Lagrangian equations are derived under the assumption that no singularity will occur; in cases of singularity, the related Lagrangian brackets are zero, so the related equation part disappears (Chobotov 1991; Xu 2008, 2013b;

Kaula 2001). On the other hand, in the case of circular and equatorial orbits, only the terms of semi-major axis $a$ and the mean anomaly $M$ exist in the Lagrangian equations of motion (cf. Xu and Xu 2013b, Eq. 10.3), so that the three-dimensional perturbations are degraded to two-dimensions (within the orbital plane), and this is not reasonable.

Analytical orbit solutions have been derived for Lagrangian equations 10.8 (Xu and Xu 2013b) disturbed by the geopotentials (see, e.g., Chap. 6 of Xu and Xu 2013b; Wnuk 1990; Shapiro 1962; Kaula 2001; Xu 2008). By integrating the differential Eqs. 10.8 (Xu and Xu 2013b), the mean value theorem for integration is used, and the coefficient functions of the long and long-periodic terms (functions of $a$, $e$, $i$, $\omega$, and $\Omega$) are considered as constants over suitably selected time intervals (Wnuk 1990; Xu et al. 2010a, b; Xu et al. 2011). In other words, the functions of $(a, e, i, \omega, \Omega)$ are considered first as constants by the indefinite integration. The time variations of the slow-changing elements $(a, e, i, \omega, \Omega)$ are taken then into account by step-wise integration of the equations.

Disturbing solutions of the second order due to the multi-body and solar radiation, as well as atmosphere perturbations, are given by Xu et al. (2010a, b, 2011) and Xu and Xu (2013a, b). Again, there exists a singularity problem in cases of circular and equatorial obits. We must emphasize that Xu's solutions in Xu and Xu 2013b (cf. also Xu et al. 2010a, b, 2011; Xu and Xu 2012, 2013a) are based on the so-called basic Lagrange and Gauss equations; therefore, the solutions are general and are further valid for the singularity-free equations discussed later in this chapter.

A so-called singularity-free theory was proposed in Xu and Xu (2012), mainly via logical derivation and in the form of indefinite integrals. A better-formulated version in the form of a differential equation was given in Xu and Xu (2013b), and these are called singularity-free Lagrange-Xu and Gauss-Xu equations. Because the equation of motion with factors $e$ and/or $\sin i$ as divisor may lead to singularity, the factors are called singular factors (Xu and Xu 2012). Xu and Xu also noted in 2012 that using $e^2$ and $\cos i$ as variables instead of $e$ and $i$ would lead to three Lagrange equations of motion singularity-free (Xu and Xu 2012, 2013b).

Two rigorous and pure mathematical proofs for the correctness of the singularity-free Lagrange-Xu and Gauss-Xu equations of motion are given in Xu et al. (2014a, b) and Xu et al. (2015a, b). Importantly, three singularity criteria are defined in Xu and Xu (2012) to give boundaries of the three cases of singularity with orbital geometric meaning. Through the study of the continuousness of the motion equations conducted by Chunhua Jiang, supervised by GXu, the optimal formulations of the singularity-free equations were found at the beginning of 2016.

Because in two-body or $N$-body problems, the same Lagrange equations are used to describe the motions of the second or the $N$th body, the derived singularity-free equations of motion lead to the conclusion that the singularity problem in the $N$-body problem is also solved [Dvorak agrees with this opinion (Xu and Xu 2013b)].

A breakthrough on the singularity problem was achieved by Xu and Xu (2012, 2013b), and the so-called singularity-free Lagrange-Xu and Gauss-Xu equations were thus derived. The following section is based on the papers of Xu et al. (2014a, b) and Xu et al. (2015a, b), as well as the research results of Xu's team at Shandong University. The key members of the team who contributed to the formulation of the optimal singularity-free equations of motion are GXu, Jia Xu, Wu Chen, Yunzhong Shen, Ta-kang Yhe, Chunhua Jiang, Nan Jiang, Jing Qiao, Wenfeng Nie, and Yujun Du.

## 12.2
## On the Singularity Problem in Orbital Mechanics

In this section, we outline the basic Lagrange and Gauss equations of motion (Xu and Xu 2012), which are important for solving equations without singularity problems. We also introduce an algorithm proposed to deal with the singularity problem in orbital mechanics (cf. Xu and Xu 2012, with minor modifications). Criteria with geometric meaning for singularity are introduced, and the so-called singularity-free Lagrange-Xu and Gauss-Xu equations of motion are derived using rigorous mathematical methods.

### 12.2.1
### Basic Lagrangian and Gaussian Equations of Motion

The Lagrangian equations of satellite motion are represented by Battin (1999), Brouwer and Clemence (1961), Van Kamp (1967), Boccaletti and Pucacco (2001), and Eberle et al. (2008).

$$\begin{aligned}
\frac{da}{dt} &= \frac{2}{na}\frac{\partial V}{\partial M} \\
\frac{de}{dt} &= \frac{1-e^2}{na^2 e}\frac{\partial V}{\partial M} - \frac{\sqrt{1-e^2}}{na^2 e}\frac{\partial V}{\partial \omega} \\
\frac{d\omega}{dt} &= \frac{\sqrt{1-e^2}}{na^2 e}\frac{\partial V}{\partial e} - \frac{\cos i}{na^2\sqrt{1-e^2}\sin i}\frac{\partial V}{\partial i} \\
\frac{di}{dt} &= \frac{1}{na^2\sqrt{1-e^2}\sin i}\left(\cos i \frac{\partial V}{\partial \omega} - \frac{\partial V}{\partial \Omega}\right) \\
\frac{\partial \Omega}{dt} &= \frac{1}{na^2\sqrt{1-e^2}\sin i}\frac{\partial V}{\partial i} \\
\frac{\partial M}{dt} &= n - \frac{2}{na}\frac{\partial V}{\partial a} - \frac{1-e^2}{na^2 e}\frac{\partial V}{\partial e}
\end{aligned}. \tag{12.1}$$

Here, $n$ is the mean angular velocity and will be omitted later on. The Keplerian elements ($a, e, \omega, i, \Omega, M, f$) are the semi-major axis, the eccentricity of the ellipse, the argument of perigee, the inclination angle, the right ascension of the ascending node, the mean anomaly, and the true anomaly, respectively. $V$ is the disturbing potential function.

The Lagrangian equations can be written as (cf. Eqs. 6.23 or 7.94 in Xu and Xu 2013b)

$$\frac{da}{dt} = \frac{2}{na}\frac{da_1}{dt} = h_1 \frac{da_1}{dt}$$

$$\frac{de}{dt} = \frac{1-e^2}{na^2 e}\frac{da_1}{dt} - \frac{\sqrt{1-e^2}}{na^2 e}\frac{de_1}{dt} = h_2 \frac{da_1}{dt} - h_3 \frac{de_1}{dt}$$

$$\frac{d\omega}{dt} = \frac{\sqrt{1-e^2}}{na^2 e}\frac{d\omega_1}{dt} - \cos i \frac{d\Omega}{dt} = h_3 \frac{d\omega_1}{dt} - h_5(\cos i/\sin i)\frac{d\Omega_1}{dt} \quad (12.2)$$

$$\frac{di}{dt} = \frac{1}{na^2\sqrt{1-e^2}\sin i}\left(\cos i \frac{de_1}{dt} - \frac{di_1}{dt}\right) = h_4 \frac{1}{\sin i}\left(\cos i \frac{de_1}{dt} - \frac{di_1}{dt}\right)$$

$$\frac{d\Omega}{dt} = \frac{1}{na^2\sqrt{1-e^2}\sin i}\frac{d\Omega_1}{dt} = h_5 \frac{1}{\sin i}\frac{d\Omega_1}{dt}$$

$$\frac{dM}{dt} = n - \frac{2}{na}\frac{dM_1}{dt} - \frac{1-e^2}{na^2 e}\frac{d\omega_1}{dt} = n - h_1 \frac{dM_1}{dt} - h_2 \frac{d\omega_1}{dt}$$

where coefficients ($h_1, h_2, h_3, h_4, h_5$) are defined, and

$$\frac{da_1}{dt} = \frac{\partial R}{\partial M}, \quad \frac{de_1}{dt} = \frac{\partial R}{\partial \omega}, \quad \frac{d\omega_1}{dt} = \frac{\partial R}{\partial e},$$

$$\frac{di_1}{dt} = \frac{\partial R}{\partial \Omega}, \quad \frac{d\Omega_1}{dt} = \frac{\partial R}{\partial i}, \quad \frac{dM_1}{dt} = \frac{\partial R}{\partial a}. \quad (12.3)$$

Equations 12.3 are called basic Lagrangian equations of motion. By integrating the Eqs. 12.2, the mean value theorem for integration is used where the functions of variables ($a, e, i$) are considered constants. The simplest way to solve the Lagrangian equations of motion 12.1 or 12.2 is to solve the basic Lagrangian Eqs. 10.10. Substituting solutions of 12.3 into 12.2, the solutions of 12.2 can be obtained. Therefore, without exception, the potential force disturbing orbit problem turns out to be the problem of solving the basic Lagrangian equations of motion 12.3. It is notable that the basic Lagrangian equations of motion are singularity-free.

The Gaussian perturbed equations of motion are (Kaula 2001)

$$\frac{da}{dt} = \frac{2}{n\sqrt{1-e^2}}[e\sin f \cdot f_\mathrm{r} + (1+e\cos f)\cdot f_\alpha]$$

$$\frac{de}{dt} = \frac{\sqrt{1-e^2}}{na}[\sin f \cdot f_\mathrm{r} + (\cos E + \cos f)\cdot f_\alpha]$$

$$\frac{d\omega}{dt} = \frac{\sqrt{1-e^2}}{nae}\left[-\cos f \cdot f_\mathrm{r} + \frac{2+e\cos f}{1+e\cos f}\sin f \cdot f_\alpha\right] - \cos i \frac{d\Omega}{dt}$$

$$\frac{di}{dt} = \frac{(1-e\cos E)\cos u}{na\sqrt{1-e^2}}\cdot f_\mathrm{h}$$

$$\frac{d\Omega}{dt} = \frac{(1-e\cos E)\sin u}{na\sqrt{1-e^2}\sin i}\cdot f_\mathrm{h}$$

$$\frac{dM}{dt} = n - \frac{2}{na}\left(\frac{1-e^2}{1+e\cos f}\right)\cdot f_\mathrm{r} - \frac{1-e^2}{nae}\left[-\cos f \cdot f_\mathrm{r} + \frac{2+e\cos f}{1+e\cos f}\sin f \cdot f_\alpha\right]$$

(12.4)

and

$$u = \omega + f, \quad \cos E = \frac{e+\cos f}{1+e\cos f}. \tag{12.5}$$

Here $(f_\mathrm{r}, f_\alpha, f_\mathrm{h})^\mathrm{T}$ is a force vector in an orbital plane coordinate system, the first two components are in the orbital plane, $f_\mathrm{r}$ is the radial component, $f_\alpha$ is the component perpendicular to $f_\mathrm{r}$ and points in the direction of the satellite motion (see Fig. 12.1, where $o$ is the focus of the orbital ellipse and $f$ is the true anomaly counted from the perigee of the orbital ellipse) and $f_\mathrm{h}$ completes a right-handed system. $E$ is the eccentric anomaly. The relation between eccentric anomaly $E$ and true anomaly $f$ is

**Fig. 12.1** Radial force vector and its perpendicular one in the orbital plane

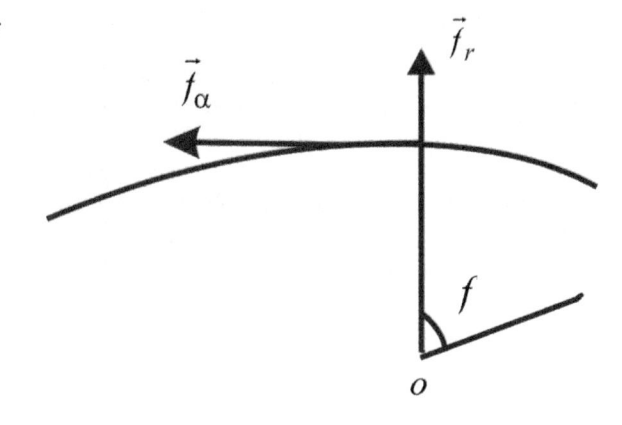

**Fig. 12.2** The eccentric and true anomalies ($f$ and $E$) of a satellite

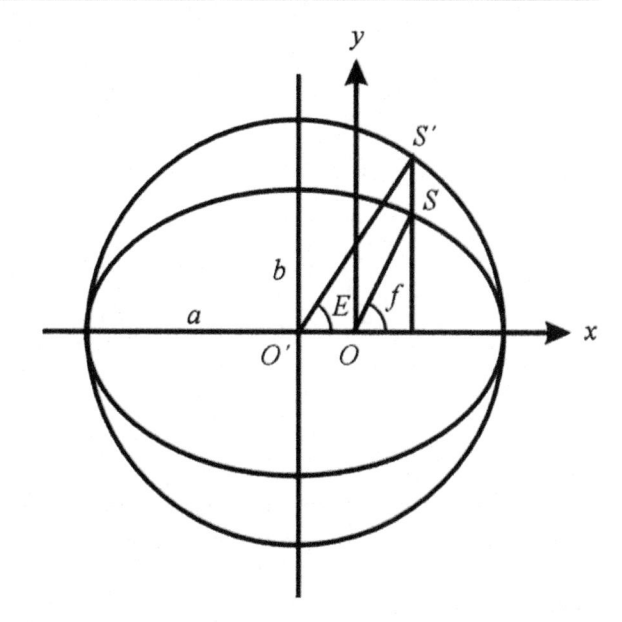

given in Fig. 12.2 (where $S'$ is the vertical projection of the satellite $S$ on the circle with a radius of $a$ (semi-major axis of the ellipse), $b$ is the semi-minor axis of the ellipse, $O'$ is the centre of the circle and $O$ is the focus of the ellipse).

The Gaussian equations can be rewritten as

$$\frac{da}{dt} = \frac{2}{n\sqrt{1-e^2}} \frac{da_1}{dt} = h_1 \frac{da_1}{dt}$$

$$\frac{de}{dt} = \frac{\sqrt{1-e^2}}{na} \frac{de_1}{dt} = h_2 \frac{de_1}{dt}$$

$$\frac{d\omega}{dt} = \frac{\sqrt{1-e^2}}{nae} \frac{d\omega_1}{dt} - \cos i \frac{d\Omega}{dt} = h_3 \frac{d\omega_1}{dt} - \cos i \frac{d\Omega}{dt} = h_3 \frac{d\omega_1}{dt} - h_7 \frac{d\Omega_1}{dt} \quad (12.6)$$

$$\frac{di}{dt} = \frac{1}{na\sqrt{1-e^2}} \frac{di_1}{dt} = h_4 \frac{di_1}{dt}$$

$$\frac{d\Omega}{dt} = \frac{1}{na\sqrt{1-e^2}\sin i} \frac{d\Omega_1}{dt} = h_5 \frac{d\Omega_1}{dt} = h_4 \frac{1}{\sin i} \frac{d\Omega_1}{dt}$$

$$\frac{dM}{dt} = n - 2\left(\frac{1-e^2}{na}\right)\frac{dM_1}{dt} - \frac{1-e^2}{nae}\frac{d\omega_1}{dt} = n - 2h_6 e \frac{dM_1}{dt} - h_6 \frac{d\omega_1}{dt}$$

where coefficients of ($h_1$, $h_2$, $h_3$, $h_4$, $h_5$, $h_6$, $h_7 = h_5 \cos i = h_4 \cos i / \sin i$) are defined, and

$$\frac{da_1}{dt} = e\sin f \cdot f_r + (1+e\cos f)\cdot f_\alpha$$

$$\frac{de_1}{dt} = \sin f \cdot f_r + (\cos E + \cos f)\cdot f_\alpha$$

$$\frac{d\omega_1}{dt} = -\cos f \cdot f_r + \frac{2+e\cos f}{1+e\cos f}\sin f \cdot f_\alpha$$

$$\frac{di_1}{dt} = (1-e\cos E)\cos u \cdot f_h \qquad (12.7)$$

$$\frac{d\Omega_1}{dt} = (1-e\cos E)\sin u \cdot f_h$$

$$\frac{dM_1}{dt} = \frac{1}{1+e\cos f}\cdot f_r$$

Equations 12.7 are called basic Gaussian equations of motion. By integrating the Eqs. 12.6 the mean value theorem for integration is used where the functions of variables $(a, e, i)$ are considered constants. For solving the Gaussian equations of motion 12.6, the simplest way is to solve the basic Gaussian Eq. 12.7. Substituting solutions of 12.7 into 12.6 the solutions of 12.6 can be obtained. Therefore, without exception, the non-conservative force disturbing orbit problem turns out to be the problem of solving the basic Gaussian equations of motion 12.7. It is notable that the basic Gaussian equations of motion are singularity-free.

In both the Lagrangian and Gaussian equations of motion 12.2 and 12.6, there exist singular problems. In the case of $e \ll 1$, the orbit is nearly circular, so that the perigee and the related Keplerian elements $f$ and $\omega$ are not defined, and the problem is singular. To overcome this problem, traditionally, let $u = f + \omega$, and a parameter set of $(a, i, \Omega, \xi = e\cos\omega, \eta = -e\sin\omega, \lambda = M + \omega)$ can be used to describe the motion of the satellite. The related Lagrangian equations of motion can be derived and the singularity of $e$ disappears. Using another set of variables $(a, h = \sin i\cos\Omega, k = -\sin i\sin\Omega, \xi = e\cos(\omega+\Omega), \eta = -e\sin(\omega+\Omega), \lambda = M+\omega+\Omega)$, both the singularities caused by $e = 0$ and $\sin i = 0$ may disappear. Furthermore, a possible variable set is $(a, h = \tan(i/2)\sin\Omega, k = \tan(i/2)\cos\Omega, \xi = e\sin(\omega+\Omega), \eta = e\cos(\omega+\Omega), \lambda = M+\omega+\Omega)$ (see, e.g., Chobotov 1991). Also often employed are the canonical transformations using, for example, Hill variables (Cui 1990, 1997; Schneider and Cui 2005) and Delaunay elements (Kaula 2001; Wnuk 1990). These are traditional methods that may be used to solve the singularity problem; however, except for Keplerian variables, they do not have a clear geometric meaning.

In orbit determination practice for an equatorial satellite, to avoid the inclinational singularity problem, the coordinate system will be rotated first so that the orbital plane will not coincide with the equatorial plane; after the orbits are determined, the results will be rotated inversely.

## 12.2.2
## Solving Algorithm for the Singularity Problem

For convenience, we call $e$ and $\sin i$ singular factors. Any singular factor as divisor in the equations of motion could lead to a singularity; however, it could also not. The partial derivatives of the potential function with respect to the Keplerian elements are not responsible for singularity (cf. Xu 2008, 2010a, b). The singularity problem (if it exists) exists in the equations from the beginning.

Rewrite Eqs. 12.1 or 12.2 as follows (Xu and Xu 2012, 2013b):

$$\frac{da}{dt} = \delta_a, \quad \frac{de}{dt} = \frac{\delta_{e1}}{ae} + \frac{\delta_{e2}}{ae}, \quad \frac{d\omega}{dt} = \frac{\delta_{\omega 1}}{ae} + \frac{\delta_{\omega 2}}{a \sin i},$$
$$\frac{di}{dt} = \frac{\delta_{i1}}{a \sin i} + \frac{\delta_{i2}}{a \sin i}, \quad \frac{d\Omega}{dt} = \frac{\delta_\Omega}{a \sin i}, \quad \frac{dM}{dt} = \delta_{M1} + \frac{\delta_{M2}}{ae}. \quad (12.8)$$

Here, $\delta$ with indices represent terms on the right-hand side of Eqs. 12.1 or 12.2. The relationship between $\delta$ and the basic Lagrangian Eqs. 12.3 can be seen by comparing Eqs. 12.2 and 12.8. $\delta$s are in the same magnitude order (comparing Eqs. 12.1 and 12.8). Keplerian elements on the left-hand side of Eq. 12.1 will be represented by $\sigma$ with indices later on. Multiplying $e$ by the second and sixth equation of 12.8, $\sin i$ by the fourth and fifth, and $e \sin i$ by the third, it follows that

$$\frac{da}{dt} = \delta_a, \quad e\frac{de}{dt} = \frac{\delta_{e1}}{a} + \frac{\delta_{e2}}{a}, \quad e \sin i \frac{d\omega}{dt} = \sin i \frac{\delta_{\omega 1}}{a} + \frac{e \delta_{\omega 2}}{a},$$
$$\sin i \frac{di}{dt} = \frac{\delta_{i1}}{a} + \frac{\delta_{i2}}{a}, \quad \sin i \frac{d\Omega}{dt} = \frac{\delta_\Omega}{a}, \quad e\frac{dM}{dt} = e\delta_{M1} + \frac{\delta_{M2}}{a}. \quad (12.9)$$

The above equations are singularity-free Lagrangian ($n$ in the sixth equation is omitted). The solutions (i.e. the indefinite integrals of 12.9) can be derived using methods described in Wnuk (1990), Xu et al. (2010a, b), (2011), and Kaula (2001), or formed using the solutions of the basic Lagrangian equations derived in this book, and have the forms

$$\Delta a = \Delta \delta_a, \quad e\Delta e = \frac{\Delta \delta_{e1}}{a} + \frac{\Delta \delta_{e2}}{a}, \quad e \sin i \Delta \omega = \frac{\sin i \Delta \delta_{\omega 1}}{a} + \frac{e \Delta \delta_{\omega 2}}{a},$$
$$\sin i \Delta i = \frac{\Delta \delta_{i1}}{a} + \frac{\Delta \delta_{i2}}{a}, \quad \sin i \Delta \Omega = \frac{\Delta \delta_\Omega}{a}, \quad e\Delta M = e\Delta \delta_{M1} + \frac{\Delta \delta_{M2}}{a}. \quad (12.10)$$

Here $\Delta \sigma$ represents the change (or delta) of $\sigma$ and $\Delta \delta$ represents indefinite integrals of $\delta$ with respect to time. Deriving the Kepler element change $\Delta \sigma$ from Eqs. 12.1 or 12.8 thus becomes a task of deriving solutions $\Delta \sigma$ from Eq. 12.10. On the right-hand side, the integrals $\Delta \delta$ are the solutions of the basic Lagrangian Eq. 12.3 multiplied by factors which can be obtained by comparing Eqs. 12.2, 12.3, and 12.9. The equations at 12.10 are called intermediate solutions. We see now why the basic Lagrangian equations are defined and solved systematically in *Orbits*

(Xu and Xu 2013b). In the case of non-singularity, dividing $e$ by the second and sixth equation of 12.10, $\sin i$ by the fourth and fifth, and $e\sin i$ by the third, it follows that

$$\Delta a = \Delta \delta_a, \quad \Delta e = \frac{\Delta \delta_{e1}}{ae} + \frac{\Delta \delta_{e2}}{ae}, \quad \Delta \omega = \frac{\Delta \delta_{\omega 1}}{ae} + \frac{\Delta \delta_{\omega 2}}{a \sin i},$$
$$\Delta i = \frac{\Delta \delta_{i1}}{a \sin i} + \frac{\Delta \delta_{i2}}{a \sin i}, \quad \Delta \Omega = \frac{\Delta \delta_{\Omega}}{a \sin i}, \quad \Delta M = \Delta \delta_{M1} + \frac{\Delta \delta_{M2}}{ae}. \quad (12.11)$$

The equations at 12.11 are the solutions of 12.8 in the case of non-singularity. The problem happens only in the case of singularity, i.e. $e$ and/or $\sin i$ ($\approx i$) approaching zero. Dividing any extremely small number could disproportionally amplify the error in the result (lead to singularity), and this must be avoided.

To summarize, the singularity-free intermediate solutions of 12.10 can be obtained by the solutions of the basic Lagrangian equations. Then, in the singularity-free case, the final solutions 12.11 of the Lagrangian equations of motion can be formed. Thus the problem now is how to form the final solutions from the intermediate solutions in cases of singularities. Several criteria are defined and introduced to distinguish the type of the singularity and to decide how to form the related final solutions in different singular cases from Eq. 12.10, which are described in Xu (2012a, b) in reasonable and rigorous detail.

### 12.2.3
### Xu's Criteria for Singularity

Define radial and equatorial biases as (Xu and Xu 2012, 2013b)

$$\varepsilon_e = ae, \quad \varepsilon_i = a \sin i. \quad (12.12)$$

Here the radial bias $ae$ is the distance between the orbital geometric centre $O$ and the geocentre $O'$ (see Fig. 12.2). The expression of the satellite radius can be expanded to order $e$ for the case of $f = 0$; it holds that (Bronstein and Semendjajew 1987; Wang et al. 1979)

$$r = \frac{a(1 - e^2)}{1 + e \cos f} \approx a(1 - e \cos f) \stackrel{f=0}{\approx} a - ae. \quad (12.13)$$

Hence, $ae$ is also the maximal radius correction (error) which could be caused by $e$. The equatorial bias $a\sin i \approx ai$ is the maximal distance of the satellite vertical to the equator which could be caused by $i$ (inclination $i$ is the maximal latitude of the satellite and is a small angle). At the initial time of integration $t_0 = 0$, $e$ and/or $i$ are approaching zero, and therefore, at the integration time $t$, the correction of $e$ is $\Delta e$ and the correction of $i$ is $\Delta i$. Thus Eq. 12.12 turns out to be

$$\varepsilon_e = ae = a\Delta e, \quad \varepsilon_i = a\sin i \approx a\Delta \sin i \approx a\Delta i. \tag{12.14}$$

An orbital bias can be defined as follows:

$$\begin{aligned}\varepsilon_o &= a(\Delta\omega + \Delta M), \quad \text{for circular orbit} \\ \varepsilon_o &= a(\Delta\Omega + \Delta\omega), \quad \text{for equatorial orbit} \\ \varepsilon_o &= a(\Delta\Omega + \Delta\omega + \Delta M), \quad \text{for circular and equatorial orbit}\end{aligned} \tag{12.15}$$

$\varepsilon_o$ has the meaning of the maximal bias in the orbital motion direction, which could be caused by $\Delta\Omega, \Delta\omega, \Delta M$. Therefore, the radial, equatorial, and orbital biases are the maximal orbit errors in the three related components. The three biases represent the maximal error area of the orbit caused by the disturbance of $(e, \omega, i, \Omega, M)$. In practice, for a required orbital precision (standard deviation or sigma) of 0.33 m, $\varepsilon_e$, $\varepsilon_i$, and $\varepsilon_o$ are ca. 1 m (three sigma), respectively. This means that $\varepsilon_e$, $\varepsilon_i$, and $\varepsilon_o$ are predefined numbers which can be used as criteria.

Similarly we can define

$$\begin{aligned}\varepsilon_M &= a\Delta M, \quad \text{for non - singular orbit} \\ \varepsilon_\omega &= a\Delta\omega, \quad \text{for non - singular orbit} . \\ \varepsilon_\Omega &= a\Delta\Omega, \quad \text{for non - singular orbit}\end{aligned}$$

They are the biases caused by changes in satellite mean anomaly, satellite perigee, and ascending node of the satellite, respectively.

## 12.2.4
## Derivation of Lagrange-Xu Equations of Motion

The perturbed equation of motion of a satellite or a planet is described by Newton's second law in an inertial Cartesian coordinate system as (Kaula 1966, 2001; Xu and Xu 2013b; Brouwer and Clemence 1961; Chobotov 1991; Herrick 1972; Vallado 2007)

$$\begin{aligned}\frac{d}{dt}x_j &= \dot{x}_j, \quad j = 1, 2, 3 \\ \frac{d}{dt}\dot{x}_j &= \frac{\partial V}{\partial x_j}, \quad j = 1, 2, 3\end{aligned} \tag{12.16}$$

Here, $V$ is the potential function of the force. The time derivatives of the position and velocity components $x$ and $\dot{x}$ in the above left-hand side can be expressed as functions of the rates of change of the Keplerian elements $(a, e, \omega, i, \Omega, M)$ (they are the semi-major axis, the eccentricity of the ellipse, the argument of perigee, the inclination angle, the right ascension of the ascending node, and the mean anomaly, and will be represented generally by $s_k$). It then yields

$$\sum_{k=1}^{6} \frac{\partial x_j}{\partial s_k} \frac{d}{dt} s_k = \dot{x}_j, \quad j = 1, 2, 3$$

$$\sum_{k=1}^{6} \frac{\partial \dot{x}_j}{\partial s_k} \frac{d}{dt} s_k = \frac{\partial V}{\partial x_j}, \quad j = 1, 2, 3$$

(12.17)

Multiplying the first equation of 12.17 by $-\frac{\partial \dot{x}_j}{\partial s_l}$ and the second by $\frac{\partial x_j}{\partial s_l}$ and then adding them together for index $j$, one has

$$\sum_{k=1}^{6} \sum_{j=1}^{3} \left( -\frac{\partial \dot{x}_j}{\partial s_l} \frac{\partial x_j}{\partial s_k} + \frac{\partial x_j}{\partial s_l} \frac{\partial \dot{x}_j}{\partial s_k} \right) \frac{d}{dt} s_k = \sum_{j=1}^{3} \left( -\frac{\partial \dot{x}_j}{\partial s_l} \dot{x}_j + \frac{\partial x_j}{\partial s_l} \frac{\partial V}{\partial x_j} \right), \quad (12.18)$$

i.e.

$$\sum_{k=1}^{6} [s_l, s_k] \frac{d}{dt} s_k = \frac{\partial F}{\partial s_l}. \quad (12.19)$$

where Lagrange brackets $[s_l, s_k]$ and the term on the right-hand side are defined by

$$[s_l, s_k] = \sum_{j=1}^{3} \left( -\frac{\partial \dot{x}_j}{\partial s_l} \frac{\partial x_j}{\partial s_k} + \frac{\partial x_j}{\partial s_l} \frac{\partial \dot{x}_j}{\partial s_k} \right)$$

$$\frac{\partial F}{\partial s_l} = \sum_{j=1}^{3} \left( -\frac{\partial}{\partial s_l}(\dot{x}_j^2/2) + \frac{\partial V}{\partial s_l} = \frac{\partial(V-T)}{\partial s_l} \right)$$

(12.20)

where $F$ is called force function and $T$ the kinetic energy

$$T = \sum_{j=1}^{3} (-\dot{x}_j^2/2). \quad (12.21)$$

Take the following relations into account (Kaula 1966, 2001, Xu and Xu 2013b)

$$\begin{pmatrix} \vec{r} \\ \dot{\vec{r}} \end{pmatrix} = R_3(-\Omega) R_1(-i) R_3(-\omega) \begin{pmatrix} \vec{q} \\ \dot{\vec{q}} \end{pmatrix} = R \begin{pmatrix} \vec{q} \\ \dot{\vec{q}} \end{pmatrix} \quad (12.22)$$

$$\vec{q} = \begin{pmatrix} a(\cos E - e) \\ a\sqrt{1-e^2}\sin E \\ 0 \end{pmatrix} = \begin{pmatrix} r\cos f \\ r\sin f \\ 0 \end{pmatrix} \quad (12.23)$$

$$\dot{\vec{q}} = \begin{pmatrix} -\sin E \\ \sqrt{1-e^2}\cos E \\ 0 \end{pmatrix} \frac{na}{1-e\cos E} = \begin{pmatrix} -\sin f \\ e+\cos f \\ 0 \end{pmatrix} \frac{na}{\sqrt{1=e^2}} \quad (12.24)$$

$$\frac{\partial \vec{r}}{\partial(\Omega, i, \omega)} = \frac{\partial R}{\partial(\Omega, i, \omega)}\vec{q} \quad \text{and} \quad \frac{\partial \dot{\vec{r}}}{\partial(\Omega, i, \omega)} = \frac{\partial R}{\partial(\Omega, i, \omega)}\dot{\vec{q}}, \quad (12.25)$$

$$\frac{\partial \vec{r}}{\partial(a, e, M)} = R\frac{\partial \vec{q}}{\partial(a, e, M)} \quad \text{and} \quad \frac{\partial \dot{\vec{r}}}{\partial(a, e, M)} = R\frac{\partial \dot{\vec{q}}}{\partial(a, e, M)}, \quad (12.26)$$

where

$$\frac{\partial \vec{q}}{\partial(a, e, M)} = \begin{pmatrix} \cos E - e & \frac{-a\sin^2 E}{1-e\cos E} - a & \frac{-a\sin E}{1-e\cos E} \\ \sqrt{1-e^2}\sin E & a\sqrt{1-e^2}\left(\frac{\sin 2E}{2(1-e\cos E)} - \frac{e\sin E}{1-e^2}\right) & \frac{a\sqrt{1-e^2}\cos E}{1-e\cos E} \\ 0 & 0 & 0 \end{pmatrix} \quad (12.27)$$

and

$$\frac{\partial \dot{\vec{q}}}{\partial(a, e, M)} = \begin{pmatrix} \frac{n\sin E}{2(1-e\cos E)} & \frac{na\sin E(e-2\cos E+e\cos^2 E)}{(1-e\cos E)^3} & \frac{na(e-\cos E)}{(1-e\cos E)^3} \\ \frac{-n\sqrt{1-e^2}\cos E}{2(1-e\cos E)} & \frac{na\left[1+e^2-2e\cos E+\sin^2 E(e\cos E-2)\right]}{\sqrt{1-e^2}(1-e\cos E)^3} & \frac{-na\sqrt{1-e^2}\sin E}{(1-e\cos E)^3} \\ 0 & 0 & 0 \end{pmatrix} \quad (12.28)$$

Here the left-hand side of 12.22 are position and velocity vectors in the Cartesian coordinate system; the left-hand sides of 12.23–12.24 are the position and velocity vectors in the orbital ellipse; $R$ with indices are the rotational matrices around the index related axis; $R$ without index is the product of the three rotational matrices. $E$ and $f$ are eccentric anomaly and true anomaly, respectively. $n$ is the mean angular velocity.

From the definition of the Lagrange bracket 12.20, one has properties of

$$[s_l, s_k] = -[s_k, s_l] \quad \text{and} \quad [s_l, s_l] = 0. \quad (12.29)$$

Noting that the Lagrange bracket $[s_l, s_k]$ is time-invariant (Kaula 2000), or

$$\frac{\partial}{\partial t}[s_l, s_k] = \sum_{j=1}^{3}\left(-\frac{\partial^2 x_j}{\partial s_l \partial t}\frac{\partial \dot{x}_j}{\partial s_k} - \frac{\partial \dot{x}_j}{\partial s_l}\frac{\partial^2 x_j}{\partial s_k \partial t} + \frac{\partial^2 x_j}{\partial s_l \partial t}\frac{\partial \dot{x}_j}{\partial s_k} + \frac{\partial x_j}{\partial s_l}\frac{\partial^2 \dot{x}_j}{\partial s_k \partial t}\right)$$

$$= \sum_{j=1}^{3}\left(\frac{\partial}{\partial s_l}\left(\frac{\partial x_j}{\partial t}\frac{\partial \dot{x}_j}{\partial s_k} - \frac{\partial \dot{x}_j}{\partial t}\frac{\partial x_j}{\partial s_k}\right) + \frac{\partial}{\partial s_k}\left(\frac{\partial x_j}{\partial s_l}\frac{\partial \dot{x}_j}{\partial t} - \frac{\partial \dot{x}_j}{\partial s_l}\frac{\partial x_j}{\partial t}\right)\right)$$

$$= \sum_{j=1}^{3}\left(\frac{\partial}{\partial s_l}\left(\dot{x}_j\frac{\partial \dot{x}_j}{\partial s_k} - \frac{\partial x_j}{\partial s_k}\ddot{x}_j\right) + \frac{\partial}{\partial s_k}\left(\frac{\partial x_j}{\partial s_l}\ddot{x}_j - \frac{\partial \dot{x}_j}{\partial s_l}\dot{x}_j\right)\right)$$

$$= \sum_{j=1}^{3}\left(\frac{\partial}{\partial s_l}\left(\frac{\partial(\dot{x}_j^2/2)}{\partial s_k}\right) - \frac{\partial}{\partial s_l}\left(\frac{\partial x_j}{\partial s_k}\ddot{x}_j\right) + \frac{\partial}{\partial s_k}\left(\frac{\partial x_j}{\partial s_l}\ddot{x}_j\right) - \frac{\partial}{\partial s_k}\left(\frac{\partial(\dot{x}_j^2/2)}{\partial s_l}\right)\right)$$

$$= \sum_{j=1}^{3}\left(-\frac{\partial}{\partial s_l}\left(\frac{\partial x_j}{\partial s_k}\frac{\partial(\mu/r)}{\partial x_j}\right) + \frac{\partial}{\partial s_k}\left(\frac{\partial x_j}{\partial s_l}\frac{\partial(\mu/r)}{\partial x_j}\right)\right)$$

$$= \sum_{j=1}^{3}\left(-\frac{\partial}{\partial s_l}\left(\frac{\partial(\mu/r)}{\partial s_k}\right) + \frac{\partial}{\partial s_k}\left(\frac{\partial(\mu/r)}{\partial s_l}\right)\right) = 0$$

(12.30)

Therefore, the Lagrange bracket $[s_l, s_k]$ can be computed, for example, at the perigee, or $E = 0$. Then non-zero Lagrange brackets are

$$\begin{aligned}
[\Omega, i] &= -na^2\sqrt{1-e^2}\sin i \\
[\Omega, a] &= \sqrt{1-e^2}\cos i\, na/2 \\
[\Omega, e] &= -na^2 e\cos i\sqrt{1-e^2} \\
[\omega, a] &= \sqrt{1-e^2}\, na/2 \\
[\omega, e] &= -na^2 e/\sqrt{1-e^2} \\
[a, M] &= -na/2
\end{aligned}$$

(12.31)

Then we have 12.19 in forms of

$$[s_l, a]\frac{d}{dt}a + [s_l, e]\frac{d}{dt}e + [s_l, \omega]\frac{d}{dt}\omega + [s_l, i]\frac{d}{dt}i + [s_l, \Omega]\frac{d}{dt}\Omega + [s_l, M]\frac{d}{dt}M + \frac{\partial F}{\partial s_l}$$
$$l = 1, 2, 3, 4, 5, 6$$

(12.32)

Of special note, a Lagrange bracket $[s_l, s_k] = 0$ means that the time derivative of the Kepler element $s_k$ has nothing to do with the partial derivative of the force function $F$ with respect to the element $s_l$.

For all $s_l$ one has the non-singular equation system (Xu et al. 2014a, b)

$$[a,\omega]\frac{d}{dt}\omega + [a,\Omega]\frac{d}{dt}\Omega + [a,M]\frac{d}{dt}M = \frac{\partial F}{\partial a}, \tag{12.33}$$

$$[e,\omega]\frac{d}{dt}\omega + [e,\Omega]\frac{d}{dt}\Omega = \frac{\partial F}{\partial e}, \tag{12.34}$$

$$[\omega,a]\frac{d}{dt}a + [\omega,e]\frac{d}{dt}e = \frac{\partial F}{\partial \omega}, \tag{12.35}$$

$$[i,\Omega]\frac{d}{dt}\Omega = \frac{\partial F}{\partial i}, \tag{12.36}$$

$$[\Omega,a]\frac{d}{dt}a + [\Omega,e]\frac{d}{dt}e + [\Omega,i]\frac{d}{dt}i = \frac{\partial F}{\partial \Omega}, \tag{12.37}$$

$$[M,a]\frac{d}{dt}a = \frac{\partial F}{\partial M}. \tag{12.38}$$

From 12.38, it yields

$$\frac{d}{dt}a = \frac{2}{na}\frac{\partial F}{\partial M}. \tag{12.39}$$

From 12.36 one has

$$\sin i \frac{d}{dt}\Omega = \frac{1}{na^2\sqrt{1-e^2}}\frac{\partial F}{\partial i}. \tag{12.40}$$

Substituting 12.39 into 12.35, one has

$$[\omega,a]\frac{2}{na}\frac{\partial F}{\partial M} + [\omega,e]\frac{d}{dt}e = \frac{\partial F}{\partial \omega}, \tag{12.41}$$

i.e.

$$e\frac{d}{dt}e = \frac{1}{na^2}\left((1-e^2)\frac{\partial F}{\partial M} - \sqrt{1-e^2}\frac{\partial F}{\partial \omega}\right), \tag{12.42}$$

or

$$\frac{d}{dt}\left(\frac{e^2}{2}\right) = \frac{1}{na^2}\left((1-e^2)\frac{\partial F}{\partial M} - \sqrt{1-e^2}\frac{\partial F}{\partial \omega}\right). \tag{12.43}$$

Reformulating 12.37 using 12.39 and 12.31, as well as 12.43, yields

$$[\Omega, a] \frac{2}{na} \frac{\partial F}{\partial M} - na^2 \cos i \frac{1}{\sqrt{1-e^2}} \frac{d}{dt}(e^2/2) + [\Omega, i] \frac{d}{dt} i = \frac{\partial F}{\partial \Omega}, \qquad (12.44)$$

i.e.

$$\sqrt{1-e^2} \cos i \frac{\partial F}{\partial M} - \cos i \left( \sqrt{1-e^2} \frac{\partial F}{\partial M} - \frac{\partial F}{\partial \omega} \right) + [\Omega, i] \frac{d}{dt} i = \frac{\partial F}{\partial \Omega}, \qquad (12.45)$$

or

$$\cos i \frac{\partial F}{\partial \omega} - na^2 \sqrt{1-e^2} \sin i \frac{d}{dt} i = \frac{\partial F}{\partial \Omega}, \qquad (12.46)$$

i.e.

$$\sin i \frac{d}{dt} i = \frac{1}{na^2 \sqrt{1-e^2}} \left( \cos i \frac{\partial F}{\partial \omega} - \frac{\partial F}{\partial \Omega} \right). \qquad (12.47)$$

or

$$\frac{d}{dt} \cos i = \frac{-1}{na^2 \sqrt{1-e^2}} \left( \cos i \frac{\partial F}{\partial \omega} - \frac{\partial F}{\partial \Omega} \right). \qquad (12.48)$$

From 12.34 and 12.40, one has

$$e \sin i \frac{d\omega}{dt} = \frac{\sqrt{1-e^2}}{na^2} \sin i \frac{\partial F}{\partial e} - \frac{\cos i}{na^2 \sqrt{1-e^2}} e \frac{\partial F}{\partial i} \qquad (12.49)$$

From 12.33 one has

$$-\sqrt{1-e^2} \frac{na}{2} \frac{d}{dt} \omega - \sqrt{1-e^2} \cos i \frac{na}{2} \frac{d}{dt} \Omega - \frac{na}{2} \frac{d}{dt} M = \frac{\partial F}{\partial a}, \qquad (12.50)$$

i.e.

$$\frac{d}{dt} M = -\sqrt{1-e^2} \frac{d}{dt} \omega - \sqrt{1-e^2} \cos i \frac{d}{dt} \Omega = \frac{2}{na} \frac{\partial F}{\partial a}. \qquad (12.51)$$

Taking into account 12.40 and 12.49, it yields

$$e \frac{d}{dt} M = ne - e \frac{2}{na} \frac{\partial F}{\partial a} - \frac{1-e^2}{na^2} \frac{\partial F}{\partial e}. \qquad (12.52)$$

Equations 12.40, 12.49, and 12.52 are the motion equations in common non-singular case.

We denote (here the energy conservative relation is used)

$$F = V - T = \frac{\mu}{r} + R - T = \frac{\mu}{2a} + R, \quad (12.53)$$

where function $R$ is called the disturbing potential, including all terms of $V$ except the central term $\mu/r$ (note that here we redefined $R$). $\mu$ is the gravitational constant of the earth. Hence, in all the above Eqs. 12.39, 12.40, 12.43, 12.48, 12.49, and 12.53, $F$ can be replaced by $R$ and become (Kepler's third law is used)

$$\begin{aligned}
\frac{da}{dt} &= \frac{2}{na}\frac{\partial R}{\partial M} \\
\frac{d}{dt}\left(\frac{e^2}{2}\right) &= \frac{1}{na^2}\left((1-e^2)\frac{\partial R}{\partial M} - \sqrt{1-e^2}\frac{\partial R}{\partial \omega}\right) \\
e\sin i \frac{d\omega}{dt} &= \frac{\sqrt{1-e^2}}{na^2}\sin i \frac{\partial R}{\partial e} - \frac{e\cos i}{na^2\sqrt{1-e^2}}\frac{\partial R}{\partial i} \\
\frac{d}{dt}\cos i &= \frac{-1}{na^2\sqrt{1-e^2}}\left(\cos i\frac{\partial R}{\partial \omega} - \frac{\partial R}{\partial \Omega}\right) \\
\sin i \frac{d\Omega}{dt} &= \frac{1}{na^2\sqrt{1-e^2}}\frac{\partial R}{\partial i} \\
e\frac{dM}{dt} &= ne - \frac{2e}{na}\frac{\partial R}{\partial a} - \frac{1-e^2}{na^2}\frac{\partial R}{\partial e}
\end{aligned} \quad (12.54)$$

Equation 12.54 is the solution to the equation system 12.33–12.38. An important contribution of Xu and Xu (2012) is that the second and fourth equations of 12.54 will never be singular. Lagrange equations of motion in common non-singular case (i.e. $\sin i \neq 0$ and $e \neq 0$) are then

$$\begin{aligned}
\frac{da}{dt} &= \frac{2}{na}\frac{\partial R}{\partial M} \\
\frac{d}{dt}\left(\frac{e^2}{2}\right) &= \frac{1}{na^2}\left((1-e^2)\frac{\partial R}{\partial M} - \sqrt{1-e^2}\frac{\partial R}{\partial \omega}\right) \\
\frac{d\omega}{dt} &= \frac{\sqrt{1-e^2}}{na^2 e}\frac{\partial R}{\partial e} - \frac{\cos i}{na^2\sqrt{1-e^2}\sin i}\frac{\partial R}{\partial i} \\
\frac{d}{dt}\cos i &= \frac{-1}{na^2\sqrt{1-e^2}}\left(\cos i\frac{\partial R}{\partial \omega} - \frac{\partial R}{\partial \Omega}\right) \\
\frac{d\Omega}{dt} &= \frac{1}{na^2\sqrt{1-e^2}\sin i}\frac{\partial R}{\partial i} \\
\frac{dM}{dt} &= n - \frac{2}{na}\frac{\partial R}{\partial a} - \frac{1-e^2}{na^2 e}\frac{\partial R}{\partial e}
\end{aligned} \quad (12.55)$$

Xu and Xu noted in 2012 that using $e^2$ and $\cos i$ as variables instead of $e$ and $i$, would lead the second and fourth Lagrange equations of motion 12.55 to be singularity-free (Xu and Xu 2012, 2013b). According to orbital geometry (cf. Fig. 3.4 in Sect. 3.1.1) and different singular cases, for circular orbit ($e = 0$), (implicit $\sin i \neq 0$), the perigee is ambiguous, and a new $M'$ which denotes the angle between satellite and ascending node should be redefined as

$$M' = M + \omega \qquad (12.56)$$

For equatorial orbit ($\sin i = 0$), (implicit $e \neq 0$), the ascending node is ambiguous, and a new $\omega'$ which denotes the angle between perigee and vernal equinox should be redefined as

$$\omega' = \Omega + \omega \qquad (12.57)$$

For circular and equatorial orbit ($e = 0$ and $\sin i = 0$), a new $M''$ which denotes the angle between satellite and vernal equinox should be redefined as

$$M'' = M + \omega + \Omega \qquad (12.58)$$

For simplicity, we define

$$h = \frac{\sqrt{1-e^2} - (1-e^2)}{e} = \left(\frac{\sqrt{1-e^2}\left(1 - \sqrt{1-e^2}\right)}{e}\right) = \left(\frac{\sqrt{1-e^2}\left(\sqrt{1 - \sqrt{1-e^2}}\right)}{\sqrt{1+\sqrt{1-e^2}}}\right). \qquad (12.59)$$

Using a mathematical expansion formula, it yields (truncation to $e^4$)

$$\sqrt{1-e^2} \approx 1 - \frac{e^2}{2} - \frac{e^4}{8}, \qquad (12.60)$$

and therefore,

$$h \approx \frac{e(4-e^2)}{8}. \qquad (12.61)$$

We note that this factor $h$ will never be singular in the case of $e = 0$. We further note that

$$\frac{1-\cos i}{\sin i} = \tan\frac{i}{2}. \qquad (12.62)$$

In cases of singularity, the singularity-free equations of motion should be as follows (Chunhua Jiang and Yan Xu contributed in part to the following derivations):

1. In the case of circular singularity ($e = 0$), (implicit $\sin i \neq 0$), we have

$$\frac{dM}{dt} + \frac{d\omega}{dt} = n - \frac{2}{na}\frac{\partial R}{\partial a} - \frac{1-e^2}{na^2 e}\frac{\partial R}{\partial e} + \frac{\sqrt{1-e^2}}{na^2 e}\frac{\partial R}{\partial e} - \frac{\cos i}{na^2\sqrt{1-e^2}\sin i}\frac{\partial R}{\partial i}$$

$$= n - \frac{2}{na}\frac{\partial R}{\partial a} - \frac{\cos i}{na^2\sqrt{1-e^2}\sin i}\frac{\partial R}{\partial i} + \frac{h}{na^2}\frac{\partial R}{\partial e}$$

(12.63)

Therefore, the motion equations in the case of circular singularity can be formed as

$$\frac{d}{dt}a = \frac{2}{na}\frac{\partial R}{\partial M}$$

$$\frac{d(e^2/2)}{dt} = \frac{1-e^2}{na^2}\frac{\partial R}{\partial M} - \frac{\sqrt{1-e^2}}{na^2}\frac{\partial R}{\partial \omega}$$

$$\frac{d\cos i}{dt} = \frac{-1}{na^2\sqrt{1-e^2}}\left(\cos i \frac{\partial R}{\partial \omega} - \frac{\partial R}{\partial \Omega}\right)$$

$$\frac{d\Omega}{dt} = \frac{1}{na^2\sqrt{1-e^2}\sin i}\frac{\partial R}{\partial i}$$

$$\frac{dM'}{dt} = n - \frac{2}{na}\frac{\partial R}{\partial a} - \frac{\cos i}{na^2\sqrt{1-e^2}\sin i}\frac{\partial R}{\partial i} + \frac{h}{na^2}\frac{\partial R}{\partial e}$$

(12.64)

which are singularity-free in the case of circular orbit and are called general Lagrange-Xu singularity-free equations of motion.

2. In the case of equatorial singularity ($\sin i = 0$), (implicit $e \neq 0$), we have

$$\frac{d\Omega}{dt} + \frac{d\omega}{dt} = \frac{\sqrt{1-e^2}}{na^2 e}\frac{\partial R}{\partial e} - \frac{\cos i}{na^2\sqrt{1-e^2}\sin i}\frac{\partial R}{\partial i} + \frac{1}{na^2\sqrt{1-e^2}\sin i}\frac{\partial R}{\partial i}$$

$$= \frac{\sqrt{1-e^2}}{na^2 e}\frac{\partial R}{\partial e} + \frac{1}{na^2\sqrt{1-e^2}}\frac{1-\cos i}{\sin i}\frac{\partial R}{\partial i}$$

$$= \frac{\sqrt{1-e^2}}{na^2 e}\frac{\partial R}{\partial e} + \frac{1}{na^2\sqrt{1-e^2}}\tan\frac{i}{2}\frac{\partial R}{\partial i}$$

(12.65)

Therefore, the motion equations in the case of equatorial singularity can be formed as

$$\begin{aligned}
\frac{d}{dt}a &= \frac{2}{na}\frac{\partial R}{\partial M} \\
\frac{d(e^2/2)}{dt} &= \frac{1-e^2}{na^2}\frac{\partial R}{\partial M} - \frac{\sqrt{1-e^2}}{na^2}\frac{\partial R}{\partial \omega} \\
\frac{d\cos i}{dt} &= \frac{-1}{na^2\sqrt{1-e^2}}\left(\cos i\frac{\partial R}{\partial \omega} - \frac{\partial R}{\partial \Omega}\right) \\
\frac{dM}{dt} &= n - \frac{2}{na}\frac{\partial R}{\partial a} - \frac{1-e^2}{na^2 e}\frac{\partial R}{\partial e} \\
\frac{d\omega'}{dt} &= \frac{\sqrt{1-e^2}}{na^2 e}\frac{\partial R}{\partial e} + \frac{1}{na^2\sqrt{1-e^2}}\tan\frac{i}{2}\frac{\partial R}{\partial i}
\end{aligned} \qquad (12.66)$$

which are singularity-free in the case of equatorial orbit and are called general Lagrange-Xu singularity-free equations of motion.

3. In the case of circular and equatorial singularity ($\sin i = 0$ and $e = 0$), we have

$$\begin{aligned}
\frac{dM}{dt} + \frac{d\omega}{dt} + \frac{d\Omega}{dt} &= n - \frac{2}{na}\frac{\partial R}{\partial a} - \frac{1-e^2}{na^2 e}\frac{\partial R}{\partial e} + \frac{\sqrt{1-e^2}}{na^2 e}\frac{\partial R}{\partial e} - \frac{\cos i}{na^2\sqrt{1-e^2}\sin i}\frac{\partial R}{\partial i} \\
&\quad + \frac{1}{na^2\sqrt{1-e^2}\sin i}\frac{\partial R}{\partial i} \\
&= n - \frac{2}{na}\frac{\partial R}{\partial a} + \frac{h}{na^2}\frac{\partial R}{\partial e} + \frac{1}{na^2\sqrt{1-e^2}}\tan\frac{i}{2}\frac{\partial R}{\partial i}
\end{aligned}$$

$$(12.67)$$

Therefore, the motion equations in the case of circular and equatorial singularity can be formed as

$$\begin{aligned}
\frac{d}{dt}a &= \frac{2}{na}\frac{\partial R}{\partial M} \\
\frac{d(e^2/2)}{dt} &= \frac{1-e^2}{na^2}\frac{\partial R}{\partial M} - \frac{\sqrt{1-e^2}}{na^2}\frac{\partial R}{\partial \omega} \\
\frac{d\cos i}{dt} &= \frac{-1}{na^2\sqrt{1-e^2}}\left(\cos i\frac{\partial R}{\partial \omega} - \frac{\partial R}{\partial \Omega}\right) \\
\frac{dM''}{dt} &= n - \frac{2}{na}\frac{\partial R}{\partial a} + \frac{h}{na^2}\frac{\partial R}{\partial e} + \frac{1}{na^2\sqrt{1-e^2}}\tan\frac{i}{2}\frac{\partial R}{\partial i}
\end{aligned} \qquad (12.68)$$

which are singularity-free in the case of circular and equatorial orbit and are called general Lagrange-Xu singularity-free equations of motion.

The above derivations of Eqs. 12.64, 12.66, and 12.68 are the so-called Lagrange-Xu singularity-free equations of motion in different singular cases (Xu and Xu 2012, 2013b). The derivations are rigorous and mathematically exact. Thus, the Lagrange-Xu singularity-free equations of motion can be summarized as follows, as Eq. 12.69:

$$\frac{da}{dt} = \frac{2}{na}\frac{\partial R}{\partial M}$$

$$\frac{d}{dt}\left(\frac{e^2}{2}\right) = \frac{1}{na^2}\left((1-e^2)\frac{\partial R}{\partial M} - \sqrt{1-e^2}\frac{\partial R}{\partial \omega}\right)$$

$$\frac{d}{dt}\cos i = \frac{-1}{na^2\sqrt{1-e^2}}\left(\cos i\frac{\partial R}{\partial \omega} - \frac{\partial R}{\partial \Omega}\right)$$

$$\frac{d\Omega}{dt} = \frac{1}{na^2\sqrt{1-e^2}\sin i}\frac{\partial R}{\partial i} \quad (\text{if } e \neq 0 \text{ and } \sin i \neq 0)$$

$$\frac{d\omega}{dt} = \begin{cases} \dfrac{\sqrt{1-e^2}}{na^2 e}\dfrac{\partial R}{\partial e} - \dfrac{\cos i}{na^2\sqrt{1-e^2}\sin i}\dfrac{\partial R}{\partial i} & (\text{if } e \neq 0 \text{ and } \sin i \neq 0) \\ \dfrac{\sqrt{1-e^2}}{na^2 e}\dfrac{\partial R}{\partial e} + \dfrac{1}{na^2\sqrt{1-e^2}}\tan\dfrac{i}{2}\dfrac{\partial R}{\partial i} & (\text{if } e \neq 0 \text{ and } \sin i = 0) \end{cases}$$

$$\frac{dM}{dt} = \begin{cases} n - \dfrac{2}{na}\dfrac{\partial R}{\partial a} - \dfrac{1-e^2}{na^2 e}\dfrac{\partial R}{\partial e} & (\text{if } e \neq 0 \text{ and } \sin i \neq 0) \\ n - \dfrac{2}{na}\dfrac{\partial R}{\partial a} - \dfrac{\cos i}{na^2\sqrt{1-e^2}\sin i}\dfrac{\partial R}{\partial i} + \dfrac{h}{na^2}\dfrac{\partial R}{\partial e} & (\text{if } e = 0 \text{ and } \sin i \neq 0) \\ n - \dfrac{2}{na}\dfrac{\partial R}{\partial a} + \dfrac{h}{na^2}\dfrac{\partial R}{\partial e} + \dfrac{1}{na^2\sqrt{1-e^2}}\tan\dfrac{i}{2}\dfrac{\partial R}{\partial i} & (\text{if } e = 0 \text{ and } \sin i = 0) \end{cases}$$

(12.69)

## 12.2.5
### Derivation of Gauss Equations from Lagrange Equations

As discussed in Sect. 12.2.4, the Lagrange equations of motion in common non-singular case (i.e. $\sin i \neq 0$ and $e \neq 0$) are

$$\frac{da}{dt} = \frac{2}{na}\frac{\partial R}{\partial M}$$

$$\frac{d}{dt}\left(\frac{e^2}{2}\right) = \frac{1}{na^2}\left((1-e^2)\frac{\partial R}{\partial M} - \sqrt{1-e^2}\frac{\partial R}{\partial \omega}\right)$$

$$\frac{d\omega}{dt} = \frac{\sqrt{1-e^2}}{na^2 e}\frac{\partial R}{\partial e} - \frac{\cos i}{na^2\sqrt{1-e^2}\sin i}\frac{\partial R}{\partial i} \qquad (12.70)$$

$$\frac{d}{dt}\cos i = \frac{-1}{na^2\sqrt{1-e^2}}\left(\cos i\frac{\partial R}{\partial \omega} - \frac{\partial R}{\partial \Omega}\right)$$

$$\frac{d\Omega}{dt} = \frac{1}{na^2\sqrt{1-e^2}\sin i}\frac{\partial R}{\partial i}$$

$$\frac{dM}{dt} = n - \frac{2}{na}\frac{\partial R}{\partial a} - \frac{1-e^2}{na^2 e}\frac{\partial R}{\partial e}$$

Here, $R$ is the disturbing potential function. The Keplerian elements $(a, e, \omega, i, \Omega, M, f)$ are the semi-major axis, the eccentricity of the ellipse, the argument of perigee, the inclination angle, the right ascension of ascending node, the mean anomaly, and the true anomaly, respectively.

The partial derivatives of the Keplerian elements can be represented as (Xu et al. 2015a, b)

$$\begin{aligned}
\frac{\partial R}{\partial a} &= \frac{r}{a} f_r \\
\frac{\partial R}{\partial e} &= -a \cos f \cdot f_r + \frac{r \sin f}{1-e^2}(2 + e \cos f) \cdot f_\alpha \\
\frac{\partial R}{\partial i} &= r \sin u \cdot f_h \\
\frac{\partial R}{\partial \Omega} &= r \cos i \cdot f_\alpha - r \sin i \cos u \cdot f_h \\
\frac{\partial R}{\partial \omega} &= r \cdot f_\alpha \\
\frac{\partial R}{\partial M} &= \frac{ae}{\sqrt{1-e^2}} \sin f \cdot f_r + \frac{a(1+e \cos f)}{\sqrt{1-e^2}} \cdot f_\alpha
\end{aligned} \quad (12.71)$$

Here, $(f_r, f_\alpha, f_h)^T$ is a force vector in an orbital plane coordinate system; the first two components are in the orbital plane, $f_r$ is the radial component, $f_\alpha$ is the component perpendicular to $f_r$ and points in the direction of the satellite motion. Furthermore, the Keplerian variables have the following relationships:

$$\begin{aligned}
r &= \frac{a(1-e^2)}{1+e \cos f} = a(1 - e \cos E) \\
r \cos f &= a(\cos E - e) \\
r \sin f &= a\sqrt{1-e^2} \sin E
\end{aligned} \quad (12.72)$$

Here $E$ is the eccentric anomaly. Substituting Eqs. 12.71 and 12.72 into 12.70, it yields

$$\begin{aligned}
\frac{da}{dt} &= \frac{2}{n\sqrt{1-e^2}}[e \sin f \cdot f_r + (1 + e \cos f) \cdot f_\alpha] \\
\frac{d(e^2/2)}{dt} &= \frac{e\sqrt{1-e^2}}{na}[\sin f \cdot f_r + (\cos f + \cos E) \cdot f_\alpha] \\
\frac{di}{dt} &= \frac{(1-e\cos E)\cos u}{na\sqrt{1-e^2}} \cdot f_h \\
\frac{d\omega}{dt} &= \frac{\sqrt{1-e^2}}{nae}\left[-\cos f \cdot f_r + \frac{2+e\cos f}{1+e\cos f}\sin f \cdot f_\alpha\right] - \frac{\cos i(1-e\cos E)\sin u}{na\sqrt{1-e^2}\sin i} \cdot f_h \\
\frac{d\Omega}{dt} &= \frac{(1-e\cos E)\sin u}{na\sqrt{1-e^2}\sin i} \cdot f_h \\
\frac{dM}{dt} &= n - \frac{1-e^2}{nae}\left[\left(-\cos f + \frac{2e}{1+e\cos f}\right) \cdot f_r + \frac{(2+e\cos f)\sin f}{1+e\cos f} \cdot f_\alpha\right]
\end{aligned}$$

$$(12.73)$$

Therefore, the Gaussian equations of motion in the common non-singular case (i.e. $\sin i \neq 0$ and $e \neq 0$) of Eq. 12.73 are derived.

## 12.2.6
### Derivation of Gauss-Xu Equations of Motion

On the basis of the derivation of Lagrange-Xu singularity-free equations of motion in Sect. 12.2.4, Gauss-Xu singularity-free equations of motion in different singular cases can be similarly derived as follows (cf. Xu et al. 2015a, b). For simplicity, we define $h$ the same as mentioned in Sect. 12.2.4. Chunhua Jiang and Yan Xu contributed part to the following derivations.

1. In the case of circular singularity ($e = 0$), (implicit $\sin i \neq 0$), we have

$$\begin{aligned}
\frac{dM}{dt} + \frac{d\omega}{dt} &= n - \frac{2}{na}\frac{\partial R}{\partial a} - \frac{\cos i}{na^2\sqrt{1-e^2}\sin i}\frac{\partial R}{\partial i} + \frac{h}{na^2}\frac{\partial R}{\partial e} \\
&= n - \frac{2(1-e^2)}{na(1+e\cos f)}\cdot f_r - \frac{\cos i(1-e\cos E)\sin u}{na\sqrt{1-e^2}\sin i}\cdot f_h \\
&\quad + \frac{h}{na}\left(-\cos f \cdot f_r + \frac{2+e\cos f}{1+e\cos f}\sin f \cdot f_\alpha\right) \\
&= n - \left[\frac{2(1-e^2)}{na(1+e\cos f)} + \frac{h}{na}\cos f\right]\cdot f_r + \frac{(2+e\cos f)h\sin f}{na(1+e\cos f)}\cdot f_\alpha \\
&\quad - \frac{\cos i(1-e\cos E)\sin u}{na\sqrt{1-e^2}\sin i}\cdot f_h
\end{aligned}$$
(12.74)

Therefore, the motion equations in the circular singularity case can be formed as

$$\begin{aligned}
\frac{da}{dt} &= \frac{2}{n\sqrt{1-e^2}}[e\sin f \cdot f_r + (1+e\cos f)\cdot f_\alpha] \\
\frac{d(e^2/2)}{dt} &= \frac{e\sqrt{1-e^2}}{na}[\sin f \cdot f_r + (\cos f + \cos E)\cdot f_\alpha] \\
\frac{di}{dt} &= \frac{(1-e\cos E)\cos u}{na\sqrt{1-e^2}}\cdot f_h \\
\frac{d\Omega}{dt} &= \frac{(1-e\cos E)\sin u}{na\sqrt{1-e^2}\sin i}\cdot f_h \\
\frac{dM'}{dt} &= n - \left[\frac{2(1-e^2)}{na(1+e\cos f)} + \frac{h}{na}\cos f\right]\cdot f_r + \frac{(2+e\cos f)h\sin f}{na(1+e\cos f)}\cdot f_\alpha \\
&\quad - \frac{\cos i(1-e\cos E)\sin u}{na\sqrt{1-e^2}\sin i}\cdot f_h
\end{aligned}$$
(12.75)

which are singularity-free in the case of circular orbit and are called general Gauss-Xu singularity-free equations of motion.

2. In the case of equatorial singularity ($\sin i = 0$), (implicit $e \neq 0$), we have

$$\begin{aligned}\frac{d\Omega}{dt} + \frac{d\omega}{dt} &= \frac{\sqrt{1-e^2}}{na^2 e} \frac{\partial R}{\partial e} + \frac{1}{na^2\sqrt{1-e^2}} \tan\frac{i}{2} \frac{\partial R}{\partial i} \\ &= \frac{\sqrt{1-e^2}}{nae}\left[-\cos f \cdot f_r + \frac{2+e\cos f}{1+e\cos f}\sin f \cdot f_\alpha\right] + \frac{(1-e\cos E)\sin u}{na\sqrt{1-e^2}}\tan\frac{i}{2}\cdot f_h\end{aligned}$$
(12.76)

Therefore, the motion equations in the case of equatorial singularity can be formed as

$$\begin{aligned}\frac{da}{dt} &= \frac{2}{n\sqrt{1-e^2}}[e\sin f \cdot f_r + (1+e\cos f)\cdot f_\alpha] \\ \frac{d(e^2/2)}{dt} &= \frac{e\sqrt{1-e^2}}{na}[\sin f \cdot f_r + (\cos f + \cos E)\cdot f_\alpha] \\ \frac{di}{dt} &= \frac{(1-e\cos E)\cos u}{na\sqrt{1-e^2}}\cdot f_h \\ \frac{dM}{dt} &= n - \frac{1-e^2}{nae}\left[\left(-\cos f + \frac{2e}{1+e\cos f}\right)\cdot f_r + \frac{(2+e\cos f)\sin f}{1+e\cos f}\cdot f_\alpha\right] \\ \frac{d\omega'}{dt} &= \frac{\sqrt{1-e^2}}{nae}\left[-\cos f \cdot f_r + \frac{2+e\cos f}{1+e\cos f}\sin f \cdot f_\alpha\right] + \frac{(1-e\cos E)\sin u}{na\sqrt{1-e^2}}\tan\frac{i}{2}\cdot f_h\end{aligned}$$
(12.77)

which are singularity-free in the case of equatorial orbit and are called general Gauss-Xu singularity-free equations of motion.

3. In the case of circular and equatorial singularity ($\sin i = 0$ and $e = 0$), we have

$$\begin{aligned}\frac{dM}{dt} + \frac{d\omega}{dt} + \frac{d\Omega}{dt} &= n - \frac{2}{na}\frac{\partial R}{\partial a} + \frac{h}{na^2}\frac{\partial R}{\partial e} + \frac{1}{na^2\sqrt{1-e^2}}\tan\frac{i}{2}\frac{\partial R}{\partial i} \\ &= n - \left[\frac{2(1-e^2)}{na(1+e\cos f)} + \frac{h}{na}\cos f\right]\cdot f_r + \frac{(2+e\cos f)h\sin f}{na(1+e\cos f)}\cdot f_\alpha \\ &\quad + \frac{(1-e\cos E)\sin u}{na\sqrt{1-e^2}}\tan\frac{i}{2}\cdot f_h\end{aligned}$$
(12.78)

## 12.2 · On the Singularity Problem in Orbital Mechanics

Therefore, the motion equation in circular and equatorial singularity case can be formed as

$$\frac{da}{dt} = \frac{2}{n\sqrt{1-e^2}}[e\sin f \cdot f_r + (1+e\cos f) \cdot f_\alpha]$$

$$\frac{d(e^2/2)}{dt} = \frac{e\sqrt{1-e^2}}{na}[\sin f \cdot f_r + (\cos f + \cos E) \cdot f_\alpha]$$

$$\frac{di}{dt} = \frac{(1-e\cos E)\cos u}{na\sqrt{1-e^2}} \cdot f_h$$

$$\frac{dM''}{dt} = n - \left[\frac{2(1-e^2)}{na(1+e\cos f)} + \frac{h}{na}\cos f\right] \cdot f_r + \frac{(2+e\cos f)h\sin f}{na(1+e\cos f)} \cdot f_\alpha$$

$$+ \frac{(1-e\cos E)\sin u}{na\sqrt{1-e^2}}\tan\frac{i}{2} \cdot f_h$$

$$(12.79)$$

which are singularity-free in the case of circular and equatorial orbit and are called general Gauss-Xu singularity-free equations of motion.

The above derivations of Eqs. 12.75, 12.77, and 12.79 are the so-called Gauss-Xu singularity-free equations of motion in different singular cases. The derivations are rigorous and mathematically exact. Thus, the Gauss-Xu singularity-free equations of motion can be summarized as follows as Eq. 12.80:

$$\frac{da}{dt} = \frac{2}{n\sqrt{1-e^2}}[e\sin f \cdot f_r + (1+e\cos f) \cdot f_\alpha]$$

$$\frac{d(e^2/2)}{dt} = \frac{e\sqrt{1-e^2}}{na}[\sin f \cdot f_r + (\cos f + \cos E) \cdot f_\alpha]$$

$$\frac{di}{dt} = \frac{(1-e\cos E)\cos u}{na\sqrt{1-e^2}} \cdot f_h$$

$$\frac{d\Omega}{dt} = \frac{(1-e\cos E)\sin u}{na\sqrt{1-e^2}\sin i} \cdot f_h \quad (\text{if } e \neq 0 \text{ and } \sin i \neq 0)$$

$$\frac{d\omega}{dt} = \begin{cases} \frac{\sqrt{1-e^2}}{nae}\left[-\cos f \cdot f_r + \frac{2+e\cos f}{1+e\cos f}\sin f \cdot f_\alpha\right] - \frac{\cos i(1-e\cos E)\sin u}{na\sqrt{1-e^2}\sin i} \cdot f_h & (\text{if } e \neq 0 \text{ and } \sin i \neq 0) \\ \frac{\sqrt{1-e^2}}{nae}\left[-\cos f \cdot f_r + \frac{2+e\cos f}{1+e\cos f}\sin f \cdot f_\alpha\right] + \frac{(1-e\cos E)\sin u}{na\sqrt{1-e^2}}\tan\frac{i}{2} \cdot f_h & (\text{if } e \neq 0 \text{ and } \sin i = 0) \end{cases}$$

$$\frac{dM}{dt} = \begin{cases} n - \frac{1-e^2}{nae}\left[(-\cos f + \frac{2e}{1+e\cos f}) \cdot f_r + \frac{(2+e\cos f)\sin f}{1+e\cos f} \cdot f_\alpha\right] & (\text{if } e \neq 0 \text{ and } \sin i \neq 0) \\ n - \left[\frac{2(1-e^2)}{na(1+e\cos f)} + \frac{h}{na}\cos f\right] \cdot f_r + \frac{(2+e\cos f)h\sin f}{na(1+e\cos f)} \cdot f_\alpha - \frac{\cos i(1-e\cos E)\sin u}{na\sqrt{1-e^2}\sin i} \cdot f_h & (\text{if } e = 0 \text{ and } \sin i \neq 0) \\ n - \left[\frac{2(1-e^2)}{na(1+e\cos f)} + \frac{h}{na}\cos f\right] \cdot f_r + \frac{(2+e\cos f)h\sin f}{na(1+e\cos f)} \cdot f_\alpha + \frac{(1-e\cos E)\sin u}{na\sqrt{1-e^2}}\tan\frac{i}{2} \cdot f_h & (\text{if } e = 0 \text{ and } \sin i = 0) \end{cases}$$

$$(12.80)$$

## 12.3
## Bridge Between Analytical Theory and Numerical Integration

In 2010, Tianhe Xu worked as a visiting scientist at the GFZ with GXu, and is the second author of two very important papers concerning extraterrestrial gravitational disturbances and atmospheric drag on satellite orbits (Xu et al. 2010a, b) using so-called basic Lagrangian and Gaussian equations of motion. He then applied these in orbital analytical solutions. However, the theoretical integrals and numerical integration showed certain large differences. Either the theoretical integrals were wrong by derivation, or the numerical integration was wrong. But both mistakes seemed to be impossible: the theoretical integrals were not wrong by derivation, and the numerical integration was not wrong. A bridge between analytical theory and numerical integration was needed so that the analytical theory could be used to obtain the same results as those using numerical integration.

It took two years for GXu to understand the reason for the difference. By numerical integration, the variables are updated by every integration step, whereas by theoretical integration they are not. With this insight, GXu discussed organising PhD studies for these topics with Wu Chen. In 2012, Jing Qiao started her study with Wu Chen at Hong Kong Polytech University and is expected to finish part of her PhD thesis with GXu at Shandong University. In 2016, Jing Qiao was able to demonstrate how to use the derived analytical solutions in Xu and Xu (2013a, b) to compute the comparable results obtained by numerical integrator. Thus, for the first time in the history of celestial mechanics, analytical solutions could be used for orbit determination and orbit optimization.

It worth to note that the navigation and remote sensing as well as celestial mechanics team of Shandong University at Weihai is making several progresses on the satellite orbit-related topics. The so-called Xu's algebra solution of the variation equation (Xu 2007 Chap. 11.5.1) is proved to have good performance in practice by Wenfeng Nie and Yujun Du. Jing Qiao contributed greatly for how to use the analytical solution to obtain comparable results with that of numerical integration. Chunhua Jiang contributed largely for the singularity-free problem. Fangzhao Zhang worked out first timely the influences of the precession and nutation as well as polar motion on the Kepler elements. Nan Jiang is using the Gauss-Xu singularity-free equations for BeiDou GEO satellite orbits maneuver detection and determination. Wenfeng Nie is working on using Xu's solar radiation and atmosphere drag models for precise orbits determination.

# References

Aarseth SJ (2003) Gravitational N-Body simulations Tools and Algorithms, Cambridge University Press

Aarseth SJ, Tout CA, Mardling RA (2008) The Cambridge N-Body Lectures, Springer Heidelsberg

Barrow-Green J (1996) Poincare and the Three Body Problem, Amer. Math. Soc.

Battin R.H. (1999) An Introduction to the Mathematics and Methods of Astrodynamics, revised version, AIAA Education Series

Bernoulli JI (1710) Extrait de la réponse de M. Bernoulli à M. Hermann, datée de Basle le 7 Octobre 1710 Joh. B. Op LXXXVI. Both letters were published in the Mémoires de l'Ac Royale des Sciences, Boudot, Paris

Boccaletti D, Pucacco G (1998) Theory of Orbits (two volumes). Springer-Verlag

Boccaletti D, Pucacco G (2001) Theory of Oribts, Vol. 1: Integrable systems and non-perturbative methods; Vol. 2: Perturbative and geometrical methods, Springer Berlin

Bronstein IN, Semendjajew KA (1987) Taschenbuch der Mathematik.B. G. Teubner Verlagsgesellschaft, Leipzig, ISBN 3-322-00259-4

Brouwer D, Clemence GM (1961) Methods of celestial mechanics. Academic Press, New York

Chobotov VA (ed) (1991) Orbital mechanics. Published by AIAA, Washington

Cui C (1990) Die Bewegung künstlicher Satelliten im anisotropen Gravitationsfeld einer gleichmässig rotierenden starren Modellerde. Deutsche Geodätische Kommission, Reihe C: Dissertationen, Heft Nr. 357

Cui C (1997) Satellite orbit integration based on canonical transformations with special regard to the resonance and coupling effects. Dtsch Geod Komm bayer Akad Wiss, Reihe A, Nr. 112, 128 pp

Delaunay (1860) Mem. De l'des Sci. 28

Delaunay (1867) Mem. De l'des Sci. 29

Diacu, F (1992) Singularities of the N-Body Problem, Les Publications CRM, Montreal

Diacu F (1996) The solution of the n-body problem, The Mathematical Intelligencer, 1996, 18, p. 66–70

Diacu F, Holmes P (1996) Celestial Encounters: The origins or chaos and stability, Princeton University Press, Princeton, NJ

Dvorak R, Lhotka C (2013) Celestial dynamics – chaoticity and dynamics of celestial systems, Wiley, Weinheim

Eberle J, Cuntz M, Musielak ZE (2008) The instability trasition for the restricted 3-body problem - I. Theoretical approach, Astronomy & Astrophysics, Vol. 489 No.3

Euler L (1767) Nov. Comm. Acad. Imp. Petropolitanae, 10, pp207-242, 11, pp152-184; Memories de IAcad. De Berlin, 11, 228-249

Goldstein H (1980) Classical Mechanics (2nd Ed), New York, Addison-Wesley

Hagihara, Y: Celestial Mechanics. (Vol I and Vol II pt 1 and Vol II pt 2.) MIT Press, 1970

Havel K (2008) N-Body Gravitational Problem: Unrestricted Solution, Brampton: Grevyt Press, 2008

Herrick S (1972) Astrodynamics, Vol. II. Van Nostrand Reinhold, London

Kaula WM (2000) Theory of satellite geodesy: applications of satellites to geodesy. Courier Corporation

Kaula WM (1966, 2001) Theory of satellite geodesy. Blaisdell Publishing Company, Dover Publications, New York

Kaula WM (2001) Theory of satellite geodesy. Blaisdell Publishing Company, Dover Publications, New York

Lagrange JL (1772) Miscellanea Taurinensia, 4, 118-243; Oeuvres, 2, pp67-121; Mechanique Analytique, 1st Ed, pp262-286; 2nd Ed, 2, pp108-121; Oeuvres, 12, pp101-114

Landau LD, Lifshitz EM (1976) Mechanics (3rd Ed), New York, Pergamon Press

Liu L, Zhao D (1979) Orbit theory of the Earth satellite. Nanjing University Press, (in Chinese)

Mittag-Leffler G. The n-body problem (Price Announcement), Acta Matematica, 1885/1886, 7
Newton I(1687): Philosophiae Naturalis Principia Mathematica, London, 1687: also English translation of 3rd (1726) edition by I. Bernard Cohen and Anne Whitman (Berkeley, CA, 1999)
Poincare (1892) Les Methodes Nouevelles de la Mechanique ce'leste Guthier villars, Paris, Chap. V, p. 250, (published in English in three volumes)
Poincare H (1992) New Methods of Celestial Mechanics, AIP
Schneider M, Cui CF (2005) Theoreme über Bewegungsintegrale und ihre Anwendung in Bahntheorien, Bayerischen Akad Wiss, Reihe A, Heft Nr. 121, 132pp, München
Shapiro II (1962) The prediction of satellite orbits, in IUTAM Symposium Paris 1962, M Roy (ed) Dynamics of Satellites
Sundman K (1906) Recherches sur le problème des trois corps, Acta Soc. Sei. Fenn. 34
Sundman K (1909) Nouvelles recherches sur le problème des trois corps, Acta Soc. Sei. Fenn. 35
Sundman KE (1912) Memoire sur le probleme de trois corps, Acta Mathematica 36 (1912): 105–179.
Tisserand F-F (1894) Mecanique Celeste, tome III (Paris, 1894), ch.III, at p. 27
Vallado David A (2007) Fundamentals of Astrodynamics and Applications (3rd Ed), Microcosm Press & Springer
Van Kamp PD (1967) Principles of Astronomy. W.H. Freemann and Company, San Francisco, CA/London
Wang QD (1991) The global solution of the n-body problem (Celestial Mechanics and Dynamical Astronomy (ISSN 0923-2958), vol. 50, no. 1, 1991, p. 73–88., URI retrieved on 2007-05-05)
Wang LX, Fang ZD, Zhang MY, Lin GB, Gu LK, Zhong TD, Yang XA, She DP, Luo ZH, Xiao BQ, Chai H, Lin DX (1979) Mathematic handbook. Educational Press, Peking, ISBN 13012-0165
Wnuk E (1990) Tesseral harmonic perturbations in the Keplerian orbital elements, Acta. Astronomica Vol.40, No.1-2, p.191-198
Xia ZH (1992) The existence of noncollision singularities in Newtonian system, Annals Math. 135(3): 411–468
Xu (2013b) Private communication with an editor of MNRAS concerning two submitted papers and the decision letters of MNRAS editorial board (for review available upon request gcxu@sdu.edu.cn)
Xu G (2008) Orbits, Springer Heidelberg, ISBN 978-3-540-78521-7, 230 pages, in English
Xu G (2010) Analytic Orbit Theory, chapter 4 in G Xu (Ed) Sciences of Geodesy - I, Advances and Future Directions, Springer, pp 105-154
Xu G (2010) (Ed.): Sciences of Geodesy - I, Advances and Future Directions, Springer Heidelberg, chapter topics (authors): Aerogravimetry (R Forsberg), Superconducting Gravimetry (J Neumeyer), Absolute and Relative Gravimetry (L Timmen), Deformation and Tectonics (L Bastos et al.), Analytic Orbit Theory (G Xu), InSAR (Y Xia), Marine Geodesy (J Reinking), Kalman Filtering (Y Yang), Equivalence of GPS Algorithms (G Xu et al.), Earth Rotation (F Seitz, H Schuh), Satellite Laser Ranging (L Combrinck), in English, 507 pages
Xu Y (2012) Studies on Antarctic GNSS Precise Positioning. Chang'an University. Xi'an, China
Xu G (2012) (Ed.): Sciences of Geodesy - II, Advances and Future Directions, Springer Heidelberg, chapter topics (authors): General Relativity and Space Geodesy (L Comblinck), Global Terrestrial Reference Systems and their Realizations (D Angermann et al), Ocean Tide Loading (M Bos, HG Scherneck), Photogrammetry (P Redweik), Regularization and Adjustment (Y Shen, G Xu), Regional Gravity Field Modelling (H Denker), VLBI (H Schuh, J Boehm), in English, 400 pages
Xu G, Xu J (2012) On the Singularity Problem of Orbital Mechanics, MNRAS, 2013, Vol.429, pp1139-1148
Xu G, Xu J (2013a) On Orbital Disturbance of Solar Radiation, MNRAS, 432 (1): 584-588 doi:10.1093/mnras/stt483
Xu G, Xu J (2013b) Orbits – $2^{nd}$ Order Singularity-free Solutions, second edition, Springer Heidelberg, ISBN 978-3-642-32792-6, 426 pages, in English

Xu G, Xu TH, Yeh TK, Chen W (2010a) Analytic Solution of Satellite Orbit Perturbed by Lunar and Solar Gravitation, MNRAS, Vol. 410, Issue 1, pp 645-653

Xu G, Xu TH, Chen W, Yeh TK (2010b) Analytic Solution of Satellite Orbit Perturbed by Atmospheric Drag, MNRAS, Vol. 410, Issue 1, pp 654-662 87.

Xu Y, Yang Y, Zhang Q, Xu G (2011) Solar Oblateness and Mercury's Perihelion Precession, MNRAS, Vol. 415, 3335-3343

Xu G, Lv ZP, Shen YZ, Yeh TK (2014) A mathematical derivation of singularity-free Lagrange equations of planetary motion, Special issue for celebration 80th birthday of academician Houze Xu, Journal of Surveying and Mapping, 2014 89.

Xu Y, Yang Y, Xu G (2014) Analysis on Tropospheric Delay in Antarctic GPS Positioning. Journal of Geodesy and Geodynamics, 34(1): pp104-107

Xu G, Chen W, Shen YZ, Jiang N, Jiang CH (2015) A mathematical derivation of singularity-free Gauss equations of planetary motion, Special issue for celebration 70th birthday of Prof Jikun Ou, Journal of Surveying and Mapping, 2015

Xu Y, Jiang N, Xu G, Yang Y, Schuh H (2015) Influence of meteorological data and horizontal gradient of tropospheric model on precise point positioning. Adv. Space Res. 56(11), pp2374-2383

# Chapter 13
# Discussions

The previous chapters of this book covered the most important material regarding static, kinematic, and dynamic GPS, including theory, algorithms, and applications. At the end of the book, the authors will emphasize, discuss, and comment on some important topics and remaining problems with GPS.

## 13.1 Independent Parameterisation and A Priori Information

*A Priori Information*

As discussed earlier with regard to the parameterisation of the GPS observation model (Sects. 9.1 and 9.2), clock errors and instrumental biases, as well as ambiguities, are partially over-parameterised or linearly correlated (related to and between themselves). Generally speaking, cancelling the over-parameterised unknowns out of the equation or modelling them first and then keeping them fixed using the a priori method (Sect. 7.8) will be equivalent. As long as one knows which parameters should be kept fixed, the a priori information used is true, and is just used as a tool for fixing the parameters to zero. If the model is not parameterised regularly, and one does not know exactly which parameters are over-parameterised, then the normal equation will be singular and cannot be solved. Again, using a priori information may make the equation solvable. However, in this case, the a priori information has the meaning of direct "measures" on the related parameters. Therefore, the a priori information used must be true and reasonable; otherwise, the given a priori information will affect the solution in unreasonable ways. If different a priori information is given, different results will be obtained. Therefore, the a priori information used should be based on true information.

## Independent Parameterisation of the Observation Model

A priori information can be obtained from external surveys or from the experience of long-term data processing that does not use a priori information. A regular (independent) parameterisation of the GPS observation model is a precondition for a stable solution of a normal equation without using a priori information. As mentioned above, parameterising the model independently and fixing the over-parameterised unknowns are equivalent. However, in order to keep some parameters fixed, one must know which parameters are over-parameterised and need to be fixed. Therefore, in any case, one must understand how to parameterise the GPS observation model in a regular manner. Fixing the over-parameterised unknowns after a general parameterisation is equivalent to a direct independent parameterisation. Therefore, regular parameterisation of the GPS observation model is important.

## Inseparability of Some of the Bias Effects

Independent parameterisation is necessary because of the linear correlation of some parameters. The linear correlation party merges the different effects so that they cannot be exactly separated from each other. The constant parts of the different effects are nearly impossible to separate without precise physical models, whereas many model parameters are presented in the GPS observation equation and have to be codetermined. The inseparability of the bias effects comes partly from the physics of the surveys and is dependent on the survey strategy. Understanding the inseparability of the bias effects is important in designing surveys. The physical models must be determined more precisely in order to separate the constant parts of the effects.

## Change in the Physical Meaning of the Parameters

Because of the linear correlation and inseparability of some parameters, the parameters that are to be adjusted may change their physical meaning. For example, the instrumental biases of the reference frequency and channel are linearly correlated with the clock errors. This indicates that these biases cannot be modelled separately so that the clock error parameters represent the summation of the clock errors and the related instrumental biases. They may be separated only through extra surveys or alternative models. If the clock errors of the reference satellite and receiver are not adjusted, then the other clock errors represent the relative errors between the other clocks and the reference clocks. If the other instrumental biases are not modelled, they will be partially absorbed into the ambiguities. In this case, the ambiguities represent not only the ambiguities, but also part of the instrumental biases, such that the ambiguities are no longer integers. The double difference may eliminate the instrumental biases so that the double-differenced ambiguities are free from the effects of instrumental biases, whereas the undifferenced ambiguities include those biases. If the instrumental errors are not modelled, the undifferenced ambiguities are no longer integers, whereas the double-differenced ambiguities are integers (no data combinations are considered here).

### Zero Setting and Fixing of the Parameters

Setting a parameter to zero or fixing the parameter to a definite value must be done carefully. Any incorrect setting or fixing is similar to a linear transformation (translation) of the linearly correlated parameters. For example, the clock errors and instrumental biases of the reference station and satellite are generally not zero. Keeping the clock errors and instrumental biases of the reference as zero is similar to carrying out a time system translation with an unknown amount, and such a translation is inhomogeneous, because the orbit data are given in the GPS time system. External surveys may help for a correct zero setting.

### Independent Parameterisation of Physical Models

Independent parameterisation of the bias parameters of the GPS observation model indicates the need for further study of the parameterisation problem. As long as the parameters of the physical models must be codetermined by the GPS observation equations, parameterisation of the physical models should be investigated with great care.

## 13.2
## Equivalence of the GPS Data Processing Algorithms

### Equivalence Principle

For definitive measurement and parameterisation of the observation model, the uncombined and combining algorithms, undifferenced and differencing algorithms, and their mixtures are equivalent. The results must be identical and the precision equivalent. The practical results should obey this principle.

The equivalence comes from the definite information contents of the surveys and the definitive parameterisation of the observation model. For better results or better precision of the results, better measurements are necessary.

### Traditional Combinations

Under traditional parameterisation, the combinations are equivalent. Under independent parameterisation, the combinations are also equivalent. However, the combinations under the traditional parameterisation and independent parameterisation are not equivalent. Because of the inexactness of traditional parameterisation, traditional combinations will lead to inexact results.

### Traditional Differencing Algorithms

Traditional differencing algorithms usually take into account only the differencing equations and leave the undifferenced part aside. In this way, the differencing part of equations includes fewer parameters and the systematic effects are reduced. Meanwhile, however, the information content of the observables is also reduced proportionally. The results of the parameters of interest remain the same.

*Equivalent Algorithms*

Equivalent algorithms are general forms of undifferenced and differencing algorithms. The observation equation can be separated into two diagonal parts. Each part uses the original observation vector (therefore the original weight matrix); however, the equation possesses only a part of the unknown parameters. The normal equation of the original observation equation can also be separated into two parts. This indicates that any solvable adjustment problem can be separated into two sub-problems.

## 13.3
## Other Comments

*Data Communication in Real-Time GNSS Positioning*

Real-time GNSS positioning technology has become a fast, efficient navigation tool that can yield survey-grade coordinates for use in a variety of applications. One of the most important rules for real-time positioning is that a robust communication link is needed for acquiring the data from the rover station or corrections to the observables at the base station (as in relative positioning methodology). When considering data communication, there are several methods to choose from. The use of radios is one option, which is robust, but its range of communication can be limited, especially in urban areas where interference and frequency usage are high. On the other hand, wireless data modems are typically CDMA (Code Division Multiple Access), GSM (Global System for Mobile communications), and GPRS (General Packet Radio Service) communication formats using TCP/IP (Transmission Control Protocol/Internet Protocol) over cellular provider networks. This will allow longer ranges in the case of good cell coverage areas. Maintaining a strong, continuous communication link for data communication in real-time GNSS positioning can still be a challenge.

*Indoor Positioning*

Indoor positioning has become a focus of research and development over the past decade, and has been widely applied in many areas, such as indoor location-based service (LBS). It is apparent that the widely used GNSS performs poorly within indoor environments due to signal outages. Technologies using FM radios, radars, cellular networks, DETC phones, WLAN, ZigBee, RFID, ultra-wideband, high-sensitivity GNSS, and pseudolite systems have been developed. The integration of different techniques in a multi-sensor positioning system is another solution in indoor positioning. However, the indoor environment lacks a system that can provide excellent performance with high accuracy, short latency, high availability, high integrity, and low user costs like GNSS in outdoor environments. Current capable indoor positioning systems have different levels of accuracy, and the provision of global indoor positioning at a low cost and with accuracy of 1 m is far from a reality. Many indoor positioning applications are still waiting for a satisfactory solution.

# Appendix A
# IAU 1980 Theory of Nutation

**Table A.1** The units of $A_i$ and $B_i$ are $0.''0001$, units of $A_i'$ and $B_i'$ are $0.''00001$ (cf. McCarthy 1996)

| Coefficients of | | | | | Values of | | | |
|---|---|---|---|---|---|---|---|---|
| l | l' | F | D | Ω | $A_i$ | $A_i'$ | $B_i$ | $B_i'$ |
| 0 | 0 | 0 | 0 | 1 | −171996 | −1742 | 92025 | 89 |
| 0 | 0 | 2 | −2 | 2 | −13187 | 16 | 5736 | −31 |
| 0 | 0 | 2 | 0 | 2 | −2274 | −2 | 977 | −5 |
| 0 | 0 | 0 | 0 | 2 | 2062 | 2 | −895 | 5 |
| 0 | −1 | 0 | 0 | 0 | −1426 | 34 | 54 | −1 |
| 1 | 0 | 0 | 0 | 0 | 712 | 1 | −7 | 0 |
| 0 | 1 | 2 | −2 | 2 | −517 | 12 | 224 | −6 |
| 0 | 0 | 2 | 0 | 1 | −386 | −4 | 200 | 0 |
| 1 | 0 | 2 | 0 | 2 | −301 | 0 | 129 | −1 |
| 0 | −1 | 2 | −2 | 2 | 217 | −5 | −95 | 3 |
| −1 | 0 | 0 | −2 | 0 | 158 | 0 | −1 | 0 |
| 0 | 0 | 2 | −2 | 1 | 129 | 1 | −70 | 0 |
| −1 | 0 | 2 | 0 | 2 | 123 | 0 | −53 | 0 |
| 1 | 0 | 0 | 0 | 1 | 63 | 1 | −33 | 0 |
| 0 | 0 | 0 | 2 | 0 | 63 | 0 | −2 | 0 |
| −1 | 0 | 2 | 2 | 2 | −59 | 0 | 26 | 0 |
| −1 | 0 | 0 | 0 | 1 | −58 | −1 | 32 | 0 |
| 1 | 0 | 2 | 0 | 1 | −51 | 0 | 27 | 0 |
| −2 | 0 | 0 | 2 | 0 | −48 | 0 | 1 | 0 |
| −2 | 0 | 2 | 0 | 1 | 46 | 0 | −24 | 0 |
| 0 | 0 | 2 | 2 | 2 | −38 | 0 | 16 | 0 |
| 2 | 0 | 2 | 0 | 2 | −31 | 0 | 13 | 0 |
| 1 | 0 | 2 | −2 | 2 | 29 | 0 | −12 | 0 |
| 2 | 0 | 0 | 0 | 0 | 29 | 0 | −1 | 0 |
| 0 | 0 | 2 | 0 | 0 | 26 | 0 | −1 | 0 |
| 0 | 0 | 2 | −2 | 0 | −22 | 0 | 0 | 0 |

(continued)

**Table A.1** (continued)

| Coefficients of | | | | | Values of | | | |
|---|---|---|---|---|---|---|---|---|
| l | l' | F | D | Ω | $A_i$ | $A_i'$ | $B_i$ | $B_i'$ |
| −1 | 0 | 2 | 0 | 1 | 21 | 0 | −10 | 0 |
| 0 | 2 | 0 | 0 | 0 | 17 | −1 | 0 | 0 |
| −1 | 0 | 0 | 2 | 1 | 16 | 0 | −8 | 0 |
| 0 | 2 | 2 | −2 | 2 | −16 | 1 | 7 | 0 |
| 0 | 1 | 0 | 0 | 1 | −15 | 0 | 9 | 0 |
| 1 | 0 | 0 | −2 | 1 | −13 | 0 | 7 | 0 |
| 0 | −1 | 0 | 0 | 1 | −12 | 0 | 6 | 0 |
| 2 | 0 | −2 | 0 | 0 | 11 | 0 | 0 | 0 |
| −1 | 0 | 2 | 2 | 1 | −10 | 0 | 5 | 0 |
| 1 | 0 | 2 | 2 | 2 | −8 | 0 | 3 | 0 |
| 0 | 0 | 2 | 2 | 1 | −7 | 0 | 3 | 0 |
| 0 | −1 | 2 | 0 | 2 | −7 | 0 | 3 | 0 |
| 0 | 1 | 2 | 0 | 2 | 7 | 0 | −3 | 0 |
| 1 | 1 | 0 | −2 | 0 | −7 | 0 | 0 | 0 |
| 1 | 0 | 2 | −2 | 1 | 6 | 0 | −3 | 0 |
| 0 | 0 | 0 | 2 | 1 | −6 | 0 | 3 | 0 |
| 2 | 0 | 2 | −2 | 2 | 6 | 0 | −3 | 0 |
| 1 | 0 | 0 | 2 | 0 | 6 | 0 | 0 | 0 |
| −2 | 0 | 0 | 2 | 1 | −6 | 0 | 3 | 0 |
| 2 | 0 | 2 | 0 | 1 | −5 | 0 | 3 | 0 |
| 1 | −1 | 0 | 0 | 0 | 5 | 0 | 0 | 0 |
| 0 | 0 | 0 | −2 | 1 | −5 | 0 | 3 | 0 |
| 0 | −1 | 2 | −2 | 1 | −5 | 0 | 3 | 0 |
| 0 | 0 | 0 | 1 | 0 | −4 | 0 | 0 | 0 |
| 1 | 0 | −2 | 0 | 0 | 4 | 0 | 0 | 0 |
| 0 | 1 | 0 | −2 | 0 | −4 | 0 | 0 | 0 |
| 1 | 0 | 0 | −1 | 0 | −4 | 0 | 0 | 0 |
| 0 | 1 | 2 | −2 | 1 | 4 | 0 | −2 | 0 |
| 2 | 0 | 0 | −2 | 1 | 4 | 0 | −2 | 0 |
| 0 | −1 | 2 | 2 | 2 | −3 | 0 | 1 | 0 |
| 3 | 0 | 2 | 0 | 2 | −3 | 0 | 1 | 0 |
| −1 | −1 | 2 | 2 | 2 | −3 | 0 | 1 | 0 |
| 1 | −1 | 2 | 0 | 2 | −3 | 0 | 1 | 0 |
| 1 | 0 | 2 | 0 | 0 | 3 | 0 | 0 | 0 |
| 1 | 1 | 0 | 0 | 0 | −3 | 0 | 0 | 0 |
| 1 | −1 | 0 | −1 | 0 | −3 | 0 | 0 | 0 |
| −2 | 0 | 2 | 0 | 2 | −3 | 0 | 1 | 0 |
| −1 | 0 | 2 | 4 | 2 | −2 | 0 | 1 | 0 |
| 0 | 0 | 2 | 1 | 2 | 2 | 0 | −1 | 0 |

(continued)

**Table A.1** (continued)

| Coefficients of | | | | | Values of | | | |
|---|---|---|---|---|---|---|---|---|
| l | l' | F | D | Ω | $A_i$ | $A_i'$ | $B_i$ | $B_i'$ |
| 3 | 0 | 0 | 0 | 0 | 2 | 0 | 0 | 0 |
| 1 | 0 | 0 | 0 | 2 | −2 | 0 | 1 | 0 |
| 2 | 0 | 0 | 0 | 1 | 2 | 0 | −1 | 0 |
| −1 | 0 | 2 | −2 | 1 | −2 | 0 | 1 | 0 |
| 1 | 1 | 2 | 0 | 2 | 2 | 0 | −1 | 0 |
| −2 | 0 | 0 | 0 | 1 | −2 | 0 | 1 | 0 |
| 0 | −2 | 2 | −2 | 1 | −2 | 0 | 1 | 0 |
| 0 | 1 | 0 | 1 | 0 | 1 | 0 | 0 | 0 |
| 0 | 0 | 2 | 4 | 2 | −1 | 0 | 0 | 0 |
| 2 | 0 | 0 | 2 | 0 | 1 | 0 | 0 | 0 |
| 1 | 0 | −2 | 2 | 0 | −1 | 0 | 0 | 0 |
| 1 | 1 | 0 | −2 | 1 | −1 | 0 | 0 | 0 |
| 0 | −1 | 2 | 0 | 1 | −1 | 0 | 0 | 0 |
| 1 | 0 | −2 | −2 | 0 | −1 | 0 | 0 | 0 |
| 0 | 1 | 0 | 2 | 0 | −1 | 0 | 0 | 0 |
| 0 | 0 | 2 | −1 | 2 | −1 | 0 | 0 | 0 |
| 0 | 0 | −2 | 0 | 1 | −1 | 0 | 0 | 0 |
| −1 | −1 | 0 | 2 | 1 | 1 | 0 | 0 | 0 |
| 0 | 1 | 2 | 0 | 1 | 1 | 0 | 0 | 0 |
| 1 | 0 | 2 | −2 | 0 | −1 | 0 | 0 | 0 |
| 3 | 0 | 2 | −2 | 2 | 1 | 0 | 0 | 0 |
| 0 | 0 | 4 | −2 | 2 | 1 | 0 | 0 | 0 |
| 1 | 0 | 0 | 2 | 1 | −1 | 0 | 0 | 0 |
| 2 | 0 | 2 | 2 | 2 | −1 | 0 | 0 | 0 |
| 2 | 0 | 2 | −2 | 1 | 1 | 0 | −1 | 0 |
| 1 | −1 | 0 | −2 | 0 | 1 | 0 | 0 | 0 |
| −1 | 0 | 4 | 0 | 2 | 1 | 0 | 0 | 0 |
| −2 | 0 | 2 | 4 | 2 | −1 | 0 | 1 | 0 |
| 1 | 0 | 2 | 2 | 1 | −1 | 0 | 1 | 0 |
| 1 | 1 | 2 | −2 | 2 | 1 | 0 | −1 | 0 |
| 2 | 0 | 0 | −4 | 0 | −1 | 0 | 0 | 0 |
| −2 | 0 | 2 | 2 | 2 | 1 | 0 | −1 | 0 |
| 1 | 0 | 0 | −4 | 0 | −1 | 0 | 0 | 0 |
| −1 | 0 | 0 | 0 | 2 | 1 | 0 | −1 | 0 |
| 0 | 1 | 2 | −2 | 0 | −1 | 0 | 0 | 0 |
| −1 | 0 | 0 | 1 | 1 | 1 | 0 | 0 | 0 |
| 0 | 1 | 0 | 0 | 2 | 1 | 0 | 0 | 0 |
| 0 | 1 | −2 | 2 | 0 | −1 | 0 | 0 | 0 |
| 0 | 0 | −2 | 2 | 1 | 1 | 0 | 0 | 0 |

(continued)

**Table A.1** (continued)

| Coefficients of | | | | | Values of | | | |
|---|---|---|---|---|---|---|---|---|
| l | l' | F | D | Ω | $A_i$ | $A_i'$ | $B_i$ | $B_i'$ |
| 2 | 1 | 0 | −2 | 0 | 1 | 0 | 0 | 0 |
| 2 | 0 | −2 | 0 | 1 | 1 | 0 | 0 | 0 |

# Appendix B
# Numerical Examples of the Diagonalisation of the Equations

As discussed in Sect. 8.3.7, a normal equation can be diagonalised and the related observation equation can be formed.

For the linearised observation equation (cf. Eq. 8.38)

$$V = L - \begin{pmatrix} A_1 & A_2 \end{pmatrix} \begin{pmatrix} X_1 \\ X_2 \end{pmatrix}, \ P, \tag{B.1}$$

the least squares normal equation can be written as (cf. Eqs. 8.39 and 8.40)

$$\begin{pmatrix} M_{11} & M_{12} \\ M_{21} & M_{22} \end{pmatrix} \begin{pmatrix} X_1 \\ X_2 \end{pmatrix} = \begin{pmatrix} W_1 \\ W_2 \end{pmatrix}, \tag{B.2}$$

where

$$\begin{pmatrix} A_1^T P A_1 & A_1^T P A_2 \\ A_2^T P A_1 & A_2^T P A_2 \end{pmatrix} = \begin{pmatrix} M_{11} & M_{12} \\ M_{21} & M_{22} \end{pmatrix} = M, \ M^{-1} = Q = \begin{pmatrix} Q_{11} & Q_{12} \\ Q_{21} & Q_{22} \end{pmatrix},$$
$$W_1 = A_1^T P L, \ W_2 = A_2^T P L, \tag{B.3}$$

The normal Eq. B.2 can be diagonalised as (cf. Eq. 8.41)

$$\begin{pmatrix} M_1 & 0 \\ 0 & M_2 \end{pmatrix} \begin{pmatrix} X_1 \\ X_2 \end{pmatrix} = \begin{pmatrix} B_1 \\ B_2 \end{pmatrix} \tag{B.4}$$

where

$$M_1 = M_{11} - M_{12} M_{22}^{-1} M_{21} \text{ and}$$
$$B_1 = W_1 - M_{12} M_{22}^{-1} W_2 \tag{B.5}$$

$$M_2 = M_{22} - M_{21}M_{11}^{-1}M_{12}$$
$$B_2 = W_2 - M_{21}M_{11}^{-1}W_1 \quad , \tag{B.6}$$

The above diagonalisation process can be repeated $r - 1$ times to the second normal equation of Eq. B.4, so that the second equation of Eq. B.4 can be fully diagonalised and Eq. B.4 can be represented as:

$$\begin{pmatrix} M_1 & 0 \\ 0 & M'_2 \end{pmatrix} \begin{pmatrix} X_1 \\ X_2 \end{pmatrix} = \begin{pmatrix} B_1 \\ B'_2 \end{pmatrix}, \tag{B.7}$$

where $M'_2$ is a diagonal matrix, $r$ is the dimension of $X_2$, and $B'_2$ is a vector.

Normal Eq. B.4 related observation equation is (cf. Eq. 8.43)

$$\begin{pmatrix} U_1 \\ U_2 \end{pmatrix} = \begin{pmatrix} L \\ L \end{pmatrix} - \begin{pmatrix} D_1 & 0 \\ 0 & D_2 \end{pmatrix} \begin{pmatrix} X_1 \\ X_2 \end{pmatrix}, \quad \begin{pmatrix} P & 0 \\ 0 & P \end{pmatrix}, \tag{B.8}$$

where

$$D_1 = (E - I)A_1, \quad D_2 = (E - J)A_2, \quad \text{and} \tag{B.9}$$

$$I = A_2 M_{22}^{-1} A_2^T P, \quad J = A_1 M_{11}^{-1} A_1^T P, \tag{B.10}$$

where $E$ is an identity matrix, and $U_1$ and $U_2$ are residual vectors which have the same property as $V$ in Eq. B.1.

By similarly repeating the above process $r - 1$ times to the observation equation of $X_2$ (i.e. the second equation of Eq. B.8), then Eq. B.8 turns out to have the form

$$\begin{pmatrix} U_1 \\ U'_2 \end{pmatrix} = \begin{pmatrix} L \\ L' \end{pmatrix} - \begin{pmatrix} D_1 & 0 \\ 0 & D'_2 \end{pmatrix} \begin{pmatrix} X_1 \\ X_2 \end{pmatrix}, \quad \begin{pmatrix} P & 0 \\ 0 & P' \end{pmatrix}, \tag{B.11}$$

where $D'_2$ is in a form of a diagonal matrix where all elements are vectors of dimension $r$, $P'$ is a diagonal matrix of $P$, $L'$ is a vector of $L$, and $U'_2$ is a residual vector that has the same property as $V$ in Eq. B.1. Equation B.11 is the observation equation of normal Eq. B.7.

Numerical examples to illustrate the diagonalisation process of the normal equation and observation equation are given below.

### 1. *The Case of Two Variables*

For the observation equation (where $\sigma$ is set to 1, which does not affect all results)

$$\begin{pmatrix} V_1 \\ V_2 \\ V_3 \end{pmatrix} = \begin{pmatrix} 1 \\ 2 \\ -1 \end{pmatrix} - \begin{pmatrix} 1 & 1 \\ 1 & 2 \\ 1 & 1 \end{pmatrix} \begin{pmatrix} X_1 \\ X_2 \end{pmatrix}, \quad P = \frac{1}{\sigma^2} \begin{pmatrix} 1 & 0 & 0 \\ 0 & 1 & 0 \\ 0 & 0 & 1 \end{pmatrix}, \tag{B.12}$$

Appendix B: Numerical Examples of the Diagonalisation of the Equations    451

the least squares normal equation is

$$\begin{pmatrix} 3 & 4 \\ 4 & 6 \end{pmatrix} \begin{pmatrix} X_1 \\ X_2 \end{pmatrix} = \begin{pmatrix} 2 \\ 4 \end{pmatrix}. \tag{B.13}$$

Because

$$M_1 = 3-4(1/6)4 = 1/3, \quad B_1 = 2-4(1/6)4 = -2/3, \quad \text{and}$$
$$M_2 = 6-4(1/3)4 = 2/3, \quad B_2 = 4-4(1/3)2 = 4/3,$$

Eq. B.13 is diagonalised as

$$\begin{pmatrix} 1/3 & 0 \\ 0 & 2/3 \end{pmatrix} \begin{pmatrix} X_1 \\ X_2 \end{pmatrix} = \begin{pmatrix} -2/3 \\ 4/3 \end{pmatrix}. \tag{B.14}$$

The solution ($X_1 = -2$, $X_2 = 2$) of Eq. B.14 is the same as that of Eq. B.13. Furthermore, to form the equivalent observation equation, there are

$$M_{11} = A_1^T A_1 = (1 \; 1 \; 1) \begin{pmatrix} 1 \\ 1 \\ 1 \end{pmatrix} = 3, \quad M_{22} = A_2^T A_2 = (1 \; 2 \; 1) \begin{pmatrix} 1 \\ 2 \\ 1 \end{pmatrix} = 6,$$

$$I = \begin{pmatrix} 1 \\ 2 \\ 1 \end{pmatrix} \frac{1}{6} (1 \; 2 \; 1) = \frac{1}{6} \begin{pmatrix} 1 & 2 & 1 \\ 2 & 4 & 2 \\ 1 & 2 & 1 \end{pmatrix}, \quad J = \begin{pmatrix} 1 \\ 1 \\ 1 \end{pmatrix} \frac{1}{3} (1 \; 1 \; 1) = \frac{1}{3} \begin{pmatrix} 1 & 1 & 1 \\ 1 & 1 & 1 \\ 1 & 1 & 1 \end{pmatrix},$$

$$D_1 = (E - I)A_1 = \frac{1}{6} \begin{pmatrix} 5 & -2 & -1 \\ -2 & 2 & -2 \\ -1 & -2 & 5 \end{pmatrix} \begin{pmatrix} 1 \\ 1 \\ 1 \end{pmatrix} = \frac{1}{3} \begin{pmatrix} 1 \\ -1 \\ 1 \end{pmatrix}, \quad \text{and,}$$

$$D_2 = (E - J)A_2 = \frac{1}{3} \begin{pmatrix} 2 & -1 & -1 \\ -1 & 2 & -1 \\ -1 & -1 & 2 \end{pmatrix} \begin{pmatrix} 1 \\ 2 \\ 1 \end{pmatrix} = \frac{1}{3} \begin{pmatrix} -1 \\ 2 \\ -1 \end{pmatrix};$$

thus, the observation equation related to Eq. B.14 is

$$\begin{pmatrix} U_1 \\ U_2 \end{pmatrix} = \left( \begin{pmatrix} 1 \\ 2 \\ -1 \\ 1 \\ 2 \\ -1 \end{pmatrix} \right) - \begin{pmatrix} \frac{1}{3} \begin{pmatrix} 1 \\ -1 \\ 1 \end{pmatrix} & 0_{3 \times 1} \\ 0_{3 \times 1} & \frac{1}{3} \begin{pmatrix} -1 \\ 2 \\ -1 \end{pmatrix} \end{pmatrix} \begin{pmatrix} X_1 \\ X_2 \end{pmatrix}, \quad \begin{pmatrix} P & 0 \\ 0 & P \end{pmatrix}. \tag{B.15}$$

The normal equation of the observation Eq. B.15 is exactly the same as Eq. B.14. This numerical example shows that the normal equation and the related observation equation can be diagonalised.

## 2. The Case of Three Variables

For the observation equation (where $\sigma$ is set to 1, which does not affect all results)

$$\begin{pmatrix} V_1 \\ V_2 \\ V_3 \\ V_4 \end{pmatrix} = \begin{pmatrix} 2 \\ 1 \\ 0 \\ -2 \end{pmatrix} - \begin{pmatrix} 1 & 1 & 1 \\ 2 & 1 & 1 \\ 1 & 1 & 2 \\ 1 & 1 & 1 \end{pmatrix} \begin{pmatrix} X_1 \\ X_2 \\ X_3 \end{pmatrix}, \quad P = \frac{1}{\sigma^2} E_{4 \times 4}, \quad (B.16)$$

the least squares normal equation is

$$\begin{pmatrix} 7 & 5 & 6 \\ 5 & 4 & 5 \\ 6 & 5 & 7 \end{pmatrix} \begin{pmatrix} X_1 \\ X_2 \\ X_3 \end{pmatrix} = \begin{pmatrix} 2 \\ 1 \\ 1 \end{pmatrix}. \quad (B.17)$$

Because

$$M_{22}^{-1} = \begin{pmatrix} 4 & 5 \\ 5 & 7 \end{pmatrix}^{-1} = \frac{1}{3} \begin{pmatrix} 7 & -5 \\ -5 & 4 \end{pmatrix}, \quad M_{11}^{-1} = \frac{1}{7},$$

$$M_1 = 7 - (5 \ 6) \frac{1}{3} \begin{pmatrix} 7 & -5 \\ -5 & 4 \end{pmatrix} \begin{pmatrix} 5 \\ 6 \end{pmatrix} = \frac{2}{3}, \quad B_1 = 2 - (5 \ 6) \frac{1}{3} \begin{pmatrix} 7 & -5 \\ -5 & 4 \end{pmatrix} \begin{pmatrix} 1 \\ 1 \end{pmatrix} = \frac{2}{3},$$

$$M_2 = \begin{pmatrix} 4 & 5 \\ 5 & 7 \end{pmatrix} - \begin{pmatrix} 5 \\ 6 \end{pmatrix} \frac{1}{7} (5 \ 6) = \frac{1}{7} \begin{pmatrix} 3 & 5 \\ 5 & 13 \end{pmatrix} \quad \text{and} \quad B_2 = \begin{pmatrix} 1 \\ 1 \end{pmatrix} - \begin{pmatrix} 5 \\ 6 \end{pmatrix} \frac{1}{7} \cdot 2 = \frac{1}{7} \begin{pmatrix} -3 \\ 5 \end{pmatrix},$$

(B.18)

Eq. B.17 is diagonalised as

$$\begin{pmatrix} 2/3 & 0 & 0 \\ 0 & 3/7 & 5/7 \\ 0 & 5/7 & 13/7 \end{pmatrix} \begin{pmatrix} X_1 \\ X_2 \\ X_3 \end{pmatrix} = \begin{pmatrix} 2/3 \\ -3/7 \\ -5/7 \end{pmatrix}. \quad (B.19)$$

The $X_2$ and $X_3$ related normal equation can be further diagonalised. Because of

$$M_1' = 3/7 - 5(1/13)(5/7) = 2/13, \quad B_1' = -3/7 - 5(1/13)(-5/7) = -2/13,$$
$$M_2' = 13/7 - 5(1/3)(5/7) = 2/3, \quad B_2' = -5/7 - 5(1/3)(-3/7) = 0,$$

Eq. B.19 is further diagonalised as

$$\begin{pmatrix} 2/3 & 0 & 0 \\ 0 & 2/13 & 0 \\ 0 & 0 & 2/3 \end{pmatrix} \begin{pmatrix} X_1 \\ X_2 \\ X_3 \end{pmatrix} = \begin{pmatrix} 2/3 \\ -2/13 \\ 0 \end{pmatrix}. \qquad (B.20)$$

The solution ($X_1 = 1$, $X_2 = -1$, $X_3 = 0$) of Eq. B.20 is the same as that of Eqs. B.17 and B.19. Furthermore, to form the equivalent observation equation of Eq. B.19, there are

$$I = \begin{pmatrix} 1 & 1 \\ 1 & 1 \\ 1 & 2 \\ 1 & 1 \end{pmatrix} \frac{1}{3} \begin{pmatrix} 7 & -5 \\ -5 & 4 \end{pmatrix} \begin{pmatrix} 1 & 1 & 1 & 1 \\ 1 & 1 & 2 & 1 \end{pmatrix} = \frac{1}{3} \begin{pmatrix} 1 & 1 & 0 & 1 \\ 1 & 1 & 0 & 1 \\ 0 & 0 & 3 & 0 \\ 1 & 1 & 0 & 1 \end{pmatrix},$$

$$J = \begin{pmatrix} 1 \\ 2 \\ 1 \\ 1 \end{pmatrix} \frac{1}{7} \begin{pmatrix} 1 & 2 & 1 & 1 \end{pmatrix} = \frac{1}{7} \begin{pmatrix} 1 & 2 & 1 & 1 \\ 2 & 4 & 2 & 2 \\ 1 & 2 & 1 & 1 \\ 1 & 2 & 1 & 1 \end{pmatrix},$$

$$D_1 = (E - I)A_1 = \frac{1}{3} \begin{pmatrix} -1 \\ 2 \\ 0 \\ -1 \end{pmatrix} \quad \text{and} \quad D_2 = (E - J)A_2 = \frac{1}{7} \begin{pmatrix} 2 & 1 \\ -3 & -5 \\ 2 & 8 \\ 2 & 1 \end{pmatrix},$$

thus the observation equation related to Eq. B.19 is

$$\begin{pmatrix} U_1 \\ U_2 \end{pmatrix} = \begin{pmatrix} L \\ L \end{pmatrix} - \begin{pmatrix} D_1 & 0 \\ 0 & D_2 \end{pmatrix} \begin{pmatrix} X_1 \\ X_2 \\ X_3 \end{pmatrix}, \quad \begin{pmatrix} P & 0 \\ 0 & P \end{pmatrix}, \quad \text{where} \quad L = \begin{pmatrix} 2 \\ 1 \\ 0 \\ -2 \end{pmatrix}.$$

(B.21)

The $X_2$ and $X_3$ related observation equation can be further diagonalised as follows. Because

$$I' = \frac{1}{7}\begin{pmatrix} 1 \\ -5 \\ 8 \\ 1 \end{pmatrix} \frac{7}{13} \cdot \frac{1}{7}(1 \quad -5 \quad 8 \quad 1) = \frac{1}{91}\begin{pmatrix} 1 & -5 & 8 & 1 \\ -5 & 25 & -40 & -5 \\ 8 & -40 & 64 & 8 \\ 1 & -5 & 8 & 1 \end{pmatrix},$$

$$J' = \frac{1}{7}\begin{pmatrix} 2 \\ -3 \\ 2 \\ 2 \end{pmatrix} \frac{7}{3} \cdot \frac{1}{7}(2 \quad -3 \quad 2 \quad 2) = \frac{1}{21}\begin{pmatrix} 4 & -6 & 4 & 4 \\ -6 & 9 & -6 & -6 \\ 4 & -6 & 4 & 4 \\ 4 & -6 & 4 & 4 \end{pmatrix}, \text{ and}$$

$$D'_{21} = A'_1 - I'A'_1 = \frac{1}{7}\begin{pmatrix} 2 \\ -3 \\ 2 \\ 2 \end{pmatrix} - \frac{1}{7 \cdot 91}\begin{pmatrix} 35 \\ -175 \\ 280 \\ 35 \end{pmatrix} = \frac{1}{13}\begin{pmatrix} 3 \\ -2 \\ -2 \\ 3 \end{pmatrix},$$

$$D'_{22} = A'_2 - J'A'_2 = \frac{1}{7}\begin{pmatrix} 1 \\ -5 \\ 8 \\ 1 \end{pmatrix} - \frac{1}{21 \cdot 7}\begin{pmatrix} 70 \\ -105 \\ 70 \\ 70 \end{pmatrix} = \frac{1}{3}\begin{pmatrix} -1 \\ 0 \\ 2 \\ -1 \end{pmatrix},$$

the observation equation related to Eq. B.20 is

$$\begin{pmatrix} U'_1 \\ U'_2 \\ U'_3 \end{pmatrix} = \begin{pmatrix} L \\ L \\ L \end{pmatrix} - \begin{pmatrix} D_1 & 0 & 0 \\ 0 & D'_{21} & 0 \\ 0 & 0 & D'_{22} \end{pmatrix} \begin{pmatrix} X_1 \\ X_2 \\ X_3 \end{pmatrix}, \quad \begin{pmatrix} P & 0 & 0 \\ 0 & P & 0 \\ 0 & 0 & P \end{pmatrix}. \quad (B.22)$$

The normal Eq. B.17 and its related observation Eq. B.16 are fully diagonalised as Eqs. B.20 and B.22, respectively. These numerical examples show that the normal equation and the related observation equation can be diagonalised as described in Sect. 8.3.7.

# References

Abidin HZ (1995) GPS and hydro-oceanographic surveying in Indonesia. Int J Geomatics 9(4):35–37

Abidin HZ et al. (2004) The deformation of Bromo volcano as detected by GPS surveys method. J. GPS 3(1-2): 16-24

Abramowitz M, Stegun IA (1965) Handbook of mathematical functions. Dover Publications, Inc., New York

Adami D, Garroppo RG, Giordano S, Lucetti S (2003) On synchronization techniques: Performance and impact on time metrics monitoring. Int. J. Comm. Syst. 16(4): 273-290

Afraimovich EL, Kosogorov EA, Leonovich LA (2000) The use of the international GPS network as the global detector (GLOBDET) simultaneously observing sudden ionospheric disturbance. Earth Planets Space 52(11):1077–1082

Akos DM (2003) The role of Global Navigation Satellite System (GNSS) software radios in embedded systems. GPS Solutions 7(1): 1-4

Al-Haifi Y, Corbett S, Cross P (1997) Performance evaluation of GPS single-epoch on-the-fly ambigu-ity resolution. J Inst Navig 44,4:479–487

Albertella A, Sacerdote F (1995) Spectral analysis of block averaged data in geopotential global model determination. J Geodesy 70,3:166–175

Andersen PH, Kristiansen O, Zarraoa N (1995) Analysis of data from the VLBI-GPS collocation ex-periment CONT94. In: GPS Trends in Precise Terrestrial, Airborne, and Spaceborne Applications: 21st IUGG General Assembly, IAG Symposium No. 115, Boulder, USA, July 3–4, 1995, Springer-Verlag, Berlin, pp 315–319

Angermann D, Becker M (2000) Untersuchungen zu Genauigkeit und systematischen Effekten in großräumigen GPS-Netzen am Beispiel von GEODYSSEA. ZfV 125(3):88–95

Angermann D, Baustert G, Klotz J (1995) The impact of IGS on the analysis of regional GPS-network. In: GPS Trends in Precise Terrestrial Airborne, and Spaceborne Applications: 21st IUGG General Assembly, IAG Symposium No. 115, Boulder, USA, July 3–4, 1995, Springer-Verlag, Berlin, pp 35–41

Arikan F, Erol CB, Arikan O (2003) Regularized estimation of vertical total electron content from Global Positioning System data. J. Geophys. Res. 108(A12): SIA20/1-12

Arnold D, Meindl M, Beutler G, Dach R, Schaer S, Lutz S, Prange L, Sosnica K, Mervart L, Jaeggi A (2015) CODE's new solar radiation pressure model for GNSS orbit determination. J Geodesy 89(8), pp775-791

Artese G, Cefalo R, Vettore A (1997) Real time kinematic GPS to bathymetry. Rep Geod 5 (28):77–87

Ashkenazi V, Beamson G, Bingley R (1995) Monitoring absolute changes in mean sea level. In: Pro-ceedings of the First Turkish International Symposium on Deformations "Istanbul-94", Istanbul, September 5–9, pp 40–46

Ashkenazi V, Park D, Dumville M (2000) Robot positioning and the global navigation satellite system. Ind Robot 27(6):419–426

Aw York Bin, Goh Pong Chai (1996) Improving cadastral survey controls using GPS surveying in Singapore. Survey rev 33:488–495

Axelsson O (1994) Iterative Solution Methods. Cambridge University Press, London/New York

Ayres F (1975) Differential- und Integralrechnung, Schaum's Outline. McGraw-Hill Book, New York

Babu R (2005) Web-based resources on software GPS receivers. GPS Solutions 9(3): 240-242

Baertlein H, Carlson B, Eckels R, Lyle S, Wilson S (2000) A high-performance, high-accuracy RTK GPS machine guidance system. GPS Solutions 3(3):4–11

Bailey Brian K (2014) GPS Modernization Update. June 2014.

Baldi P, Bonvalot S, Briole P, Marsella M (2000) Digital photogrammetry and kinematic GPS applied to the monitoring of Vulcano Island, Aeolian Arc, Italy. Geophys J Int 142(3):801–811

Balmino G, Schrama E, Sneeuw N (1996) Compatibility of first-order circular orbit perturbations theories: Consequences for cross-track inclination functions. J Geodesy 70(9):554–561

Banyai L, Gianniou M (1997) Comparison of Turbo-Rogue and Trimble SSI GPS receivers for iono-spheric investigation under anti-spoofing. ZfV 3:136–142

Bar-Sever YE (1996) A new model for GPS yaw attitude. J Geodesy 70:714-723

Barthelmes F (1996) Die Wavelet-Transformation zur Zeitreihenanalyse. Erste Geodätische Woche, Stuttgart, 7.–12. Oktober 1996, 15 Blatt

Bastos L, Landau H(1988) Fixing cycle slips in dual-frequency kinematic GPS-application using Kalman filtering. Manuscr Geodaet 13:249–256

Bastos L, Osorio J, Hein G (1995) GPS derived displacements in the Azores Triple Junction Region. In: GPS Trends in Precise Terrestrial Airborne, and Spaceborne Applications: 21st IUGG General Assembly, IAG Symposium No. 115, Boulder, USA, July 3–4, 1995, Springer-Verlag, Berlin, pp 99–104

Baugh CM (2006) A Primer on Hierarchical Galaxy Formation: the Semi-Analytical Approach, Reports on Progress in Physics, Vol 69, Issue 12, p3101-3156

Bause F, Toelle W (1993) Programmieren mit C++, Version 3. Vieweg & Sohn, Verlagsgesellschaft mbH, Braunschweig

Becker M, Angermann D, Nordin S, Reigber C, Reinhart E (2000) Das Geschwindigkeitsfeld in Südostasien aus einer kombinierten GPS Lösung der drei GEODYSSEA Kampagnen von 1994 bis 1998. ZfV 125(3):74–80

Berrocoso M, Garate J, Martin J (1996) Improving the local geoid with GPS. In: Proceedings of the Techniques for local geoid determination, session G7 European Geophysical Society XXI[st] General Assembly The Hague, The Netherlands, 6–10 May, 1996, Masala, pp 91–96

Beutler G (1994) GPS trends in precise terrestrial, airborne, and space borne applications. Springer-Verlag, Heidelberg

Beutler G (1996) GPS satellite orbits. In Kleusberg A, Teunissen PJG (eds) GPS for geodesy. Springer-Verlag, Berlin

Beutler G (1996) The GPS as a tool in global geodynamics. In: Kleusberg A, Teunissen PJG (eds) GPS for geodesy. Springer-Verlag, Berlin

Beutler G, Brockmann E, Hugentobler U (1996) Combining consecutive short arcs into long arcs for precise and efficient GPS orbit determination. J Geodesy 70,5:287–299

Beutler G, Schildknecht T, Hugentobler U, Gurtner W (2003) Orbit determination in satellite geodesy. Adv. Space Res. 31(8): 1853-1868

Beyerle G, Wickert J, Schmidt T, Reigber C (2004) Atmos. sounding by global navigation satellite system radio occultation: An analysis of the negative refractivity bias using CHAMP observations. J. Geophys. Res. 109(D01106): 1-8

Bian S (1996) Topography supported GPS leveling. ZfV 121,3:109–113

Bian S, Jin J, Fang Z (2005) The Beidou satellite positioning system and its positioning accuracy. Navigation 52(3): 123-129

Bilitza D, Altadill D, Zhang Y, et al. (2014) The International Reference Ionosphere 2012 – a model of international collaboration. J. Space Weather Space Clim., 4, A07

Bisnath S, Wells D, Howden S, Dodd D, Wiesenburg D (2004) Development of an operational RTK GPS-equipped buoy for tidal datum determination. Int. Hydrogr. Rev. 5(1): 54-64

Blanchard D (2012) Galileo Programme Status Update. ION GNSS 2012, November 20, pp553-587

Blomenhofer H (1996) Untersuchungen zu hochpräzisen kinematischen DGPS-Echtzeitverfahren mit besonderer Berücksichtigung atmosphärischer Fehlereinflüsse. Neubiberg, 166 S

Bock H, Dach R, Jaeggi A, Beutler G (2009) High-rate GPS clock corrections from CODE: support of 1 Hz applications. J Geodesy 83:1083

Bock Y (1996) Reference systems. In: Kleusberg A, Teunissen PJG (eds) GPS for geodesy. Springer-Verlag, Berlin

Bock Y (1996) Medium distance GPS measurements. In: Kleusberg A, Teunissen PJG (eds) GPS for geodesy. Springer-Verlag, Berlin

Bock Y, Beutler G, Schaer S, Springer TA, Rothacher M (2000) Processing aspects related to permanent GPS arrays. Earth Planets Space 52(10):657–662

Bock Y, Prawirodirdjo L, Melbourne TI (2004) Detection of arbitrarily large dynamic ground motions with a dense high-rate GPS network. Geophys. Res. Lett. 31(L06604): 1-4

Boehme S (1970) Zum Einfluß eines Quadrupolmoments der Sonne auf die Bahnlage der Planeten, Astron. Nachr., Bd. 292, H. 1

Boey SS, Coombe LJ, Gerdan GP (1996) Assessing the accuracy of real time kinematic GPS positions for the purposes of cadastral surveying. Aust Surveyor 41,2:109–120

Bona P (2000) Precision, cross correlation, and time correlation of GPS phase and code observations. GPS Solutions 4(2):3–13

Boomkamp H, Dow J (2005) Use of double difference observations in combined orbit solutions for LEO and GPS satellites. Adv. Space Res. 36(3): 382-391

Borge TK, Forssell B (1994) A new real time ambiguity resolution strategy based on polynomial indentification, In: Proceedings of the International Symposium on Kinematic Systems in Geod-esy, Geomatics and Navigation, Banff, Canada, 30 August–2 September, pp 233–240

Borre K (2003) The GPS Easy Suit-Matlab code for the GPS newcomer. GPS Solutions 7(1): 47-51

Borre K, Akos DM, Bertelsen N, Rinder P, Jensen SH (2007) A Software-Defined GPS and Galileo Receiver, A Single-Frequency Approach, Birkhaüser, Boston, M.A. ISBN 978-0-8176-4390-4

Bosco M (2011) The European GNSS Programmes EGNOS and Galileo International Challenges Ahead. November 23, 2011.

Bottke WF, Cellino A, Paolicchi P, Binzel RP (ed 2002) Asteroids III, Space Science Series, Univetrsity of Arizona Press

Bouin M-N, Vigny C (2000) New constraints on Antarctic plate motion and deformation from GPS data. J Geophys Res 105(B12):28279–28293

Boulton WJ, 1983, The effect of solar radiation pressure on the orbit of a cylindrical satellite, Planet. Space Sci. Vol. 32, No. 3, pp. 287-296Bauer M (1994) Vermessung und Ortung mit Satelliten. Wichmann Verlag, Karslruhe, Germany

Brodin G, Cooper J, Walsh D, Stevens J (2005) The effect of helicopter rotors on GPS signal reception. J. Navig. 58(3): 433-450

Broederbauer V, Weber R (2003) Results of modelling GPS satellite clocks. Osterr. Z. Vermess. Geoinf. 91(1): 38-47

Brumberg VA (1995) Analytical Techniques of Celestial Mechanics, Springer

Brunner FK (1998) Advances in positioning and reference frames. Springer-Verlag, Heidelberg

Brunner FK, Gu M (1991) An improved model for the dual frequency ionospheric correction of GPS observations. Manuscr Geodaet 16:205–214

Brunner FK, Welsch WM (1993) Effect of the troposphere on GPS measurements. GPS World 4:42–51

Burns JA, Lamy PL, Soter S (1979) Radiation force on small particles in the solar system, Icarus, Vol 40 p1-48

Bust GS, Coco D, Makela JJ (2000) Combined Ionospheric Campaign 1: Ionospheric tomography and GPS total electron content (TEC) depletions. Geophys Res Lett 27(18):2849–2852

Cai C, Gao Y (2013) Modeling and assessment of combined GPS/GLONASS precise point positioning. GPS Solutions 17(2), 223-236.

Campbell J, Goerres B, Siemens M, Wirsch J, Becker M (2004) Zur Genauigkeit der GPS Antennenkalibrierung auf der Grundlage von Labormessungen und deren Vergleich mit anderen Verfahren. Allgemeine Vermessungs-Nachrichten 111(1): 2-11

Campbell L, Moffat JW (1983) Quadrupole Moment of the Sun and the Planetary Orbits, Astrophysical Journal, 275:L77-L79

Campbell L, Mcdow JC, Moffat JV, Vincent D (1983) The Sun's quadrupole moment and perihelion precession of Mercury, Nature, Vol. 305, P508

Campos MA, Krueger CP (1995) GPS kinematic real-time applications in rivers and train. In: GPS Trends in Precise Terrestrial, Airborne, and Spaceborne Applications: 21$^{st}$ IUGG General Assem-bly, IAG Symposium No. 115, Boulder, USA, July 3–4, 1995. Springer-Verlag, Berlin, pp 222–225

Cangahuala L, Muellerschoen R, Yuan D-N (1995) TOPEX/Poseidon precision orbit determination with SLR and GPS anti-spoofing data. In: GPS Trends in Precise Terrestrial Airborne, and Spaceborne Applications: 21$^{st}$ IUGG General Assembly, IAG Symposium No. 115, Boulder, USA, July 3–4, 1995. Springer-Verlag, Berlin, pp 123–127

Cannon E, Weisenburger S (2000) The use of multiple receivers for constraining GPS carrier phase ambiguity resolution. Lighthouse 57:7–18

Cannon ME, Skone S, Karunanayake MD, Kassam A (2004) Performance analysis of the real-time Canada-Wide DGPS Service (CDGPS). Geomatica 58(2): 95-105

Cardellach E, Behrend D, Ruffini G, Rius R (2000) The use of GPS buoys in the determination of oceanic variables. Earth Planets Space 52(11):1113–1116

Casotto S, Zin A (2000) An assessment of the benefits of including GLONASS data in GPS-based precise orbit determination – I: S/A analysis. Advances in the Astronautical Sciences 105 (1):237–256

Castleden et al. (2004) First results from Virtual Reference Station (VRS) and precise point positioning (PPP) GPS research at the Western Australian Centre for Geodesy. J. GPS 3(1-2): 79-84

Celebi M (2000) GPS in dynamic monitoring of long-period structures. Soil Dyn Earthq Eng 20 (5–8): 477–483

Celleti A (2010) Stability and Chaos in celestial mechanics, Springer

Chang C-C (2000) Estimation of local subsidence using GPS and leveling data. Surveying and Land Information Systems 60(2):85–94

Chang C-C, Sun Y-D (2004) Application of a GPS-based method to tidal datum transfer. Hydrogr. J. 112: 15-20

Chen CS, Chen Y-J, Yeh T-K (2000) The impact of GPS antenna phase center offset and variation on the positioning accuracy. Bull Geod Sci Affini 59(1):73–94

Chen D (1994) Development of a fast ambiguity search filtering (FASF) method for GPS carrier phase ambiguity resolution. Reports of the Department of Geomatics Engineering of the University of Calgary, Vol. 20071

Chen D, Lachapelle G (1994) A comparison of the FASF and least-squares search algorithms for ambiguity resolution on the fly. In: Proceedings of the International Symposium on Kinematic Systems in Geodesy, Geomatics and Navigation, Banff, Canada, August 30–September 2, pp 241–253

Chen H, Dai L, Rizos C, Han S (2005) Ambiguity recovery using the triple-differenced carrier phase type approach for long-range GPS kinematic positioning. Mar. Geod. 28(2): 119-135

Chen W et al. (2004) Kinematic GPS precise point positioning for sea level monitoring with GPS buoy. J. GPS 3(1-2): 302-307

Chen X, Langley RB, Dragert H (1995) The Western Canada Deformation Array: An update on GPS solutions and error analysis. In: GPS Trends in Precise Terrestrial Airborne, and Spaceborne Applications: 21$^{st}$ IUGG General Assembly, IAG Symposium No. 115, Boulder, USA, July 3–4, 1995. Springer-Verlag, Berlin, pp 70–74

Chen Y-Q, Wang J-L (1996) Reliability measures for correlated observations. ZfV 121,5:211–219

Chen Y-Q, Ding XL, Huang DF, Zhu JJ (2000) A multi-antenna GPS system for local area deformation. Earth Planets Space 52(10):873–876

China Satellite Navigation Office (2013) Report on the Development of BeiDou (COMPASS) Navigation Satellite System (V2.2). December 2013.

Christou AA, Asher DJ (2011) A long-lived horseshoe companion to the Earth, MNRAS 414, 2965 - 2969

Clark TA (1995) Low-cost GPS time synchronization: The "Totally Accurate Clock". In: GPS Trends in Precise Terrestrial,Airborne, and Spaceborne Applications: 21$^{st}$ IUGG General Assembly,IAG Symposium No. 115, Boulder, USA, July 3–4, 1995. Springer-Verlag, Berlin, pp 325–327

Collins GW (2004) The foundations of celestial mechanics, Pachart Publishing House

Colombo OL (1984a) Altimetry, orbits and tides. NASA Technical Memorandum 86180

Colombo OL (1984b) The global mapping of gravity with two satellites. Netherlands Geodetic Commission, Delft, The Netherlands Publications on Geodesy, Vol. 7, No. 3, 253pp

Colombo OL, Rizos C, Hirsch B (1995) Testing high-accuracy, long-range carrier phase DGPS in Australasia. In: GPS Trends in Precise Terrestrial,Airborne, and Spaceborne Applications: 21$^{st}$ IUGG General Assembly, IAG Symposium No. 115, Boulder, USA, July 3–4, 1995. Springer-Verlag, Berlin, pp 226–230

Colombo OL, Hernández-Pajares M, Juan JM, Sanz J, Talaya J (1999) Resolving carrier-phase ambigu-ities on the fly, at more than 100 km from nearest reference site, with the help of ionospheric topography. ION GPS 99 14–17, September 1999, pp 1635–1642

Conway BA (Ed, 2010) Spacecraft Trajectory Optimization, Cambridge University Press

Cooray A, Sheth R (2002) Halo models of large scale structure, Physics Reports, Vol 372, Issue 1, P1-129

Corbett SJ, Cross PA (1995) GPS single epoch ambiguity resolution. Survey rev 33(257):149–160

Cross PA, Ramjattan AN (1995) A Kalman filter model for an integrated land vehicle navigation sys-tem. In: Proceedings of the 3$^{rd}$ international workshop on high precision navigation: High preci-sion navigation 95. University of Stuttgart, April 1995, Bonn, pp 423–434

Cui C, Lelgemann D (1995) Analytical dynamic orbit improvement for the evaluation of geodetic-geodynamic satellite data. J Geodesy 70:83–97

Cui X, Yang Y (2006) Adaptively Robust Filtering with Classified Adaptive Factors. Progress in Natural Science 16(8):846-851

Dach R, Boehm J, Lutz S, Steigenberger P, Beutler G (2011) Evaluation of the impact of atmospheric pressure loading modeling on GNSS data analysis. J Geodesy 85(2), pp75-91

Dach R, Dietrich R (2000) Influence of the ocean loading effect on GPS derived precipitable water vapor. Geophys Res Lett 27(18):2953–2956

Dam T van, Larson KM, Wahr J, Gross S, Francis O (2000) Using GPS and gravity to infer ice mass changes in Greenland. EOS Trans. AGU 81(37):421, 426–427

Davis JL, Cosmo ML, Elgered G (1995) Using the Global Positioning System to study the atmosphere of the Earth: Overview and prospects. In: GPS Trends in Precise Terrestrial, Airborne, and Spaceborne Applications: 21$^{st}$ IUGG General Assembly, IAG Symposium No. 115, Boulder, USA, July 3–4, 1995. Springer-Verlag, Berlin, pp 233–242

Davis P, Rabinowitz P (1984) Methods of numerical integration, 2$^{nd}$ Ed. Academic Press, INC

Davis PJ (1963) Interpolation and approximation. Dover Publications Inc., New York

Denker H (1995) Grossräumige Höhenbestimmung mit GPS- und Schwerefelddaten. Schriftenreihe des Deutschen Vereins für Vermessungswesen, Bd. 18, Stuttgart, pp 233–258

Desai SD, Haines BJ (2003) Near-real-time GPS-based orbit determination and sea surface height observations from The Jason-1 mission. Mar. Geod. 26(3-4): 383-397

Desmars J, Arlot S, Arlot JE, Lainey V, Vienne A (2009) Estimating the accuracy of satellite ephemrides using the bootstrap method, Astronomy & Astrophysics, Vol. 499 No.1

Dick G (1997) Nutzung von GPS zur Bahnbestimmung niedrigfliegender Satelliten. GPS-Anwendungen und Ergebnisse '96: Beiträge zum 41. DVW-Fortbildungsseminar vom 7. bis 8. November 1996 am Geo-Forschungszentrum Potsdam, pp 241-249

Dick G, Gendt G (1997) GPS-Anwendungen und Ergebnisse '96: Beiträge zum 41. DVW-Fortbildungs-seminar vom 7. bis 8. November 1996 am Geo-Forschungszentrum Potsdam. Geodesia: Nederl. geod. t., Stuttgart

Dicke RH (1970) The solar oblateness and the gravitational quadrupole moments, Ap. J. 159, 1-23

Dicke RH, Kuhn JR, Libbrecht KG (1987) Is the solar oblateness variable? Measurements of 1985, Astrophysical Journal 318: 451-458

Dierendonck AJ Van, Hegarty C (2000) The new L5 civil GPS signal. GPS World 11(9):64-71

Dietrich R (1997) Untersuchung von vertikalen Krustendeformationen wegen wechselnder Eislauflasten in Grönland. GPS-Anwendungen und Ergebnisse '96: Beiträge zum 41. DVW-Fortbildungsseminar vom 7. bis 8. November 1996 am Geo-Forschungszentrum Potsdam, pp 94-102

Dietrich R, Rulke A, Scheinert M (2005) Present-day vertical crustal deformations in West Greenland from repeated GPS observations. Geophys. J. Int. 163(3): 865-874

Diggelen F (1998) GPS accuracy: Lies, damm lies, and statistics. GPS World 9,1:41-44

Diggelen F, Martin W (1997) GPS + GLONASS RTK: A quantum leap in RTK performance. Int J Geo-matics 11(11):69-71

Ding X, Coleman R (1996) Multiple outlier detection by evaluating redundancy contributions of observations. J Geodesy 708:489-498

Ding X, Coleman R (1996) Adjustment of precision metrology networks in three dimension. Survey rev 33,259:305-315

Ding XL et al. (2005) Seasonal and secular positional variations at eight co-located GPS and VLBI stations. J. Geod. 79(1-3): 71-81

Dittrich J, Kuehmstedt E, Richter B, Reinhart E (1997) Accurate positioning by low frequency (ALF) and other services for emission of DGPS correction data in Germany. Rep Geod 6 (29):97-108

Dodson AH, Shardlow PJ, Hubbard LCM (1995) Wet tropospheric effects on precise relative GPS height determination. J Geodesy 70(4):188-202

Douša J (2004): Precise orbits for ground-based GPS meteorology: processing strategy and quality assessment of the orbits determined at geodetic observatory Pecny. J. Meteor. Soc. Japan 82 (1B): 371-380

Dow JM, Romay-Merino MM, Piriz R (1993) High precision orbits for ERS-1: 3-day and 35-day repeat cycles. In: Proceedings of the Second ERS-1 symposium: Space at the service of our environment, Hamburg, 11-14 October 1993, Vol. 2, Jan. 1994, Noordwijk, pp 1349-1354

Dragert H, James TS, Lambert A (2000) Ocean loading corrections for continuous GPS: A case study at the Canadian coastal site Holberg. Geophys Res Lett 27(14):2045-2048

Drewes H (1996) Kinematische Referenzsysteme für die Landesvermessung. ZfV 121(6):277-285

Drewes H (1997) Realisierung des geozentrischen Referenzsystems für Südamerika (SIRGAS). GPS-Anwendungen und Ergebnisse '96: Beiträge zum 41. DVW-Fortbildungsseminar vom 7. bis 8. No-vember 1996 am Geo-Forschungszentrum Potsdam, pp 54-63

Du RL, Qiao XJ, Wang Q, Xing CF, You XZ (2005) Deformation in the Three Gorges Reservoir after the first impoundment determined by GPS measurements. Progress Natural Sci. 15(6): 515-522

El-Mowafy A (2013) GNSS multi-frequency receiver single-satellite measurement validation method. GPS Solutions 18(4), pp553-561

Eissfeller B, Ameres G, Kropp V, et al. (2007) Performance of GPS, GLONASS and Galileo. Dieter Fritsch, Wichmann, pp185-199

Eissfeller B, Teuber A, Zucker P (2005) Untersuchungen zum GPS-Satellitenempfang in Gebaeuden. Allgemeine Vermessungs-Nachrichten 112(4): 137-145

Elosequi P, Davis JL, Jaldehag RTK (1995) Geodesy using the global positioning system: The effects of signal scattering on estimates of site position. J Geophys Res 100(B6):9921–9934

Elsobeiey M, AI-Harbi S (2015) Performance of real-time Precise Point Positioning using IGS real-time service. GPS Solutions, DOI: 10.1007/s10291-015-0467-z

Emardson TR, Jarlemark POJ (1999) Atmospheric modelling in GPS analysis and its effect on the estimated geodetic parameters. J Geodesy 73:322–331

Enge P (2003) GPS Modernization: Capabilities of New Civil Signals. Australian International Aerospace Congress, Brisbane, 29 July-1 August 2003.

Engel F, Heiser G, Mumford P, Parkinson K, Rizos C (2004) An open GNSS receiver platform architecture. J. GPS 3(1-2): 63-69

Engelhardt G, Mikolaiski H (1996) Concepts and results of the GPS data processing with Bernese and GIPSY Software. In: German Contributions to the SCAR 95 Epoch Campaign, 1996: The Geodetic Antarctic Project GAP95, Muenchen, pp 37–51

Ephishov II, Baran LW, Shagimuratov II, Yakimova GA (2000) Comparison of total electron content obtained from GPS with IRI. Phys Chem Earth 25C(4):339–342

ESA (2015) http://www.esa.int/Our_Activities/Navigation/The_future_-_Galileo/What_is_Galileo.

ESA nauipedia (2014) http://www.navipedia.net/index.php/BeiDou_General_Introduction.

Euler H-J (1995) Statische/Kinematische Echtzeitvermessung mit GPS. Schriftenreihe des Deutschen Vereins für Vermessungswesen, Bd. 18, Stuttgart, pp 271–286

Euler HJ, Seeger S, Takac F (2004) Analysis of biases influencing successful rover positioning with GNSS-network RTK. J. GPS 3(1-2): 70-78

Even-Tzur G, Agmon E (2005) Monitoring vertical movements in Mount Carmel by means of GPS and precise leveling. Surv. Rev. 38(296): 146-157

Exertier P, Bonnefond P (1997) Analytical solution of perturbed circular motion: Application to satellite geodesy. J Geodesy 71(3):149–159

Farguhar R (2011) Fifty Years on the Space Frontier: Halo Orbits, Comets, Asteroids, and more, Outskirts Press, Inc. Denver, Colorado

Faruqi FA, Turner KJ (2000) Extended Kalman filter synthesis for integrated global positioning/iner-tial navigation systems. Appl Math Comput 115(2–3):213–227

Featherstone W, Dentith M, Kirby J (1998) Strategies for the accurate determination of orthometric heights from GPS. Survey rev 34(267):278–296

Featherstone WE (2004) Evidence of a north-south trend between AUSGeoid98 and the Australian height datum in southwest Australia. Survey Rev. 37(291): 334-343

Feltens J (1991) Nicht gravitative Störeinflüsse bei der Modellierungen von GPS-Erdumlaufbahnen. DGK, Reihe C, Heft 371, Verlag der Bayerischen Akademie der Wissenschaften

Feng Y, Kubik K (1997) On the internal stability of GPS solutions. J Geodesy 72:1–10

Feng Y (2005) Future GNSS performance. Predictions using GPS with a virtual Galileo constellation. GPS World 16(3): 46-52

Fivian M, Hudson H, Lin RP, Zahid J (2008) A Large Excess in Apparent Solar Oblateness Due to Surface Magnetism, Science, Vol. 322, 24 Oct 2008

Fivian M, Hudson H, Lin RP, Zahid J (2009) Response to Comment on "A Large Excess in Apparent Solar Oblateness Due to Surface Magnetism", Sciences, Vol. 324, 1143, 29 May 2009

Flores A, Escudero A, Sedo MJ, Rius A (2000) A near real time system for tropospheric monitoring using GPS hourly data. Earth Planets Space 52(10):681–684

Fotopoulos G, Kotsakis C, Sideris MG (2003) How accurately can we determine orthometric height differences from GPS and geoid data. J. Surv. Eng., ASCE 129(1): 1-10

Forsberg R, Olesen AV, Timmen L, Xu GC, Bastos L, Hehl K, Solheim D (1998) Airborne gravity in Skagerrak and elsewhere: The AGMASCO project and a nordic outlook. In: Proceedings NKG meeting Gvle, May 1998

Forsberg R, Keller K, Nielsen CS, Gundestrup N, Tscherning CC, Madsen SN, Dall J (2000) Elevation change measurements of the Greenland Ice Sheet. Earth Planets Space 52(11):1049–1053

Freda P, Angrisano A, Gaglione S, Troisi S (2015) Time-differenced carrier phases technique for precise GNSS velocity estimation. GPS Solutions 19(2), pp335-341

Fry WG (1997) GPS flies high in Midwest flood study: The Mississippi River Project demonstrates viability of large-area airborne GPS-controlled mapping. EOM: mag. geogr, mapp, Earth inf 6(1):28–31

Gabor MJ, Nerem RS (2004) Characteristics of satellite-satellite single difference widelane fractional carrier-phase biases. Navigation 51(1): 77-92

Galas R, Reigber C (1997) Status of the IGS stations provided by GFZ. International GPS Service for Geodynamics: 1996 annual report, Pasadena, pp 393–396

Galas R, Reigber C, Baustert G (1995) Permanent betriebene GPS-Stationen in globalen und regionalen Netzen. ZfV 1209:431–438

Gallimore J, Maini A (2000) Galileo: The public-private partnership. GPS World 11(9):58–63

Gao Y, McLellan J, Schleppe J (1998) Integrating GPS with barometry for high-precision real-time kinematic seismic survey, Survey. Land Inf Syst 58(2):115–119

Gao Y, Wojciechowski, Chen K (2005) Airborne kinematic positioning using precise point positioning methodology. Geomatica 59(1): 29-36

Ge LL, Han SW, Rizos C (2000) Multipath mitigation of continuous GPS measurements using an adap-tive filter. GPS Solutions 4(2):19–30

Ge L (2003) Integration of GPS and radar interferometry. GPS Solutions 7(1): 52-54

Ge M, Calais E, Haase J (2000) Reducing satellite orbit error effects in near real-time GPS zenith tropospheric delay estimation for meteorology. Geophys Res Lett 27(13):1915–1918

Ge M, Gendt G, Dick G, Zhang FP, Reigber C (2005) Impact of GPS satellite antenna offsets on scale changes in global network solutions. Geophys. Res. Lett. 32(L06310): 1-4

Geiger A, Hirter H, Cocard M (1995) Mitigation of tropospheric effects in local and regional GPS networks. In: GPS Trends in Precise Terrestrial, Airborne, and Spaceborne Applications: 21$^{st}$ IUGG General As-sembly, IAG Symposium No. 115, Boulder, USA, July 3–4, 1995. Springer-Verlag, Berlin, pp 263–267

Gendt G (1997) Analysen der IGS-Daten und Ergebnisse, GPS-Anwendungen und Ergebnisse '96: Beiträge zum 41. DVW-Fortbildungsseminar vom 7. bis 8. November 1996 am Geo-Forschungs-zentrum Potsdam, 1997, Stuttgart, pp 43–53

Gendt G, Dick G, Reigber C (1995) Global plate kinematics estimated by GPS data of the IGS core network. In: GPS Trends in Precise Terrestrial, Airborne, and Spaceborne Applications: 21$^{st}$ IUGG General As-sembly, IAG Symposium No. 115, Boulder, USA, July 3–4, 1995. Springer-Verlag, Berlin, pp 30–34

Gianniou M (1996) Genauigkeitssteigerung bei kurzzeit-statischen und kinematischen Satelliten-messungen bis hin zur Echtzeitanwendung. DGK, Reihe C, Heft 458, Verlag der Bayerischen Akademie der Wissenschaften

Gili JA, Corominas J, Rius J (2000) Using Global Positioning System techniques in landslide monitoring. Eng Geol 55(3):167–192

Gilvarry JJ, Sturrock PA (1967) Sloar Oblateness and the Perihelion Advances of Planets, Nature, Vol. 216, December 30, 1967

Gleason DM (1996) Avoiding numerical stability problems of long duration DGPS/INS Kalman filters. J Geodesy 70(5):263–275

Goad C (1996) Single-site GPS models. In: Kleusberg A, Teunissen PJG (eds) GPS for geodesy. Springer-Verlag, Berlin

Goad C (1996) Short distance GPS models. In: Kleusberg A, Teunissen PJG (eds) GPS for geodesy. Springer-Verlag, Berlin

Goad C, Yang M (1997) A new approach to precision airborne GPS positioning for photogrammetry. Photogramm Eng Rem S 63(9):1067–1077

Goad C, Dorota A, Brzezinska G, Yang M (1996) Determination of high-precision GPS orbits using triple differencing technique. J Geodesy 70:655–662

Godier S, Rozelot JP (1999) Quadrupole moment of the Sun. Gravitational and rotational potentials, A&A 350, 310-317

Godier S, Rozelot JP (2000) The solar oblateness and its relationship with the structure of the tachocline and of the Sun's subsurface, A&A 355, 365-374

Goerres B, Campbell J (1998) Bestimmung vertikaler Punktbewegungen mit GPS. ZfV 123 (7):222–230

Goodhue J (1997) Experiments aloft: Balloon-borne payloads reach near space. GPS World 8 (9):34–42

GPS.gov, National Coordination Office for Space-Based Positioning, Navigation and Timing (2015) GPS Modernization, http://www.gps.gov/systems/gps/modernization/

Grafarend EW (2000) Mixed integer-real valued adjustment (IRA) problems: GPS initial cycle ambi-guity resolution by means of the LLL algorithm. GPS Solutions 4(2):31–44

Grafarend E, Ardalan AW (1997) An estimate in the Finnish Height Datum N60, epoch 1993.4, from twenty-five GPS points of the Baltic Sea Level Project. J Geodesy 71(11):673–679

Grejner-Brzezinska D, Toth C, Yi YD (2005) On improving navigation accuracy of GPS/INS systems. Photogramm. Eng. Remot Sens. 71(4): 377-389

Grejner-Brzezinska DA et al. (2004) An analysis of the effects of different network-based ionosphere estimation models on rover positioning accuracy. J. GPS 3(1-2): 115-131

Grewal MS, Weill LR, Andrews AP (2001) Global Positioning System, Inertial Navigation, and Integration. John Wiley & Sons, Inc., New York: 392 p.

Griffiths J, Ray JR (2009) On the precision and accuracy of IGS orbits. J Geodesy 83(3), pp277-287

Groten E (1979) Geodesy and the Earth's gravity field, Vol. I: Principles and conventional methods. Dümmler-Verlag, Bonn

Groten E (1980) Geodesy and the Earth's gravity field, Vol. II: Geodynamics and advanced methods. Dümmler-Verlag, Bonn

Guinn J, Muellerschoen R, Cangahuala L (1995) TOPEX/Poseidon precision orbit determination us-ing combined GPS, SLR and DORIS. In: GPS Trends in Precise Terrestrial, Airborne and Spaceborne Applications: 21$^{st}$ IUGG General Assembly, IAG Symposium No. 115, Boulder, USA, July 3–4, 1995. Springer-Verlag, Berlin, pp 128–132

Guo JF, Ou JK, Ren C (2005) Partial continuation model and its application in mitigating systematic errors of doubled-differenced GPS measurements. Progress in Natural Sci. 15(3): 246-251

Gurtner W, Mader G (1990) Receiver independent exchange format version 2. GPS Bulletin 3 (3):1–8

Gurtner W, Boucher C, Bruyninx C (1997) The use of the IGS/EUREF permanent network for EUREF densification campaigns. In: Symposium of the IAG Subcommission for Europe (EUREF), Sofia, Bulgaria, June 4–7, 1997, 3 pp

Guo Q (2015) Precision comparison and analysis of four online free PPP services in static positioning and tropospheric delay estimation. GPS Solutions 19(4), pp537-544

Haines BJ, Christensen EJ, Guinn JR (1995) Observations of TOPEX/Poseidon orbit errors due to gravitational and tidal modeling errors using the Global Positioning System. In: GPS Trends in Precise Terrestrial,Airborne, and Spaceborne Applications: 21$^{st}$ IUGG General Assembly,IAG Sym-posium No. 115, Boulder, USA, July 3–4, 1995. Springer-Verlag, Berlin, pp 133–138

Haines B, Bar-Server Y, Bertiger W, Desai S, Willis P (2004) One-centimeter orbit determination for Jason-1: New GPS-based strategies. Mar. Geod. 27(1-2): 299-318

Hajj GA, Kursinski ER, Bertiger WI (1995) Initial results of GPS-LEO occultation measurements of Earth's atmosphere obtained with the GPS-MET experiment. In: GPS Trends in Precise Terrestrial, Airborne, and Spaceborne Applications: 21$^{st}$ IUGG General Assembly, IAG Symposium No. 115, Boulder, USA, July 3–4, 1995. Springer-Verlag, Berlin, pp 144–153

Hamilton GS, Whillans IA (2000) Point measurements of mass balance of the Greenland Ice Sheet using precision vertical Global Positioning System (GPS) surveys. J Geophys Res 105(B7): 16295–16301

Han S (1997) Quality-control issues relating to instantaneous ambiguity resolution for real-time GPS kinematic positioning. J Geodesy 71(6):351–361

Han S, Rizos C (1996) Validation and rejection criteria for integer least-squares estimation. Survey rev 33(260):375–382

Han S, Rizos C (2000) GPS multipath mitigation using FIR filters. Survey rev 35(277):487–498

Han S, Rizos C (2000) An instantaneous ambiguity resolution technique for medium-range GPS kinematic positioning. J Inst Navig 47(1):17–31

Han S, Rizos C (2000) Airborne GPS kinematic positioning and its application to oceanographic mapping. Earth Planets Space 52(10):819–824

Hariharan R, Krumm J, Horvitz E (2005) Web-enhanced GPS. Lect. Notes Comput. Sci. 3479: 95-104

Harwood NM, Swinerd GG (1995) Long-periodic and secular perturbations to the orbit of a spherical satellite due to direct solar radiation pressure, Celestial Mechanics and Dynamical Astronomy 62: 71-80

Hatanaka Y, Tsuji H, Iimura Y, Kobayashi K, Morishita H (1995) Application of GPS kinematic method for detection of crustal movements with high temporal resolution. In: GPS Trends in Precise Ter-restrial, Airborne, and Spaceborne Applications: 21$^{st}$ IUGG General Assembly, IAG Symposium No. 115, Boulder, USA, July 3–4, 1995. Springer-Verlag, Berlin, pp 105–112

Hatch RR (1996) The promise of a third frequency, GPS World, 7(1996)5, 55–58

Hatch RR (2004) Those scandalous clocks. GPS Solutions 8(2): 67-73

Hatch RR, Sharpe RT (2004) Recent improvements to the SDtar Fire global DGPS navigation software. J. GPS 3(1-2): 143-153

Hauschild A, Montenbruck O (2009) Kalman-filter-based GPS clock estimation for near real-time positioning. GPS Solutions 13(3), pp173-182

Hay C, Wong J (2000) Enhancing GPS: Tropospheric delay prediction at the Master control Station. GPS World 11(1):56–62

He HB, Li JL, Yang YX, Xu JY, Guo HR, Wang AB (2014) Performance assessment of single- and dual-frequency BeiDou/GPS single-epoch kinematic positioning. GPS Solutions 18(3), pp393-403

He JK, Cai DS, Li YX, Gong ZS (2004) Active extension of the Shanxi rift, north China: does it result from anticlockwise block rotations? Terra Nova 16(1): 38-42

He K, Xu G, Xu T, Flechtner F (2014) GNSS navigation and positioning for the GEOHALO experiment in Italy. GPS Solutions, DOI: 10.1007/s10291-014-0430-4.

He XF, Guang Y, Ding XL, Chen YQ (2004) Application and evaluation of a GPS-multi-antenna system for dam deformation monitoring. Earth Planets and Space 56(11): 1035-1039

Heck B (1995a) Grundlagen der SatellitenGeodaesie. Schriftenreihe des Deutschen Vereins für Vermessungswesen, Bd. 18, Stuttgart, pp 10–31

Heck B (1995b) Grundlagen der erd- und himmelsfesten Referenzsysteme. Schriftenreihe des Deutschen Vereins für Vermessungswesen, Bd. 18, Stuttgart, pp 138–153

Hefty J, Rothacher M, Springer T, Weber R, Beutler G (2000) Analysis of the first year of Earth rotation parameters with a sub-daily resolution gained at the CODE processing center of the IGS. J Geodesy 74(6):479–487

Hehl K, Xu G, Fritsch J (1995) Results from field tests of an airborne gravity meter system. In: Proceed-ings of IUGG XXI General Assembly, IAG meeting at Boulder,Colorado, USA,July 1995, IAG Sympo-sium G4, pp 169–174

Hein GW (2000) From GPS and GLONASS via EGNOS to Galileo. Position and navigation in the third millennium. GPS Solutions 3(4):39–47

Hein GW, Riedl B (1995) High precision deformation monitoring using differential GPS. In: GPS Trends in Precise Terrestrial,Airborne, and Spaceborne Applications: 21$^{st}$ IUGG General Assembly, IAG Symposium No. 115, Boulder, USA, July 3–4, 1995. Springer-Verlag, Berlin, pp 180–184

Hein GW, Eisfeller B, Pielmeier J (1995) Developments in airborne "high precision" digital photo flight navigation in "realtime". In: GPS Trends in Precise Terrestrial, Airborne, and Spaceborne Applications: 21$^{st}$ IUGG General Assembly, IAG Symposium No. 115, Boulder, USA, July 3–4, 1995. Springer-Verlag, Berlin, pp 175–179

Hein GW et al. (2003) Galileo frequency & signal design. GPS World 14(6): 30-37

Heinrich G et al. (2004) HIGAPS. A highly integrated Galileo/GPS chipset for consumer applications. GPS World 15(9): 38-47

Heitz S (1988) Coordinates in geodesy. Springer-Verlag, Berlin

Hernández-Pajares M, Juan JM, Sanz J, Colombo OL (2000) Application of ionospheric tomography to real-time GPS carrier-phase ambiguities resolution, at scales of 400–1 000 km and with high geomagnetic activity. Geophys Res Lett 27(13):2009–2012

Hernandez-Pajares M, Juan JM, Sanz J, Colombo OL (2003) Impact of real-time ionospheric determination on improving precise navigation with GALILEO and next-generation GPS. Navigation 50(3): 205-218

Herring T (2003) MATLAB Tools for viewing GPS velocities and time series. GPS Solutions 7 (3): 194-199

Hess D, Keller W (1999) Gradiometrie mit GRACE Teil I, Fehleranalyse künstlicher Gradiometerdaten. ZfV 5:137–144

Hess D, Keller W (1999) Gradiometrie mit GRACE Teil II, Simulationsstudie. ZfV 7:205–211

Highsmith D, Axelrad P (1999) Relative state estimation using GPS flight data from co-orbiting space-craft. ION GPS '99, 14–17 September 1999, pp 401–409

Hill HA, Clayton PD, Patz DL, Healy AW, Stebbins RT, Oleson JR, Zanoni CA (1974) Solar oblateness, Excess Brightness, and Relativity. Physical Review Letters, Vol. 33 No. 25, 1497-1500

Hilla S (2004) Plotting pseudorange multipath with respect to satellite azimuth and elevation. GPS Solutions 8(1): 44-48

Hiller W, Lauterbach P, Wlaka M (1997) Seeking sovereignty: A European navigation satellite system. GPS World 8(9):56–60

Hirahara K (2000) Local GPS tropospheric tomography. Earth Planets Space 52(11):935–939

Hofmann-Wellenhof B, Legat K, Weiser M (2003) Navigation, Principles of Positioning Guidance. Springer-Verlag, New York, NY: xxix+427p

Hong Y, Ou JK (2006) Algebra solution of the variation equation and its numerical validation. (in Chinese)

Horvath I, Essex EA (2000) Using observations from the GPS and TOPEX satellites to investigate night-time TEC enhancements at mid-latitudes in the southern hemisphere during a low sunspot number period. J Atmos Sol-Terr Phy 62(5):371–391

Hostetter GH (1987) Handbook of digital signal processing. Engineering Applications, Academic Press, San Diego, CA

Hotine M (1991) Differential geodesy. Springer-Verlag, Berlin

Hsu R, Li S (2004) Decomposition of deformation primitives of horizontal geodetic networks: application to Taiwan's GPS network. J. Geod. 78(4-5): 251-262

Hu GR, Khoo HS, Goh PC, Law CL (2003) Development and assessment of GPS virtual reference station for RTK positioning. J. Geod. 77(5-6): 292-302

Hugentobler U, Ineichen D, Beutler G (2003) GPS satellites: Radiation pressure, attitude and resonance. Adv. Space Res. 31(8): 1917-1926

Hughes S (1977) Satellite Orbits Perturbed by Direct Solar Radiation Pressure: General Expansion of the Disturbing Function, Planet. Space Sci. Vol 25 pp 809-815

Hünerbein K von, Hamann HJ, Rüter E, Wiltschko W (2000) A GPS-based system for recording the flight paths of birds. Naturwissenschaften 87(6):278–279

Ifadis IM (2000) A new approach to mapping the atmospheric effects for GPS. Earth Planets Space 52(10):703–708

IGS, RTCM-SC104 (2015) RINEX- The Receiver Independent Exchange Format (Version 3.03). ftp://igs.org/pub/data/format/rinex303.pdf

Ihde J (1996) Geoidbestimmung unter Nutzung von GPS und Nivellement. Erste Geodätische Woche, Stuttgart, 7.–12. Oktober 1996, 6 Blatt

Ince CD, Sahin M (2000) Real-time deformation monitoring with GPS and Kalman Filter. Earth Plan-ets Space 52(10):837–840

ION, The Institute of Navigation. Proceedings of ION GPS-91, 92, 93, 94, 95, 96, 97, 98, 99, 00, 01, 02, 03, 04, 05, 06, 07, 08, 09, 10, 11, 12, 13, 14, 15

Iorio L (2005) On the possibility of measuring the solar oblateness and some relativistic effects from planetary ranging, A&A 433, 385-393

Jaggi A, Beutler G, Hugentobler U (2005) Reduced-dynamic orbit determination and the use of accelerometer data. Adv. Space Res. 36(3): 438-444

Jakowski N, Mayer C, Hoque MM, Wilken V (2011) Total electron content models and their use in ionosphere monitoring. Radio Science, 46(6)

Jakowski N, Sardon E, Engler E (1995) About the use of GPS measurements for ionospheric studies. In: GPS Trends in Precise Terrestrial, Airborne, and Spaceborne Applications: $21^{st}$ IUGG General As-sembly, IAG Symposium No. 115, Boulder, USA, July 3–4, 1995. Springer-Verlag, Berlin, pp 248–252

Jerde CL, Visscher DR (2005) GPS measurement error influences on movement model parameterization. Ecological Appl. 15(3): 806-810

Jekeli C, Garcia R (1997) GPS phase accelerations for moving – base vector gravimetry. J Geodesy 71:630–639

Jeyapalan K (2004) Local geoid determination using global positioning systems. Surv. Land Information Sci. 64(1): 65-75

Jiang N (2013) Studies on error analysis of positioning and real-time precise point positioning. Chang'an University. Xi'an, China

Jiang N, Xu T (2013) An improved velocity determination method based on GOCE kinematic orbit. Geodesy and Geodynamics 4(2): pp47-52

Jiang N, Xu T, Xu Y (2012) An Improved Method for Determination of GOCE Orbital Velocity of the Geometric Method. Bulletin of Surveying and Mapping, 11: pp7-10

Jiang N, Xu T, Xu Y (2013) A real-time precise point positioning method without precise clock bias. Journal of Central South University (Science and Technology), 44(11): pp4520-4526

Jiang N, Xu T, Xu Y (2013) Influence of the Receiver Antenna random to GPS Positioning Precision. Geomatics and Information Science of Wuhan University, 38(5): pp566-570

Jiang N, Xu T, Xu Y (2013) Multipath error estimation and its improved algorithm for Chinese IGS stations. Journal of Geodesy annd Geodynamics, 33(2): pp143-146

Jiang N, Xu T, Xu Y (2013) Real-time estimation of satellite clock and PPP precision analysis based on IGS regional net. Journal of Geodesy and Geodynamics, 33(5): pp44-48

Jiang N, Xu Y, Xu T, Xu G, Sun Z, Schuh H (2016) GPS/BDS short-term ISB modelling and prediction. GPS Solutions, DOI: 10.1007/s10291-015-0513-x

Jin SG, Zhu WY (2003) Active motion of tectonic blocks in East Asia: Evidence from GPS measurements. Acta Geologica Sinica 77(1): 59-63

Jin X-X (1995) A recursive procedure for computation and quality control of GPS differential corrections. Delft University of Technology, Faculty Geod. Engin., Delft Geodetic Computing Centre, Delft, 83 S

Jin X-X, Jong K de, Cees D (1996) Relationship between satellite elevation and precision of GPS code observations. Leipzig, 13 S

Jong K de (1999) The Influence of Code Multipath on the Estimated Parameters of the Geometry-Free GPS Model. GPS Solutions, Vol. 3, No. 2, 11-18

Jong K de (2000) Minimal detectable biases of cross-correlated GPS observations. GPS Solutions 3(3): 12–18

Jong K de, Teunissen PJG (2000) Minimal detectable biases of GPS observations for a weighted ionosphere. Earth Planets Space 52(10):857–862

Jonkman NF, Jong K de (2000) Integrity monitoring of IGEX-98 data, Part I: Availability. GPS Solu-tions 3(4):10–23

Jonkman NF, Jong K de (2000) Integrity monitoring of IGEX-98 data, Part II: Cycle slip and outlier detection. GPS Solutions 3(4):24–34

Jonkman NF, Jong K de (2000) Integrity monitoring of IGEX-98 data, Part III: Broadcast navigation message validation. GPS Solutions 4(2):45–53

Joosten J (2000) The GPS integer least-squares statistics. Phys Chem Earth 25(A9–A11):687–692

Joosten P, Tiberius C (2000) Fixing the ambiguities. Are you sure they're right? GPS World 11 (5):46–51

Kaczorowski M (1995) Calculation of the Green's loading functions. Part 1: Theory. Artificial Satellites, Journal of Planetary Geodesy 30(1):77–93

Kälber S, Jäger R, Schwäble R (2000) A GPS-based online control and alarm system. GPS Solutions 3(3):19–25

Kammeyer P (2000) A UT1-like quantity from analysis of GPS orbit planes. Celest Mech Dyn Astr 77(4):241–272

Kamp PD van (1967) Principles of astrometry. W. H. Freemann and Company, San Francisco and London

Kang Z, Nagel P, Pastor R (2003) Precise orbit determination for GRACE. Adv. Space Res. 31(8): 1875-1881

Kaniuth K, Kleuren D, Tremel H (1998) Sensitivity of GPS height estimates to tropospheric delay modelling. Allgemeine Vermessungsnachrichten 105(6):200–207

Kaniuth K, Kleuren D, Tremel H, Schlueter W (1998) Elevationabhängige Phasenzentrums-variationen geodätischer GPS-Antennen. ZfV 10:320–325

Karslioglu MO (2005) An interactive program for GPS-based dynamic orbit determination of small satellites. Comput. and Geosci. 31(3): 309-317

Kashani I, Wielgosz P, Grejner-Brzezinska D (2003) Datum definition in the long range instantaneous RTK GPS network solution. J. GPS 2(2): 100-108

Kechine MO, Tiberius CCJM, van der Marel H (2004) An experimental performance analysis of real-time kinematic positioning with NASA's Internet-based Global Differential GPS. GPS Solutions 8(1): 9-22

Keshin MO (2004) Directional statistics of satellite-satellite single-difference widelane phases biases. Artificial Satellites 39(4): 305-324

Kezerashvili RY, Vazquez-Poritz J (2009) Solar Radiation Pressure and Deviations from Keplerian Orbits, Physics Letters B 675, 18-21

Khan SA, Scherneck HG (2003) The M-2 ocean tide loading wave in Alaska: vertical and horizontal displacements, modeled and observed. J. Geod. 77(3-4): 117-127

Khan SA, Tscherning CC (2001) Determination of semi-diurnal ocean tide loading constituents using GPS in Alaska. Geophys Res Lett 28(11):2249–2252

Khazaradze G, Klotz J (2003) Short- and long-term effects of GPS measured crustal deformation rates along the south central Andes. J. Geophys. Res. 108(B6): ETG5/1-15

Kelley K, Bologlu A (1995) DGPS on the waterfront: tracking cargo and equipment in maritime terminals. GPS World 6(9):62–71

Keong J, Lachapelle G (2000) Heading and pitch determination using GPS/GLONASS. GPS Solutions 3(3):26–36

Khodabandeln A, Teunissen PJG (2015) An analytical study of PPP-RTK corrections: precision, correlation and user-impact. J Geodesy 89(11), pp1109-1132

Kim D, Langley RB (2000) A search space optimization technique for improving ambiguity resolution and computational efficiency. Earth Planets Space 52(10):807–812

Kim D, Langley RB (2003) On ultrahigh-precision GPS positioning and navigation. Navigation 50 (2): 103-116

King-Hele D (1964) Theory of satellite orbits in an atmosphere, Butterworths Mathematical Texts, Butterworths & Co. Publ., London

King MA, Penna NT, Clarke PJ, King EC (2005) Validation of ocean tide models around Antarctica using onshore GPS and gravity data. J. Geophys. Res. 110(B08401): 1-21

King M, Coleman R, Morgan P (2000) Treatment of horizontal and vertical tidal signals in GPS data: A case study on a floating ice shelf. Earth Planets Space 52(11):1043–1047

Kislik MD (1983) On the solar oblateness, Sov. Astron. Lett. 9(5)

Kistler M, Geiger A (2000) GPS am Seil herunterlassen: das Global Positioning System im Dienste des Seilbahnwesen. Vermess Photogramm Kulturtech 98(7):441–445

Kleusberg A (1995) Mathematics of attitude determination with GPS. GPS World, 6(9):72–78

Kleusberg A, Teunissen PJG (eds) (1996) GPS for geodesy. Springer-Verlag, Berlin

Klotz J, Angermann D, Reinking J (1995) Großräumige GPS-Netze zur Bestimmung der rezenten Kinematik der Erde. ZfV 120(9):449–460

Knickmeyer ET, Knickmeyer EH, Nitschke M (1996) Zur Auswertung kinematischer Messungen mit dem Kalman-Filter. Schriftenreihe des Deutschen Vereins für Vermessungswesen, Bd. 22, Stuttgart, pp 141–166

Knudsen P, Andersen O (1997) Global marine gravity and mean sea surface from multi mission satel-lite altimetry. Scientific Assembly of the International Association of Geodesy in conjunction with 28$^{th}$ Brazilian Congress of Cartography; Rio de Janeiro, 3–9 September 1997, 4 pp

Koch KR (1980) Parameterschätzung und Hypothesentests in linearen Modellen. Dümmler-Verlag,Bonn

Koch KR (1996) Robuste Parameterschätzung. Allgemeine Vermessungsnanchrichten 103(1):1–18

Komjathy A, Langley RB, Vejrazka F (1995) Assessment of two methods to provide ionospheric range emror corrections for single-frequency GPS users. In: GPS Trends in Precise Terrestrial, Airborne, and Spaceborne Applications: 21$^{st}$ IUGG General Assembly, IAG Symposium No. 115, Boulder, USA, July 3–4, 1995. Springer-Verlag, Berlin, pp 253–257n

Komjathy A, Zavorotny VU, Axelrad P, Born GH, Garrison JL (2000) GPS signal scattering from sea surface: Wind speed retrieval using experimental data and theoretical model. Remote Sens Environ 73(2):162–174

Konig R, Schwintzer P, Bode A (1996) GFZ-1: A small laser satellite mission for gravity nfield model improvement. Geophys Res Lett 23(22):3143–3146

Konig R, Reigber C, Zhu SY (2005) Dynamic model orbits and earth system parameters from combined GPS and LEO data. Adv. Space Res. 36(3): 431-437

Kraus JD (1966) Radio astronomy. McGraw-Hill, New York

Kristiansen O (1995) Experiences with high precision GPS processing in Norway. Rep Finnish Geod Inst 4:77–84

Krivov AV, Sokolov LL, Dikarev VV (1996) Dynamics of Mars-Orbiting Dust: Effects if Light Pressure and Planetary Oblateness, Celestial Mechanics and Dynamical Astronomy 63: 313-339

Kroes R, Montenbruck O (2004) Spacecraft formation flying: Relative positioning using dual-frequency carrier phase. GPS World 15(7): 37-42

Kroes R, Montenbruck O, Bertiger W, Visser P (2005) Precise GRACE baseline determination using GPS. GPS Solutions 9(1): 21-31

Kuang D, Rim HJ, Schutz BE (1996) Modeling GPS satellite attitude variation for precise orbit determination. J Geodesy 70(9):572–580

Kuang D, Rim HJ, Schutz BE, Abusali PAM (1996) Modeling GPS satellite attitude variation for precise orbit determination. J Geodesy 70:572–580

Kubo-oka T, Sengoku A (1999) Solar Radiation Pressure Model for the Relay Satellite of SELENE, Earth Planets Space, 51, 979-986

Kudak VI, Klimik VU, Epishev VP (2010) Evalution of disturbances from solar radiation in orbital elements of geosychronous satellite based on harmonics, Astrophysical Bulletin, Vol. 65, No. 3, pp. 300-310

Kudryavtsev SM (2007) Long-term harmonic development of lunar ephemeris, Astronomy & Astrophysics, Vol. 472 No.2

Kuhn JR, Bush RI, Scheick X, Scherrer P (1998) The Sun's Shape and Brightness, Nature, Vol. 392, P155

Kuhn JR, Emilio M, Bush R (2009) Comment on "A Large Excess in Apparent Solar Oblateness Due to Surface Magnetism, Sciences, Vol. 324, 1143, 29 May 2009

Kumar M (1997) Time-invariant bathymetry: A new concept to define and survey it using GPS. In: Proceedings of Fourteenth United Nations Regional Cartografic Conference for Asia and the Pa-cific, Bangkok, 3–7 February 1997. Bangkok, 4 pp

Kwon JH, Grejner-Brzezinska D, Bae TS, Hong CK (2003) A triple difference approach to Low Earth Orbiter precision orbit determination. J. Navig. 56(3): 457-473

Lachapelle G (1995) Post-mission GPS absolute kinematic positioning at one-metre accuracy level. Int J Geomatics 9(1):37–39

Lachapelle G, Cannon ME, Qiu W, Varner C (1996) Precise aircraft single-point positioning using GPS post-mission orbits and satellite clock corrections. J Geodesy 70:562–571

Lachapelle G (2004) GNSS indoor location technologies. J. GPS 3(1-2): 2-11

Lachapelle G, Kuusniemi H, Dao DTH, Macgougan G, Cannon ME (2004) HSGPS signal analysis and performance under various indoor conditions. Navigation 51(1): 29-43

Lambert A, Pagiatakis SD, Billyard AP, Dragert H (1988) Improved ocean tide loading correction for gravity and displacement: Canada and northern United States. J Geophys Res 103 (B12):30231–30244

Landau H (1988) Zur Nutzung des Global Positioning Systems in Geodaesie und Geodynamik: Modell-bildung, Software-Entwicklung und Analyse. Universität der Bundeswehr Müchen, Studiengang Vermessungswesen, Schriftenreihe, Heft 36

Landspersky D, Mervart L (1997) A contribution to the study of modelling of the troposphere biases of GPS observations with high accuracy. In: Proceedings of the EGS symposium G14 'Geodetic and Geodynamic programmes of the CEI': 22 General Assembly of the EGS, Vienna, Austria, 21– 25 April 1997. Warszawa, pp 207–211

Langley RB (1997a) GLONASS: Review and update. GPS World 8(7):46–51

Langley RB (1997b) The GPS error budget. GPS World 8(3):51–56

Langley RB (2000) GPS, the ionosphere, and the solar maximum. GPS World 11(7):44–49

Langley RB (2003) Getting your bearings. The magnetic compass and GPS. GPS World 14(9): 70 +

Lapucha D (1994) Real-time centimeter-accuracy positioning with on-the-fly carrier phase ambiguity resolution. Rep Geod 1:52–59

Larson WJ, Wertz JR (1995) Space mission analysis and design, $2^{nd}$ Ed, Microcosm, Inc. California and Kluwer Academic Publishers Boston

Lechner W (1995) Telemetriekonzepte für die GPS-unterstützte Echtzeitvermessung. Schriftenreihe des Deutschen Vereins für Vermessungswesen, Bd. 18, pp 260–286

Lee HK, Hewitson S, Wang J (2004) Web-based resources on GPS/INS integration. GPS Solutions 8(3): 189-191

Lee J-T, Mezera DF (2000) Concerns related to GPS-derived geoid determination. Survey rev 35 (276):379–397

Lee YC, O'Laughlin DG (2000) Performance analysis of a tightly coupled GPS/Inertial system for two integrity monitoring methods. J Inst Navig 47(3):175–189

Leinen S (1997) Hochpräzise Positionierung über große Entfernungen und in Echtzeit mit dem Global Positioning System. DGK, Reihe C, Heft 472,Verlag der Bayerischen Akademie der Wissenschaften

Lelgemann D (1983) A linear solution of equation of motion of an Earth-orbiting satellite based on a Lie-series. Celestial Mech 30:309

Lelgemann D (1996) Geodaesie im Weltraumzeitalter. Dtsch Geod Komm 25:59–77
Lelgemann D, Petrovic S (1997) Bemerkungen über den Höhenbegriff in der Geodaesie. ZfV 122 (11): 503–509
Lemmens R (2004) Book review: GPS – Theory, Algorithms and Applications, Xu G 2003, International J. Applied Earth Observation and Geoinformation 5 (2004) 165-166
Leroy E (1995) GPS real-time levelling on the world's longest suspension bridge. Int J Geomatics 9(8):6–8
Levin E (1968) Solar radiation pressure perturbations of earth satellite orbits, AIAA Journal, vol. 6, issue 1, pp. 120-12
Levine J (2001) GPS and the legal traceability of time. GPS World 12(1):52–58
Li H, Xu G, Xue H, Zhao H, Chen J,Wang G (1999) Design of GPS application program. Science Press, Peking, ISBN 7-03-007204-9/TP.1049, 337 p (in Chinese and in C)
Li X (2004) The advantage of an integrated RTK-GPS system in monitoring structural deformation. J. GPS 3(1-2): 191-199
Li X, Zhang X, Ren X, Fritsche M, Wickert J, Schuh H (2015) Precise positioning with current multi-constellation Global Navigation Satellite Systems: GPS, GLONASS, Galileo and BeiDou. Scientific reports 5, 8328.
Licandro J, Alvarenz-Candal A, Leon Jde, Pinilla-Aloso N, Lazzaro D, Hampins H (2008) Spectral properties of asteroids in cometary orbits, Astronomy & Astrophysics, Vol. 481 No.3
Lightsey EG, Blackburn GC, Simpson JE (2000) Going up: A GPS receiver adapts to space. GPS World 11(9):30–34
Linkwitz K, Hangleiter U (eds) (1995) High precision navigation 95. Dümmler-Verlag, Bonn
Liu D, Liu J, Liu G (1993) The three-dimensional combined adjustment of GPS and terrestrial surveying data. Acta Geod Cartogr Sinica 41–54 (select. papers Engl. ed.)
Liu YX, Chen YQ, Liu JN (2000) Determination of weighted mean tropospheric temperature using ground meteorological measurement. J Wuhan Techn Univ Survey Mapp 25(5):400–403
Liu Z (2011) A new automated cycle slip detection and repair method for a single dual-frequency GPS receiver. J Geodesy 85(3), pp171-183
Lowe S, Zuffada C, Chao Y, Kroger P, Young J (2002b) 5-cm-precision aircraft ocean altimetry using GPS reflections. Geophys. Res. Lett., 29(10)
Ludwig R (1969) Methoden der Fehler- und Ausgleichsrechnung. Vieweg & Sohn, Braunschweig
Lynden-Bell D (2009) Analytical orbits in any central potential, MNRAS Vol. 402 Issue 3 (p1937-1941)
Mackenzie R, Moore P (1997) A geopotential error analysis for a non planar satellite to satellite tracking mission J Geodesy 71(5):262–272
Mackie JB (1985) The elements of astronomy for surveyors. Charles Griffin & Company Ltd.
Mader GL (1995) Kinematic and rapid static (KARS) GPS positioning: Techniques and recent experiences. In: GPS Trends in Precise Terrestrial, Airborne, and Spaceborne Applications: 21$^{st}$ IUGG General Assembly, IAG Symposium No. 115, Boulder, USA, July 3–4, 1995. Springer-Verlag, Berlin, pp 170–174
Madsen FB, Madsen F (1994) Realization of the EUREF89 reference frame in Denmark. Report on the Symposium of the IAG Subcommission for the European Reference Frame (EUREF) held in Warsaw 8–11 June 1994. Reports of the EUREF Technical Working Group, Muenchen, pp 270–274
MalekiL, Prestage J(2005) Applications of clocks and frequency standards: from the routine to tests of fundamental models.Metrologia42(3):S145-S153
Mander A, Bisnath S (2013) GPS-based precise orbit determination of Low Earth Orbiters with limited resources. GPS Solutions 17(4), pp587-594
Manning J, Johnston G (1995) A fiducial GPS network to monitor the motion of the Australien plate. In: Proceedings of the First Turkish International Symposium on Deformations "Istanbul-94", Istanbul, Sept. 5–9, 1995, Istanbul, pp 85–89
Mannucci A J, Wilson B D, Yuan D N, et al. (1998) A global mapping technique for GPS-derived ionospheric total electron content measurements. Radio Science, 33(3): 565-582

Mansfeld W (2004) Satellitenortung und Navigation, 2nd Edition, Wiesbanden: Vieweg Verlag, 352 p.
McCarthy DD (1996) International Earth Rotation Service. IERS conventions, Paris, 95 pp. IERS Technical Note No. 21
McCarthy DD, Capitaine N (2002) Practical Consequences of Resolution B1.6 "IAU2000 Precession-Nutation Model", Resolution B1.7 "Definition of Celestial Intermediate Pole", and Resolution B1.8 "Definition and Use of Celestrial and Terrestrial Ephemeris Origin". In: Proceedings of the IERS Workshop on the Implementation of the New IAU Resolutions, Paris, France, April 18-19, 2002 (Capitaine N et al, ed.) IERS Technical Note No. 29
McCarthy DD, Luzum BJ (1995) Using GPS to determine Earth orientation. In: GPS Trends in Precise Terrestrial, Airborne, and Spaceborne Applications: 21$^{st}$ IUGG General Assembly, IAG Symposium No. 115, Boulder, USA, July 3–4, 1995. Springer-Verlag, Berlin, pp 52–58
McCarthy DD , Petit G (Ed) (2003) International Earth Rotation Service. IERS conventions (2003), IERS Technical Note No. 32
McInnes CR (1999) Solar Sailing: Technology, Dynamics and Missions Applications, Springer Berlin
Mertikas SP, Rizos C (1997) On-line detection of abrupt changes in the carrier-phase measurements of GPS. J Geodesy 71(8):469–482
Mervart L (1995) Ambiguity resolution techniques in geodetic and geodynamic applications of the Global Positioning System. Dissertation an der Philosophisch- naturwissenschaftlichen Fakultät der Universität Bern
Mervart L, Beutler G, Rothacher M (1995) The impact of ambiguity resolution on GPS orbit determination and on global geodynamics studies. In: GPS Trends in Precise Terrestrial, Airborne, and Spaceborne Applications: 21$^{st}$ IUGG General Assembly, IAG Symposium No. 115, Boulder, USA, July 3–4, 1995. Springer-Verlag, Berlin, pp 285–289
Michel GW, Becker M, Angermann D, Reigber C, Reinhart E (2000) Crustal motion in E- and SE-Asia from GPS measurements. Earth Planets Space 52(10):713–720
Mickler D, Axelrad P, Born G (2004) Using GPS reflections for satellite remote sensing. Acta Astronautica 55(1): 39-49
Milani A, Rossi A, Vokrouhlicky D, Villani D, Bonanno C (2001) Gravity field and rotation state of Mercury from the BepiColombo Radio Science Experiments, Planetary and Space Science 49, pp 1579-1596
Milbert D(2005) Correction to `Influence of pseudorange accuracy on phase ambiguity resolution in various GPS modernization scenarios'.Navigation52(3): 121-121
Milbert D(2005) Influence of pseudorange accuracy on phase ambiguity resolution invarious GPS modernization scenarios.Navigation52(1):29-38
Miller KM (2000) A review of GLONASS. Hydrographic Journal 98:15–21
Minovitch M (1961) A method for determining interplanetary free-fall reconnaissance trajectories. Jet Propulsion Laboratory Technical Memo TM-312-130, pages 38-44 (23 August 1961)
Mirgorodskaya T (2013) GLONASS Government Policy, Status and Modernization Plans. IGNSS 2013, Gold Coast, Queensland, Australia, July 16, 2013.
Mireault Y, Kouba J, Lahaye F (1995) IGS combination of precise GPS satellite ephemerides and clock. In: GPS Trends in Precise Terrestrial, Airborne, and Spaceborne Applications: 21$^{st}$ IUGG General Assembly, IAG Symposium No. 115, Boulder, USA, July 3–4, 1995. Springer-Verlag, Berlin, pp 14–23
Mitchell S, Jackson B, Cubbedge S (1996) Navigation solution accuracy from a spaceborne GPS receiver, GPS World, 7(1996)6, 42, 44, 46–48, 50
Montenbruck O (2003) Kinematic GPS positioning of LEO satellites using ionospheric-free single frequency measurements. Aerospace Sci. Technol. 7(5): 396-405
Montenbruck O, Gill E, Kroes R (2005) Rapid orbit determination of LEO satellites using IGS clock and ephemeris products. GPS Solutions 9(3):226-235
Montenbruck O, Kroes R (2003) In flight performance analysis of the CHAMP BlackJack GPS receiver. GPS Solutions 7(2): 74-86

Montenbruck O, van Helleputte T, Kroes R, Gill E (2005) Reduced dynamic orbit determination using GPS code and carrier measurements. Aerospace Sci. Technol.9(3): 261-271

Moore T, Zhang K, Close G, Moore R (2000) Real-time river level monitoring using GPS heighting. GPS Solutions 4(2):63–67

Moreau MC, Axelrad P, Garrison JL, Long A (2000) GPS receiver architecture and expected performance for autonomous navigation in high earth orbits. J Inst Navig 47(3):191–204

Mostafa MMR(2005) Airborne GPS augmentation alternatives. Potogramm. Eng. Remote Sens.71 (5): 545+

Mostafa MMR(2005) Direct georeferencing - Airborn GPS augmentation alternatives - Part II satellite-based correction service. Potogramm. Eng. Remote Sens.71(7): 783-783

Mueller II (1964) Introduction to satellite geodesy. Frederick Ungar Publishing Co.

Murakami M (1996) Precise determination of the GPS satellite orbits and its new applications: GPS orbit determination at the Geographical Survey Institute. J Geod Soc Japan 42(1):1–14

Murray CD, Dermott SF (1999) Solar system dynamics, Cambridge University Press

Musen P (1960) The Influence of the Solar Radiation Pressure on the Motion of an Artificial Satellite, Journal of Geophysical Research, Vol. 65, No. 5, p.1391-1396

Musman S (1995) Deriving ionospheric TEC from GPS observations. In: GPS Trends in Precise Terrestrial, Airborne, and Spaceborne Applications: 21$^{st}$ IUGG General Assembly, IAG Symposium No. 115, Boulder, USA, July 3–4, 1995. Springer-Verlag, Berlin, pp 258–262

Mysen E (2009) On the predictability of unstable satellite motion around elongated elestial bodies, Astronomy & Astrophysics, Vol. 506 No.2

Niell AE (2000) Improved atmospheric mapping functions for VLBI and GPS. Earth Planets Space 52(10):703–708

Ning F, Sun Y (1992) The center form of the n-body problem generalized force potential (I): Basic equations and theoretical analysis, Astronomy, 1992 vol 2 page 8

Ning F, Sun Y (1992) The center form of the n-body problem generalized force potential (II) - 2 ≤ n ≤ 4 cases Center conformation number and shape, Astronomy, 1992 vol 3

Obana K, Katao H, Ando M (2000) Seafloor positioning system with GPS-acoustic link for crustal dynamics observations. A preliminary result from experiments in the sea. Earth Planets Space 52(6):415–423

Odijk D, Marel H van der, Song I (2000) Precise GPS positioning by applying ionospheric corrections from an active control network. GPS Solutions 3(3):49–57

O'Keefe K, Stephen J, Lachapelle G, Gonzales RA (2000) Effect of ice loading of a GPS antenna. Geomatica 54(1):63–74

Ou J (1995) On atmospheric effects on GPS surveying. In: GPS Trends in Precise Terrestrial, Airborne, and Spaceborne Applications: 21$^{st}$ IUGG General Assembly, IAG Symposium No. 115, Boulder, USA, July 3–4, 1995. Springer-Verlag, Berlin, pp 243–247

Ou JK, Wang ZJ (2004) An improved regularization method to resolve integer ambiguity in rapid positioning using single frequency GPS receivers. Chinese Sci. Bull. 49(2): 196-200

Pachelski W (1995) GPS phases: Single epoch ambiguity and slip resolution. In: GPS Trends in Precise Terrestrial, Airborne, and Spaceborne Applications: 21$^{st}$ IUGG General Assembly, IAG Symposium No. 115, Boulder, USA, July 3–4, 1995. Springer-Verlag, Berlin, pp 295–299

Pal A (2009) An analytical solution for Kepler's problem, MNRAS Vol. 396 Issue 3 (p1737-1742)

Pan M, Sjoeberg LE (1995) Unification of regional vertical datums using GPS. In: GPS Trends in Precise Terrestrial, Airborne, and Spaceborne Applications: 21$^{st}$ IUGG General Assembly, IAG Symposium No. 115, Boulder, USA, July 3–4, 1995. Springer-Verlag, Berlin, pp 94–98

Parkinson BW, Spilker JJ (eds) (1996) Global Positioning System: Theory and applications, Vol. I, II. American Institute of Aeronautics and Astronautics, Progress in Astronautics and Aeronautics, Vol. 163

Parkinson RW, Jones HM, Shapiro II (1960) Effects of Solar Radiation Pressure on Earth Satellite Orbits, Science, V. 131, Issue 3404, pp. 920-921

Pavlis EC, Beard RL (1995) The Laser Retroreflector Experiment on GPS-35 and 36. In: GPS Trends in Precise Terrestrial,Airborne, and Spaceborne Applications: 21$^{st}$ IUGG General Assembly,IAG Symposium No. 115, Boulder, USA, July 3–4, 1995. Springer-Verlag, Berlin, pp 154–158

Perozzi E, Ferraz-Mello S (Ed, 2010) Space manifold dynamics, Springer

Petovello MG, Lachapelle G (2000) Estimation of clock stability using GPS. GPS Solutions 4 (1):21–33

Pireaux S, Rozelot JP (2003) Solar quadrupole moment and purely relativistic gravitation contributions to Mercury's perihelion Advance, Astrophysics and Space Science, Vol, 284, No. 4, P1159-1194

Pireaux S, Barriot JP, Rosenblatt P (2006) (SC)RMI: A (S)emi-(C)lassical (R)elativistic (M)otion (I)integrator, to model the orbits of space probes around the Earth and other planets, Acta Astronautica, Vol, 59, Issue 4, P517-523

Pitjeva EV (2005) Relativistic Effects and Solar Oblateness from Radar Observations of Planets and Spacecraft, Astronomy Letters, Vol. 31 No. 5, p340-349

Plumb J, Larson KM, White J, Powers E (2005) Absolute calibration of a geodetic time transfer. IEEE Trans. Ultrason.Ferr.Freq. Contr.52(11):1904-1911

Psiaki ML, Powell SP, Kintner PM Jr. (2000) Accuracy of the Global Positioning System-derived acceleration vector. J Guid Control Dynam 23(3):532–538

Puglisi G, Briole P, Bonforte A (2004) Twelve years of ground deformation studies on Mt. Etna volcano based on GPS surveys. Geophys. Monogr. 143: 321-341

Rabaeijs A, Grosso D, Huang X, Qi D (2003) GPS receiver prototype for integration into system-on-chip. IEEE Trans. on Consumer Electronics 49(1): 48-58

Rajal BS, Madhwal HB (1997) Kinematic Global Positioning System survey as the solution for quick large scale mapping. Survey rev 34(265):159–162

Reigber C (1997) Geowissenschaftlicher Kleinsatellit CHAMP. GPS-Anwendungen und Ergebnisse '96: Beiträge zum 41. DVW-Fortbildungsseminar vom 7. bis 8. November 1996 am Geo-Forschungs-zentrum Potsdam, pp 266–273

Reigber C (1997) IERS und IGS: Stand und Perspektiven. GPS-Anwendungen und Ergebnisse '96: Beitraege zum 41. DVW-Fortbildungsseminar vom 7. bis 8. November 1996 am Geo-Forschungs-zentrum Potsdam, pp 34–42

Reigber C, Feissel M (1997) IERS missions, present and future. International Earth Rotation Service (ed) Report on the 1996 IERS workshop, Paris, 50 p (IERS technical note 22)

Reigber C, Schwintzer P, Luehr H (1996) CHAMP – a challenging mini-satellite payload for geoscientific research and application. Erste Geodaetische Woche, Stuttgart, 7.-12. Oktober 1996, 4 p)

Reigber C et al. (2003) Global gravity field recovery using solely GPS tracking and accelerometer data from CHAMP. Space Sci. Reviews 108(1-2): 55-66

Reigber C et al. (2005) An Earth gravity field model complete to degree and order 150 from GRACE: EIGEN-GRACE25.J. Geodyn. 39(1): 1-10

Reinhart E, Franke P, Habrich H, Schlueter W, Seeger H, Weber G (1997) Implications of permanent GPS-arrays for the monitoring of geodetic reference frames. Sixth United Regional Cartographic Conference for the America, New York, 2–6 June 1997, 10 pp

Reinking J (2009) Book review: GPS – Theory, Algorithms and Applications, Xu G 2007 2nd Ed. ZfV - Zeitschrift für Geodäsie, Geoinformation und Landmanagement, 2009, Vol. 134 (2): 122-123

Reinking J, Angermann D, Klotz J (1995) Zur Anlage und Beobachtung grossräumiger GPS-Netze für geodynamische Untersuchungen. Allgemeine Vermessungsnachrichten 102(6):221–231

Remondi BW, Brown G (2000) Triple differencing with Kalman filtering: Making it work. GPS Solu-tions 3(3):58–64

Remondi BW (2004) Computing satellite velocity using the broadcast ephemeris. GPS Solutions 8 (3): 181-183

Remondi BW, Brown RG (2004) A comparison of a Hi/Lo GPS constellation with a populated conventional GPS constellation in support of RTK: A covariance analysis. GPS Solutions 8(2): 82-92

Retscher G, Chao CHJ (2000) Precise real-time positioning in WADGPS networks. GPS Solutions 4(2):68–75

Revnivykh S (2010) GLONASS Status and Progress. ION GNSS 2010, Portland, Oregon, September 21-24, 2010.

Revnivykh S (2007) GLONASS Status, Development and Application. International Committee on Global Navigation Satellite Systems (ICG), Bangalore, India, September 4-7, 2007.

Rizos C, Han S, Chen HY (2000) Regional-scale multiple reference stations for carrier phase-based GPS positioning: A correction generation algorithm. Earth Planets Space 52(10):795–800

Rizos C, Han S, Ge L, Chen HY, Hatanaka Y, Abe K (2000) Low-cost densification of permanent GPS networks for natural hazard navigation: First tests on GSI's GEONET network. Earth Planets Space 52(10):867–871

Roberts GW, Dodson AH, Ashkenazi V (2000) Experimental plan guidance and control by kinematic GPS. Proc. Institution of Civil Engineers, Civil Engineering 138(1):19–25

Rochus P (2010) Book Review: Orbits, Xu G 2008. MatheSciNet, American Mathematical Society, MR2494776 (2010a:70033) 70M20 (74F05 74F15)

Rochus P (2008) Private communication, for review upon request under gcxu@sdu.edu.cn

Rodolpho Vilhena De Moraes (1981) Combined solar radiation pressure and drag effects on the orbits of artificial satellites, Celestial Mechanics, V. 25, Issue 3, pp.281-292

Rothacher M, Gurtner W, Schaer S (1995) Azimuth- and elevation-dependent phase center corrections for geodetic GPS antennas estimated from calibration campaigns. In: GPS Trends in Precise Terrestrial, Airborne, and Spaceborne Applications: 21$^{st}$ IUGG General Assembly, IAG Symposium No. 115, Boulder, USA, July 3–4, 1995. Springer-Verlag, Berlin, pp 333–338

Rothacher M, Mervart L (1996) Bernese GPS Software Version 4.0. Astronomical Institute of Univer-sity of Bern

Rothacher M, Schaer S (1995) GPS-Auswertetechniken. Schriftenreihe des Deutschen Vereins für Vermessungswesen, Bd. 18, pp 107–121

Rozelot JP and Damiani C (2011) History of solar oblateness measurements and interpretation, The European Physical Journal H 36, 407-436

Rozelot JP, Damiani C, Lefebvre S, Kilcik A, Kosovichev AG (2011) A brief history of the solar oblateness. A Review. Journal of Atmospheric and Solar-Terrestrial Physics 73, 2-3 (2011) 241-250

Rozelot JP, Godier S, Lefebvre S (2001) On the Theory of the Oblateness of the Sun, Solar Physics 198, 223-240

Rozelot JP, Pireaux S, Lefebvre S, Ajabshirizadeh A (2004) Solar Rotation and Gravitational Moments: Some Astrophysical outcomes, Proceedings of the 14/GONG 2004 Workshop, New Haven, Connecticut, USA

Roßbach U (2006) Positioning and Navigation Using the Russian Satellite System GLONASS, Universität der Bundeswehr München, URN: de:bvb:707-648

Rummel R, Gelderen M van (1995) Meissl scheme: Spectral characteristics of physical geodesy. Manuscr Geodaet 20(5):379–385

Rummel R, Ilk KH (1995) Height datum connection – the ocean part. Allgemeine Vermessungsnachrichten 102(8/9):321–330

Rush J (2000) Current issues in the use of the global positioning system aboard satellites. Acta Astronaut 47(2–9):377–387

Rutten JAc (2004) Book review: GPS – Theory, Algorithms and Applications, Xu G 2003, European Journal of Navigation Vol. 2 Number 1(2004) 94

Saad Nadia A, Khalil Kh I, Amin Magdy Y (2010) Analytical Solution for the Combined Solar Radiation Pressure and Luni-Solar Effects on the Orbits of High Altitude Satellites, The Open Astronomy Journal, vol. 3, issue 1, pp. 113-122

Salazar D, Hernandez-Pajares M, Juan-Zornoza JM, Sanz-Subirana J, Aragon-Angel A (2011) EVA: GPS-based extended velocity and acceleration determination. J Geodesy 85(6), pp329-340

Sandlin A, McDonald K, Donahue A (1995) Selective availability: To be or not to be? GPS World 6(9): 44–51

Satirapod C, Wang J, Rizos C (2003) Comparing different Global Positioning System data processing techniques for modeling residual systematic errors. J. Surv. Eng. ASCE 129(4): 129-135

Schaal RE, Netto NP (2000) Quantifying multipath using MNR ratios. GPS Solutions 3(3):44–48

Schaffrin B (1995) On some alternative to Kalman filtering. In: Sanso F (ed) Geodetic theory today. Springer-Verlag, Berlin, pp 235–245

Scherrer R (1985) The WM GPS primer. WM Satellite Survey Company, Wild, Herrbrugg, Switzerland

Schildknecht T, Dudle G (2000) Time and frequency transfer: High precision using GPS phase measurements. GPS World 11(2):48–52

Schmid R, Rothacher M, Thaller D, Steigenberger P(2005) Absolute phase center corrections of satellite and receiver antennas. Impact of global GPS solutions and estimation of azimuthal phase center variations of the satellite antenna. GPS Solutions 9(4): 283-293

Schneider M (1988) Satellitengeodaesie. Wissenschaftsverlag, Mannheim

Schneider M, Cui CF (2005) Theoreme über Bewegungsintegrale und ihre Anwendung in Bahntheorien, Bayerischen Akad Wiss, Reihe A, Heft Nr. 121, 132pp, München

Schoene T, Reigber C, Braun A (2003) GPS offshore buoys and continuous GPS control of tide gauges. Int. Hydrogr. Rev. 4(3): 64-70

Scholl H, Marzari F, Tricarico P (2005) Dynamics of Mars Trojans, Icarus 175, 397-408

Schutz BE (2000) Numerical studies in the vicinity of GPS deep resonance. Advances in the Astronautical Sciences 105(1):287–302

Schwarz K-P, El-Sheimy N (1995) Multi-sensor arrays for mapping from moving vehicles. In: GPS Trends in Precise Terrestrial, Airborne, and Spaceborne Applications: 21st IUGG General Assembly, IAG Symposium No. 115, Boulder, USA, July 3–4, 1995. Springer-Verlag, Berlin, pp 185–189

Schwintzer P, Kang Z, Reigber C (1995) GPS satellite-to-satellite tracking for TOPEX/Poseidon precise orbit determination and gravity field model improvement. J Geodyn 20(2):155–166

Seeber G (1996) Grundprinzipien zur Vermessung mit GPS. Vermessungsingenieur 47(2):53–64

Seeber G (2003) Satellite Geodesy: Foundations, Methods, and Applications, Berlin: Walter de Gruyter, xx+589

Seeger H, Franke P, Schlueter H, Weber G (1997) The significance and results of permanent GPS arrays. In: Proceedings Fourteenth United Nations Regional Cartografic Conference for Asia and the Pacific, Bangkok, 3–7 February 1997. 10 p

Seitz K, Urakawa MJ, Heck B, Krueger C(2005) Zu jeder Zeit an jedem Ort – Studie zur Verfuegbarkeit und Genauigkeit von GPS-Echtzeitmessungen im SAPOS-Service HEPS.Z. Geod.Geoinf.Landmanag.130(1):47-55

Shank C (1998) GPS navigation message enhancements. GPS World 7(4):38–44

Shapiro II (1999) A century of relativity, Reviews of Modern Physics, Vol. 71, No. 2

Shapiro II (1963) The prediction of satellite orbits, in Maurice Roy (ed) Dynamics of satellites, pp257-312

Shaw M (2011) GPS Modernization: On the Road to the Future GPS IIR / IIR-M and GPS III. UN/ UAE/ US Workshop On GNSS Applications, Dubai, 2011.

Shen YZ, Chen Y, Zheng DH (2006) A Quaternion-based geodetic datum transformation algorithm, J. Geodesy 80:233-239

Shen Y, Li B, Chen Y (2011) An iterative solution of weighted total least-squares adjustment. J Geodesy 85(4), pp229-238

Shi J, Xu C, Li Y, Gao Y (2015) Impact of real-time satellite clock errors on GPS precise point positioning-based troposphere zenith delay estimation. J Geodesy 89(8), pp747-756

Sien-Chong Wu, William GM (1993) An optimal GPS data processing technique for precise positioning. IEEE Transactions on Geoscience and Remote Sensing, 31:146–152, January 1993.

Sigl R (1978) Geodätische Astronomie. Wichmann Verlag, Karlsruhe

Sjoeberg LE (1998) On the estimation of GPS phase ambiguities by triple frequency phase and code data. ZfV 1235:162–163

Skaloud J, Schwarz KP (2000) Accurate orientation for airborne mapping systems. Photogramm Eng Rem S 66(4):393–401

Smith AJE, Hesper ET, Kuijper DC, Mets GJ, Visser PN, Ambrosius BAC, Wakker KF (1996) TOPEX/ Poseidon orbit error assessment. J Geodesy 70:546–553

Snow KB, Schaffrin B (2003) Three-dimensional outlier detection for GPS networks and their densification via the BLIMPBE approach. GPS Solutions 7(2): 130-139

Spilker JJ (1996) GPS navigation data. In: Parkinson BW, Spilker JJ (eds) Global Positioning System: Theory and applications, Vol. I, Chapter 4

Springer TA, Beutler G, Rothacher M (1999) Improving the orbit estimates of GPS satellites. J Geodesy 73:147–157

Steigenberger P, Boehm J, Tesmer V (2009) Comparison of GMF/GPT with VMF1/ECMWF and implications for atmospheric loading. J Geodesy 83:943

Sterle O, Stopar B, Preseren PP (2015) Single-frequency precise point positioning: an analytical approach. J Geodesy 89(8), pp 793-810

Stoew B, Elgered G (2004) Characterization of atmospheric parameters using a ground based GPS network in north Europe. J. Meteor. Soc. Japan 82(1B): 587-596

Stowers D, Moore A, Iijima B, Lindqwister U, Lockhart T, Marcin M, Khachikyan R (1996) JPL-supported permanent tracking stations. International GPS Service for Geodynamics: 1996 annual report, Nov. 1997, Pasadena, pp 409–420

Sturrock PA, Gilvarry JJ (1967) Sloar Oblateness and Magnetic Field, Nature, Vol 216

Sun HP, Ducarme B, Dehant V (1995) Effect of the atmospheric pressure on surface displacements. J Geodesy 70:131–139

Tapley BD, Schutz BE, Eanes RJ, Ries JC, Watkins MM (1993) Lageos laser ranging contributions to geodynamics, geodesy, and orbital dynamics. In: Contributions of Space Geodesy to Geodynamics: Earth Dynamics, Geodyn. Ser. 24:147–174

Testoyedov N (2015) Space Navigation in Russia: History of Development. United Nations / Russian Federation Workshop on the Applications of Global Navigation Satellite Systems, Krasnoyarsk, May 18-22, 2015.

Tetreault P, Kouba J, Heroux P, Legree P (2005) CSRS-PPP: An Internet service for GPS user access to the Canadian Spatial reference Frame. Geomatica 59(1): 17-28

Teunissen P (1996) An analytical study of ambiguity decorrelation using dual frequency code and carrier phase. J Geodesy 70(8):515–528

Teunissen P (1997) Closed form expressions for the volume of the GPS ambiguity search spaces. Artificial Satellites, Journal of Planetary Geodesy 32(1):5–20

Teunissen P (2005) GNSS ambiguity resolution with optimally controlled failure-rate. Artificial Satellites40(4):219-227

Teunissen P (1996) GPS carrier phase ambiguity fixing concepts. In: Kleusberg A, Teunissen PJG (eds) GPS for geodesy. Springer-Verlag, Berlin

Teunissen P (1998) Minimal detectable biases of GPS data. J Geodesy 72:630–639

Teunissen P (2004) Penalized GNSS Ambiguity resolution. J. Geod. 78(4-5): 235-244

Teunissen P (2003) Towards a unified theory of GPS ambiguity resolution. J. GPS 2(1): 1-12

Teunissen P, Jonge P, Tiberius C (1997) Performance of the LAMBDA method for fast GPS ambiguity resolution. J Inst Navig 44(3):373–383

Teunissen PJG, Kleusberg A (1996) GPS observation equations and positioning concepts. In: Kleusberg A, Teunissen PJG (eds) GPS for geodesy. Springer-Verlag, Berlin

Theakstone WH, Jacobsen FM, Knudsen NT (2000) Changes of snow cover thickness measured by conventional mass balance methods and by global positioning system surveying. Geografiska Annaler 81(A4):767–776

Tiberius CCJM, Kenselaar F (2000) Estimation of the stochastic model for GPS code and phase observables. Survey rev 35(277):441–454

Tiberius C (2003) Standard Positioning Service: Handheld GPS receiver accuracy. GPS World 14 (2): 46-51

Timmen L, Ye X (1997) SAR-Interferometrie unterstützt durch GPS zur Überwachung von Erdoberflächendeformationen. GPS-Anwendungen und Ergebnisse '96: Beiträge zum 41. DVW-Fortbildungsseminar vom 7. bis 8. November 1996 am GeoForschungszentrum Potsdam, pp 104–114

Timmen L, Bastos L, Boebel T, Cunha S, Forsberg R, Gidskehaug A, Hehl K, Meyer U, Nesemann M, Olesen AV, Rubek F, Xu G (1998) The European Airborne Geoid Mapping System for Coastal Oceanography (AGMASCO). Progress in Geodetic Science at GW 98. In: Proceedings of the Geodetic Week 1998. University of Kaiserslautern, Germany, Shaker press, Aachen

Touma JR, Tremaine S, Kazandjian MV (2009) Gauss's methods for secular dynamics, softened, MNRAS Vol. 394 Issue 1 (p 1085-1108)

Tsai C, Kurz L (1983) An adaptive robustifing approach to Kalman filtering. Automatica 19:279–288

Tscherning C, Rubek F, Forsberg R (1997) Combining airborne and ground gravity using collocation, Scientific Assembly of the International Association of Geodesy in conjunction with 28th Brazilian Congress of Cartography; Rio de Janeiro, 3–9 September 1997, 6 pp

Tsujii T, Harigae M, Inagaki T, Kanai T (2000) Flight tests of GPS/GLONASS precise positioning versus dual frequency KGPS profile. Earth Planets Space 52(10):825–829

Tregoning P, van Dam T (2005) Atmospheric pressure loading corrections applied to GPS data at the observing level. Geophys. Res. Lett.32(L22310):1-4

Urlichich Y, Subbotin V, Stupak G, et al. (2010) GLONASS Developing Strategy. ION GNSS 2010, the 23$^{rd}$ International Technical Meeting of the Institute of Navigation, Portland, Oregon, September 21-24, 2010.

Urlichich Y, Subbotin V, Stupak G, et al. (2011) Innovation: GLONASS Developing strategies for the Future, GPS World, April, pp42-49

UrschlC, Dach R, Hugentobler U, Schaer S, Beutler G(2005) Validating ocean tide loading models using GPS.J. Geod.78(10):616-625

van Dam T, Francis O (eds) (2004) The state of GPS vertical positioning precision: Separation of Earth processes by space geodesy. Centre Europeen de Geodynamique et de Seismologie, Luxembourg, Cahiers 23: xxii+176.

van Sickle J (2003) GPS for Land Surveyors (2nd ed.) Chelsea, MI, Sleeping Bear Press: xii+284

Verhagen S (2004) Integer ambiguity validation: An open problem? GPS Solutions 8(1): 36-43

Visser PNAM, IJssel J van den (2000) GPS-based precise orbit determination of the very low Earth-orbiting gravity mission GOCE. J Geodesy 74(7/8):590–602

Vittorini LD, Robinson B (2003) Receiver frequency standards. Optimizing indoor GPS performance. GPS World 14(11): 40-42, 44, 46-48

Vokrouhlický D, Milani A (2000) Direct Solar Radiation Pressure on orbits of small near-Earth asteroids: observable effects? A&A, v.362, p.746-755

Vokrouhlicky D, Farinella P, Mignard F (1993) Solar radiation pressure perturbations for Earth satellites, I: A complete theory including penumbra transitions, A&A v. 280, p. 295-312

Vokrouhlicky D, Farinella P, Mignard F (1994) Solar radiation pressure perturbations for Earth satellites, II. an approximate method to model penumbra transitions and their long-term orbital effects on LAGEOS, A&A V. 285, 333-343

Vollath U, Birnbach S, Landau H (1998) An analysis of three-carrier ambiguity resolution (TCAR) technique for precise relative positioning in GNSS-2. Proceedings of ION GPS-98, The 11$^{th}$ International Technical Meeting of the Satellite Division of the Institute of Navigation, Nashville, Tennessee, USA, September 15-18, pp. 417-426

Wagner C, Klokocnik J (2003) The value of ocean reflections of GPS signals to enhance satellite altimetry: data distribution and error analysis. J. Geod. 77(3-4): 128-138

Wagner J, Bauer M (1997) GPS-Vermessung mit Echtzeitauswertung (RTK-Vermessung): ein Beitrag zur Einschätzung der Praxistauglichkeit und Praxisrelevanz. Vermessungsingenieur 48 (2):87–92

Wagner JF (2005) GNSS/INS integration: still an attractive candidate for automatic landing systems? GPS Solutions9(3):179-193

Wang CM, Hajj G, Pi XQ, Rosen IG, Wilson B (2004) Development of the Global Assimilative Ionospheric Model. Radio Sci. 39(1) RS1S06: 1-11

Wang G, Wang H, Xu G (1995) The principle of the satellite altimetry. Science Press, Peking, ISBN 7-03-004499-1/P.797, 390 p, (in Chinese)

Wang J (2000) An approach to GLONASS ambiguity resolution. J Geodesy 74(5):421–430

Wang J, Rizos C, Stewart MP, Leick A (2001) GPS and GLONASS integration-Modeling and Ambiguity Resolution Issues. GPS Solutions 5(1), 55-64.

Wang J, Steward MP, Tsakiri M (1999) Adaptive Kalman filtering for integration of GPS with GLONASS and INS. Presentation in the XXII$^{th}$ IUGG, Birmingham, England

Wang J, Stewart MP, Tsakiri M (2000) A comparative study of the integer ambiguity validation procedures. Earth Planets Space 52(10):813–817

Wang JG (1997) Filtermethoden zur fehlertoleranten kinematischen Positionsbestimmung. Neubiberg, 135 S

Wanninger L (1999) Der Einfluss ionosphärischer Störungen auf die präzise GPS-Positionierung mit Hilfe virtueller Referenzstationen. ZfV 10:322–330

Wanninger L (2003) Permanent GPS-Stationen als Referenz fuer praezise kinematische Positionierung. Photogramm. Fernerkund. Geoinf. 7(4): 343-348

Wanninger L (2003) Virtuelle GPS-Referenzstationen fuer großraeumige kinematische Anwendungen. Z.f. Verm.wessen 128(3): 196-202

Ware RH, Fulker DW, Stein SA, Anderson DN, Avery SK, Clark RD, Droegemeier KK, Kuettner JP, Minster J, Sorooshian S (2000) Real-time national GPS networks: Opportunities for atmospheric sensing. Earth Planets Space 52(11):901–905

Warnant R, Pottiaux E (2000) The increase of the ionospheric activity as measured by GPS. Earth Planets Space 52(11):1055–1060

Wayte R (2010) On the Estimated Precession of Mercury's Orbit, submitted to PMC Physics A

Weber G (1994) Initial operational capability für das GPS: aktuelle Entwicklungen der US Satellitennavigation. SATNAV 94: Satellitennavigationssysteme – Grundlagen und Anwendungen, DGON-Seminar, Hamburg 24.–26. Oktober 1994, pp 1–14

Weber R (1996) Monitoring Earth orientation variations at the Center for Orbit Determination in Europe (CODE). Oesterr Z Vermess Geoinf 84(3):269–275

Weiland C (2010) Computational space flight mechanics, Springer

Wickert J, Schmidt T, Beyerle G, Konig R, Reigber C (2004) The radio occultation experiment aboard CHAMP: Operational data analysis and validation of vertical atmospheric profiles. J. Meteor. Soc. Japan 82(1B): 381-395

Wicki F (1998) Robuste Schätzverfahren für die Parameterschätzung in geodätischen Netzen. Technische Hochschule Zürich

Wieser A, Brunner FK (2000) An extended weight model for GPS phase observations. Earth Planets Space 52(10):777–782

Williams SDP (2003) Offsets in Global Positioning System time series. J. Geophys. Res 108(B6): ETG12/1-13

Witte TH, Wilson AM (2004) Accuracy of non-differential GPS for the determination of speed over ground. J. Biomechanics 37(12): 1891-1898

Wolverton M (2004) The depths of space: the Pioneer planetary probes, Joseph Henry Press, ISBN 0-309-09050-4

Won J-H,Lee J-S(2005) A note on the group delay and phase advance phenomenon associated with GPS signal propagation through theionosphere.Navigation52(2):95-97

Wu CC, Kuo HC, Hsu HH, Jou BJD (2000) Weather and climate research in Taiwan: Potential application of GPS/MET data. Terr Atmos Ocean Sci 11(1):211–234

Wu J, Lin SG (1995) Height accuracy of one and a half centimetres by GPS rapid static surveying. Int J Remote Sens 16(15):2863–2874

Wuebbena G (1991) Zur Modellierung von GPS-Beobachtungen für die hochgenaue Positionsbestimmung. Unversität Hannover

Wuebbena G, Seeber G (1995) Developments in real-time precise DGPS applications: concepts and status. In: GPS Trends in Precise Terrestrial, Airborne, and Spaceborne Applications: 21$^{st}$ IUGG General Assembly, IAG Symposium No. 115, Boulder, USA, July 3–4, 1995. Springer-Verlag, Berlin, pp 212–216

Xia Y, Michel GW, Reigber C, Klotz J, Kaufmann H (2003) Seismic unloading and loading in northern central Chile as observed by differential Synthetic Aperture Radar Interferometry (D-INSAR) and GPS. Int. J. Remote Sensing 24(22): 4375-4391

Xu CJ, Liu JN, Song CH, Jiang WP, Shi C (2000) GPS measurements of present-day uplift in the Southern Tibet. Earth Planets Space 52(10):735–739

Xu G (1984) Very long baseline interferometry and tidal theories. The Institute of Geodesy and Geophysics, Chinese Academy of Sciences, M.Sc. Thesis No. 84011, (in Chinese)

Xu G (1999): KGsoft – Kinematic GPS Software – Software User Manual, Version of 1999, Kort & Matrikelstyrelsen (National Survey and Cadastre – Denmark), ISBN 87-7866-158-7, ISSN 0109-1344, 35 pages, in English

Xu G (2003a) A diagonalization algorithm and its application in ambiguity search. J. GPS 2(1): 35-41

Xu G (2011): GPS – Theory, Algorithms and Applications, 1st Ed in Chinese, Tschinghua University Press Peking, ISBN 978-7-302-27164-2, 332 pages, translated from Xu (2007): GPS – Theory, Algorithms and Applications, 2nd Ed, Springer, in translation series of Space Technology of Springer, translation organised by the Peking Institute of Satellite Controlling and Telecommunication, translated by Qiang Li, Guangjun Liu, Hailiang Yu, Bo Li, and Haiying Luo, proved by Xurong Dong.

Xu, G. (2014): GPS – Theory, Algorithms and Applications, 1st Ed in Persian, Iran Technical University Press, ISBN 978-7-302-27164-2, 332 pages, translated from Xu (2007): GPS – Theory, Algorithms and Applications, 2nd Ed, Springer, translation organized by the Iran Technical University.

Xu G, Hehl K, Angermann D (1994) GPS software development for use in aerogravimetry: Strategy, realisation, and first results. In: Proceedings of ION GPS-94, pp 1637–1642

Xu G, Timmen L (1997) Airborne gravimetry results of the AGMASCO test campaign in Braunschweig. Geodetic Week Berlin '97, Oct. 6–11, 1997, electronic version published in http://www.geodesy.tu-berlin.de

Xu P, Ando M, Tadokoro K (2005) Precise, three-dimensional seafloor geodetic deformation measurements using differential techniques. Earth, Planetsand Space57(9): 795-808

Xu P (1995) Estimating the state vector in observable and singular hybrid INS/GPS systems without the knowledge of initial conditions. Bull Geod Sci Affini 54(4):389–406

Xu Y, Yang Y, Xu G, Jiang N (2013) Ionospheric delay in the Antarctic GPS positioning. Journal of Beijing University of Aeronautics and Astronautics, 39(10), pp1370-1375

Xu Y, Yang Y, Xu G (2012) Precise determination of GNSS trajectory in the Antarctic airborne kinematic positioning. In: Proceedings / China Satellite Navigation Conference (CSNC) 2012: revised selected papers, (Lecture Notes in Electrical Engineering; vol. 159), Springer, pp 95-105

Yang M (2005) Noniterative method of solving the GPS doubled-differenced pseudorange equations. J. Surv. Eng. ASCE 131(4):130-134

Yang Y (1993) Robust estimation and its applications. Bayi Publishing House, Peking

Yang Y (1994) Robust estimation for dependent observations. Manuscr Geodaet 19:10–17

Yang Y, Gao W (2010) Robust Kalman filtering with constraints: a case study for integrated navigation. J Geodesy 84(6), pp373-381

Yang YX, Xu TH, Song LJ (2005) Robust estimation of variance components with application in global positioning system network adjustment. J. Surv.Eng.ASCE131(4): 107-112

Yang Y, Zha M, Song L, Wei Z et al. (2005) Combined adjustment project of national astronomical geodetic networks and 2000' national GPS control network. Progress Natural Sci.15(5):435-441

Yeh TK, Chen CS (2006) Clarifying the relationship between the clock errors and positioning precision of GPS receiver, VI Hotline-Marussi Symposium of Theoretical and Computational Geodesy Wuhan.

Yeh TK, Chen CS, Lee CW (2004) Sensing of precipitable water vapor in atmosphere using GPS technology. Boll. Geod. Sci. Affini 63(4): 251-258

Yi ZH, Li GY, Luo YJ, Xia Y, Zhao HB (2010) Common orbital plane restricted three-body problem and its application - Science in China: G Series, 2010

Yonetoku D, Murakami T, Gunji S, et al (2010) Gamma-Ray Burst Polarimeter – GAP – aboard the Small Solar Power Sail Demonstrator IKAROS, arXiv: 1010.5305v1 [astro-ph.IM] 26 Oct 2010

Yoon JC, Lee BS, Choi KH (2000) Spacecraft orbit determination using GPS navigation solutions. Aerosp Sci Technol 4(3):215–221

Yuan YB, Ou JK (2003) Preliminary results and analyses of using IGS GPS data to determine global ionospheric TEC. Progress Natural Sci. 13(6):446-450

Yuan YB, Ou JK (2004) Ionospheric eclipse factor method (IEFM) for determining the ionospheric delay using GPS data. Progress Natural Sci. 14(9): 800-804

Yunck TP, Melbourne WG (1995) Spaceborne GPS for earth science. In: GPS Trends in Precise Terrestrial, Airborne, and Spaceborne Applications: 21$^{st}$ IUGG General Assembly, IAG Symposium No. 115, Boulder, USA, July 3–4, 1995. Springer-Verlag, Berlin, pp 113–122

Zehentner N, Mayer-Guerr T (2015) Precise orbit determination based on raw GPS measurements. J Geodesy, DOI: 10.1007/s00190-015-0872-7.

Zhang Q, Qiu H (2004) A dynamic path search algorithm for tractor automatic navigation. Trans. ASAE 47(2): 639-646

Zhang Q, Moore P, Hanley J, Martin S (2006) Auto-BAHN: Software for Near Real-time GPS Orbit and Clock Computation.

Zhang X, Forsberg R (2009) Assessment of long-range kinematic GPS positioning errors by comparison with airborne laser altimetry and satellite altimetry. J Geodesy 81(3), pp201-211

Zhang W, Cannon M, Julien O, Alves P (2003) Investigation of combined GPS/GALILEO cascading ambiguity resolution schemes. Proceedings of ION GPS/GNSS 2003, The 16$^{th}$ International Technical Meeting of the Satellite Division of the Institute of Navigation, Portland, OR, USA, September 9-12, pp. 2599-2610

Zhao CM, Ou JK, Yuan YB (2005) Positioning accuracy and reliability of GALILEO, integrated GPS-GALILEO system based on single positioning model. Chinese Sci. Bull.50 (12): 1252-1260

Zhao CM, Yuan YB, Ou JK, Chen JP(2005) Variation properties of ionospheric eclipse factor and ionospheric influence factor. Progress Natural Sci.15(6): 573-576

Zheng DW, Zhong P, Ding XL, Chen W(2005) Filtering GPS time-series using aVondrak filter and cross-validation.J.Geod.79(6-7): 363-369

Zhou J (1989) Classical theory of errors and robust estimation. Acta Geod Cartogr Sinica 18:115–120

Zhu J (1996) Robustness and the robust estimate. J Geodesy 70(9):586–590

Zhu SY (1997) GPS-Bahnfehler und ihre Auswirkung auf die Positionierung. GPS-Anwendungen und Ergebnisse '96: Beiträge zum 41. DVW-Fortbildungsseminar vom 7. bis 8. November 1996 am GeoForschungszentrum Potsdam, pp 219–226

Zhu S, Reigber C, Koenig (2004) Integrated adjustment of CHAMP, GRACE, and GPS data. J. Geod. 78(1-2): 103-108

Zhu WY, Fu Y, Li Y (2003) Global elevation vibration and seasonal changes derived by the analysis of GPS height. Science in China, Ser. D 46(8): 765-778

Ziebart M, Cross P (2003) LEO GPS attitude determination algorithm for a micro-satellite using boon-arm deployed antennas. GPS Solutions 6(4): 242-256

Zizka J, Vokrouhlicky D (2011) Solar radiation pressure on (99942) Apophis, Icarus, V. 211, Issue 1, p. 511-518

# Index

**A**

A priori constraint, 182, 183, 220, 221, 226, 257, 258
A priori datum, 173, 182, 183, 187, 220, 221, 226, 301, 304
A priori information, 219, 220, 225, 257, 265, 274, 437, 438
Accuracy, 4, 5, 7–9, 11, 13, 79, 88, 94, 102–104, 108, 110, 111, 114, 121, 122, 150, 192, 212, 236, 293, 306, 311, 321, 326, 329, 331, 332, 337, 338, 391, 394, 396, 397, 440
Adams algorithms, 391, 395, 397
Adaptively robust Kalman filter, 187, 212, 217, 225
Additional perturbation, 365–367
Adjustment, 1–3, 136, 149, 170, 183, 187, 191–196, 198, 201, 203, 204, 208, 209, 212, 218, 220, 221, 224, 225, 231, 232, 234–238, 271, 276, 296, 300, 306, 308, 330, 331, 360–362, 365, 379, 383, 440
Airborne, 91, 103, 104, 117, 120, 287, 288, 313, 324, 329, 337
Aircraft, 120, 224, 225, 323, 325, 329–331, 334, 337, 339
Altimetry, 117, 118, 120, 329, 337
Ambiguity, 243, 244
  criterion, 224, 231, 232, 236, 238, 239, 241, 243–245, 247, 248, 330
  fixing, 168, 171, 196, 231, 236, 241, 243, 245–247, 251–253, 255, 257, 309, 311, 330
  function, 134, 229, 232, 247–250
  ionospheric equation, 147, 252, 256, 257, 291, 295, 299
  resolution, 231, 232, 245–247, 251, 252, 255, 257–259, 311, 337
  search, 224, 229, 231, 232, 234–236, 238, 239, 241–245, 247, 257, 330, 339

Antenna phase centre, 122, 125, 126
Anti-spoofing (AS), 63, 121, 232
Apogee, 43
Apparent sidereal time, 25, 27
Argument
  of latitude, 7, 48, 50
  of perigee, 43, 44, 369, 413, 419, 429
Ascending node, 3, 7, 9, 26, 27, 30, 40, 43, 44, 50, 365, 369, 370, 413, 419, 420, 426, 430
Astronomical coordinate system, 23
Atmospheric
  drag, 49, 134, 362, 364, 397, 406, 434
  pressure, 81, 84
Attraction, 353
Azimuth, 23, 70, 71, 74, 87, 108, 139, 140, 331

**B**

Barycentre, 33
Barycentric Dynamic Time (TDB), 33
Baseline, 89, 103, 110, 115, 154, 155, 158, 160, 178, 181, 249, 251, 252, 255, 266–269, 271, 274, 277, 278, 287–289, 327, 331, 334, 338, 379
BeiDou Navigation Satellite System (BDS), 1, 12, 11, 12, 35, 53
Block-wise least squares adjustment, 187, 196, 198, 203, 276, 296
Broadcast ephemerides, 3, 49, 52, 122
Broadcast ionospheric model, 70, 72

**C**

Carrier phase, 13, 55, 57, 66, 116, 133, 253, 294, 298, 321
Cartesian coordinates, 119, 343, 351, 399
Celestial Reference Frame (CRF), 24
Central force, 37, 39, 49, 350, 410

Clock
  bias, 56, 111, 265
  drift, 303, 304
  error, 51, 56, 57, 59, 61, 99, 110–114, 116, 119, 138, 139, 153, 156, 172, 176, 179, 184, 225, 266, 267, 270, 271, 274, 278, 280, 291–297, 299, 301, 302, 307, 318, 438
  frequency, 97, 304
  offset, 52, 99
  parameter, 265, 295
Code
  delay, 63
  phase combination, 200, 246
  pseudorange, 133, 161, 292, 298
  smoothing, 152
Cofactor matrix, 189, 276
Combining algorithms, 163–171, 264, 265, 275, 276, 279, 439
Conditional least squares adjustment, 187, 193, 195, 219, 232, 234
Constrained adjustment, 187, 219
Conventional International Origin (CIO), 17, 24
Conventional Terrestrial System (CTS), 17
Coordinate
  system, 11, 13, 17, 21–24, 28, 38, 47, 49, 51, 97, 135, 307, 345, 360, 361, 365, 368, 371, 398, 414, 430
  transformation, 28, 32, 135, 317, 373
Correlation, 55, 94, 121, 136, 155, 160, 178, 273, 438
Covariance
  matrix, 141, 147, 155, 157, 160, 164, 170, 208, 213, 306
  propagation, 133, 141, 207, 307
Cowell algorithms, 387, 394, 397
Cycle slip
  detection, 151, 152, 229, 230, 291

**D**

Data
  combination, 2, 133, 142, 144, 171, 298
  condition, 271, 274, 275, 277, 278
  differentiation, 133, 152, 280
  processing, 2, 87, 109, 111, 116, 160, 168, 170–172, 182, 198, 224, 225, 277, 291, 304, 320, 329, 439
Diagonalisation, 205, 206, 225, 448
Differencing algorithms, 133, 160, 172, 263, 265, 275, 278, 439, 440

Differential
  Doppler, 133, 150, 151, 305, 317
  equation, 42, 106, 342, 385, 396, 411
  GPS, 104, 122, 182, 184, 296, 301, 326, 327
  phases, 152, 230
  positioning, 301, 326
Dilution of Precision (DOP), 13, 307
Disturbed Satellite Motion, 49
Disturbing
  force, 49, 355, 357, 394, 397, 402
  potential, 350, 365, 413, 429
Doppler
  data, 111, 148, 230, 302, 304
  effect, 60
  frequency shift, 60
  integration, 133, 150, 151, 230
Double difference, 114, 156–158, 160, 179, 182, 253, 266, 301, 328, 438
Drag, 291, 341, 348, 364, 407
Dry component, 81
Dual frequency, 322
Dynamic time, 33

**E**

Earth
  Centred–Earth–Fixed (ECEF), 17
  Centred–Ecliptic–Inertial Coordinate System (ECEI), 24
  Centred–Inertial system (ECI), 24
  Centred–Space–Fixed (ECSF), 24, 38, 47, 49, 352, 357, 361, 365, 369
  gravitational field, 134, 348, 352
  potential, 350
  rotation, 20, 25, 27, 28, 32, 57, 302, 366
  rotational effect, 97, 98
  tide displacement, 99, 101, 104
Eccentric anomaly, 44, 46, 99, 369, 414, 430
Eccentricity, 43, 44, 50, 99, 363, 413, 429
Ecliptic, 24, 26, 369, 370
Electronic density, 65, 66, 170
Elevation, 70, 71, 88, 92, 331
Ellipsoidal
  coordinates, 18
  height, 18
  mapping function, 75, 85
Emission time, 55, 57, 58, 111, 122, 137, 302
Ephemerides
  of satellite, 3
  of the Sun, the Moon and planets, 102, 316, 341, 368, 369

Equatorial
  plane, 17, 32, 33, 39, 43, 416
  system, 25
Equivalence
  theorem, 2, 274, 276, 287
Equivalent criterion, 231, 232, 241, 242, 244
Equivalent observation equation, 172–175, 187, 204, 206, 225, 240, 281, 283, 284, 290, 451

**F**

Fictitious observations, 219
Flight state monitoring, 324, 331
Float
  ambiguity, 251, 257
  solution, 234, 235, 237, 239, 242, 255
Frequency
  drift, 112, 303
  effect, 97
  offset, 52, 114
  reference, 114
Fundamental frequency, 3, 97, 253

**G**

Gain matrix, 207, 209
Galileo, 1, 2, 9, 10, 13, 14, 22, 35, 53, 246, 313
Galileo carrier phase ambiguity resolution, 245
Gaussian equations, 410, 412, 415, 416, 431, 434
Gauss-Xu equations
  derivation, 411, 431
General criterion, 224, 231, 237–240, 242, 243, 245, 246, 330
General relativity, 94, 96, 98, 99, 341
Geocentric
  distance, 50, 98, 100, 101, 354, 356, 358, 361, 363
  latitude, 18, 74, 363
Geometric co-mapping function, 86
Geometric dilution of precision (GDOP), 307
Geometric mapping function, 74, 75
Geometry-free combination, 145, 147–149, 164, 165, 167, 280
GLONASS, 1, 7–9, 13, 14, 20, 22, 35, 52, 53, 112, 322
GNSS, 1, 7–9, 13, 80, 90, 112–114, 171, 218, 281, 287, 296, 308, 309, 322, 440, 441
GPS, 1–10, 13, 14, 22–24, 33–35, 37, 49, 51–53, 55–60, 63, 64, 66, 67, 69–71, 76, 77, 80, 87, 93, 94, 96–99, 102–104, 109–112, 114–122, 125–127, 133, 134, 136, 137, 140, 142, 146, 147, 149, 151, 152, 154, 155, 158, 160, 161, 163, 168, 170, 172, 181–184, 187, 198, 208, 217, 224, 231, 232, 234, 239, 240, 242, 243, 245–248, 250–253, 255
GPS
  altimetry, 118, 120
  modernization, 4, 5
  observation equation, 133, 142, 145, 146, 161, 168, 169, 172–175, 177–181, 183, 187–191, 193, 195, 196, 198–200, 202, 204–206, 219, 231, 234, 239, 240, 250, 264, 275, 276, 280, 299, 300, 380, 383, 385, 397, 438, 439
  software, 2, 114, 121, 232, 255, 318, 320, 321, 323, 330, 338
  time, GPST, 33, 35
  week, 34, 35
Gravitational
  constant, 37, 50, 97, 99–102, 106, 343, 350, 354, 356, 371, 372, 402, 425
  field, 23, 33, 96, 134, 348, 351, 352
  force, 348, 409
  potential, 97, 367
Greenwich
  hour angle, 33
  meridian, 17, 25, 27, 33
  sidereal, 32, 33, 51, 373
Greenwich Apparent Sidereal Time (GAST), 25, 27, 32
Greenwich Mean Sidereal Time (GMST), 27
Group delay, 71

**H**

Helmert transformation, 22
Hopfield model, 83, 84, 88
Horizontal dilution of precision (HDOP), 308
Horizontal plane, 22
Hour angle, 33, 101, 354

**I**

IAU 2000, 28, 31
Inertial navigation system (INS), 332, 336, 337
Initial
  state, 342, 383–385
  value, 135, 208, 225, 292, 299, 301, 304, 306, 385–387, 391, 393, 394, 396, 401

value problem, 385–387, 391
Instrumental bias, 57, 63, 126, 231, 295, 437–439
Integer ambiguity, 116, 224, 229, 231–236, 238, 239, 241, 244, 245, 253, 257, 258, 330
Integration, 39, 40, 42, 44, 46, 60, 65, 78, 80, 81, 133, 230, 251, 255, 311, 349, 356, 375, 378, 383, 385, 391, 393, 396, 397, 411, 413, 416, 418, 434, 440
Intelligent Kalman filter, 218
International
  Atomic Time, TAI, 33
  Earth Rotation Service, IERS, 31–33, 51, 369
  GPS Service for Geodynamics, IGS, 51, 52, 56, 109, 122, 296
International Astronomical Union (IAU), 26–28, 31, 443
International Earth Rotation Service (IERS), 20, 26, 28, 29, 31–33, 50, 51, 99, 101–103, 109, 369
International GNSS Service (IGS), 114
International GPS Service for Geodynamics (IGS), 3, 51, 52, 56, 109, 114, 122, 125, 248, 277, 291, 292, 296, 297, 302, 304, 311, 326, 360, 379, 397
International Terrestrial Reference Frame (ITRF), 20
Interpolation, 52, 79, 89, 91, 310, 341, 387, 392, 397
Ionosphere-free combination, 67, 68, 144, 145, 162–164, 252–258, 291, 294, 295, 297, 299, 330
Ionospheric
  ambiguity correction, 252, 253, 255–257
  effect, 57, 59, 63, 65–67, 69, 71, 72, 76, 116, 122, 134, 135, 144, 145, 150, 249, 252–257, 294, 298, 330, 331, 339
  model, 3, 69, 71–74, 76, 78–80, 122, 146, 148, 254, 291, 294, 311
  residual, 133, 149, 150, 230, 253–255

**J**

Julian Date (JD), 24, 34

**K**

Kalman filter, 113, 137, 151, 187, 206, 208, 209, 212, 213, 216, 218, 225, 305, 306, 321

Keplerian
  elements, 47–49, 344–346, 366, 370, 373, 378–382, 410, 413, 416, 417, 419, 429
  ellipse, 343
  equation, 44, 46, 51, 369
  motion, 37, 49, 368
Kinematic
  ionospheric model, 3, 69, 71–73, 76, 78–80, 122, 254, 291, 311
  positioning, 1, 2, 91, 119, 151, 212, 213, 287, 297, 313, 321, 323, 324, 326–332, 335, 339

**L**

Lagrange
  interpolation, 52, 397
  polynomial, 51, 53
Lagrange-Xu equations
  derivation, 419
Lagrangian equations, 410–413, 416–418
Least squares adjustment (LS Adjustment), 187, 189, 191, 193, 195, 198, 201, 204, 208, 210, 219, 224, 225, 232, 234
Least squares ambiguity search criterion (LSAS criterion), 242, 243, 245
Linear
  combination, 53, 66, 143, 144, 246, 252, 311
  correlation, 438
  transformation, 133, 141, 147, 157, 171, 172, 263, 266, 269, 276, 439
Linearization, 133, 135, 136, 299, 331, 397
Loading tide, 57, 59, 63, 99, 105, 106, 108–110, 134, 141, 291, 327, 338, 341, 402
Local coordinate system, 22, 23, 307
Lock-wise least squares adjustment, 224
Loss of lock, 151, 229, 230, 232

**M**

Mapping function, 66, 71, 73–76, 81, 82, 85–90, 146, 168, 170, 254
Mean anomaly, 26, 45, 46, 369, 411, 413, 415, 419, 420, 430
Minimum spanning tree, 277
Modified Julian Date (MJD), 34
Multipath, 57, 59, 61, 63, 114–117, 120, 121, 134, 292, 297
Multiple static references, 327, 334, 335, 337

## N

Narrow lane, 167
Navigation message, 11, 49, 52, 53, 56, 99, 122, 292, 302
Normal equation, 173, 176–180, 183, 187, 189–191, 196, 197, 199, 201, 202, 204–206, 219–222, 224–226, 239, 240, 264–266, 272, 275, 283, 289, 290, 305, 306, 437, 438, 440, 447, 448, 450, 452
Numerical
 differentiation, 150, 385
 eccentricity, 74
 integration, 230, 250, 341, 383
 Integration, 387, 434
Nutation, 24–27, 30–32, 367, 369, 443

## O

Observational Model, 135
Ocean loading tide displacement, 105
Ocean tide, 49, 105, 107, 108, 297, 327, 341, 348, 354, 356, 357
Optimal baseline, 154, 277, 278
Orbit
 correction, 49, 341, 379, 380
 determination, 1, 2, 12, 49, 56, 77, 123, 134, 208, 217, 218, 291, 299, 321, 322, 341, 360, 361, 383, 384, 397, 409, 416, 434
Orbital
 coordinate system, 47
 mechanics, 1, 412
 plane, 3, 7, 9, 11, 39, 40, 42, 44–46, 50, 345, 368, 381, 411, 414, 416, 430

## P

Parameterisation, 2, 84, 85, 87, 110, 163, 168, 170, 175, 229, 252, 253, 263–267, 271, 273–276, 279, 280, 287, 383, 437–439
Partial derivatives, 133, 137–140, 299, 341, 345, 346, 352, 355, 366, 373, 380–384, 388, 389, 398, 401, 402, 404, 406, 407, 417, 430
Path delay, 74, 76, 80, 81, 83–85, 87, 146
Path range effect, 98
P-code, 116

Perigee, 43, 44, 46, 47, 344, 365, 368, 369, 413, 414, 416, 419, 422, 426, 429
Perturbation
 of point mass, 341, 353
 of tide, 354–356, 402
 order, 366, 372, 402
Perturbation force, 348, 351, 352, 367, 402, 404
Perturbed
 equation of satellite motion, 341, 342, 345
 orbit, 2, 375, 379
Phase
 advance, 63, 71
 centre, 63, 122, 125
 code combination, 67, 69, 148, 150, 264, 272–274
 combination, 146, 150, 246
 difference, 11, 114
 model, 59, 111
Point positioning, 111, 119, 120, 293, 294, 296, 297, 317
Polar motion, 18, 25, 27, 32, 33, 318
Post processing, 3, 77, 121, 322
Precession, 24, 25, 31, 367
Precipitable Water Vapor (PWV), 93
Precise ephemerides, 3, 51
Precise point positioning
 ambiguity fixing, 257, 311
Pre-processing, 291, 317, 318
Projection mapping function, 73, 74, 85
Pseudorange, 3, 8, 55, 57, 59, 72, 114, 116, 133, 292, 298, 321

## Q

Quasi-stable datum, 187, 222, 226

## R

Radial
 component, 102, 414, 430
 velocity, 59
Range rate, 61, 380, 383
Real-time
 positioning, 77, 112, 308, 440
Receiver
 clock, 56, 58, 114, 119, 266, 278, 292, 293, 297, 301, 307

independent exchange format (rinex), 291
Receiver independent exchange format (RINEX), 53, 291, 322
Reference
 frequency, 10, 438
 satellite, 156, 183, 225, 267, 271, 274, 287, 288, 301, 438
 station, 8, 91, 92, 154, 226, 258, 259, 270, 274, 289, 308–310, 326, 327, 333, 439
Refractive index, 64, 65, 80
Relative positioning, 301
Relativistic effect, 94, 97, 99
Relativity, 1
Right ascension, 40, 44, 48, 50, 364, 369, 413, 420, 430
Robust Kalman filter, 187, 209, 218, 225
Rotational
 angle, 21
 matrix, 21
Runge-Kutta algorithms, 387, 388, 391, 393, 396, 397

## S

Saastamoinen model, 81, 84, 88, 91
Sagnac effect, 98
Satellite
 antenna, 122, 125
 clock, 3, 51, 53, 56, 97, 112, 122, 172, 175, 184, 218, 266, 299, 302, 311
 motion, 33, 37, 40, 44, 49, 342, 345, 348, 412, 430
 orbit, 1, 2, 37, 43, 52, 99, 134, 296, 318, 342, 410, 434
Selective availability (SA), 4, 63, 111, 122
Sequential Least Squares Adjustment, 113, 187, 191, 212, 290
Shadow, 125, 358, 360, 368
Sidereal time, 25, 33
Single
 difference, 153–156, 175, 177, 179, 269, 328
 point positioning, 3, 111, 287, 292–297, 320
Singularity
 criteria, 409, 411
 criterion, 412
 free, 438
Solar radiation, 49, 123, 134, 291, 341, 345, 348, 358, 360–362, 368, 404, 411
Special relativity, 94, 96, 97

Speed of light, 56, 58–60, 64, 67, 71, 94, 95, 99, 110, 134, 161, 175, 179, 292, 297
Spherical coordinate, 18, 249, 343, 348, 351, 352, 355, 356
Spherical harmonics, 89, 349
Standard deviation, 110, 142–145, 161, 171, 189, 194, 210, 223, 233, 236, 238, 239, 248, 249, 307, 329, 335, 338, 419
State vector, 46–49, 134, 135, 137, 213, 215, 342, 343, 365, 366, 380, 381, 383, 384
Static reference, 103, 293, 303, 323, 325–329, 334, 335, 337, 338

## T

Terrestrial
 Dynamic Time, TDT, 33
 Time, TT, 28, 33
Terrestrial dynamic time (TDT), 33, 102
Tidal
 deformation, 102, 354
 effect, 103, 134, 140, 152, 327
 potential, 100–102, 354, 355
Time dilution of precision (TDOP), 307
Time system, 2, 7, 13, 25, 33, 35, 52, 439
Total electron content (TEC), 66, 73, 78, 79
Transition matrix, 207, 208, 384, 385
Transmitting time, 3, 51, 55–58, 95, 98, 110, 119, 134
Triple difference, 133, 158–160, 172, 173, 181, 182, 266, 280, 301
Triple-frequency combination, 67
Tropospheric
 delay, 80, 81, 83, 84, 87, 88, 91–93, 114, 139, 297
 effect, 2, 57, 59, 63, 80, 82, 84, 87, 135, 139, 153, 156, 159, 252
 model, 81, 82, 87, 88, 91, 110, 139, 291, 323, 329, 339
True anomaly, 43, 46, 344, 365, 369, 413, 414, 421, 430

## U

Uncombined algorithm, 164, 167
Undifferenced algorithm, 265, 271, 274
Unified equivalent algorithm, 183
Universal
 Time (UT), 27, 33
 Time Coordinated (UTC), 13, 33, 35
Universal Time (UT), 27, 33

Universal Time Coordinated (UTC), 33, 35
UT1, polar motion corrected UT, 27

## V

Variational equation, 385
Velocity
    determination, 302–304, 320
    vector, 46, 60, 98, 111, 302, 303, 305, 331, 362, 364, 368, 371, 381, 383, 386, 402, 405, 421
Vernal equinox, 24, 32, 33, 40, 47, 426
Vertical dilution of precision (VDOP), 308

## W

Water vapor, 93, 94

Weight matrix, 136, 161, 172–174, 176, 177, 179, 181–183, 188, 200, 204, 206, 209, 210, 212, 215, 216, 220, 221, 225, 226, 233, 256, 257, 263, 272, 281, 286, 290, 293, 295, 296, 299, 300, 303, 304, 306, 307, 440
Wide lane, 246
World Geodetic System (WGS), 20

## Y

Yang's Filter, 212

## Z

Zenith delay, 74, 88, 89, 146, 254, 321